FORAGE RESOURCE

Development and Utilization

The Author

Dr. Kumar Amrendra Singh, Ph. D. (Agronomy) from Indian Agricultural Research Institute, New Delhi and Post Doctorate in Forestry from OFI, Oxford University (UK), had been in forage research for 39 years with a remarkable credit of 223 research papers, reviews, bulletins and technical reports in the area of forage resource, agroforestry and farming systems. Dr. Singh was conferred with "Fakhruddin Ali Ahmed Award" for his outstanding Research in Agricultural Sciences for the Hilly Tribal Region of North East India. He is also recipient of several awards and the Fellows by professional societies (Indian Society of Agronomy and Range Management Society of India). He had occupied research management positions for about 15 years as Director, ICAR- Indian Grassland and Fodder Research Institute; Joint Director, ICAR Research Complex for NEH Region; and Head, Cropping Systems, ICAR Research Complex for Eastern Region. During his professional pursuit, he contributed as a member and chairman in various advisory committees of Planning Commission; Department of Animal Husbandry, Dairying & Fisheries; Agroforestry, Research, Education, Training, Capacity building and Information management for the Environment & Forests Sector; and Ministry of Science and Technology, Government of India.

FORAGE RESOURCE
Development and Utilization

by

Kumar Amrendra Singh

2018

Daya Publishing House®

A Division of

Astral International Pvt. Ltd.

New Delhi – 110 002

ISBN: 978-93-5124-947-4 (HB)

Published by : **Daya Publishing House®**
A Division of
Astral International Pvt. Ltd.
– ISO 9001:2015 Certified Company –
4736/23, Ansari Road, Darya Ganj
New Delhi-110 002
Ph. 011-43549197, 23278134
E-mail: info@astralint.com
Website: www.astralint.com

Digitally Printed at : **Replika Press Pvt. Ltd.**

Foreword

Indian livestock production is interwoven with crop production *viz.* field, fruit and vegetables, commercial crops in mixed farming systems with the intricate fabric of the sociocultural, religious and economic ways. Livestock rearing is an integral part of rural livelihood. Milk contributes 60 per cent of livestock economy well comparable to the contributions of wheat, rice and sugarcane. In general, livestock contributes 35 per cent of agrarian economy. The poor productivity of Indian livestock cannot be attributed to low genetic potential only but to a large extent poorly and underfed livestock. Livestock main feed resource is crop residues and monsoon bound natural vegetation which is low in nutrition and seasonal in availability. Forage production is not given due importance in animal husbandry practices. Somehow forage has fallen between two stools, the compartmentalized research in the divisions of crop science and animal sciences in Indian Council of Agricultural Research and central and state departments of Agriculture and Animal Husbandry. For improving productivity of livestock particularly ruminants, forage needs its due place in policy, plan and implementation by various agencies. They must realize the importance of green forage particularly cultivated forage under irrigated lands for quality milk and health of livestock and human beings. Green forage had been proven to be adequate feed for dairy livestock producing six to seven litres milk in a day. A new paradigm for biological production systems in coming years will be growing healthier food. There is an intricate linkage between agricultural practices and health of livestock and human beings. Green forage is a great natural resource for overall wellbeing of not only livestock but to improve soil productivity, conserving biodiversity and playing significant role in mitigating climate change.

The author has attempted in this book a holistic approach to forage resource development and utilization. Various chapters in the book cover current status of forage resources and animal husbandry in India; ecological perspectives and environmental impacts of forage based livestock production under changing climates; history of global forage resource development, forage resource development and utilization in hilly and plain regions of India, managing natural calamities; scope of forage replacement in non-ruminants; organic forage production; strategies to develop forage resources by appropriate policy support; organisational changes in structure of central and state animal husbandry departments; investment in forage resource development; road map and appropriate technological interventions for forage resource development and utilization with a chapter on glossary of terminology in this book suitably entitled "Forage Resource Development and Utilization". He also describes plant adaptation and the complexity of forage management by integrating climate, soil and herbivory factors. The book is a step forward to capture insight of the different types of forage resources and their utilization to evolve holistic approach to forage based livestock production to promote healthy livestock for healthy human beings. Policy makers, livestock development agencies and wild life conservation people may find it useful. I wish all success in this endeavour.

Dr. R.M. Acharya PhD (USA)

Former Deputy Director General (Animal Sciences),
Indian Council of Agricultural Research and
FAO/UNDP Consultants
House No. 784, Sector – 9
Faridabad (Haryana), India

Preface

With domestication of animals and increasing population of human beings with evolution of human civilization, natural grasslands changed into intensively and more intensive managed grasslands as they are today, distinguishing gradually between grassland developed by man and natural grasslands. The world history of forage crops dates back 1300 BC when alfalfa was cultivated in Turkey. In northern India, sorghum (*Sorghum bicolor* (*L.*) *Moench*) was recorded to be used as forage crop long back in the second century BC. Major advances in forage crops happened during 13th to 19th centuries. Now, forage crops are an important source of feed for domestic animals, although the history of their use is shorter than that of the use of native grasslands. The multiple benefits of forage resource to humans is not limited only to livestock production but its contribution to food crop production, soil conservation and amelioration; landscape and wildlife conservation; improvement and protection of the environment; revegetation and ecological repair of degraded lands. Thus, forage resource is a valuable plant species ranging from herbaceous to trees, grasses, legumes and forbs covering the largest baskets of annual and perennials to serve the human society, if it is managed wisely.

In India, livelihood of people is interwoven with livestock rearing in crop-livestock mixed farming systems. In recent few decades, the revolution in livestock sector was witnessed with the increasing demand of livestock products, propelled by changing dietary habits, urbanization and income growth. Growing demand promoted establishment of peri-urban dairy, modern dairy and livestock farms. This has brought new challenges to provide nutritious forage resources to feed livestock and upscale livestock productivity. The low productivity cannot be attributed only to low genetic potential. The low productivity of Indian livestock can be largely attributed to underfeeding and providing low nutrition feed *i.e.* crop residues and monsoon bound natural vegetation. The forage plant adaptation and the complexity of forage management by integrating soil, climate, and herbivory factors with forage production goals and livestock utilization need to be well understood for successful animal husbandry practices. Looking into the growing demand and current scarcity of forage, there is need to comprehend livestock production; grazing, or harvesting and feeding forages; and the overall cost-price-return factors with enabling policy environment, organisational structure and basic infrastructures required in its totality. Large scale institutional and policy change needed pilot approach evidenced, which need to be documented and the experiences gained shared and built upon successes and stimulating policy dialogues for larger application. A new paradigm for agriculture in coming years will be growing healthier food. There is an intricate linkage between agricultural practices and health of livestock and human beings. Growing healthy crops require fertile, well-nourished soils, which in turn provide feed materials which are balanced in all nutrients to rear healthy livestock. The livestock products obtained from these animals ensure wholesome milk and meat products to human beings, free from any contaminants and pesticide residues.

The book "**Forage Resource: Development and Utilization**" is a step forward to capture current insight of animal husbandry practices in the country, structural bottlenecks in the concerned government organisations at state and central governments, perception of stakeholders including NGO's, forage innovation at international levels, issues and policy support for investment for development of forage resources in the hilly and plain agro-eco regions of the country. The book also take into account forage technology including organic forage production, proper forage conservation and its optimum utilization, mitigating natural calamities like drought and flood with emerging challenges of climate change and reducing availability of land. It also describes the potential use of forage in non-ruminants. The 11 chapters in the book are an attempt to provide a comprehensive approach to enhance forage resource and their utilization in the country.

I express my deep sense of gratitude to all my professional colleagues and friends for suggestions to improve various chapters in the book. I specially express my gratitude to IGFRI colleagues especially Drs. D. R. Malviya, A. K. Roy, P. K. Pathak, Sunil Seth, Sunil Kumar, S. B. Maity, J. P. Singh for valuable inputs to enable this endeavour and technical assistance by Shri G. P. Nigam, K. P. Rao and Shailendra Kumar. Dr. Anil Kumar, Principal Scientist (Livestock Production and Management) has provided valuable inputs for the chapter on natural calamities. The knowledge and experience gained during my service with Indian Council of Agricultural Research enabled me to make this endeavour possible. I duly acknowledge it. I hope this book will achieve the intended purpose to address the needs of today's students and all stakeholders of forage resource management and animal husbandry.

Kumar Amrendra Singh

Contents

Chapter 1

History of Global Development of Forage Resources

Neolithic tribes first discovered that some animals can be tamed and plants can be grown to alleviate the uncertainty of the hunting chase and food searching expedition (Choi, 2014 and Gordon King, 2017). This was a major step in agriculture evolution. In this endeavour, some animals were domesticated as livestock to provide commodities for consumption by the group of humanbeings or family with occasional surpluses for barter or sale. The domestication of animals necessitated taking responsibility for care and ensure foraging material for the livestock. Later on, keeping livestock for food began in tropical or semitropical regions of the Middle East. Innovations of milk let down reflex and selection of individual livestock for milk led to development of dairying. Also, the discovery of methods for incubating birds' eggs without nesting hens led to increased availability of poultry products. During 1880-1950, application of genetic selection for livestock improvement and employing transport of dairy and meat products using cold chain and storage revolutionized livestock production. Consequently, this progression made livestock totally dependent on people for care and feeding. Now, history of animal husbandry worldwide records 100 different animal species kept by humans for multi-purpose use (Groenefeld and Glodek, 2000). Most of the 40 different mammalian species (*e.g.* cattle, yaks, banteng, mithun, buffalo, sheep, goat, horse, donkey, pig, reindeer, camel, cat, dog, rabbit, guinea pig, liamas, elephant, rat, mice), 24 different bird species (*e.g.*, chicken, geese, ducks, turkeys, doves, guinea fowl, ostriches), 18 different fish species (*e.g.*, carp, trout, salmon) and several insect species (*e.g.*, earthworms, bees, silkworms) are kept on the farms and have been domesticated at different stages of history. In the last two centuries, animal production had focussed on improved feeding, health care, stable keeping and breeding high yielding livestock. Worldwide, 863 sheep breeds, 783 cattle breeds, 313 goat breeds, 357 horse breeds, 263 pig breeds, 78 donkey breeds and 62 buffalo breeds are known. Most important grazing animals *i.e.* cattle, sheep and goats believed to be domesticated about 8000, 11 000 and 9000 years ago, respectively. This followed domestication of buffaloes, horses, asses and camels about 5000 years ago. Although the practice of tilling the soil, growing of crops and rearing of livestock was common throughout parts of Asia, Europe and Africa and even in the Americas for millennia, it had a much shorter history in some other regions. Grazing animals required forage as basal food. The husbandry of grazing animals required that humans make forage available to the animals.

Domestication of livestock led to pastoralism which thrived in semiarid areas, such as central Asia, the Sudanic belt south of the Sahara desert in Africa, and the savannah zone of East and South Africa (Grigg, 1974, Lambert, 1988, Knapp, 1991 and Masahiko Hirata, 2009). The nomadic, herding way of life had tended to produce independent and hardy peoples. Major grassland areas, such as the prairies of the United States, the pampas of South America, the steppes of central Asia, and the velds of Africa are estimated to be developed about 20-25 million years ago. Forage plants occurring in grassland, rangeland and fallow land were an arbitrary gift of nature. The undisturbed grasslands produced deep, fertile topsoil because of the natural life and death cycle over long periods of time. Before human activity began to alter grasslands they were naturally maintained as productive grasslands by favourable climate, grazing wildlife, and natural fires. As a consequence of domestication of grazing animals, forage had been a major concern in animal husbandry. Evidently its importance had been found in the Old Testament "God is praised for his creation as: "You let the Earth produce grass for cattle, plants for our food (Psalms 104:14)"; Moses, in the thirteenth century BC, promised "He will send rain at the right seasons, so you will have more than enough food, wine, and olive oil, and there will be plenty of grass for your cattle (Deuteronomy 11:14-15)" and "Our cattle wander aimlessly, moaning for lack of pasture, and sheep are suffering" (Joel 1:18)." Early recognition of importance of forage had led to forage conservation in the form of hay and silage to optimise the stage of forage growth for use during lean period when it is unavailable. Hay making and scythe date back from 750 BC and silage dates back to 2000-1500 BC. Paintings found in Egypt, dating from the period about 1500-1000 BC, suggest that the ancient Egyptians were familiar with ensiling (the silage making process). Silos were found in the ruins of Carthage, which indicates that forage was ensiled there in about 1200 BC. With domestication of animals, evolution of human civilization and increasing population of human beings, natural grasslands evolved into intensively and more intensive managed grasslands as they are today, distinguishing gradually between grassland developed by man and natural grasslands.

Forage crops have served as an important source of feed of domestic animals for a long period, although the history of their use is shorter than that of the use of native forages (Hartley, 1963, Vasey, 1992 and Masahiko Hirata, 2009). Nearly 200 plant species had been on record as forage crops. The benefit of forage crops to human beings is not a feed source but they also contribute to food crop production and many other aspects of human life through soil conservation and amelioration; landscape and wildlife conservation; improvement and protection of the environment from pollution; reclamation, revegetation and ecological repair of degraded land and potential conversion of biomass to energy. Bolton *et al.* (1972) reported that *Medicago sativa* (lucerne) was first forage crop to be domesticated and cultivated for over three millennia as the "Queen of the Forage Crops." Hittite brick tablets (1400-1200 BC) discovered in Turkey indicated that animals were fed alfalfa all through the winter season and alfalfa was regarded as a highly nutritious animal feed. There was also historical evidence that testified to the wide distribution of alfalfa in Media (NW Persia) in the first millennium BC. The name "alfalfa" can be traced to the old Iranian word meaning "horse fodder." The crop was brought to Greece by the invading Median armies in the fifth century BC. After its introduction, alfalfa soon gained some importance in Greek agriculture. Many writers, such as Aristophanes (450 to 385 BC), Aristotle (384-322 BC), Theophrastus and others, either mentioned it or discussed it at length. For the next 200 or 300 years, however, no further references to alfalfa were found until the Roman era. The Romans were credited with being the fathers of forage culture. It was due to their advanced knowledge about seeding, management, and hay-making practices of forage crops. In the second century BC, Romans accessed alfalfa from the Greek civilization (Bolton *et al.* (1972). The new crop thrived and quickly spread throughout Italy (27 BC-395 AD). Several Roman writers mention several forage crops. For example, the culture and management of alfalfa was discussed in detail by Varro (116-27 BC), Pliny (23/24-79 AD) and Colum Ella (in the first half of the first century AD). Varro suggested a rate of seeding (about 38 kg ha^{-1}). Pliny recommended planting in a well-drained soil, the use of chalk (liming), and cutting at early flowering. Colum Ella cautioned his readers to feed sparingly at first, lest bloating occur and the blood supply greatly increased. He knew that alfalfa improves the soil, and had a high regard for its persistence and feed value. Besides alfalfa, crops such as turnip, lupines and vetches were mentioned by Cato (234-149 BC) and Varro, and cowpea (*Vigna unguiculata* (L.) Walp. ssp. *unguiculata*) by Pliny and Colum Ella. Colum Ella explained how the Gauls used turnips as feed for animals during the winter period and rape (*Brassica napus L.*) for cattle feed. The early use of forage crops was also seen in other parts of the world. Simultaneously with its arrival in Italy in the second century BC, alfalfa spread eastward into China, and in succeeding years, it became established throughout northern China as an important forage crop. By this time, sorghum (*Sorghum bicolor* (L.) Moench) had been used for forage in northern India. At beginning of Christian era, wide spread use of forage crops happened in terms of both geographical distribution and number of forage plant species used. In the first century AD, alfalfa was grown in Andalusia in southern Spain and in the Lake Lucerne region of central Switzerland and the name lucerne was derived from that Swiss lake region where the crop gained its popularity. During same period, it reached southern France, and was probably making inroads into North Africa. Red clover (*Trifolium pratense* L.), originated in Asia Minor and south-eastern Europe and found to be cultivated in southern Europe in the third and fourth centuries AD and used as cut and fed to animals as green forage or hay. A wild type of alsike clover (*Trifolium hybridum* L.) was grown in Sweden in the 10th century. Despite the use of forage crops in these times, their contribution to the husbandry of domestic grazing animals was still very limited (Masahiko Hirata, 2009). The animals were mostly dependent on naturally growing forages or by-products of food crops. Reflecting this, for instance, in northern Europe the scarcity of winter feed resulting from the relatively short growing season of native forages limited the number of animals which could be maintained over the winter.

Use of forage crops as feed significantly declined with the fall of the Roman Empire and the advent of the Dark Ages in Europe. After the barbaric invasions in the fifth century, alfalfa almost disappeared from Italy until Islamic ages when forage crops were rediscovered in southern Europe. With the expansion of Islam during the seventh and eighth centuries, its triumphant armies movement to westward across Africa to the Atlantic, then northward across the Strait of Gibraltar, throughout Spain, and as far north as the Pyrenees, alfalfa was possibly reintroduced into Spain via North Africa by the invading Moors in the early eighth century. The Spanish acceptance of the Arabic word "alfalfa" in preference to the Roman words, "medica" or "lucerne", gives credence to this route. The crop was then reintroduced into Italy from Spain. During the 11th and 12thcenturies, the recognition of the importance of forage crops by the Muslims was mentioned in agricultural manuscripts in Spain. Ibn al-Awwâm, highly valued alfalfa for its persistence and feed value, as the "Queen of the Forage Crops" as well as mention of as lupines, vetches and red clover primarily as rotation crops that improve soil fertility. In northern Europe, the role of forage crops in the animal husbandry remained unimportant, with few forage crops being grown up to 12th century. Little happened between the fifth and twelfth centuries, but the thirteenth to nineteenth centuries saw great advances in forage crops (Masahiko Hirata, 2009). It is, however, assumed that a wild type of alsike clover (*Trifolium hybridum* L.) used to be grown in Sweden as early as the 10th century. Between 1000 to 1300 AD, grazing land was converted to arable land due to increasing population in Europe (The European Agricultural Revolution). Consequently, livestock population reduced so the reduced supply of manure to cereal crops. As result cereal yields declined in 13th century.

The modern era dates from the beginning of the 12th century. They contributed to the development of mixed farming in the European agricultural revolution, and spread globally with the expansion of Europeans into other continents. The use of forage crops developed also with the rise of industrialized agriculture, which involved the use of new techniques and industrial inputs, the application of science and the rise of a forage seed industry.

Development of mixed farming in Low Countries, first in Flanders and Brabant and later in Zeeland and Friesland led to european agricultural revolution. Two major systems were devised. First, the improved three-course system by planting of forage crops replaced fallows. Vetches had been grown in Flanders by 1278. Turnip and red clover were grown in the 14th century. Second, the ley farming system practiced by growing two years of cereals, one year of fallow followed by three to six years of forage crops at Ghent by 1323. Initially, ley was in grass but later on red clover and to some degree white clover (*Trifolium repens* L.) was also used. The improved three-course system and the ley farming were common practices in the Low Countries by the 15th and 17th centuries, respectively. In the cereal-livestock mixed farming, planting of forage not only improved supply of feed to livestock but provided the soil a rest from cereal to increase the soil fertility and soil physical structure by roots. Ultimately productivity of cereal and forage crops increased. High amount of forage produced fed to livestock augmented the supply of manure to cereal crops. In the 16th and 17th centuries, the use of forage crops was rapidly adopted into England, where the alternate system of cereal crops and indigenous grasses had been long and widely practiced. Richard Weston, an Englishman, learned the use of turnip and clovers as rotation crops when he travelled in Flanders in 1644 (Masahiko Hirata, 2009). An English writer, Walter Blith, dealt with clovers, sainfoin (*Onobrychis viciifolia* Scop.), alfalfa and turnip as new crops and recommended their culture in 1652. By the late 17thcentury, turnip and clovers were grown in parts of southeastern England, and in the early 18th century the classic four-course system, *i.e.* a four-year rotation of wheat, turnip, barley and red clover, was established in Norfolk by Charles Townshend. The ley farming system enabled the first whole-year stall-feeding of cattle. Thus, use of forage crops enabled mixed farming integrating cereal and livestock production. Origin of ley farming in Flanders in the 13th century, consequently matured in 18th century in England at the beginning of the European agricultural revolution, and further spread to the rest of the Western Europe. It was said that the influence of red clover on civilization and European agriculture was greater than that of the potato and much greater than that of any other forage plant.

Among 10,000 species of grasses, only about 40 - 50 were used on grasslands developed by man (Hoveland, 1987). Often, grasses from various forest edges and woodlands around the world became more important forage species than the species that were considered native or natural grasslands. Sometimes very important forage species were carried intentionally or unintentionally by immigrants to their new habitat. For example, alfalfa, one of the most important forages in the United States of America, originated in the area of modern-day Iran and Turkey. Bermuda grass, very important forage in the southern United States of America, was apparently brought to the country inadvertently by African slave ships. Kentucky bluegrass, well known pasture grass, was apparently introduced to America from England as part of the sweepings of cattle boats.

In Indian context, first record of growing forage crop was found in second century BC. Sorghum (*Sorghum bicolor* (L.) Moench) used to be grown as forage crop. In ancient India, livestock were largely dependent on grazing lands with supplementary feeding of edible crop residues as mixed farming and livestock rearing forms an integral part of socio-cultural-religious and economical ways of living. During 20th century, animal husbandry departments set up a section/

demonstration centres across India. Systematic development of forage crops and grassland was focussed in 20th century. Grassland survey in sixties led to establishment of Indian Grassland and Fodder Research Institute in early seventy and All India Coordinated Research on Forage Crops in late eighty in State Agricultural Universities across the country during 20th century AD. Systematic forage crop breeding programmes at the research institutions under the Indian Council of Agricultural Research (ICAR) and the State Agricultural Universities (SAUs) have led to the development and release of a large number of improved varieties in different forage crops suitable for different agro-ecological zones. Biodiversity in forage resources of Indian sub-continent is one of the world's mega centres of crop origin and crop plant diversity, as it presents a wide spectrum of eco-climate ranging from humid tropical to semi-arid and temperate to alpine (Joshi *et al.*, 2009). The Indian gene centre possesses a rich genetic diversity in native grasses and legumes. There are reports of 245 genera and 1,256 species of *Gramineae* of which about 21 genera and 139 species are endemic. One-third of Indian grasses are considered to have fodder value. Most of the grasses belong to the tribes *Andropogoneae* (30 per cent), *Paniceae* (15 per cent), and *Eragrosteae* (9 per cent). Similarly, out of about 400 species of 60 genera of *Leguminosae*, 21 genera are reported to be useful as forage. The main centres of genetic diversity are peninsular India (for tropical types) and North-Eastern Region (for sub-tropical types) besides some micro-centres for certain species. Major forage genera exhibiting forage biodiversity include legumes like *Desmodium, Lablab, Stylosanthes, Vigna, Macroptilium, Centrosema, etc.*; grasses like *Bothriochloa, Dichanthium, Cynodon, Panicum, Pennisetum, Cenchrus, Lasiurus, etc.* and browse plants such as *Leucaena, Sesbania, Albizia, Bauhinia, Cassia, Grewia, etc.* These genera besides many others form an integral part of feed and forage resources of the country. The country is further endowed with the rich heritage of traditional know-how of raising, maintaining and utilizing forage, feed and livestock resources. The Table 1.1 highlights key events in the world history of agriculture and forage production to get an appropriate historical perspective of forage production and management.

Further, evolutionary changes in the principles and practices for producing forage crops began with parallel to historical developments in agricultural technology in general. The development of forage crops has been dramatically accelerated by two main factors. The first is the enormous impact of modern science and technology. Forage crop scientists started the process by eventually using the agriculture technology to discover and breed high yielding and high-quality forage crops. Progress in plant improvement has led to the development of many new cultivars and the dramatic growth of the seed industry, and thus to the global dispersion of many new forage plants. The second is the diversification in the value of forage crops in a changing society. On one hand, the value of forages as livestock feed has been continuously well recognized, with their production and utilization being strengthened by industrialized agriculture. On the other hand, many other uses of forage crops have been highlighted with the rising concern of

Table 1.1: Some Major Highlights in the Early Evolution of Animal Production and Global Development of Forage Resources

Time Period	*Events*
Long long Ago	Small numbers of hunter-gatherers.
circa 8,000 BCE	Earliest domestication of animals and cultivation of plants. Beginnings of mixed farming.
16th century	Colonization begins and European livestock breeds transported to "New World."
About 20-25 million years ago.	Major grassland areas, such as the prairies of the United States, the pampas of South America, the steppes of central Asia, and the velds of Africa developed
Over three millennia	*Medicago sativa* L. (lucerne) "Queen of the Forage Crops." is first forage crop to be domesticated and cultivated.
2000-1500 BC.	Paintings found in Egypt, dating from the period about 1500-1000 BC, suggest that the ancient Egyptians were familiar with ensiling (the silage making process). Silos were found in the ruins of Carthage, which indicates that forage was ensiled there in about 1200 BC.
750 BC	Hay making.
Second century BC	1. Sorghum (*Sorghum bicolor* (L.) Moench) grown in India.
	2. Alfalfa with its arrival in Italy spread eastward into China, and established in northern China as an important forage crop.
234-149 BC	Turnip, lupines and vetches, cowpea (*Vigna unguiculata* (L.) Walp. ssp. *unguiculata*), rape (*Brassica napus L*) were grown as forage crops
First century AD	Alfalfa was grown in Andalusia in southern Spain, central Switzerland, Southern France, and probably into North Africa.
Third and fourth centuries AD	Red clover (*Trifolium pratense* L.), originated in Asia Minor and south-eastern Europe and cultivated in southern Europe
Fifth to seventh century AD	Use of forage crops as feed declined with the fall of the Roman Empire in Europe.Alfalfa almost disappeared from Italy until Islamic ages when forage crops were rediscovered in southern Europe.
Eighth century AD	With the expansion of Islam during the seventh and eighth centuries. Alfalfa was possibly reintroduced into Spain via North Africa in the early eighth century. The Spanish acceptance of the Arabic word "alfalfa" in preference to the Roman words, "medica" or "lucerne", gives credence to this route.
10th century AD	A wild type of alsike clover (*Trifolium hybridum* L.) grown in Sweden
11th and 12th century AD	Alfalfa was reintroduced into Italy from Spain.
Up to 12th century AD	In northern Europe, the role of forage crops in the animal husbandry remained unimportant, with few forage crops being grown. Little happened between the fifth and 12th centuries.
10th to 13th century AD	Grazing land was converted to arable land due to increasing human population in Europe. Consequently, cereal yields declined in 13th century due to reduced livestock population reducing supply of manure to cereal crops.
13th and 19th centuries	Great advances in forage crops happened. Planting of forage crops replaced fallows. The use of forage crops developed also with the rise of industrialized agriculture, which involved the use of new techniques and industrial inputs, the application of science and the rise of a forage seed industry and development of mixed farming in Low countries, first in Flanders and Brabant and later in Zeeland and Friesland which led to European agricultural revolution.
Early 1600's AD	Sir Richard Weston introduces red clover into pastures in England.

Contd...

Time Period	Events
Mid 1600's	Forage grasses introduced in North America.
Late 1600's	Native Americans raise cattle from colonists.
	Bermuda grass imported from Northern Africa'
17th century	Formation of scientific societies and animal breed associations resulted in enhancing the importance of forage resource.
Mid 1800's	Johnson grass introduced in southern United States of America.
1862	Morrill Act established United States of America Land Grant Universities
1873	Construction of first silo in United States of America
1886	Discovery of nitrogen fixation in legumes by Hellriegel and Willfarth
1931	Tall fescue discovered in eastern Kentucky
1940	Development of self-tying hay bale

Sources: Modified Hartley (1963), Bolton *et al.* (1972), Grigg, 1974, Hoveland, 1987, Lambert, 1988, Knapp, (1991), Groenefeld and Glodek (2000), Masahiko Hirata (2009), Choi (2014), Gordon King (2017).

environmental issues. This emerging diversity of functions points to the future roles of forage crops in human existence and development.

Much research has been carried out to improve the quality and availability of forage resources, including work on sown forages, forage conservation, the use of multi-purpose trees, fibrous crop residues and strategic supplementation. Given the prevalence of mixed crop-livestock systems in many parts of the world, closer integration of crops and livestock in such systems can give rise to increased productivity and increased soil fertility (McIntire *et al.,* 1992, Herrero *et al.,* 2009). However, ruminants in pastoral and extensive mixed farming systems in many developing countries still continue to suffer from permanent or seasonal nutritional stress (Bruinsma, 2003). Moreover, the hazards of factory style livestock production and environmental pollution have thrown many issues related to wholesome food production taking into account the ethical issues and the emergence of new diseases besides polluting the environment by livestock (Otte *et al.,* 2007, Gura, 2008). Many medicinal plants including bamboo will come to the aid of safeguarding the health of livestock. Some species of bamboo contain immune stimulatory compounds that the farmers can utilize for their animals (Thornton, 2010). Bio-active forages, like bamboo, offer low cost and environmental friendly solutions for feeding all species of domestic animals. The linkage between healthy soil-crop-livestock was emphasized over a century ago by Sir Albert Howard, need revisit. He had reported that when livestock were fed on plants grown on fertile healthy soil, even the FMD disease failed to show any symptoms on these animals although they frequently came in contact with the diseased animals. There is an intricate linkage between agricultural practices and health of livestock and human beings. Growing healthy crops require fertile, well-nourished soils, which in turn provide nutrient balanced forage to livestock. The livestock products obtained from these animals ensure wholesome milk and meat products to human beings. Good health affects agriculture by boosting peoples' capacity for work and thus increasing how much they can produce. Understanding plant adaptation and the complexity of forage management by integrating soil, climate, and herbivory factors with production goals need to be further focussed. Systems analysis approach at farm level will provide a method of putting forage and livestock variables together with the appropriate cost factors, and for dealing with cause-effect relationships rather experimental techniques used to obtain biological evaluation (Van Keuren *et al.*, 1974). Another important facet of the total forage-animal system is the development of new planting, harvesting, handling, and feeding equipment needed in the system.

In the future, forage production will increasingly be affected by competition for natural resources, particularly land and water, competition between food and forage production and by the need to operate in a carbon-constrained economy (Brunori *et al.*, 2008, Thorton *et al.*, 2009, Thornton, 2010, Thornton and Gerber 2010, Thorton *et al.*, 2009). Improved forage based feeding practices can reduce methane emissions from livestock to an optimum level of productivity (Smith *et al.*, 2007). Besides, many specific agents and dietary additives had been proposed to reduce methane emissions, including certain antibiotics, compounds that inhibit methanogenic bacteria, probiotics such as yeast culture and propionate precursors such as fumarate or malate that can reduce methane formation whether these various agents and additives are viable for practical use or not, and what their ultimate impacts could be on greenhouse gas mitigation, are areas that need further research. A new paradigm for agriculture in coming years will be growing healthier food. Therefore, demand for livestock products in the future could be heavily moderated by socio-economic factors such as human health concerns and changing socio-cultural values (Brunori *et al.*, 2008). There is considerable uncertainty as to how these factors will play out in different regions of the world in the coming decades. Little research has been conducted on integrating agriculture-livestock-human health in promoting the wellbeing of people and livestock as a whole.

Chapter 2

Status of Forage Resource, Animal Husbandry and Dairying in India

The country had largely unorganized primitive subsidiary livestock production systems that do not allow scientific and technological interventions to permeate into them. Animal husbandry continues to be a subsidiary appendage to crop production except peri-urban dairies and some organized dairies in private sectors. Recently, there had been a rapid change in agricultural scenario in the country. It was predicted that meat and milk consumption will grow at 2.8 and 3.3 per cent per annum, respectively, in developing countries like India. Peri-urban livestock farming and emerging forage markets are indicators of fast changing economic scenario in livestock sector. Livestock population is around 500 million and is expected to grow at the rate of 1.23 per cent in the coming years (Table 2.1). There is need to meet the demand of increasing number of livestock and also enhance their productivity for which availability of forage and feed resources have to be increased (Table 2.2).

Table 2.1: Projected Livestock Population Estimates (Million adult cattle unit)

Year	*Cattle*	*Buffalo*	*Sheep*	*Goat*	*Equine*	*Camel*	*Total*
1995	180.5	82.8	4.0	9.2	0.5	0.9	278.0
2000	187.1	87.7	4.1	9.9	0.4	1.0	290.0
2005	192.2	92.6	4.2	10.5	0.3	1.0	301.0
2010	197.3	97.5	4.3	11.2	0.3	1.0	312.0
2015	202.3	102.4	4.4	11.8	0.1	1.1	322.0
2020	207.4	107.3	4.5	12.5	0.1	1.1	333.0
2025	212.5	112.2	4.6	13.2	0.1	1.1	344.0
2030	217.7	117.3	4.7	13.9	0.1	1.1	354.8

Source: Based on 10th and 11th five year plan document projections made.

In present days, there is emerging emphasis on intensive and diversified cropping systems, transformation of sustenance farming into market oriented contract farming, emergence of corporate world in input supply and output procurement, processing and marketing through organized retail chains, linking production to consumption under changing food habits towards livestock products, quality consciousness and standards. In other words, livestock production is changing from sustenance to agri-business in the form of enterprises and industries based on milk, meat, egg and other livestock products.

Table 2.2: Projected Availability, Requirements and Deficit of CP and TDN (including concentrate) (million tonnes)

Year	*Requirement*		*Availability*		*Per cent Deficit*	
	CP	*TDN*	*CP*	*TDN*	*CP*	*TDN*
2000	44.49	321.29	30.81	242.42	30.75	24.55
2005	46.12	333.11	32.62	253.63	29.27	23.86
2010	47.76	344.93	34.18	262.02	28.44	24.04
2015	49.39	356.73	35.98	273.24	27.15	23.41
2020	51.04	368.61	37.50	281.23	26.52	23.70
2025	52.68	380.49	39.31	292.45	25.38	23.14
2030	54.37	392.70	41.20	304.11	24.22	23.32

Source: Based on 10th and 11th five year plan document projections made.

To increase milk production from 101 million at present to 160 million tonnes by 2020, the demands on fodder will be to the tune of 520, 850 and 90 million tonnes in respect dry forage, green forage and concentrates. Crop residues constitute the single largest bulk feed material, especially for the ruminants. Several agro-industrials by-products such as vegetable and fruit processing residues and bagasse are also fed to animals. Across ecosystems, dry matter availability through greens is highest in the rain fed ecosystem, followed by irrigated and hill/mountain ecosystems. The dry forage (roughage) is highest in the irrigated and rain fed areas. Paddy straw is the major crop residue in all the ecosystems, except arid zones where wheat straw predominates. Coarse cereals are also available in rain fed and arid ecosystems. There is a wide gap in dairy feed concentrates looking to the actual requirement of livestock. It is because of low demands from livestock owners as many feed homemade concentrate mixtures and there is lack of awareness among them of feeding balanced feed. However, in advanced milk shed districts there is a trend of increased demands for scientifically formulated cattle feed that enhances milk yield per animal by 20-30 per cent.

Forage crops possess certain unique problems. It is intermediary inputs for livestock production which is quite different from the food, grain, and horticultural crops. Besides multiplicity of the crop species and their nature, the forage crops are usually area and region specific. The preference for forages varies from season to season and locality to locality. Non-commercial nature of the crop does not encourage forage crop cultivation. In the most cases the degraded and marginal lands are usually given for forage production with minimum inputs in term of fertilizers, water and operational energy input. The regional and seasonal deficits are more important than the national deficit in case of forage as it is not economical to transport over long distances.

Present Status and Future Projection

The desired annual growth of agriculture sector at the rate of 4 per cent can also be accomplished by enhancing productivity from the livestock sector. This would require a steady supply of forage for supporting the livestock population. Forage is derived from both arable and non-arable lands, including common property resources, fringe forests and other grazing areas. Forage is cultivated only on 4.60 per cent of total cropping area which has resulted in a severe deficit of green forage and roughage from crop residues and agricultural by-products. However, certain states like Haryana, Punjab, Gujarat, Rajasthan and Maharashtra have about 10 per cent or more area under forage cultivation. Green forage production is high in only five states *i.e.* Rajasthan followed by Maharashtra, Gujarat, Punjab and Haryana. About 50 per cent of the cultivated area under forage is occupied by forage sorghum and berseem. The other major crops include lucerne, maize, pearl millet and oats. The permanent pastures and forests occupy 3.6 and 22.6 per cent of the geographical area of the country that offer forage for the livestock. However, only the forest areas outside the purview of environmental protection acts are permitted for grazing. There are areas under community control in many villages that provide forage to animals, either in grazing or cut and carry mode. These categories of range resources are in degraded state and need improvement for enhancing their productivity and contribute fully to forage basket. Some 70 million rural farmers largely feed their animals by allowing them to graze on common lands outside their villages including the land alongside of railways, road side's irrigation canals *etc.* and forest areas and supplementing agro-industrial products like oil cakes and other such concentrates.

There is a need for re-structuring the land use strategy to elevate the overall percentage of cultivable lands for forage production to not less than 10 per cent. Miscellaneous top feeds and weeds cover an area of 1.2 per cent of geographical area while cultivable waste in the country is about 4.5 per cent, which needs to be developed for forage production. Crop residues are the single largest bulk feed material available at the doorsteps of the farmers for feeding livestock, specifically, the ruminants. They include coarse straws, fine straws, leguminous straws, pulses straws and sugar cane tops. There are several agro-industrial by-products such as vegetable and fruit processing residues and bagasse, which are available in abundance to be fed to livestock. Permanent pastures productivity and carrying capacity are declining, though these lands support grazing ruminants such as cattle, sheep and goats in large numbers. There are large chunks of common property and community lands which are under the public domain, but becoming drastically reduced for livestock grazing. In India, the forest cover is to the tune of

22.6 per cent of which over 85 per cent area is under protection and conservation. These lands used to be a major source of feeds and forage for the livestock rearing communities dwelling within and nearby the forests. Forage deficit can mainly be attributed to our limitations in increasing the area under forage crops, lack of availability of quality seeds of improved varieties/hybrids, poor quality of dry forage like paddy/wheat straw *etc.* besides, low priority accorded to investment in forage production, lack of post-harvest management for surplus forage, crop residues, poor management of grazing/pasture lands and inadequate, extension and manpower support have only aggravated the situation. Forage based cheaper feeding strategies are required to reduce the cost of quality livestock product as the feed alone constitutes 70 per cent of the milk production cost. There is tremendous pressure of livestock on available total feed and forage, as land available for forage production has been almost constant. At present, the country faces a net deficit of 61.1 per cent green forage, 21.9 per cent dry crop residues and 64 per cent feeds (Table 2.3). Supply and demand scenario of forage and roughage is presented in Table 2.2. To meet the current level of livestock production and its annual growth in population, the deficit in all components of forage, dry crop residues and feed has to be met from either increasing productivity, utilizing untapped forage resources, increasing land area (not possible due to human pressure for food crops). In animal feed supply, coarse cereals have a major role and four major cereals *viz.*, maize, barley, sorghum and pearl millet account for about 44 per cent of the total cereals. Production of these cereals is stagnating at around 30 million tonnes. To meet the feed requirements there is a need to improve productivity of these cereals.

Table 2.3: Supply and Demand Scenario of Forage and Roughage (in million tonnes)

Year	*Supply*		*Demand*		*Deficit as per cent of Demand*	
	Green	*Dry*	*Green*	*Dry*	*Green*	*Dry*
1995	379.3	421	947	526	59.95 (568)	19.95 (105)
2000	384.5	428	988	549	61.10 (604)	21.93 (121)
2005	389.9	443	1025	569	61.96 (635)	22.08 (126)
2010	395.2	451	1061	589	62.76 (666)	23.46 (138)
2015	400.6	466	1097	609	63.50 (696)	23.56 (143)
2020	405.9	473	1134	630	64.21 (728)	24.81 (157)
2025	411.3	488	1170	650	64.87 (759)	24.92 (162)
2030	416.7	503.4	1207.1	670.6	65.45 (790)	24.90 (167)

Figure in parenthesis indicates actual deficit.

Source: Based on 10th and 11th five year plan document.

Among the concentrate feed ingredients, coarse grains occupy the primary position. The production of coarse grains such as maize, sorghum, pearl millet and other millets amount to 37.76 million tonnes of which not more than 10 per cent is presently used for livestock feeding. Cereal by-products such as brans and polish and various oil meals including groundnut cake, mustard cake, coconut cake, soybean meal, cotton seed meal and sesame cake amount to about 15.71 million tonnes while various agro-by-products, unconventional products such as molasses, silkworm pupae, distillery waste, wastes and by-products from meat, milk and fish, chiefly meat scrap, whey powder and fish meal are put into use for livestock feeding. Total concentrate ingredients availability is about 35.32 million tonnes. Judging by the present requirements and availability of forage, the deficits may persist and even aggravate unless adequate measures are undertaken to augment these resources. Assuming the requirement of 250 g of concentrate per animal/day for 300 days in a year for cattle and buffaloes in milk; the total concentrate requirement works out to be 5.3 million tonnes. Among the concentrate feed ingredients, coarse grains *viz.*, maize, sorghum, pearl millet and other millets occupy the primary position. Out of total production of such coarse grains (37.76 million t) only about 10 per cent is presently used for livestock feeding. Soybean meal is largely exported. Maize substituted by sorghum and pearl millet for poultry is not preferred for lack of yellowness of yolk in eggs, provided by only maize. The high milk yielding dairy animals is fed more concentrates. The demand for total compound feed is estimated at 42 million tonnes and is expected to increase to 68 million tonnes in near future. The present production is in the range of 6-8 million tonnes only by the members of Compound Livestock feed Manufacturing Association (CLFMA) and other non-members (Table 2.4). This, however, does not include 5.3 million t of concentrate used in the rural backyard livestock.

Table 2.4: Compounded Cattle and Poultry Feed Manufactured by CLFMA Members (in'000 tonnes)

Year	*Cattle and Others*	*Poultry*	*Total*
1990	1325	834	2159
1991	1479	943	2422
1992-93	1475	863	2338
1993-94	1371	877	2248
1994-95	1465	1075	2540
1995-96	1543	1268	2811
1996-97	1551	1361	2912
1997-98	1430	1340	2770
1998-99	1490	1690	3180
1999-00	1630	1900	3530
2000-01	1275	1519	2794
2001-02	1199	1363	2562
2002-03	1176	1477	2653
2003-04	1175	1457	2632

Source: CLFMA, 2005.

Generally, feeding livestock on home grown forage, especially leguminous or a mixture of leguminous and non-leguminous is considered more economical as compared to feeding them on concentrates or crop residues. Liberal feeding of greens or ensiled forage may cut down the costs on concentrates by 33 per cent or even more, if leguminous forage is also added up in the greens/silage. Silage and hay will provide feed particularly during lean periods - summer in plains and winter in hilly areas and balance the availability of forage round the year. Increased demands for seeds of cultivated forages, range grasses and legumes are coming from many parts of the country where livestock rearing is seen as an enterprise and livelihood. Based on the area under cultivation, current demand for seeds of cultivated fodder is estimated at 3, 55, 000 tonnes per annum. The availability of quality seed is only to the tune of 5-6 per cent. The estimated requirements in respect of major forage crops *viz.*, maize, berseem, oat, lucerne, sorghum and pearl millet in near future is likely to increase. It indicates that there is considerable scope of improving seed production through Breeder Seed -Foundation Seed- Certified Seed/truthfully levelled seed chain. The activities in the area of seed production of range grasses are not properly organized. The limited quantity of seed of range grasses and legumes are produced by different agencies being primarily used by the forest departments. There are demands for import of forage seeds, especially in respect of hybrid sorghum (from Australia and USA) and berseem (from Egypt) in the recent years due to absence of institutional arrangement of multiplication of breeder-foundation-certified/truthfully levelled seed, economically unviable venture of seed production of forage crops as well as absence of organized marketing channels

The Weakness

Conclusively, there is lack of statistics/databases on the area under forage crops and forage plant species grown. The area, production and cost of cultivation data is not collected in respect of forage crops by NSSO. Similarly, the database on the availability of forage in different seasons in various regions *vis* a *vis* livestock's requirement are not available. Timely availability of quality forage seeds is one of the important limiting factors. The seed chain from breeder to certified seeds does not exist at national level (NSC) and also the institutional mechanism in most of the states. Despite huge demand of forage seed, indented breeder seed is not fully lifted due to inefficient marketing. Forage section at centre and in most of the states is not adequately manned and in many cases they are vacant since 5 to 15 years. There is lack of awareness about forage production, utilization and marketing aspects among the farmers as well as extension workers. The adoption of the technologies for conservation of surplus green forage during monsoon, efficient utilization of existing forage resources and its value addition is poor. The aspects related to proper livestock care like housing, water provisioning *etc.* receive little attention. In any livestock improvement programme such issues are very important. The nutritional aspects like mineral deficiency do not receive proper attention, resulting in loss of productivity. The flow of funds for forage production and forage related research and extension is low and not institutionalized for up scaling to the level of requirement. The allocation of funds for forage development is very low out of the total fund allocation for the animal husbandry sector. Linkage among various organizations related with forage in some or the other way is missing. A holistic viewpoint what is needed is not in existence. For example seed for forage crops are being dealt by DAC and state agriculture department but functional linkage with DAHF of DAC and States Agriculture Departments, is completely missing. Forage is an intermediate input in livestock production but green forage market is highly unorganized, except peri-urban areas and around cities. The price fluctuation in dry forage prices is quite high, going up to five times of the average price during acute shortage because of natural calamities like drought/flood. Lack of marketing opportunity to sell surplus forage at remunerative prices is also an important factor in extending the area under forage production like other commercial agricultural crops. This aspect needs a regional focus. In spite of the shortages in quality forage supplies for the livestock of the country, the area under forage crops is not expected to increase on account of limited land resources and other requirement for land.

The natural resource base for forage is under severe pressure and already is in a degraded state. The agencies (like forest department) managing these resource bases do not have adequate concern towards forage for the livestock, particularly in forest margin. Transportation costs of planting materials, feed ingredients, bulky crop residues and green forage *etc.* make it economically unviable.

Livestock in National Economy

India has the largest livestock population in the world. Distributed over 100 million households in approximately 6,00,000 villages, Indian farmer's stock animals as varied as the little known Yak and Mithun to the seemingly insignificant backyard poultry. Of the total household in the rural area, about 73 per cent own some form of livestock. More importantly, small and marginal farmers account for three quarters of these households. The importance of livestock in India's economy can be gauged from the fact that 90 million farming families rear over 90 million milch animals. An estimated 70 million rural milch animal households are engaged in milk production. The focus of the poor is on small animals. In India, the livestock production and agriculture are intrinsically linked, each one being dependent on the other and both crucial for the overall food security of the people. It forms an important livelihood activity for most of the farmers, supporting agriculture in the form of critical inputs, contributing to the health and nutrition of the household, supplementing incomes, offering employment opportunities, and finally being a dependable "bank on hooves" in times of need. This is the sector where the poor contribute to growth directly instead of getting benefit from growth generated elsewhere. The rural women play a significant role in animal husbandry and are directly involved in most of the operations relating to feeding, breeding, management and health-care of the livestock.

The overall growth rate in livestock sector is steady and is around 4-5 per cent and this has been achieved despite the fact that investment in this sector was not substantial (Table 2.5). Despite the importance of livestock sector in the Indian economy, especially towards livelihood of resource poor farmers and landless labourers, government policy for the sector has suffered from the lack of clear and strong thrust and focus. One of the indicators of priority to a sector could be judged from budgetary allocation under various plan periods to the sector. Over the past ten five-year plans, the allocation of funds to Animal Husbandry and Dairying has come down from about 1.2 per cent initially to about 0.2 per cent in the ninth plan.

Table 2.5: Share of Agriculture and Livestock Sector in GDP

Year	*Agriculture Sector*	*Livestock Sector*
1993-94	28.39	6.49
1994-95	27.82	6.38
1995-96	25.62	6.20
1996-97	26.16	6.00
1997-98	24.26	5.88
1998-99	24.34	5.73
1999-00	22.93	5.30
2000-01	21.84	5.29
2001-02	21.98	5.17
2002-03	19.44	5.21
2003-04	19.86	4.85
2004-05*	18.47	4.81
2005-06	17.91	4.61
2006-07	16.96	4.39
2007-08	16.34	4.19

* At 1999-2000 prices.

Structure of Livestock Farming in India

Mixed crop-livestock farming and pastoralism are the two common production systems found across rain fed agriculture zones. In the former, farmers derive their livelihood somewhat equally from agriculture and livestock; in the latter, people's livelihoods depend primarily upon their livestock, which are exclusively maintained on grazing. Dry land and alpine-temperate hilly regions traditionally harbour the 'grasslands' of India, providing pasture/grass for some parts of the year. In these harsh climates with minimal precipitation, sustained agriculture through the year is extremely difficult and it is livestock, which has historically played an important role in people's livelihoods. Livestock are important in their livelihood culture and they have limited alternative opportunities for employment. Studies have shown that development of small holders' mixed crop - livestock production is one of the most effective methods of poverty alleviation.

In contrast, the resource-rich section of the rural population also utilizes livestock development for optimizing their wealth. This section has the means and ability to obtain the desired variety of livestock and the inputs and services needed to harness them. From the point of view of the national economy as well as exports, output of this resource-rich segment of the population is also important.

The peri-urban production system has developed around cities and towns that have high demand for milk. The peri-urban dairy farmers rely mainly on purchased feed. They are commercially oriented and respond to improved technical input supply and marketing services. Buffalo has been a favoured milch animal in the system but these farmers have also improved productivity of their cattle by taking to cross breeding

Presently, only a very small fraction of the livestock sector exists as industrialized system. Examples include commercial poultry farms, dairy farms and a few commercial goat and pig farms. Since current trends indicate that increasing share of the supply will be met by industrialised production given economies of scale, increasing labour and declining capital costs, there is a need to introduce favourable policy changes so that small producers are able to benefit from this demand and compete with organized industries. These policy changes include vertical integration of small producers with livestock food processors through contract farming, improving the efficiency of their operations and the productivity of their animals, which largely depends upon improved research and availability of services.

This will require overcoming feed and forage scarcity and improvements in delivery of animal health and breeding services. Technology will be a key driver of growth and concerted efforts will be needed to generate and disseminate yield-enhancing and yield-saving technologies.

Review of Five Year Plans

The efforts made during the 10th five-year plan in raising the feed and forage resources for the livestock were not very successful. The performance of central forage development organization of Government of India was evaluated by the Centre for Management Development, Thiruvanthapuram and found to be unsatisfactory.

Small ruminants play an important role in the rural economy. They are primarily maintained by poorer section of the rural community providing them a source of livelihood. However, this sector appears to have been neglected. During the last four decade, there has not been much increase in sheep population. On the contrary, despite any serious developmental programme, the goat population has increased at a very fast rate making it an important species of animal for meat production.

Market opportunities have opened up for the livestock sector following the economic liberalization. There are expectations of faster growth in demand for livestock products due to expected increase in income combined with the high-income elasticity of demand for livestock products. Export earnings from livestock products rose from Rs.1500.93 crores in 2001-02 to 2253.33 crores during 2004-05. Meat and poultry products accounted for 83 per cent of total export earnings. Export of milk and milk products also improved and accounted for 17 per cent of the export of livestock products. The sector's ability to capitalize on new market opportunity is constrained by the availability and quality of support services. Although, the possibility of export of livestock products is immense but it is not fully harnessed because India is unable to adjust effectively to the open trade regime under the WTO particularly in respect of requirements under Sanitary and Phyto-sanitary (SPS) regime.

The goals for the 11th five year plan with an outlay of 8,174 crore for the livestock sector was i) to achieve an overall growth between 6 per cent to 7 per cent per annum for the sector as a whole with milk group achieving a growth of 5.0 per cent per annum and meat and poultry group achieving a growth of 10 per cent per annum, ii) the benefit of growth should be equitable, benefiting mainly the small and marginal farmers and landless labourers and should benefit poorly endowed areas like drought prone, arid and semi-arid areas iii) the sector should generate additional employment opportunity to people in the rural areas especially to the female population, iv) livestock should provide major source of income in the selected areas having potential for mixed crop-livestock farming system, and v) the growth in the sector should result in the improvement of environment specially in the rural areas. During 11th plan, growth of value of output was about 4.8 per cent.

In 12th plan (2012-17), an allocation of Rs. 14,179 crore was approved. In this plan, National Livestock Mission had provision of Rs. 2800 crore, having sub-mission component "Feed and Fodder" for promoting high yielding forage varieties, optimum land utilisation, perennial forage grasses as well as legume planting and intercroping forage crops with food crops. The concept of fodder bank and its spread throughout the country, was also included in the 12th plan.

At present, Government is the main provider of support services. The quality of the services is however not satisfactory and these services are not available at the doorsteps of the producers. There is a need to restructure service delivery mechanism to become conducive to the requirement of the rural livestock producers. Lack of credit for livestock production has been another major problem

The Department of Animal Husbandry and Dairying is managing large infrastructure of livestock farms and forage production stations. Many of the infrastructures are out dated and have not kept pace with the development of science and technology. An exercise on restructuring the existing infrastructure needs to be taken up on priority basis.

Adequate availability of quality forage is essential for enhancing livestock productivity. For this, there is need to target at least ten per cent of the cultivable land for growing forage crops. Since major limitation to increasing forage production is insufficient availability of forage seeds, a programme on forage seed production through registered growers was launched in collaboration with State

Agricultural Universities and State Seed Corporations. Assistance was provided for better use of crop residues through industrial manufacture of feed blocks using crops residues, briquettes, and utilization of agro-industrial by-products. The projects on development of Common Property Resource (CPR) and fodder banks was implemented as public private partnership initiative (PPP).

Major policy reforms are required to provide support to small ruminant development programme. The grazing policy for livestock in forest including joint forest management with particular reference to high altitude forest needs to be developed and jointly implemented by Animal Husbandry and Forest Departments. Regeneration and development of common property resources and wasteland with involvement of *Panchayats* and NGOs need to be given greater attention. It was proposed to launch this programme in 40 backward districts in 6 states having high population of sheep and goats.

Animal Welfare during natural calamities and disaster will require attention since such calamities can drive the poor into destitution. A disaster management fund for livestock was created to support livestock owners during calamities. NGOs working for livestock welfare were strengthened so that they can ensure and promote animal care and wellbeing.

Innovations in Public Sector Technology Dissemination – ATMA Model

The traditional system of extension in the State Governments suffered from several constraints. The linkages between research-extension and farmer remained weak or non-existent. Media and information management was not suitably addressed. The system was supply-driven with scarcely any accountability to farmers. Technology generation systems were also largely in a top-down mode with very weak mechanisms for feedback from the client farmers.

Under the current extension system in most of the states, the Departments of Animal Husbandry (DAH) have no extension machinery below the district level. The veterinary officers at taluka/block level have a clinical approach and are involved in the treatment and surgery of the animals, which are brought to them rather than in the delivery of extension services at the farmers' door-step.

Despite the growing influence of the private sector, public extension would continue to remain central to technology dissemination and small and marginal farmers and economically backward regions will need to be serviced by it. This implies that public extension functionaries will have to be placed in new decentralized institutional arrangements which are demand-driven, farmer-accountable, bottom-up and have a farming systems approach. States have before them the model of the Agriculture Technology Management Agency (ATMA), which was pilot project, tested under the National Agriculture Technology Project. The ATMA model has now been adopted in half the total districts in the country. With supplementation from the private sector, media and information technology, the public extension service would be made leaner and more professional. Under this model, the public extension functionaries in the form of subject matter specialists will function only up to the block level, below the block extension activities will be carried out by the private service providers, community-based organizations, farmers' cooperatives, agri-businesses, agri-clinics and private service providers, para-vet-professionals.

Research-Extension Linkages: Validation of Technologies

Currently, there is no formal system, which can be considered for livestock farmers' feedback. This requires establishing a system by which the research in field is tested by various resource group livestock owners. This will bring out more location and resource specific scientific recommendations to suit the needs of the various livestock farmers. Livestock technology can only be passed on to the farmers with success, if the transfer of technology is undertaken under the supervision of qualified and trained livestock extension professionals. There should be institutional arrangement at each KVK to strengthen livestock component with placement of animal science qualified persons trained in extension methods to carry out the field trials on livestock production. In the absence of a proper formal mechanism for technology validation, the technologies developed by the scientists are being also self-validated by them, giving ample scope for exaggeration of their efficacy and impact for quick gains and popularity for self, which may be detrimental in long run. A system needs to be put in place where the technologies developed are validated independently through cross validation across organizations.

Technology Assessment and Refinement (TAR) through Institute-Village Linkage Programme (IVLP) in 70 centres located at the ICAR research institutes and SAUs did appreciable work in agriculture extension during the X plan. It should be continued with emphasis on technology related needs of the livestock farmers by organizing farmer participatory meetings and farmer-scientists interface.

Impact of WTO Agreements on the Indian Livestock Sector

India's trade in milk products – both exports and imports – is highly influenced by the world trade policies. High production and export subsidies provided by several developed countries distort international dairy market to the disadvantage of many developing countries including India. It was expected that the implementation of Agreement on Agriculture (AoA) would substantially reform the trade of agricultural products. This has, however, not happened and high market access barriers are being maintained. Export subsidies also continue to be a significant factor in world dairy trade. The quantity of dairy products eligible for export subsidies, even after reduction of commitments, is close to 60 per cent of estimated world trade in all products. The global prices are significantly influenced by these subsidies.

Extension and Technology Transfer

Whereas public extension played a major role in technology and knowledge transfer in the crop sector, in the livestock sector the concept of extension service delivery has been very weak. Despite its growing importance, livestock extension is a field neglected both by policy makers and by researchers. Agricultural extension services have developed around crop production, and remain tied largely to the seasonal nature of cropping. Such system is less useful for livestock production, with a longer time-scale and a lack of synchronization of different animals and herds. The institutional arrangement in the state departments of animal husbandry which are mainly run by veterinarians have primarily emphasized the clinical and diagnostic aspects of animal's health rather than the preventive and extension aspects.

Growth Potential

The Planning Commission had suggested a growth rate of 8 per cent in livestock sector during the 11th Five Year Plan period. High growth especially in milk was not achieved due to requirement of huge investment in the processing, value addition and market development including exports. The two growth scenario with an overall 7 per cent and 6 per cent growth rate in livestock sector were presented in the Table 2.6.

Table 2.6: Growth Potential in Livestock Sector (Annual Growth in Per cent)

Sector/Sub-Sector	*Annual Growth 1981-2000*	*Potential Growth*	
		Situation I (High Growth)	*Situation II (Moderate Growth)*
Overall for Sector	4.5	7	6
Milk Group	4.7	6	5
Meat Group			
Buffalo Meat	6.7	10	10
Poultry Meat	9.6	10	10
Goat Meat	3.9	5	4
Pig Meat	8.0	8	8
Eggs	6.7	10	10
Wool	1.5	2	0

Institutional Restructuring

The key pillars of Institutional structure providing services in Animal Husbandry and Dairy Sector are; i) Government, ii) Private Sector, iii). Producer Institution, iv) Autonomous Bodies and v) NGOs. These institutions do not have a well-defined role and their functions overlap each other making assessment of their performance difficult. There is a need to define the role of various institutions to make them more effective. A role restructuring is indicated below:

Government (Central and State Governments) should concentrate on the following:

1. Policy formulation and implementation.
2. Provision of public goods including physical infrastructure.
3. Regulation of private activity and promotion of competition.
4. Technology development.
5. Overall governance and promotion of credible and secure legal environment.
6. International interfacing.

Private Sector should concentrate on following:

1. Production, marketing, input supply and service support.
2. Technology development.
3. Compliance with regulatory structure.

Producers Institutions should focus on following:

1. Representation of producer interests in policy making and resource allocation decisions.
2. Promotion of producer interests in emerging markets.

Autonomous Bodies should focus on following:

1. Information generation and dissemination.
2. Policy analysis and promotion of stakeholder dialogue.
3. Technology development.
4. Capacity building.

NGOs should keep their focus on following:

1. Information generation and dissemination.
2. Complimenting government's development role.
3. Equitable participation.
4. Capacity building.
5. Watchdog function.

Important sectors where PPP can be promoted on priority are listed below:

1. Public Private Partnership for Joint Management of Public Infrastructure and Resources.
2. Role of PPP and NGOs in Delivery of Livestock Services;
3. PPP in Integrated Small Ruminant Production and Marketing;
4. Management of Community Pastures through PPP;

5. Establishment of Community Fodder Banks linked with Complete Feed Production;
6. Disease surveillance, prevention, control and eradication;
7. Disease investigation and diagnosis;
8. Extension and technology transfer;
9. Research and Development;
10. Data Collection, Information Management and Sharing.

Feed and Forage Resources

Assessment of On-Going Schemes

The schemes and programmes relating to feed, forage and pasture development in the country are quite limited. The Department of Animal Husbandry and Dairying, Government of India, is the principal governmental agency in this regard. In addition, the Departments of Animal Husbandry, Agriculture, Forestry, Environment, Rural Development, State and Central Agricultural Universities, ICAR Institutions, National Seed Corporation and various local bodies such as *Panchayats*, municipalities and corporations are engaged in the task.

A Central Fodder Development Organization was formed in IX Plan, merging the 7 regional stations for forage productions in the country, a Central Fodder Seed Production Farm and a Central Mini kit programme. The objectives included introduction of forage crops, formulation of forage calendars, organization of farmers' field days, production of forage crop foundation seeds, conduct of training programmes and distribution of forage seed mini kits and testing their performance in the field.

In addition to the central schemes, there are some centrally sponsored state level schemes which are envisaged to implement programmes for strengthening the State forage/forage seed production farms, grassland development, silvipasture development, forage seed production through registered growers and enrichment of straws and other cellulosic wastes. During the Xth plan, assistance to forage block making units, grassland development, forage seed production, biotechnology research interventions had been taken up. Though about Rs. 45 crores each for central schemes and centrally sponsored State schemes were outlaid during the Xth plan, financial expenditure was to the tune of not more than Rs.28 crores in each of these groups of schemes.

The Centre for Management Development, Thiruvanthapuram made an impact assessment of the Scheme of Central Forage Development Organization. The study revealed that the overall performance of the Central Forage Development Organization is not producing result as envisaged and the impact of these activities on the farming community and the society as a whole was too little, especially when compared to the extent of the funds spent. The seed production performance of Regional Seed for Forage Production and Development (RSFP&D) and Central

Forage Seed Production Farm (CFSPF) was far from desired and extension service failed to popularize the high yielding varieties of forage. The R&D to identity new varieties and develop package of practices suitable to the region remains neglected.

The study also reveals that the Central Mini Kit Testing Programme could not create any tangible impact on the mandated objectives of popularizing high yielding varieties of forage, attracting new entrants to the forage development activities and bridging the gap in demand and supply of forage, mainly due to the organizational and operational issues.

The evaluation report suggests that commercial seed production should be organized through certified seed producers. The Regional Forage Stations should have the focus objectives on producing foundation seeds and organizing certified seed production through registered seed growers. The station should perform as nodal agencies for transfer of advanced technology in forage seed production. This would call for discontinuation of the present farmer's training/farmer's fair and focusing on 'Trainer's Training' and Research and Development on forage seed production.

Study suggested that the Central Mini Kit Testing Program should be redrafted, restricting area of operation to five intensively farming states in the first round and then shifting to another five states in the next round. Dairy cooperative would be ideal state partner for implementing the scheme.

Feed and Forage Development

Enhancing Forage Production

Adequate availability of quality forage is essential for enhancing livestock productivity. Swaminathan Second Report on "Farmers commission" (2005) suggested that forage and feed availability should be ensured by promoting integrated management of grazing lands, forage production and stall feeding, creation of forage and feed bank and by establishing Livestock Feed Corporation. With proper planning and programme, the forage production, particularly, the cultivated green forage can be doubled. For this, there is a need to target at least 10 per cent of the net cultivable land for growing forage crops.

Appropriate strategies are needed for effective utilization of existing grasslands based on carrying capacity and nutritive status. Institutional mechanisms have to be developed for active community participation and adoption of optimal system of rotational and deferred grazing. Grasslands can be better utilized by rotational grazing of cattle, sheep and goats. In fact, it will be most valuable to restrict cattle and goats to in-house rearing conditions since these species are most amenable to stall-fed maintenance. Large scale reseeding of pasture lands with legumes and non-legumes, introduction of tree fodder and forage shrubs in the grasslands and wastelands, use of orchard interspaces for forage production and forage leaf meal production are among the strategies which need to be taken up extensively. There is a need for greater collaboration among forest, agriculture, animal husbandry departments and local bodies for development and implementation of these strategies.

However, there are formidable limitations in exploiting the scope for enhancement of green forage production. In cultivable lands, green forage production is marginal, subsidiary and non-commercial unlike food, horticultural and other field crops. Lands utilized for forage cultivation are invariably uncultivable, semi-cultivable or arid which, though may be reasonable from land use perspective, and would naturally limit the forage production and productivity. Lack of data and inventory on forage production scenario is a hindrance to planning and formulation of schemes. Forage production is primarily a land use activity. It however does not attract crop scientists for research and technology development since forage crops are still considered as low priority crops.

Forage Seed Production

The demand for forage seeds is increasing and the availability is very low, meeting only 15–20 per cent of the requirement. A number of promising forage varieties have been developed which can be incorporated into the seed production chain. At present, forage seed market is not very organized. Timely availability of quality seeds is also a major constraint. Often seeds produced by agencies against specific requisitions are also not lifted, creating difficulty in utilizing these seed materials. One other important handicap is that the quality standards for forage seeds and database on requirement and availability have not been adequately developed.

Forage Preservation

Forage preservation methods such as wet preservation in the form of silage and dry preservation in the form of hay are well known, established, time–tested technologies adopted all over the world. The objective of preserving green forage is either to save the surplus material for use during the lean period or to harvest the green forage in right time to recover maximum nutrients and preserve them so that the preserved material can be used separately or as part of a total mixed ration. Under Indian conditions, the adoption of these preservation techniques is by and large limited to organized government/few private dairy farms. The non-adoption of these methods are due to several reasons, the chief among which being, cultivation of forage is scanty and hardly any surplus is available and small and scattered land holdings with few herds of livestock do not enthuse the farmers to undertake such laborious and manual tasks, particularly, when the livestock rearing is not a primary occupation with most farmers. However, large scale commercial production of forage, their timely harvest and preservation into hay or silage will help in the manufacture of feed blocks and total mixed rations by the industry for delivery of the same to the doorsteps of intensive, commercial livestock units which are emerging of late, in many parts of the country. In many parts of the country, however, large quantities of paddy straw, wheat straw and sorghum stalks are stored in dry condition and used throughout the year.

Crop Residues

Paddy straw and wheat straw are abundantly available in the country within the farm holdings. They constitute the basic, bulk, roughage materials for large ruminants like cattle and buffaloes. Other cereals and millets also yield substantial quantities of residues of which sorghum and pearl millet straws are very important. Various pulse crops and oilseeds also provide residual material in the form of their stems and stalks to form part of the roughage material. Sugarcane, one of the important agricultural crop yields both sugarcane tops and *bagasse* which are quite edible for the ruminant stock. Availability of these crop residues depends by and large on the acreage under the main products. It must be kept in mind that competitive demands for crop residues from paper, cement and packaging industries exist and may increase. Most of the edible biomass in future may be diverted to bio-fuel production, which would require the livestock industry to formulate alternate futuristic strategies. Crop residues can be improved in their nutritive value through ammonization and urea treatment. Baling, densification, making of blocks, pellets, briquettes *etc.* and nutritional enrichment by microbial means and incorporation of other ingredients to crop residues to make balanced rations are the various feed technology measures available for their better utilization.

Biotechnology Interventions

Biotechnology measures have become a handy tool for improvement of nutritive value of feeds and better nutrient utilization. Feeds, particularly crop residues are improved through microbial, chemical and physical processes by splitting of ligno-cellulosic bonds of the fibrous matter. This improves digestibility and better availability of cellulose and other digestible fibre for rumen microbial degradation. Microbial processing also enriches the crude fibrous material with microbial protein. The rumen metabolism can be manipulated with advantage by using anaerobic fungal probiotics, resulting in better fibre digestion and microbial protein synthesis. Microbial fermentation invariably results in some methane production, to the tune of 5–7 per cent of the calorific value of the dry matter consumed. This is wasteful and environmentally detrimental. Reduction in particulate size, higher concentrate or grain feeding and introduction of propionic bacteria into the rumen along with the feed can control methane production. Use of ionophores and unsaturated fatty acids also helps in reduction in methane production. Biotechnological means can also be employed for multiplication of plant material through tissue culture and genetic improvement of forage plant varieties.

Regulatory Mechanism for Feed Quality Control

Presently, standards and specifications for several feed ingredients and compounded feeds for livestock and poultry are available under the Bureau of Indian Standards. However, there is no control or regulatory mechanism available for imposing these standards as mandatory. CLFMA has evolved its own standards for cattle feeds. However, there is an urgent need for formulation of standards and specifications for large number of feed ingredients and feed formulations for different categories of livestock and poultry. Accredited laboratories have to be established and the equipment and procedures adopted in these laboratories must be uniform and highly sophisticated. There is also a need for development of adequate human resources to undertake quality control work through, chemical, physical, biological and micro-biological methods. Since the whole exercise encompasses the monitoring of food chain to ensure sanitary and phyto-sanitary safety, the quality control measures should invariably include several pollutants, contaminants and toxicants either natural or those introduced into the food chain.

Farming community and households must be actively involved in establishment of intensive farming units, forage seed production, production of coarse grains, cultivable and commercial forage production and storage and utilization of farm by-products such as straws, stovers, rice brans, oil meals, vegetable by-products and other edible biomass. Public institutions, governmental agencies and local bodies must be collectively engaged in the monumental task of grassland and community land development and development of forest lands for forage production, establishment of fodder banks, forage seed production, agro-processing infrastructure, support for the establishment of commercial feed supply units and intensive livestock/poultry farm units.

Industrial participation and entrepreneurship building are most valuable to undertake and execute several innovative schemes, which incorporate within them latest technologies on feeds and forage development. Their tasks are related to scaling up the production of concentrate mixtures, total mixed rations, complete feeds, high-energy ration and pellets. New technologies such as bypass nutrients, probiotics and feed enzymes can be taken up on industrial production scale. Production of location specific mineral mixtures, trace element supplements, urea-molasses-mineral blocks, granules and salt licks have to be enhanced by the participation of the industry. Manufacture of feed blocks using crop residues, briquettes, utilization of agro-industrial by-products, rice bran, oil meals, coarse cereals, vegetable, fruit by-products, sugarcane, distillery, meat, milk by-products in large organized scale is possible only through industrial involvement. Even forage seed production, production of novel feeds like *azola*, leaf meal and incorporation of preserved forage such as hay and silage in total mixed rations can be taken up through intense entrepreneurship and organized industrial efforts.

R&D institutions have great stakes in feeds, forage and pasture development. Research and technology interventions are needed in the aspects of feed technology and processing, determination of nutrient requirements, evaluation of economic feed formulations, nutrient utilization studies, post-harvest measures for feeds, forage and seeds, feed supplementation, feed additives, novel feeds and agro-industrial by-products. Extensive evaluation and performance studies have to be carried out under in-house rearing systems of livestock. Field level pilot projects, training and demonstration activities have to be undertaken. It is also important to take up studies of feeds-livestock-environment relationship.

Public Private Partnership Initiative

The strategy for development of Common Property Resources (CPR) and forage banks has not been successful primarily because such projects were implemented without public participation.

Development of Community Pasture for Forage Production

The Common Property Resources (CPR) consisting of revenue wastelands, Panchayat land and the forest land which constitute the most important input for livestock production and subsistence for the poor. These are under depletion and degeneration affecting the livelihood security of the poor. There are innumerable successes in CPR development and management through community participation. Based on these models, broad frame work/norms for investment of resources, community contributions and inputs (*e.g.* labour), use of outputs, distribution of income, stocking rate, breed improvement, interagency coordination, *etc.* should be agreed by the local community and formalized. The Panchayat, and the Departments of Forests, Revenue and Animal Husbandry should endorse these and cooperate. Local NGOs/Civil Society Organizations should be encouraged and trained to facilitate.

The common land should be delineated on the ground and tenurial, usufruct and long term management rights given to the user groups/associations. All communities dependant on CPRs including the landless, livestock and non-livestock owners should actively participate. CPRs available should be divided into sections, some fenced and developed for regeneration and the others for immediate continued/controlled grazing until the first part starts yielding. Community controlled social norms and codes should be followed to generate revenue through distribution of produce for sustainability. Continuous degradation of CPRs are already reaching a point of no return, jeopardizing the livelihoods of lakhs of landless. It is high time that government took steps to reverse the trend at costs, which are not very high.

CPR user groups/associations need to be organized through facilitation by local NGOs and Panchayats. The Government support would be mainly for (i) Delineation of the land and assigning user rights, (ii) Facilitation support through the three Departments. and (iii) Initial financial support for land development, live fencing, pasture regeneration, watch and ward, institutional support and costs of Civil Society Organization *etc.* until reasonable yield begins to emerge (in about 4 years). Further, sustainability would be on contributions from user groups as labour inputs, payment for produce *etc.*

Community, NGO-Government Partnership for Management of Fodder Banks

Acute forage scarcity is a common phenomenon in arid zones in different parts of the country, where forage prices rise steeply during drought periods and small and marginal farmers are compelled to get rid of their livestock due to inability to feed them. In such areas, establishment of community fodder banks can be a boon to sustain their livelihood even under distress conditions. Dr. M. S. Swaminathan Committee Second Report (2005) also suggested setting up a "Livestock Feed Corporation of India" on the pattern of Food Corporation of India. Suitable strategy should be developed to sustain the activities of fodder banks in arid regions.

The operation of fodder bank on economically viable scale is essential. The responsibility of fodder bank management should be entrusted to local community or dairy farmers' organizations or NGOs who are active in this field. With fodder bank, the production of complete feed can also be undertaken as it has good demand in the local market as well as in peri-urban areas, even during normal seasons. Depending on the availability of crop residues, the surplus forage can be converted into complete feed, while avoiding the wastage and recovering the cost of operation. The organisations setting up fodder banks can also establish their procurement units in forage surplus areas, where agriculture by-products are burnt due to lack of demand.

Livestock is the main source of livelihood for small farmers and landless in drought prone areas. However, in the absence of adequate forage supply, particularly during scarcity, these farmers are compelled to either migrate or discard their livestock, which will force them into chronic poverty. Therefore, it is necessary to support them through supply of forage at affordable price.

In the proposed initiative the organizations/District level Dairy Farmers' cooperative or NGOs would set-up the fodder bank cum complete feed production units. The District Rural Development Agencies would provide finance to cover the shortfall for operating the forage banks. The Forest and Revenue Departments will facilitate collection and transportation of forage from various sources. The Animal Husbandry Department will monitor the quality of the feed, provide information on demand and supply of forage and assist in disbursement of surplus forage in other scarcity areas.

Recommendations

The suggestions in the major areas *viz.*, General Organization of activities; Pasture development with Community involvement; Forage seed production; Regulatory measures and Forage research are :

Area of Activities/Recommendations

General Organization of Activities

- ☆ Bring more area under forage production as presently practiced in Punjab, Gujarat and few other states through food – forage cropping systems.
- ☆ Producing high yielding varieties of forage seed to make forage production more remunerative. Encourage forage seed production in states by making it economically viable.

- ☆ Rehabilitation of degraded lands, management of permanent pasture utilization of forest resources, forest marginal lands and village common property lands, silvipastoral systems for forage production
- ☆ Enrichment and judicious use of crop residues. Stop practice of burning stovers in certain surplus zones such as Punjab, Haryana *etc.*
- ☆ Establishment of fodder banks in flood/drought prone areas by bringing crop residues from surplus zones to meet the forage requirement during natural calamities and scarcity.
- ☆ NSC should give due priority to seed production of forage crops and have a functional relationship with AHD and State AHDs.
- ☆ Need to promote research-cum-capacity building programme in drought prone areas using watershed development as a core strategy to mitigate the impact of drought.
- ☆ Evolve mechanism to review the forage production situation including seed industries in the country.
- ☆ Technical linkages between IGFRI, Jhansi and state/central forage production and demonstration farms and AHDF need to be established.
- ☆ Development of low cost feed formulations complete feed based on locally available ingredients including agricultural by-products and top feed.
- ☆ Replacement of concentrate mixture by leaf meal to the extent possible for economizing the cost of production of concentrate.
- ☆ Use of mineral mapping in the country and area specific mineral mixture.
- ☆ Alternate land use system like agro-forestry and silvipasture on wasteland will integrate the concerns for productivity, conservation of resources and environment and profitability, They should be promoted to produce more biomass per unit area and time.
- ☆ Efforts should be made to develop forage crops based feeding systems for different agricultural crops grown in different zones of the country.
- ☆ Creation of end product market – sink - The dairy system or suitable markets for end product like milk, wool, meat etc should be created which will encourage the development of dairies and thereby better quality forage and seed production.
- ☆ Intensive forage production system in resource rich areas - Use of Intensive forage production system in achieving maximum sustainable harvest of nutritive herbage per unit area and time by efficient utilization of land and other farm inputs without jeopardizing the productive capacity of the soil. It includes Multiple cropping,/Overlapping Cropping, Parallel cropping, Mixed/Intercropping systems for quality herbage production.
- ☆ Promoting peri-urban forage markets by increasing infrastructure facility and price regulation like food crops (MSP)

Pasture Development with Community Involvement

- ☆ Improvement of degraded grasslands, reclamation of wastelands, rehabilitation of problematic soils like sodic, acidic and heavy soils through edible vegetation cover of grasses and silvipasture/horti pasture.
- ☆ Govt. of India (Department of Animal Husbandary, Dairying and Fisheries) provides 100 per cent subsidy for rehabilitation of degraded pastures/ grasslands to states which should be fully utilized for forage resource development.
- ☆ Need be focused on programme on regeneration, promotion of silvi-pasture or *Gochar*, ravines and wastelands, community village lands. This will not only meet shortage of forage but will provide equal access to poor livestock owners.
- ☆ Collaboration of village community, NGO and state government is needed for exploitation of common property resources and community lands.
- ☆ Controlled grazing practices are promoted. Regulating grazing through local community based institutions in tune with production capacity will ensure sustainability of traditional pasture. Pastoralist be educated about benefits of regulated grazing.
- ☆ The grazing resources should be developed as per the grazing habits and pasture requirements of particular animal species.
    1. For the browser animal species like goats, grazing land can be predominated with nutritionally superior shrubs and tree species.
    2. For grazing livestock like sheep and cattle proper balances of superior quality grasses and legumes may be the major component of the pasture along with tree fodder for lean period.
- ☆ According to the carrying capacity and availability of forage plant species, grazing of mixed herd like cattle, sheep and goats should be adjusted intermittently so that quality forage can be made available round the year.
- ☆ The improved grazing management systems like, rotational, deferred rotational systems can be practiced to improve the quantity and quality of pasture from degraded land.
- ☆ Supplementation of energy, protein and mineral during lean period at individual level or in combination of various proportions may be exploited for economic livestock production under grazing condition.

Forage Seed Production

- ☆ Development of forage seed production chain from nucleus seed, breeder seed , foundation seed to certified seed/TFL as existing in cereal and other important field crops.
- ☆ Strengthening Breeder's seed and linking with production of foundation and certified seed production with state and central seed agencies for production of major forage crops

- ☆ Mandating National Seed Corporation (NSC) for forage seed production and distribution network.
- ☆ Develop suitable network for production of forage and range grasses seed and its delivery up to block level by involving agencies like NSC, Joint Forest Management (JFM) *etc.* It will help in ensuring availability and market for seeds.
- ☆ Encouragement and incentive to farmers (*e.g.* buy back guarantee, minimum support price *etc.*) with small farm holdings for forage seed production in a participatory manner.
- ☆ Registered forage seed growing enterprises may be encouraged.
- ☆ The forage seed multiplication farm in each directorate of State Animal Husbandry setup is required in line with the milk unions of Gujarat, Bihar, Andhra Pradesh *etc.*
- ☆ Establishment of a nodal agency to coordinate production and marketing of quality forage seeds.
- ☆ Preparation of database at each state level about availability and requirement of forage seed.
- ☆ Forage seed reserves should be developed for forage crop security as National Seed Reserve Policy
- ☆ Area specific seed production targets should be given and closely monitored.
- ☆ Improved varieties to be introduced with target at different seed production units.
- ☆ Perennial grass as well as range legume seeds should be incorporated to improve rangeland productivity.
- ☆ Quality standards of forage seed to be developed and enforced.

Regulatory Measures

- ☆ **Feed and Forage Development Authority:** A separate distinct authority may be established with in the Ministry of Agriculture with the necessary professional manpower to undertake the inter-departmental/agency coordination in forage production, forage seed production, forage conservation *etc.*
- ☆ **Legislation to regulate the quality of compounded feed:** It is felt that legislation to regulate the quality of compounded feed sold in the market be put in place. At initial level, packaged feed should contain the information on the composition and the ingredients used.
- ☆ **Establishment of reference forage quality and seed quality laboratories:** For effective enforcement of legislative provisions,

development of critical parameters in respect of feed quality control must be introduced. Rapid screening methods have to be evolved and applied in practice for moisture, fibre, protein, silica and various intrinsic and extraneous contaminants. Such regulatory measures may require both legislative and administrative sanction by the Centre and the State Governments.

- ☆ National Sample Survey Organisation (NSSO) should take up creating database on various feeds and forage resources, feeding practices: Inadequate information is available on feed and forage availability and consumption pattern by the livestock under various field conditions in villages/towns/metro cities in different regions. In the absence of this information all kinds of estimates about feed and forage consumed can at best be treated as informed guesses. Therefore, there is an urgent need of a central sector scheme for establishing a database on various feeds and forage resources, feeding practices and consumption pattern in various agro climatic regions of the country, with the provision to upgrade it periodically. This would be helpful in optimum utilization and reducing regional imbalances.

Forage Research

- ☆ Mission mode approach should be developed by forming a consortium of IGFRI with other National Agricultural Reasearch System (NARS) institutions and organizations in private sector to gear up the forage research in the country.

Present Technology Needs and Priorities in Relation to Agro-climatic Regions of India

Western and Central Himalayan Region

- ☆ Development of suitable technology for terrace cultivation on sloppy land, Clover management for increased herbage yield and quality
- ☆ Introduction of quality forage crops/varieties
- ☆ Technology for wetlands of low elevation areas
- ☆ Soil conservation practices to minimize the soil erosion
- ☆ Integrated effort to supplement grazing with cultivated forage and terrace risers/bunds and hill slopes planted with perennial forage; enrichment and forage conservation, residues of paddy straw and other agro-wastes
- ☆ Assessment of the high altitude grazing areas in terms of productivity and carrying capacity
- ☆ Meeting the forage requirement of migratory herds of livestock at the foot hills during winter

Northern Plain Region

- ✩ Standardization of forage production technology for quality forage
- ✩ Incorporation of forage crops in existing food based cropping systems
- ✩ Introduction of short duration forage crop during fallow period (summer) in irrigated area
- ✩ Emphasis on dual purpose cultivars of pearl millet, sorghum and stay green maize
- ✩ Need to select adaptive plant type to harness the positive associative effects of cereal-legume mixed cropping situations for forage production
- ✩ Development of suitable agro-forestry/agro-silvipasture systems for wastelands and problem soils

Central Plateau Region

- ✩ Intercropping of forages in prevalent cropping systems
- ✩ Introduction of more productive and nutrient rich in quality forage crops and tree fodder
- ✩ Improvement and management of degraded waste lands through forage based system
- ✩ Multi-tier system of forage and agro forestry
- ✩ Multifaceted management strategy including *in situ* moisture conservation and rainwater harvesting technology

Eastern Plateau and Chhota Nagpur Region

- ✩ Area expansion under quality forages
- ✩ Integration of suitable forages in the cropping practices of the region
- ✩ Forage production strategies for fallow and cultivable waste lands
- ✩ Suitable forage production technology for acidic soils
- ✩ Watershed based management strategies
- ✩ Forage based agro-forestry for augmenting feed resources during lean periods

Western Semi-arid and Arid Region

- ✩ Identification of alternate forage resources for weather aberration/ drought periods *viz.*, top feeds, bushes and perennials
- ✩ Inclusion of short duration legumes in food-forage cropping systems
- ✩ Raising shelterbelts through fodder trees
- ✩ Efficient water harvesting and recycling strategies using drip/micro irrigation practices
- ✩ Contingent crop planning for forage during drought

- ✰ Introduction of underutilized forage plant species suitable for arid climate *e.g.* edible cactus; promotion of trees like *Khejri, Anjan, Ardu etc.*
- ✰ Increase the pasture productivity in dry areas by replacing low yielding annual grasses with high yielding perennial grasses adapted to the region. and their optimal utilization to prevent overgrazing

West Coast Plains and Hills Region

- ✰ Round the year forage production technology
- ✰ Incorporation of legumes and forage bushes for high protein herbage
- ✰ Intensive management of forage crops/sequences
- ✰ Integration of forage crops, grasses and legumes and forage tree/shrubs/ bushes under multitier systems

Island Region

- ✰ Introduction of fast growing perennial grasses suitable for tropical humid environment
- ✰ Forage production strategies in rice fellow
- ✰ Incorporation of cultivated forage crops and grasses in coconut plantation
- ✰ Supplementation of grazing by cultivated forage

Table 2.7: Specific Training Needs for Various Stake Holders in the Areas of Forage Production and Utilization

Stakeholder Category	*Training Needs*
Farming Community and Households	(i) Establishment of intensive livestock farming units
	(ii) Commercial forage production technologies on cultivated land
	(iii) Forage seed production
	(iv) Production of coarse grains to economize concentrate
	(v) Grazing land management/silvipasture
	(vi) Effective Utilization of farm by products such as straws, stover, brans, oil meals, vegetable by products and other edible biomass
	(vii) Forage preservation schemes
Pastoralists	(i) Grassland managements – benefits of controlled regulated grazing
	(ii) Ideal pasture for mixed grazing system
	(iii) Grazing routes and grazing systems
	(iv) Designing appropriate production system for migratory graziers
Central, state and Non-Government Agencies	(i) Establishment of commercial feed supply units and intensive livestock farm units.
	(ii) Grassland and community land development.
	(iii) Forge production from forest fringes and JFM areas.
	(iv) Forage seed production and its marketing.
	(v) Establishment of fodder banks.

Contd...

Stakeholder Category	*Training Needs*
	(vi) Effective Utilization of agro-by-products and industrial wastes as forage resource.
	(vii) Standards and specifications for feed ingredients and compound feeds.
	(viii) Forage and feed standards and quality control works.
	(ix) Monitoring of food chain to ensure sanitary and phyto-sanitary safety.
	(x) Enrichment of crop residues
	(xi) Forage preservation schemes
	(xii) Bailing, densification, making of blocks and balanced rations.
Industry and entrepreneurs	(i) Establishment of commercial feed supply units and intensive livestock farm units.
	(ii) New technologies such as bypass nutrients, probiotics and feed enzymes.
	(iii) Location specific mineral mixtures, trace element supplements, urea-molasses-mineral blocks, granules and salt licks.
	(iii) Complete feed blocks
	(iii) Fodder seed production
	(iv) Novel feeds (*Azolla*, leaf meal *etc.*)
	(vi) Effective Utilization of farm by products such as straws stover, brans, oil meals, vegetable by-products and other edible biomass.
	(vii) Post harvest technology for forage conservation schemes
Researchers	(i) Feed technology and processing.
	(ii) Economic feed formulations based on livestock nutrient requirements and the targeted quality of livestock products
	(iii) Feed supplementation, feed additives, novel feeds and agro-industrial by products
	(iv) New technologies such as bypass nutrients, probiotics and feed enzymes
	(v) Grassland and community land development in a participatory mode

Chapter 3

Ecological Perspective of Forage Based Livestock Production

The changes in subtropical and tropical farming systems after settled agriculture started from long fallow to short fallow systems, short fallow systems to permanent land use, rain fed farming to irrigation farming, low intensity cropping to high intensity of cropping, natural grazing to cultivated forage, arable farming to perennial cropping, natural regeneration of soil, fertility to intensive system of manuring and fertilization, manual hoe cultivation to animal-tractor as farm power source, traditional methods to increased use of modern high technology-high yielding varieties, genetically modified crops, drip and sprinkler irrigation, bio-control of pest, disease and weeds, precision farming, hydroponics, controlled environment agriculture *etc.* which included the methods involving an increasing volume of purchased inputs. Recently, there has been a major shift from productivity enhancement to a wider goal of increasing livelihood potential of agriculture system/technologies as well as organic production and system approach in agricultural research and development.

Sustainable Agriculture

Sustainable agriculture systematically pursue the following goals:

- ☆ Through integration of natural processes such as nutrient recycling, nitrogen fixation, increasing role of microbes in crop and livestock nutrition, bio-agents/pest predator relationships for crop protection into agricultural production systems, so ensuring profitable and efficient food production.
- ☆ Minimizing the use of those external and nonrenewable inputs with the potential to damage the environment or harm the health of farmers and

consumers and a targeted use of the remaining inputs used with a view to minimizing costs.

- ☆ The full participation of farmers and other rural people in all process of problem analysis and technology development, adaption and extension leading to an increase in self-reliance amongst farmers and rural communities.
- ☆ A greater productive use of local knowledge and practices, including innovative approaches not yet fully understood by scientists or widely adopted by farmers.
- ☆ The enhancement of public good of the country as a whole.

There is a need to follow procedure of ecosystem analysis before we design new system besides critically examining the new system for pursuing sustainable farming practices (Figure 3.1).

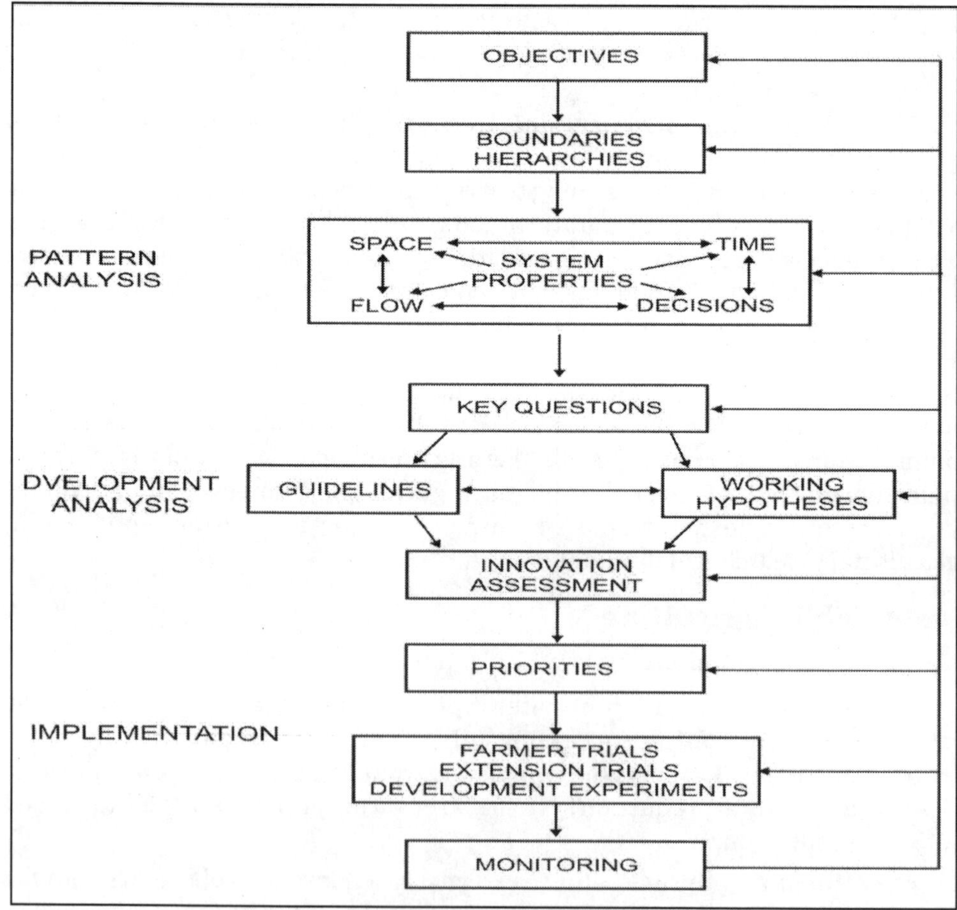

Figure 3.1: The Procedure of Agro-ecosystem Analysis for Development (Conway *et al.*, 1985).

History of Farming System's Perspective in Research

There is need to look at evolution of farming practices and their understanding for developing appropriate/alternative new farming systems. Chronological sequences of important events are as follows:

1834: The application of science to agriculture began when Boussingault laid the foundation of agricultural chemistry. Previously, all the improvements in farming practice resulted from the labourers of a few exceptional men whose innovations were afterwards copied by their neighbors.

1840: Liebigs classical monograph on agricultural chemistry. In his onslaught on the humus theory, he did not call in nature to verify his conclusions. Humus itself might be right. His failure has cast its shadow on the scientific investigation of the next 100 years. Rothamsted long term experiment started in 1843.

During 1840-1900, agricultural science was a branch of chemistry, being enlarged and embraced to chemistry, physics, bacteriology, and geology

1840-1915: Rediscovery of Mendel Law by Corren, the unit species and recognition of its importance in improvement by selection by Johannsen, Marquis and Farrer in wheat, Beaven in malting barley, Barber in sugarcane

1889: Voelcker writing of Indian agriculture quite explicitly made statement on importance of understanding what farmers do.

1925: Chayanov viewed that peasant economy really demanded a different system of economics.

1950: The adaptation of farm management ideas to the circumstances of tropical farmers was pioneered. (Meller 1966; Norman 1974-83; Norman 1983; Norman *et al.*, 1981, 1982, 1984; Collison 1982, 1983; Dillon *et al.,* 1978).

1978-83: The idea of system grew out of wartime operational research later applied to industrial problems. Systems ideas also entered technology-based agriculture in rich countries where enterprise, whole farm and production unit models developed usually with a maximizing/optimize objective in view of crop and animal production system. Collinson idea latter developed into operational research project on farm research/farming system perspective.

Simultaneously, farming system research was initiated by CFIAR (IRRI, Philippine and other research centers). Some farming system models/modules were developed. Few attempts are underway to develop new farming systems by National Agriculture Research System and Innovative Farmers/Farmers Group/ Corporate Groups in one or other way.

Diagnosis and design process of new farming systems development (NFSD) noticed at its starting point that many tropical farming systems are already so stressed that radical restructuring rather than stepwise change is necessary. The testing/refinement of inventions and exploitation of new refined systems/models are essential. My experiences with On Farm Research (OFR)/Micro watershed based Farming System Research Project (FSRP) in humid subtropical climate of hilly region of Sikkim, Meghalaya and Arunachal Pradesh to semiarid plateau of *Bundelkhand* (part of Uttar Pradesh) suggested that there is a need for revolution rather than evolution to build radically new systems/modes *ab initio*. It is essentially a practical oriented agriculture research and must obviously be founded on at least some FSP. It differs from OFR in degree rather than nature. Both seek to generate beneficial change but NFSD has a lesser OFR component and upstream process. There are no rules for how to invent NFS because each case must be *sui generis*, a sum authoritarian element. Therefore, it needs cash pull element and a degree of autocracy in the management. It has an adaptive phase. Full understanding of a system implies complete numerical specification, which did not happen in tropical agriculture research (Figure 3.2). Only farming system perspective cannot develop the appropriate farm models.

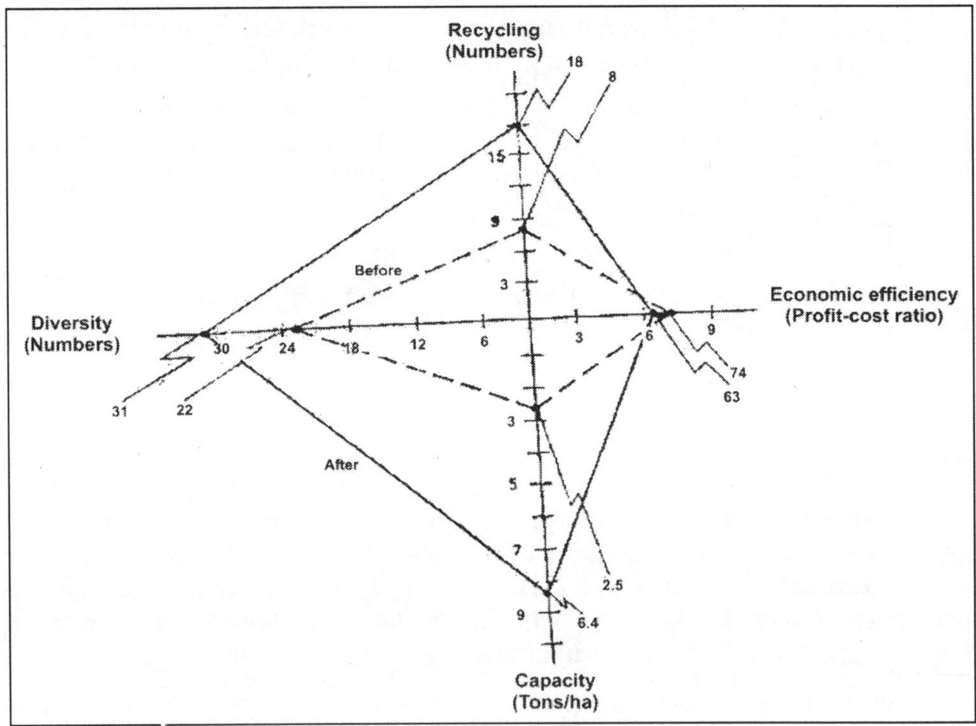

Figure 3.2: Farming Systems Performance Indicator Kites before and after Integration (Bimabao *et al.*).

A look on current scenario in context of livestock production systems in India reveals that the rapidly increasing trend towards commercialization is leading to intensive livestock environment interactions. The natural resources constituting vegetation, water, air and biodiversity as a whole are at risk due to increasing livestock population and changes in climate. The key issues are:

- Introduction of improved forage plant species with high nutrient demand under low/no fertilizer input systems is adversely affecting the soil nutrient balance.
- Only 25-30 per cent of organic matter, nitrogen and potassium contained in the total feed of livestock returned to the soil through compost of dung and urine.
- Micronutrient deficiency arising out of imbalanced fertilizer application and increased mining of soil nutrients without sufficient replacement, have adversely affected fertility and health of livestock in some of the intensively cropped areas in the country.
- Peri-urban dairy systems area generating animal wastes far exceeding adsorptive capacity of the surrounding area, causing air and water pollution.
- Intensive cultivation and emerging markets for green forage and feed also exports nutrients out of the sites, thereby increasing risk of poor soil health in the long run.
- Mineral nutrient deficient forage is also affecting livestock health and productivity.
- Imbalanced feeding systems largely based on crop residues under crop livestock mixed farming systems disturb rumen ecology with resultant enhanced emission of methane, a greenhouse gas (Table 3.1).

Table 3.1: Total Methane Emission in Atmosphere and Methane Emitted in India by different Sources (mT)

Source	*India*	*India's Share to World Production*
Solid waste	1.8	4.1
Livestock	10.0	13.2
Hard coal	0.8	5.2
Rice field	18.0	27.3
Pipeline leakage	0.2	0.3
Total	30.8	12.1

- Uncontrolled grazing is adversely affecting natural resources through vegetation loss and land degradation in arid and semi-arid regions.
- A vicious cycle of accelerated natural resource degradation has set in due to recurrent droughts resulting into changing preference towards small ruminants.

Growing and Eating Healthy Food

A new paradigm for agriculture in coming years will be growing healthier food. Non-availability of organic milk and meat besides insufficient calorie intake and lack of essential micronutrients like Fe, Zn, Vitamins and other essential nutrients are responsible for poor mental and physical development, disability, disease and death of millions of poor people. Little research has been conducted on integrating agriculture-livestock-human health in promoting the wellbeing of people as a whole. Till date, public health, agriculture and livestock development personnel are working in isolation.

A study in Vietnam indicated that the system of backyard garden, fishponds and keeping a few livestock made a significant impact on the nutrition of the family, increased the intake of iron in young children, besides promoting their education. There was a trade-off between incomes; the farmers would gain from selling their home produced, nutrient rich foods and the health benefits they would gain. The benefits of consumption led production focused agricultural intervention, which improved the overall nutrition level of the family.

In Indian context, the primary concern is to increase the availability of livestock products that requires our livestock to be fed adequately with balanced ration. Looking into the acute feed and forage scarcity in the country, there is a need for comprehensive policy and programmes coupled with some changes in existing one to develop forage resources in totality from grassland and arable lands.

Healthy Livestock for Healthy Human being

There is an intricate linkage between agricultural practices and health of livestock and human beings. Growing healthy crops require fertile, well-nourished soils, which in turn provide feed materials to livestock, which are balanced in all nutrients. The livestock products obtained from these animals ensure wholesome milk and meat products to human beings, free from contaminants and pesticide residues. Good health of people has a direct bearing on agriculture by boosting peoples' capacity to work and thus increasing the productivity. The revolution in livestock sector unfolding on Indian horizon being witnessed in terms of surge in demand of livestock products, propelled by changing dietary habits, urbanization and income which growth has brought to the fore the issues related to the health aspects of the livestock products being delivered to the consumers.

Zoonosis

The transfer of diseases between human beings and livestock, sometimes take epidemic proportions. The recent outbreak of bird flu and swine fever in several parts of the world caused panic among the people as well as governments, resulting in destroying millions of birds to check the spread of the disease. The scare of Mad Cow disease (BSE) in the last decade in Europe is still afresh in the minds of people. A recent controversy in United Kingdom, regarding killing of a temple bull suffering from tuberculosis by the authorities, in spite of the vociferous protest by the devotees, underlines the risk of disease transfer between animals and human beings. Therefore, stringent quality control needs to be in place, right from farm practices to the consumer table.

Pesticide Residues and Heavy Metal in the Food Chain

According to one estimate 90 per cent of the foodstuff being sold in the organized sector, is adulterated. Indiscriminate use of synthetic pesticides over three decades has started taxing upon the health of soil, livestock and human beings. Several villages in Punjab have their soil and ground water polluted to the extent that the children begin greying after three years of age, bypassing the joy of youth. Pesticides creep into livestock feed through crop residues and concentrate ingredients, get accumulated in animal body and further passed on to the consumers. The presence of residues or any toxic substance in milk or other food commodity is undesirable not only for the safety of human health but also form export point of view, since residues act as a non-tariff trade barrier. In many places like Punjab, Lucknow, Bangalore *etc.*, feed and forage like wheat grain, straw, cattle feed, oilseed cakes were found to contain residues of two commonly used insecticides (DDT, and BHC) till they were banned, having concentrations exceeding 5mg/kg. Analysis of feed and forage in Bundelkhand region have revealed that over 40 per cent of the samples were contaminated with organochlorides (HCH, DDT, endosulfan, organophosphates (chlorpyriphos, quinalphos, Malathion*etc.*), carbamates and synthetic pyrethroids. Feed samples of mustard, linseed, cottonseed cake *etc.* were more contaminated. Use of additives like oxytocin also induces abortions and sterility in human beings.

Industrialization and urbanization have also seriously polluted the quality of water bodies. Several states in India are now suffering from heavy metal load in drinking water such as arsenic, cadmium, lead, nickel, mercury *etc.* One of the major reasons for this problem is the over exploitation of the ground water resource for irrigation and other purposes, which has changed the character of aquifers. The earlier we start reversing the trend, the better, or tomorrow would be too late.

Biological Efficiency of Milk Production

The best biological efficiency for milk production in tropical areas is reported to be about 4000-4500 litres per lactation in crossbred cattle. Any attempt to push milk yield beyond this level, will result into low efficiency of forage and feed resource utilization. However, maximum sustained milk yield from well managed tropical pasture were estimated to be 3000-3200 litres per lactation with high dependence on tropical grasses, legumes and crop residues based feeding systems. Moreover, the best-sustained milk yield would be only 1800-2200 litres of milk. Therefore, more efforts should be given to raise livestock productivity from low to intermediate level than to provide genetic potential for high productivity, which cannot be supported economically and might seriously hamper sustainable use of natural resources. Matching animal to the levels of availability of feed and forage resources and developing feeding systems should be our endeavour for increasing productivity of various livestock. The crop livestock integrated mixed farming system, is an ideal choice to sustain the land and livestock resources in predominantly marginal and small land holdings farms in the country.

Therefore, there is a need of readymade energy protein rich forage duly fortified with essential minerals, scientific management and recycling of animal wastes to the soil for sustained livestock productivity and reduced emission of greenhouse gases particularly methane and nitrous oxide. It will be possible though holistic approach in forage resource development, post-harvest processing, storage, transport to balance supply to commensurate with demand and developing local resource based feeding systems.

Strategies to Mitigate GHG Emission from Livestock

Forage based feeding systems have a lot of potential to reduce methane production in ruminants (Table 3.2). Forages rich in soluble sugars and cell contents change the fermentation pattern to produce more propionic acid and less methane on feeding to ruminants. Selection of forage plant species and varietal development targeting such traits will be helpful. Mixing berseem with wheat/paddy straw reduce methane production up to 30 per cent depending on the proportion of berseem mixed. Increased use of tree leaves will reduce GH emission. Feeding of oat, wheat straw combinations to ruminants reduce methane production by 8-23 per cent. Forage type affects methane production as green sorghum (4.2 per cent) produces less methane than green maize (5.3 per cent) and green bajra (6.2 per cent). Grassland provides more nutritious and balanced feed resource to animal resulting in efficient rumen fermentation and ultimately reduces methane. Grassland provide mixed sward of heterogeneous biochemical entities that decreases methane emission than grazing on mono sward. Excretion in open leads to drying-off, of dung and urine in the grazing fields resulting in inhibition of methanogenesis process.

Table 3.2: Total Gas and Methane Production from Feeds and Forages

Forages	*CH_4 as per cent of Total Gas*
Forages	
Non Legume	34.39
Legume	31.70
Crop residue	32.33
Tree leaves	
Non Legume	21.56
Legume	21.94

Livestock Water Productivity

Studies on livestock water productivity have been sporadic and lacking in Indian context (Table 3.3). Efficient use of water in livestock sector calls for studying the livestock production system under different agro-climatic zones of the country. There is need to develop crop varieties requiring less water being drought tolerant. Water use efficiency for forage production can be improved by employing water conservation technologies in rain fed production system with greater emphasis on water harvesting in grassland and silvi/hortipasture production systems.

Table 3.3: Estimated Water Requirement to Produce unit Product (Milk and Meat)

Rearing System	*Species*	*Product*	*Water Required (Liters/kg milk or meat)*
Intensive	Buffalo	Milk	3751
Semi- Intensive	Buffalo	Milk	2451
Extensive	Buffalo	Milk	475
	Buffalo	Meat	9680
	Sheep	Meat	8215
	Goat	Meat	9338

A relative synergy exists among forage, grassland and livestock for ecological balance which could be further strengthened by developing appropriate technologies meeting dual objective of high livestock water productivity which reduce methane emission coupled with new method of animal waste management.

Reducing Negative Environmental Impacts

Loss of vegetation due to overgrazing has resulted in increased soil erosion, down slope sedimentation and reduced water infiltration. Research has proven that low to moderate grazing pressure has little negative impact on hydrology. Managing animal in ways that reduce land and water degradation, for example, by restricting animal access to certain areas and more integrated management of grazing land will help to reduce negative environmental effects.

IGFRI Efforts in this Direction

Some of the efforts made at Indian Grassland and Fodder Research Institute are listed below

- Screening of tree leaves and shrubs for the anti-protozoan and antimicrobial activities.
- Exploitation of phenols and tannins and other anti-nutritional agents as potential methane inhibitor or possible rumen fermentation manipulator.
- Chemicals of plant origin and plant extracts particularly from the unconventional resources tested for both for their defaunation and anti-methanogenic properties.
- Evaluation of forage specific processing techniques for reduced methane production.
- Development of total mixed rations through varied proportions of crop residues, tree leaves, legume and non-leguminous crops based on their precise chemical evaluation for varied carbohydrate and protein fractions, their utilization and interaction with anti-nutritional constituents to harness dual objective of eco-friendly higher animal productivity.

Ecological Approach to Livestock Production

The hazards of factory style livestock production has thrown many issues related to wholesome food production taking into account the ethical issues and the emergence of new diseases like BSE and avian influenza, besides polluting the environment. National Institute of Livestock and Grassland Science, Tsukuba, Japan has reported that producing a kilogram of beef has a warming potential equivalent to 36.4 kilograms of carbon dioxide. The growing health consciousness has created a demand for organic livestock products the world over and more and more farms are getting into this production system. Organic farming delivers the highest quality, best tasting food, produced without artificial chemicals or genetic modification, and respect for animal welfare and the environment, while helping to maintain the rural landscape and rural communities. The organic livestock produce will not only fetch premium market price but also create new avenues for the rural people. Many medicinal plants including bamboo will come to the aid of safeguarding the health of livestock. Some species of bamboo contain immune stimulatory compounds that the farmers can utilize for their animals. Bioactive forages, like bamboo, offer low cost and environmental friendly solutions for feeding all species of domestic animals. The linkages between healthy soil-crop-livestock was emphasized over a century ago by Sir Albert Howard (also known as Father of organic agriculture) while working at the Imperial Agricultural Research Institute, Pusa. He had reported that when livestock were fed on plants grown on fertile healthy soil, even the FMD disease failed to show any symptoms on these animals although they frequently came in contact with the diseased animals.

Grasslands as a Hub of Organic Livestock Production

Grasslands, wherever they exist in India remain a neglected resource, be it in Gujarat, Rajasthan or the alpine Himalayan pastures. The traditional livestock rearing practices on these grasslands have been going on for thousands of years with little external inputs. While these unique production systems need to be studied in detail taking into account the soil-vegetation-climate aspects together with studying the society that are traditionally pastoralists, they offer an opportunity to brand the livestock products coming out of these grasslands as organic with Geographic Indication such as *Kangayam Milk* (Tamil Nadu), *Banni Milk* (Gujarat), *Alpine Milk/Meat etc.* Such products will fetch premium price, enhancing income and setting a movement to preserve these ecosystem in its traditional form. This requires characterizing the livestock production system backed with suitable legislation by the state.

The IGFRI has developed models for 'organic livestock production' by introducing the prized indigenous cattle breeds *Tharparker, Gir* and Sahiwal and developed standard for organic milk production. The soil-crop-livestock interrelationship is being studied under the 'crop residue project'. A project on 'periurban livestock production system' is attempted to address the issues related with it. Yet another project on 'pesticide residues in feed and forage' is tried to identify the chain and quantify the pesticide load, the consumers are being exposed to. Only healthy, wholesome diet will make foundation for a healthy India.

Environmental Impact Assessment (EIA) guidelines should include grassland as a mandatory component. Considering a wide range of activities and programmes under each Ministry and R&D Institutions, it would be extremely important to identify some of the cross cutting themes and launching the Integrated Research and Development Programmes in the grasslands. Long Term Ecological Research (LTER) needs to be initiated in representative biogeography zones, which includes new dimensions of impact of climate change and land use practices within and around grassland. There is need to increase grasslands ecosystems in Protected Area System, especially in the Thar Desert and the cold desert of Ladakh and Sikkim which are grossly under-represented. The Ministry of Environment and Forests Government of India should start a division or a section to look after the grasslands issues, on the pattern of Wetland Division headed by a Joint Secretary. Other relevant issues concerning importance and development of grasslands are: grasslands as hub of organic livestock production, winter sojourn of transhumant grazers in the foothills of Himalayas is creating a zone of conflict because of vanishing CPRs *etc.* Benefit of carbon sequestration in the grassland and keepers of livestock gene pool need to be extended to the communities making a living out of grassland, who are also keepers of a vast pool of genetic resources of livestock breeds.

Forage based Feeding System for Efficient Livestock Production

The livelihood of a large segment of people in India is intrinsically woven with the livestock sector. The livestock sector in India accounts for 24.72 per cent of the agricultural GDP and about 4.36 per cent of the total GDP. This is derived from a livestock population of about 185 million cattle, 98 million buffaloes, 61 million sheep, 124 million goats, 13 million pigs and 458 million poultry including 11 million other livestock. India ranks first in milk production (over 100 million tonnes), third in egg production and eighth in the world's meat production (4.8 million tonnes). The milk production in the world is expected to increase from 568 million tonnes to 700 million tonnes between 2000 and 2020; and the meat production from 233 million tonnes to 300 million tonnes. With increasing income and changing dietary habits, the demand for products from animal origin is increasing. To keep pace with the demand, the livestock sector has to gear itself by increasing productivity through cost effective measures. To achieve the projected increase in milk and meat, there is need to promote feeding systems, which is primarily forage based systems. This would result in reducing the total food grain requirement of the country.

Of the total world cereal consumption as animal feed (660 million tonnes) in 1997, developed countries used about 425 million tonnes (64 per cent) which is expected to increase to the tune of 511 million tonnes by 2020. Livestock production in developing countries including India is mostly based on grazing and crop residues (mixed farming system) with little supplementation of agro-industrial by-products.

Poultry are the major consumer of cereals like maize, barley, sorghum, pearl millet *etc.* In India, poultry enterprises consume around 50 per cent of maize, 30 per cent of barley, sorghum and other cereals produced. A part of this cereal grain could be saved for human consumption by adopting alternative approaches. Studies have shown that use of *Stylosanthes* leaf meal in poultry feed can be economical besides increasing the acceptability of eggs because of more yellowness of the yolk. Other non-ruminant's species can also be given higher levels of green forage in their ration. Pigs can be fed up to 20 per cent green forage like succulent sorghum, guinea grass, hybrid napier, *etc.* Another 20 per cent of their requirement can also be met through tuber crops like sweet potato, turnip, sugar beet *etc.* A recent study by ACIAR has concluded that when pigs are fed with *Stylosanthes guianensis*, the average daily gain increased from 100g to 250g for virtually no extra cost, under village conditions in South-East Asia and the pigs could be sold in six rather than 10 months.

The cost of feeding accounts for 60-70 per cent of the total cost of animal production in ruminants like cattle, buffalo, sheep and goats. Livestock production based on green forage is not only economical but also good for the general health and wellbeing of the animals, which in turn adds to the income of the livestock owners on a sustainable basis. Ruminants require 7-10 kg feed to produce one kg of meat. Although, poultry are very efficient feed converters, requiring 2 - 2.5 kg of feed per kg meat and even less per kg for egg production and pigs require 2.5-4 kg dry matter per kg pig meat but they cannot utilize the crop residues and green forage in large quantities. Forages like oats, sorghum, maize, napier grass, lucerne and berseem are the cheaper source of digestible crude protein and total digestible nutrients than concentrate. The cost of DCP and TDN from oats, berseem and lucerne is around Rs. 20 per kg and from sorghum and maize is around Rs. 25 per kg while from concentrate sources it varies from Rs. 40-59 per kg.

The experiments with all forage diets have shown that 400-500 g average daily gain in heifers and milk yield of about 8-10 litters in lactating buffaloes and cows can be achieved. The well-managed green forage having a dry matter digestibility of 53 per cent, are able to provide all the required nutrients (1.0 kg DCP, 7.9 kg TDN, 51 g calcium and 37 g phosphorus) for a cow/buffalo of 400-500 kg body weight, yielding 8-10 litters of milk. The results suggest that the first 8 litters of milk in buffaloes and 10 litters of milk in cows can be produced exclusively on high quality forage diet without any concentrate supplementation. For additional milk production required energy may be provided from energy rich concentrate feeds.

There is need to change the production practices to take advantage of natural process, by-passing the energy intensive grain-fed operations that have dominated in the developed countries livestock production systems for last several decades. The food products of green forage/grass-fed ruminants (mutton, lamb, beef, milk) are nutritionally superior and contain much more omega–3 fatty acids and conjugated linoleic acid (CLA) than those products obtained from grain-fed animals. In fact, grass/forage fed animals can produce 300-500 per cent more omega–3 fatty acids and CLA than cows fed with the typical diet of 50 per cent hay and silage with 50 per cent grain. On the other hand, green forage and crop residue combination based feeding system is able to reduce the methane emission to the extent of 8-30 per cent depending upon the type of forage used and its proportion in the diet.

There is great potential to meet the feed and forage shortage through 19.34 million hectares of land under permanent pasture and grazing lands, 108.42 million hectares under degraded forest and 18.13 million hectares under agricultural land inside notified forest area. Even if 50 per cent of this land is brought under grassland and silvipasture system, there is potential to produce 200 million tonnes of dry matter to meet shortage of forage, being faced now.

In the country like India, adoption of forage based feeding regimens for livestock production would ensure lesser methane emission from enteric fermentation as well as nutritionally superior animal products for human consumption besides ensuring a better environment to live in. Moreover, the cost of production of milk and meat will come down providing better income to the millions of farmers.

Ecological Management of a Traditional Grazing Land in a Rain-Shadow Belt of Kangayam Region

The livelihoods of poor livestock keepers primarily depend on the productivity of edible biomass available from common property resources like village commons, the roadsides, along railway tracts, canals, bunds *etc.* for livestock. In 4,000 square kilometre of a rain-shadow belt of Kangayam region in Tamil Nadu, grass is the main crop for feeding livestock. The pattern of ownership of livestock in the Kangayam grassland indicate that 70 per cent of the farmers own one to four cattle or buffalo, 65 per cent own 15 or more sheep (*Curumbar* breed for wool and *Shyambliar* breed a hairy one) and 85 per cent of the farmers do not own any goats. The population of crossbred cattle is 43 per cent of the total cattle in the region. Sheep are the choice of animal reared in paddocks which are far off and where daily milk collection is not possible. Goats are not a preferred animal because they demand constant attention of their keepers and damage the live fences of *Balasmodendron berryi.* There were very few herds of goats in the region being grazed along the roadsides and they mostly belonged to the landless people.

The Kangayam region is a drought prone being located in the rain shadow region of south India, where pasture grass is the main crop and livestock cattle and buffaloes rearing is the major farming practices. The annual rainfall is 666 mm in the Kangayam grassland. The amount of rain and its distribution is hardly sufficient for raising the traditional food crops. The soil of the grassland is red loam with a '*Kanker gravel*' sub-soil rich in calcium and phosphorus, adding soil fertility and providing aeration and greater absorption of moisture. The soil is broken periodically by cultivating a crop, once in a rotation cycle of four to five years. The rainfall in the grassland is sufficient to encourage healthy growth of grasses. There were eight species of perennial grasses, six annual grasses, nine legumes and 16 forbs present in the grassland. Importance value index of herbaceous vegetation indicated that *Aristida setacea* had the highest IVI (126.9) followed by *Cenchrus ciliaris* (49.2), *Blumea* sp. (37.9) and others. The average yield of green grass was 28.48±3.06 tonnes/ha with dry matter of 9.8 tonnes/ha. This yield was sufficient for a carrying capacity of six to seven ACU which consisted of two cattle and 25 - 30 sheep for 8 months from July to February. The grassland has usually two flushes of regrowth, one occurs after the rain in May and the other major flush in September-October. Re-seeding of grassland is done by broadcasting seed *of Cenchrus ciliaris*, to boost the forage yield in subsequent years. One month after rain in May as well as in September, livestock are withdrawn from the grazing paddocks to allow new flush to grow and avoid overgrazing. From mid-June to mid-September and from mid-October to January, livestock's are allowed to graze without any supplementary feeding. Withholding grazing animals after rains is practiced in paddocks to stop over grazing so that the regrowth come up well in subsequent years. Enrichment of the *Cenchrus* dominated grassland with incorporation of *Phaseolus trilobus* legume in many areas of the Kangayam grasslands are done by progressive farmers by a way of ploughing in alternate years and sowing the legume seeds in October. The grazing continues from mid-December till mid-January. During lean season, the crop residues of food crops, tapioca leaves, collected *Acacia* pods and the grass-legume hay dried after harvesting are fed. In periods of severe drought, leaves of *palm* tree are lopped and fed to animals.

The *Kangayam* grassland in its present state of expanse has been in place for the last one hundred and fifty years. The grazing lands are in private ownership and there are no common property grazing lands. The unoccupied wastelands were taken on *puttah* (lease) by a native people between 1855 and 1881 and live fences of *B. berryi* were raised along the field boundaries. A system of rotational grazing of livestock in the paddocks was introduced which required minimal labour input. Rotational management of the grazing paddocks was strictly followed along with supplementary feeding during lean period. The sustainable management of the grassland over a century also had some positive social spinoffs such as moderate growth in human population and a healthy female to male ratio. In fact, the necessity to check human population growth may also have arisen to prevent the fragmentation of land holdings, making them unsustainably small. It had also observed that as far as common property grazing land is concerned, there is everything against it but when the grazing was regulated by not overstocking grazing livestock and allow manure to remain in the pasture, sustainable grazing was possible. Over a century, the grasslands in the *Kangayam* region have evolved into an ecologically sustainable, productive and economically viable system leading to a secured livelihood in a natural resource poor region.

Building Empire on Hay: A Successful Enterprise on Forage Resource

The story of Abbott goes back to pre-independence days when the first war of independence was fought by the Indians against the British rule. A young British boy,

William Lumsden Abbott (1828-1880) from Scotland had joined the British army and came to India in 1850. During sepoy mutiny Mr. William was severely wounded in Sitapur. Thereafter, they went back to England in 1858 but Mr. William again came back to India along with his family as an engineer in the Indian Railway and settled in Morar near Gwalior of Madhya Pradesh state. In 1880, William Lumsden Abbott died of Cholera in Morar. J. H. Abbott, his eldest child took charge of livelihood of his family after Cantonment was closed down. The young J.H. Abbott who was then 18 years old had the responsibility of his mother and 3 siblings. In 1881, the Abbott family put their belongings in a horse cart and came to Jhansi. They had settled in front Jhansi hotel (existing as on today). J.H. Abbott had only 8 annas in hand. J.H. Abbott had a difficult time to full fill the responsibility for the family.

In 1885, he struck upon an idea of supplying hay to the army. The army at that time used to employ a large number of animals including horses, elephant, mules *etc.* to cater to the transport need. For feeding their animals army used to purchase hay locally but the total quantum required was not substantial. J.H. Abbott contacted the army and offered to supply all the requirements of hay which was agreed upon. Jhansi during those days had plenty of grassland, called 'Rund' which we still find today in names of places as 'RundKarari'. In the beginning Abbott organized the supply of hay locally through contractors. The army also found it convenient that their requirement was being met without any difficulty. Moreover, J.H. Abbott was a man of word. He kept his word in meeting the supply of hay in time. With passage of time army cantonments from elsewhere started placing their requirement of hay to Mr. Abbott. This provided him an opportunity to expand his business and incorporate innovative ways of baling and transporting hay to distant places. Orders for hay started pouring from places as far as Quetta, Lahore and Peshawar in Pakistan.

Transporting hay loose was not feasible. Therefore, J.H. Abbot purchased an old steam engine which was put up for auction by the railways. Wood was used to fuel the engine to produce steam which propelled the baler. Each bale of grass weighed approximately one mound (40 kg). The first such plant was established at Ganeshpura (Ronija) near Karari in Jhansi. Each bale was accurately weighed before being dispatched. The Imperial Gazetteer of India (1908) also mentions that "Grass is also exported from Jhansi and hay was baled in large quantities for the Military Department during the *Tirah* expedition of 1895 and South African War". Similarly, the *Journal of Agriculture and Livestock* published during British period mentions that Mr. Abbott of Jhansi was the *"Pioneer of Central India Hay trade".* He had attended the Cattle Conference held in Lucknow on 4-8-1910 and put forth his views in a forceful way. He was also a member of the Management Committee of Imperial Agricultural Research Institute, Pusa, Bihar. With the demand for baled hay pouring from faraway places; he commissioned another plant for baling of grasses during 1895 at Markundi in Manikpur, Uttar Pradesh. J.H. Abbott used to collect the grasses which grew in abundance in the grasslands of Jhansi and Manikpur. Theses grasses were harvested and collected in October and November, dried and staked near the baling machines. The baling operations used to start from December which continued up to February. The baling of grass was done with the help large balers who were run with the help of an old steam engine procured from the Railways. This engine is now displayed at Cant area in Jhansi. Initially wood was used to power the engine to produce steam. But later on crude oil was used for the purpose. Out of four sons and two daughters, J.H. Abbott employed three of his sons at different places to look after the business of grass baling and marketing of bailed hay. While the eldest son Roy Leslie Abbott looked after agriculture in Sagar district, Mr. Neil Abbott was in charge of the baling plant at Karari. Charles Abbott looked after the baling operations at Markundi, Manikpur. Bruce Abbott was stationed at Lahore, who looked after the distribution of bales dispatched from Karari (Jhansi) and Makundi (Manikpur). Thus, the extent of the hay business of Abbotts grew and covered almost whole of the North India. It also had the strategic supplies of baled hay for feeding animals used for transport of men and weapons during the war.

J.H. Abbott was a man of vision and an astute businessman. He used to keep promise in making timely delivery of the baled hay which in turn brought him more orders from far off places. Within a short span of time the business of baled hay spread far and wide and J.H. Abbott became a rich man. He started diversifying his activity. He opened the Jhansi Hotel in 1905 and ventured in the contract business. J.H. Abbott also took up agriculture in a big way. In 1932, the Raja of Jabalpur Raja Gokul Das started selling villages under his suzerainty in the districts of Sagar, Damoh and Jhansi. J.H. Abbott bought almost all of the villages sold by the Raja. With the wealth earned from his business of hay bailing, Abbott acquired a 100 acre plot in Pithoragarh, called Abbot Mound now. In Jhansi, he had established low cost housing colonies like Royganj, Charliganj, and Masihaganj. which still retains the family names today. The life and work of J.H. Abbott is of great relevance to forage resource management in India which demonstrates economic value of the resource. It was given the shape of a business which we now dream of creating.

Epilogue

Agriculture has never been static and it will go on changing and developing substitutes for the inputs/techniques/technologies on which it is currently dependent, with changing needs and accumulating knowledge of agricultural innovations.

Chapter 4

Organizational Change, Policy Support, Investment and Road Map for Accelerated Forage Development and Utilization

Sustainable growth in livestock sector is crucial for the overall growth of agriculture. Livestock also provide livelihood through income and employment to weaker section of rural population. Livelihood security during extremes of weather *viz.* floods and droughts comes from livestock wealth. Over the years, there has been a definite shift towards resilience on livestock for income and employment generation by a large segment of rural people. Where dairy co-operatives have been successful, the farmers have stayed back in the villages relying on dairy animals for the major share of their income. In peri urban areas, educated unemployed youths keep 4-5 milking buffaloes/cattle for livelihood to sustain their families. It is not inappropriate to say that cattle and buffaloes are now sustaining families in peri urban areas also.

Livestock production system is undergoing a unique change towards intensification. The factors driving this change are increase in demand for livestock products because of changing dietary habits of the people brought about by urbanization, increase in income and population growth. But are we equipped to meet the surge in demand with due consideration for ensuring the livestock owners a fair price and simultaneously, making them affordable for the consumers? The major constraint in the productivity of livestock is non-availability of adequate quantity and quality of forage beside improved breed of livestock. No adequate attention has been given on the development of forage and feed resources, which account for 60-70 per cent of the total cost of milk and meat production. Now, it has been established that improving feeding system based on green forage can significantly lower the cost of milk production and replacing ingredients of concentrate with leaf meal and enriched nutrient complete feed block *etc.* Livestock production based on green forage is not only economical but also good for the general health and wellbeing of livestock besides being environmental friendly. Feeding concentrates at higher level by dairy owners without sufficient green forage supplementation adversely affect the digestive system of animals in long run and thus having negative impact upon the productivity of livestock. Feeding green forage, changes the rumen eco-system enhancing the utilization of nutrients by rumen microbes. With favourable climate in the Indian subcontinent, we should take advantage by adopting round the year forage production and conserving the excess for lean period, feed duly supplemented with leaf meal of leguminous tree fodder and range legumes. However, forage crop being an intermediary input for livestock production, possesses certain unique problems, which is quite different from the food and horticultural crops. Besides multiplicity of the forage crop species and their nature, they are usually area, region and season specific. The preference for forages varies from season to season and locality to locality. Non-commercial nature does not encourage forage crop cultivation. In most of the cases the degraded and marginal lands are usually allocated for forage production with minimum inputs of fertilizers, water and operational energy. There is a lack of awareness among the livestock owners regarding benefit of green forage, balanced feeding and nutrition of livestock. In demand and supply scenario, the regional and seasonal deficits are more important than the national deficit of green forage. In case of forage, it is not economical to transport them over the long distances. Out of 55 micro-region of the country, only 12 regions have surplus forage, while the remaining 43 have seasonal or annual deficit. Therefore, it becomes difficult to take policy decisions and measures for their optimum utilization and tackling the problems of regional imbalances and also in developing strategies during natural disasters. To increase milk production from 101 million tonnes at present to 160 million tonnes by 2020, the demands will be to the tune of 520, 850 and 90 million tonnes in respect of dry forage, green forage and concentrates. Policy provides conducive environment and programme to accelerate the pace of development. Corrective measures are also required in existing organisational structure for livestock and enhancing investment to develop forage resources with the changing national scenario in livestock sector.

Department of Animal Husbandry, Dairying and Fishery, Ministry of Agriculture, Government of India

The Department of Animal Husbandry and Dairying (AH&D) - now renamed as Department of Animal Husbandry Dairying and Fisheries (DAHD&F) is one of the Departments in the Ministry of Agriculture. it came into existence with effect from 1st February, 1991, by merging two divisions of the Department of Agriculture and Cooperation, namely Animal Husbandry and Dairy Development into a separate Department. The Department is responsible for matters relating to livestock production, preservation and protection from disease and improvement of stocks and dairy development, and also for matters relating to the Delhi Milk Scheme and the National Dairy Development Board including fisheries. Animal Husbandry Commissioner heads the technical department. The overall monitoring of feed and forage unit has been assigned to Joint Commissioner (CBF) from January, 2005 in addition to cattle breeding and fisheries activities. The personnel of feed and forage unit in the Department of Animal Husbandry and Dairying, Govt. of India are forage experts having degree either in Agriculture Science or Plant Science (Table 4.1).

Table 4.1: The Personnel of Feed and Forage Unit in the Department of Animal Husbandry and Dairying, Govt. of India for Forage Development Programme

Post	*No.*	*Place of Posting*
Dy. Commissioner (Feed and Fodder)	1	HQ
Assistant Commissioner (Feed and Fodder)	1	HQ
Director	7	7 Regional stations
Director	1	Seed Farm, Hessarghata
Agronomist	7	7 Regional stations
Agronomist	1	Seed Farm, Hessarghata
Senior Technical Assistant	1	HQ
Senior Technical Assistant	1	Seed Farm, Hessarghata
Senior Technical Assistant	7	7 Regional stations

The Department advises the State Governments/Union Territories in the formulation of Policies and programmes in the field of animal husbandry, dairy development and fisheries. The main focus on forage resource development of the department is through the two Central Sector scheme (1)Regional Stations for Forage Production and Demonstration and (2) Central Minikit Testing Programme on Forage Crops.

The increased demand for protein foods in the country is the main driver of growth in livestock sector. The growth has characteristic feature of more inclusiveness as in the ownership of livestock, small holders and landless farmers has major share. The current challenges of animal husbandry sectors are effective control of animal diseases, shortage of feed and forage, breed improvement while preserving diverse genetic resources and dissemination of technology, skills and quality services to farmers for improving productivity, which need to be addressed. The National Livestock Mission in 12th five year plan had an important component to promote availability of feed and forage to substantially reduce the gap between availability and demand. The thrust of 12th Plan was to promote optimal utilization of land, promote high yielding varieties of forage, higher income from cropping pattern, systems incorporating forage crops, improvement of grazing lands including barren and other common property forage resources, promote improved perennial forage grasses and legumes in forest area, conservation and utilization of edible crop residues and other by-products, fodder bank development and strengthening extension activities. The various schemes targeted forage seed production and distribution, hand and electric chaff cutter machines, grass reserve and grazing lands development, strengthening laboratories for testing nutritive values, establishment of silage and forage block making units, azolla production and demonstration units, bypass protein production units, area specific mineral production units and fodder banks.

Currently, DAHDF is implementing two schemes *viz.* 1) Central Sector Scheme namely Central Forage Development Organization and 2) Centrally Sponsored Scheme - Centrally Sponsored Forage and Feed Development Scheme – under which assistance is provided to States for Feed and Forage Development to overcome the shortage of feed and forage and to improve the nutritive value. It is comprised of three components namely (i) seven Regional Stations for Forage Production and Demonstration, (ii) one Central Forage Seed Production Farm, Hessarghatta, Bangalore and (iii) Central Minikit Testing Programme on Forage Crops. The seed production farm, Hessarghatta is also engaged in production of breeder's seed, foundation seed and certified seeds of high yielding varieties of different forage crops, grasses and legumes. The component wise details on physical achievement Forage and Feed Development Scheme during the year 2013-14 is given in the Table 4.2.

Organizational Structure Related to Feed and Forage Sector in National Dairy Development Board (NDDB)

In NDDB, The feed and forage production activities are coordinated by a senior manager (Production and Enhancement) who happens to be a forage specialist duly assisted by two officer (one senior scientist and one scientist) who happens to be animal nutritionist/biochemist. The densifications of crop residues *etc.* are monitored by another senior manager (P and E), an agricultural engineer. The forage development activities at 08 regional offices situated at New Delhi, Mumbai, Kolkata, Bangalore, Hubli, Chandigarh, Jaipur and Vijayawada are monitored by Deputy Manager (Forage) who happens to be an agronomist. The feed activities are monitored by senior manager level at Mumbai and New Delhi regional offices and deputy manager level at Bangalore, Chandigarh and Vijaywada regional offices. They are animal nutritionist/biochemist/feed technologist. National Dairy Development Board plays a vital role in development of feed and forage in tune with dairy farmer's demand in different zones of the country. At present, its main activity towards feed and forage revolves round the year green forage supply for balanced cattle feed and certified forage seed, demonstration, and financial assistance for forage seed multiplication.

Table 4.2: Component-wise Physical Achievement under Centrally Sponsored Forage and Feed Development Scheme during the Year 2013-14

Sl.No.	Name of Components	State/Coverage	Beneficiaries	Pattern of Assistance	Physical Achievement upto 31.03.14
1	Establishment of Forage Block Making Unit	All States	Public or Private entrepreneurship including Cooperatives and Self Help Group.	50:50	—
2	Grassland Development including Grass Reserves	All States	Department of Animal Husbandry and State Department of Forest. However, NGOs/Village Panchayats would be involved for development of grass land on Panchayat land and other common property resources. On an average Rs.7.00 lakh per unit of 10 hectare.	100:00	3,126 (hectare)
3	Forage Seed Procurement and Distribution	All States	Department of Animal Husbandry of the States. The State Govt. May involve the SIA/Cooperative/NGO for implementation of the project. Average rate of Rs.5000 per quintal of forage seeds	75:25	82,953.5 (qtls)
4	Strengthening of Feed Testing Laboratories	All States	Milk Federations/Research Institutes/Existing Soil Testing Laboratories to take up Feed testing, Laboratories working in the field of feed and forage research and development.	50:50	3 Nos.
5a	Introduction of hand driven chaff cutter	All States	Members of Milk Cooperatives	75:25	14196 Nos.
5b	Introduction of power driven chaff cutter	All states	Members of Milk Cooperatives	75:25	10450 Nos.
6	Establishment of silage making Units	All States	Milk Cooperatives	100:00	2090 Nos.

Contd...

Sl.No.	Name of Components	State/Coverage	Beneficiaries	Pattern of Assistance	Physical Achievement upto 31.03.14
7	Demonstration of *Azolla* cultivation and production units	All States	NGOs/Self Help Group/Milk Cooperatives	50:50	2270 Nos.
8	Establishment of by-pass protein production units	W.B, M.P., U.P, Punjab, Haryana, Kerala, Maharashtra, Andhra Pradesh, Karnataka and Tamil Nadu	Dairy Federations/Private Entrepreneurs.	25:75	—
9	Establishment of Area Specific Mineral Mixture/Feed Pelleting/ Feed Manufacturing Unit	All States	Public or Private entrepreneurship including Milk Cooperatives and Self Help Group.	25:75	—

Forage Specialist in the Directorate of Animal Husbandry in the States

Out of 29 states and seven union territories in India, very few states namely Punjab, West Bengal, Sikkim, Maharashtra, Uttar Pradesh, Tripura, Nagaland *etc.* have realized the importance of forage in animal husbandry in true sprit as evidenced by their organizational structure set up (Annexure 4.1). Department of Animal Husbandry in Punjab can be considered as an exemplary strong organization so far feed and forage specialist's deployment in the organisational structure is concerned. Moreover, all the livestock farms in Punjab have activities of green forage production which is being monitored by Deputy Director (Fodder). This organisational structure is almost in resemblance with Bhutan, a small country with effective organisational structure. The organizational structure of Animal Husbandry Department in Bhutan deploys Forage Development Officer up to block level. Out of six farms, seed of forage crops is grown at four farms. Guidance of Forage Development Officers at block level has boosted the forage production in the state. This has resulted in total green forage availability of 23-kg/animal/day. Forage Development Officers is able to organize field level demonstrations in a large number of villages of the block both for forage and forage seed production. Farmers are also educated for enrichment of crop residues of wheat and paddy straw available through ammonisation. The state was able to increase total area under forage production from 7.4 lakh hectares in 1999-2000 to 8.6 lakh hectares to enhance green forage production. Punjab state is also concerned about forage conservation techniques like converting green forage into baled hay for sale and silage making to be practiced at village level and also looking ahead for fodder bank. In West Bengal, the entire forage production activities are looked after by one officer of agriculture background and are headed by Joint Director (Forage) and 04 Deputy Directors (Forage). Three forage farms in the state are looked after by Deputy Director (Forage) while 07 Assistant Directors (Forage) are looking after activities in 07 farms. Forage development activities at each of the 18 districts are also looked after by Assistant Director (Forage) while 42 forage supervisors (Assistant Superintendent) are also deployed at block level in the state. Department of Animal husbandry in Uttar Pradesh, is lacking forage specialists as one Joint Director (Forage) and just one Deputy Director (Forage) is posted at head quarter to monitor the entire forage activities in the state. All the 08 state government farms are headed by the Farm Managers who are of agricultural background. It has 07 Forage Development Officer's/Senior Forage Inspectors and 35 Supervisors. In Sikkim, a relatively small state, has got a strong organizational structure in feed and forage sector as visualized by one Joint Director, 04 Deputy Directors, 04 Forage Development Officers and 10 forage supervisors. Similar is the situation in Tripura. In the animal husbandry department of Tripura, sufficient number of forage professionals are posted as Assistant Director (Forage) in each of their livestock farms. Posting of 05 Assistant Director (Forage), 04 officers at four districts in Tripura and 13 block level officers showed a balanced organizational structure in the state for forage development. In Nagaland, the forage development activity is monitored by one Deputy Director (Forage) and one Assistant Research Officer (Animal Nutrition). Forage Development Officers' are veterinary graduates not forage specialist. Andhra Pradesh, Assam, Bihar, Gujarat, Meghalaya, Madhya Pradesh, Chhattisgarh, Haryana, Arunachal Pradesh, Himachal Pradesh, Mizoram, Orissa, Jammu and Kashmir, wherein all designated forage posts in the Officer cadre up to Deputy Director level and even a few states such as Meghalaya up to Forage Development Officer level is filled up by veterinary graduates and not by forage specialists because of the service rules framed for Animal Husbandry department of the respective state governments. It is a matter of concern that post of forage specialist and Forage Development Officer occupied by the staff with veterinary background. Lack of forage specialists in the Animal Husbandry Department has resulted into poor implementation of forage development programme. There is need to bring changes in service rule and create posts of forage specialist.

Role of Dairy Cooperative Societies in the States for Forage Resource Development

Co-operative dairy farming has been found successful in many states of India. National Dairy Development Board is an apex body promoting dairy cooperatives. Dairy cooperatives pave the way for market oriented milk production. In this context, dairy federations through their milk unions in respective states need to come forward to provide technical guidance on forage production technology and production inputs particularly seed of improved varieties of forage to milk producers for dairy development. A survey of the milk federations conducted by Indian Grassland and Fodder Research Institute, Jhansi showed that in majority of states in the country forage resource development activities can be promoted through milk cooperatives. In Gujarat, Bihar, Madhya Pradesh, Maharastra, Uttar Pradesh, Andhra Pradesh and Nagaland, dairy cooperatives are giving much emphasis on promotion of feed and forage to enhance milk production. It is a matter of great concern that milk federation situated at eastern and north eastern parts of the country namely West Bengal, Orissa, Assam, Meghalaya, Jharkhand are very weak in feed and forage development programmes. Gujarat state is one of the major milk-shed of the country having reasonably balanced cattle feed production but not strong in green forage production. NDDB, Anand is also giving financial support to milk unions for distribution of forage seed. Some state like Gujarat has started promoting enrichment of crop residues through urea and demonstration to farmers. Keeping in view of large quantity of crop residues available in the state, complete feed block processing units need to be established by the dairy federations or state government directorate. This will facilitate the transport of forage from surplus zone of the state to deficit zone such as Saurashtra and Kutch regions at affordable cost. Bihar dairy cooperative federation (COMFED) has also some activity in feed and forage sector through its five subunits. Each unit is monitored by managing director and executed by one manager (P&I), one deputy manager (P&I), one assistant manager (Forage) duly supported by one Fodder Development Officer. COMFED also organizes demonstration on forage and urea enrichment of straw at farmer's field. The federation operates three cattle feed plants, producing 45,500-tonnes/annum. In Maharashtra, forage activities are monitored by managing director at Katraj dairy and coordinated by Manager (Inputs) assisted by one Forage Development Officer. Other milk cooperatives in Maharashtra lack forage developmental activities as well as have no forage specialist in their organizational structure.

Andhra Pradesh Dairy Development Cooperative Federation ranks fourth in milk production in the country. General Manager (P&I) is assisted by manager (Coordination), forage specialist, who monitors the feed and forage related programmes. Nine out of the 11 district milk unions have got feed and forage developmental activities. These officers are veterinarian/agriculturist/dairy technologist. In addition, there are nine Field Development Officers in dairy cooperatives with agricultural background. Nagaland, being a non- traditional dairy state, beginning has been made with the establishment of Kohima Milk Union Limited (KOMUL). It promotes dairying including feed and forage development. Project Officer (Development) looks after the forage developmental activities, who happen to be either dairy technologist or agricultural graduate. Tamil Nadu Cooperative Milk Producers Federation operates 11 district milk cooperative unions. Only one position in Group 'A' and 11 positions in Group 'B' category are available for the forage specialist duly supported by 11 technical staff. Forage development programmes are carried out at Madurai, Salem, Villipuram, Coimbatore and Vellore, Erode, Sivagangai, Dharmapuri and Thanjavur Milk Cooperative Unions. Karnataka Cooperative Milk Producers Federation Limited (KMF) has 13 milk cooperative unions in the state. Forage related facilities include supply of forage seed, seedlings, cattle feed and mineral mixture to the dairy farmers through milk unions as well as dairy cooperative societies. The KMF has one forage seed production farm and three forage demonstration farms. In Haryana Dairy Development Cooperative Federation, the Manager (P&I), a specialist in forage is responsible for forage development. Forage development programmes at each of the milk cooperative unions are executed by Forage Development Officer with the assistance of 18 Forage Supervisors. Punjab Milk Federation has sufficient forage development personnel's in its organisational structure. The forage activities are coordinated by manager (P&I), manager (Seed Unit) and deputy manager (Seed Production), each of them supported by two Forage Supervisors. The Milk Federation owns 11 milk unions; each milk union has forage production as well as seed production programmes. Deputy Manager is responsible for milk procurement, its sale and also forage development in each milk union.

In Dairy Cooperative of Chhattisgarh, the forage development programmes are monitored by one senior manager (Field Operation) and one deputy manager duly assisted by three assistant managers (forage production). Assistant Manager (forage production) is assisted by four Field Officers and six senior village extension officers and 19 village extension officers, in the field for forage development and milk procurement.

Tripura Cooperative Milk Producer's Union headed by a Managing Director is assisted by Manager (Forage) in forage development and one Forage Development Officer. Procurement and animal husbandry assistants also assist in forage development activities. The forage development activities are round the year forage production demonstrations, supply of seed mini-kits and rooted slips of hybrid Napier and Signal grass. Besides, one feed mixing plant is also operated under its aegis and cattle feed is supplied to farmers on no loss and profit basis. In Uttar Pradesh Milk Cooperative Federation, forage development is responsibility of General Manager (P&I), a forage specialist who is assisted by three Assistant General Managers (Forage). It has forage seed processing plant at Aligarh. It has a chain of supervisory staff that along with milk procurement also looks after forage production in the farmer's field. Its main activity is to encourage forage seed production at farmer's field and then its distribution after it's processing. Madhya Pradesh Cooperative Dairy Federation (MPCDF) has five milk cooperative unions. Feed and forage development activities are coordinated by Deputy General Manager (Field Operation) with assistance of five General Managers (either veterinarian or dairy technologist), heading five milk cooperative unions, and five Managers in the state. Besides, one Manager at each cattle feed plants is posted at Indore and Bhopal. There are no forage development activities except the supply of mini-kit forage seed to farmers. The balanced cattle feed is supplied to dairy owners on cost basis. Orissa Milk Federation (OMFED), which comprises of 15 milk cooperative unions has no forage activities due to absence of forage specialist in its organizational structure. However, assistant manager (P&I) look after distribution of forage seed/rooted slips, imparting training and conducting forage demonstrations.

West Bengal, Dairy Federation owning 15 milk unions and Assam Dairy Cooperative have no designated staff for forage development. Similar is the condition of Sikkim Milk Cooperative Federation with two milk unions. One Assistant manager (P&I) look after forage programmes with the assistance of 10 Field Supervisors in addition to milk procurement duties. The forage development is restricted to distribution of forage seeds and tree fodder saplings to farmers. Meghalaya, Jammu and Kashmir, Himachal Pradesh, Pondicherry and Jharkhand have neither forage designated staff in their organizational structure nor any feed and forage development programme.

Issues and Policy Perspectives

It is proven fact that in emerging scenario of demand and supply, its own momentum propelled the growth in livestock sector but the lack of investment and proper linkage and policy change have restricted realization of its large potential. Long back, it has been demonstrated under All India Coordinated Project on Economics of Milk Production that 6-7 litres of milk per day could be obtained from bovine by feeding green forage alone and the feeding based on green forage is most economical and eco-friendly. Moreover, cost of production of milk and other livestock products increased with high levels of concentration feeding which adversely affected the producer and consumer both. Economical production of livestock products can be achieved by lowering the cost of feeding, encouraging feeding of balanced green forage and processed forage products. Accelerated growth of livestock sector can be achieved only if feed and forage can be made available at reasonable price to the livestock farmers for which greater investment and administrative support is urgently required in a mission mode approach as in other sector of agriculture.

There are many issues, which need to be addressed through appropriate policy interventions, administrative reforms and structural changes in the department of animal husbandry and dairy at central and various states, enhanced investment, changes in the priorities and establishing linkage among research and development institutions to accelerate the growth of feed and forage resource development in India. Ultimately, forage resource development is largely driven by urbanization and markets rather than by availability of forage technologies as found by Ergano *et al.* (2010). The success of this emerging paradigm relies much on its acceptance, premised on a reasonable understanding and a willingness to change. Large scale institutional and policy change need an evidence base through piloting the approach in different contexts, documenting and sharing experiences, building on successes and stimulating policy dialogue.

Organizational Structure

Majority of the directorates of state Animal Husbandry and Dairy Federations including Government of India- Department of Animal Husbandry, Dairy and Fisheries, have very poor organizational structure for feed and forage development. There is a need of forage line staff with agriculture background in these organizations as in Punjab. The post of Joint Commissioner (Feed and Forage) which existed in the past needs to be revived in the DAHDF along with three Dy. Commissioner for cultivated forage, rangeland and CPR and forage seed production in Government of India. In most of the states, the forage development programme has been relegated to the lowest priority. There are hardly any specialist in feed and forage department. Even the sanctioned post in this department remains unmanned in most of the states including non-existent in Department of Animal Husbandry, Dairy and Fisheries in Government of India. Very few states have forage development officers with adequate training in this field. Many of the centrally sponsored schemes are not taken up in right spirit due to lack of staff in line department. On the other hand, an example exists in the state like Punjab where development of feed and forage is looked after by a well-organized institutional mechanism with a joint director, 07 deputy directors and 121 assistant forage development officers to ensure extensive forage cultivation by the livestock owners. A small country like Bhutan has also a very exemplary well organized structure for feed and forage development.

A vibrant institutional mechanism comprising of manpower with required expertise will facilitate the accelerated development of feed and forage subsector.

Long back in1937, the need for such a department was felt during the 'Cattle Conference' held in Shimla. The conference adopted a resolution to 'set up a forage and grazing committee in every province on the lines of agriculture department'. While the agriculture flourished over the years, the forage resource development in livestock sector remained non-attended entity. India has got favourable climatic conditions to grow forage round the year, which will give it an edge over grain based livestock production followed in western world leading to conflict and competition between food and feed. The success of co-operative sector in the dairy sector in different states is very well known. They have been successful in many states where, the government departments have failed to deliver. The members of the milk co-operatives are increasingly asking the cooperatives to supply them quality seeds of forage crops. Will it be more effective to promote forage production programmes through co-operatives, which have a direct stake in their development, need to be explored?

Lack of Statistics and Need of Database for Policy Decision

There is no mechanism of collecting data on area and production on forage resource at regular interval. Proper statistics on these aspects will help for policy and research intervention as well as assess the adoption of technologies. Therefore, the National Sample Survey (NSS) may also be entrusted to take up the survey as is being done in case of food and other crops. Efforts in this context could not achieve success as NSS is under Department of Agriculture and Cooperation, unwilling to take up additional task. Neither NSSO nor any other agency is involved in realistic quantification on area, production, productivity and availability of feed and forage. NSSO need to take up survey on regular basis on area, production and productivity of cultivated forage and forage from CPR's, forest fringe *etc.* for planning and policy decisions. Forage need of milch dairy animals should be separated from dry large ruminants and small ruminants for targeted approach. This also needs different data base.

Need of Investment on Forage Resource Development

The livestock sector has remained for a long time, subsidiary to agriculture in India. It moved to next stage by establishment of milk cooperatives in various states under the guidance and assistance by National Dairy Development Board and emerged as a profitable enterprise. Now, livestock keeping is being taken up not only by the poor and landless for livelihood but also on commercial scale by large farmers and in clusters of peri-urban dairy around the cities. Allocating about 07 per cent of the agricultural outlay, livestock sector has contributed 24.72 per cent (Rs. 1239 billion, 2004-05) to total GDP of agriculture and allied sector. There is a need to enhance investment in forage resource development at national and state level. In 11th Plan, out of total approved allocation (Rs.4903 crores) for Animal Husbandry and Dairy Sector (DAHDF, GOI) only 2.88 per cent (141.4 crores) has been allocated to feed and forage development. Similarly, 22 states in 10th plan had expended an average of 0.77 per cent (0.09 per cent in Andhra Pradesh to 6.63 per cent in Mizoram) of total out lay for State Animal Husbandry and Dairy sector for feed and forage development. Allocation of more funds for forage development programmes, including research, extension and human resource development is required. Simultaneously, more funds are needed to be allocated to the programmes like grassland improvement and increasing the productivity of common property resources of villages with scientific management. Investment in common property resources to improve their productivity should be seen as lending a helping hand to resource poor livestock farmers who graze their animals along roadsides, bunds and common property resources.

Credit Facility for Forage Production and Marketing

Extending credit facility for livestock and forage production in peri-urban markets is required. Farmers seeking loan against livestock and producing forage crops are facing difficulty in getting credit facility. Some provision for livestock related activities has been made in virtual fund of GOI executed by NAABARD. Developments of a forage marketing system and extending credit system for livestock and forage production are required for accelerated growth of livestock sector.

Policy Related to Breeding for Crop Varieties

In the pursuit to maximize the yield of grains of cereal crops, the yield of straw component has been reduced. Any breeding programme for cereal crops must realize that crop residues are equally important in Indian context, where they are the mainstay of livestock feeding. Therefore, a minimum straw to grain ratio must be ensured before any new crop variety is released. Central Variety Release Committee (CVRC), DAC, Ministry of agriculture, Government of India should consider straw yield and proximate analysis for feeding quality as one of the parameter for release of food crop varieties in order to assess the nutritional value of edible crop residues of the newly released food crop varieties.

Requirement of Quality Forage Seed, National Seed Reserve, Support Price, Marketing and Import of Seed

The states generally do not have trained manpower to look after the seed production activities. Moreover, there are few farms to cater to the needs of quality seeds of forage crops. It has been estimated that about 1.64 million tonnes of certified forage seeds would be required during 2010-2015 (Annexure 4.2a and b). But the quantity of total forage seed produced by the Central farms is meagre approximately 1627 tonnes. Forage seed production programmes are non-existent in the most of the states and the farmers are forced to depend on the private vendors where quality is often unreliable. Therefore, there is an urgent need to take up forage seed programme on a large scale involving National Seed Corporation. The production and distribution of quality forage seeds need to be taken up on priority by National Seed Corporation and State seed corporations. Mechanism should be developed to take up seed production expeditiously so that seeds of improved varieties of forage crops are accessible to the livestock owners. Import of berseem and forage sorghum seed need to be reviewed (Tables 4.3 and 4.4).

Table 4.3: Forage Seed Import Status in Berseem

Year	*Recommended Qty (tonnes)*
2004-05	2062
2005-06	2930
2006-07	7912
2007-08	7622
2008-09 (Till Jan 09)	7520

Note: Pending (deferred) requirements from 43 proposals (companies) is 29,228 tonnes.

Ref. : 135 EXIM Committee meeting (3 March 2009).

Table 4.4: Forage Seed Import Status in Forage Sorghum

Cultivars/Varieties	*Qty Sanctioned*
Cv Pacific BMR, Superdan, PAC/F 8411, Sweet Jumbo	40 kg (10 kg for each variety) from Australia
Cv 23402, 24213, 23431,22050, 22053	60 kg (10 kg for each variety) from USA

Ref.: 135 EXIM Committee meeting (25 March 2009).

Need of 50,000 tonnes for 2 million hectares area under berseem could be met without import if seed production chain is created as breeder- foundation-certified/truthfully levelled. As compared to imported seed, national varieties are well adapted to specific agro-eco zones, hence gives better forage yield than imported one. As per import data, 7520 MT was imported till January 2009. 43 proposals were pending for 29228 MT. Import of forage sorghum seed is limited to hybrid (40 kg from Australia and 60 kg from USA).

National seed reserve policy and programme is at final stage. Forage seed must find a place in this project. Presently, indents of breeder seed are less than the production. Hence, there is a need to upscale breeder seed requirement and create awareness about new varieties of forage crops released through various media of communication. Considering low seed yield as compared to many field crops, pricing should be fixed in a way to encourage farmers to grow forage seed.

There may be a system of support price in case of seed of forage crops considering their low seed yields as compared to food crops. Its pricing may be fixed in a way that encourages farmers to grow forage seed for own use and marketing. There is a need to develop forage seed marketing system. Table 4.5 quantify estimated seed requirement of forage crops (IGFRI Perspective Plan 2025)

Table 4.5: Estimated Requirement of Seed of Range Plant Species ('000 q)

Type of Seed	*2010-15*	*2020-25*
Grasses	305.9	315.3
Legumes	303.3	312.7
Fodder Trees	18.8	18.3

Cenchrus species, Dichanthium annulatum, Laisurus indicus, Panicum maximum, Penisetum pedicellatum, Brachiaria species, *Dactylis* species, miscellaneous species).

Issues Related to Edible Crop Residues

Crop residues provide 53 per cent of the feed requirement of the livestock and it will continue to be the mainstay of livestock feeding in coming future. The straw of crops in many parts of the country is burnt after the harvest of grain. The burning of crop residues not only result in environmental pollution but also have impacts on the forage availability for the livestock. There is a need for policy level intervention to prevent the burning of crop residues. Large quantities of edible crop residues are also being diverted for packaging, for making bricks as source of fuel in kilns and other non-agricultural purpose. Diversion of edible crop residues towards packing industry should be checked. Only non-edible crop residues may be allowed for packaging. Edible crop residues for livestock need not be diverted for other uses through policy interventions. A futuristic scenario is also emerging, wherein diversion of crop residues for bio-fuel production is foreseen. Consequently, prices of crop residues and grain are coming closer, which needs proper understanding and policy solutions. During 2009-10, National Consultation on Forage Resource Development and Utilization held under the chairmanship of Secretary, Department of Agriculture Research and Education at Indian Grassland and Fodder Research Institute also recommended promoting integrated crop-livestock production systems and other means of minimizing conflict of demands on crop residues through better understanding of farming systems (PACA, 2008). Therefore, there is a need of policy support to stop diversion of edible crop residues for power, packaging and other uses.

Policy on Export of Oil Seed Cake

Oilseed cakes are important ingredients of feed concentrate. Hence its export needs to be restricted. Currently about 25 million tonnes of Oil Seed Cake are produced in India, of which more than 50 per cent is exported. Largely, groundnut cake (5.3 million tonnes) and soybean cake (8.6 million tonnes) are exported. A policy guideline is needed to restrict the export of these oil seed cakes, which will pay more dividends by using as protein supplement in feed ingredients for livestock production than the export earnings.

Regulation on Pesticide Residues Contamination in Livestock Products

Oil seed cakes in the concentrate feeds are also an important source of pesticide contamination in milk. Therefore, the use of pesticide in crops like groundnut, mustard, soybean, sunflower and cotton crops need to be regulated.

Attitudes towards Pastoralist

In social evolutionary thinking, the nomadic lifestyle has traditionally been treated as less civilized, less productive and more degrading than a settled lifestyle. Such a cultural bias seems to have many policy implications for the Himalayan pastoralists. Due to the problem of their cultural stereotyping, small population and migratory lifestyle, the Himalayan pastoralists are ignored in the various policy-level decisions. Ignorance of their due rights and status in the Indian state has seriously marginalized these communities. Today, policy and decision-makers perceive Himalayan pastoralists as an environmental threat to the Himalayas and the local pastoral groups are incessantly blamed for overgrazing and livestock increase in population. There is little interest in a detailed objective analysis of the condition of the environment. At a local level, especially Forestry Department officials hold these attitudes. For example, due to recent notification of Great Himalayan National Park in Himachal Pradesh all the pastoralists who used to occupy the vast alpine pastures of the Park during the summer months have been deprived of access to approximately 300 square kilometres of pastures without being allotted grazing rights in alternative regions.

Common Property Resources (CPR)

In India current land use provides 12 million hectares area under permanent pasture and grazing lands which is lying as wasteland. Out of 75 million hectares area under forest, half is lying at the last stage of degradation. Similarly, different surveys have pointed out 155 to 175 million hectares as wastelands. These CPR, are important sources of livelihood and income for poor people in all the states (Table 4.6). CPRs contribute and allow considerable access to all users, but not all the sections of the rural community are equally attracted by these potentials and opportunities. The proportion of poor households dependent on fuel, forage and food items from CPR's ranged between 84-100 per cent in different villages. In general, the rural poor obtain the bulk of forage from CPR's. CPR's complement private farming by providing crop inputs, sustaining animals and are especially predominant in drought prone area. These land resources are vital for fuel and forage production. Thus, there is great potential to meet feed and forage shortage through this natural resource. Even if 50 per cent of this land is brought under silvipastural system, there is potential to produce 200 million tonnes of dry matter to meet shortage of forage.

Table 4.6: Extent of Dependence of Poor and Wealthy Households on CPRs in Dry Land of India

State	Household Category	CPR Contributions			
		Income Per cent	Fuel Supplies Per cent	Animal Grazing Per cent	Days of Employment per Household
Andhra Pradesh	Poor	17	84	0	139
	Wealthy	1	13	0	35
Gujarat	Poor	18	66	82	196
	Wealthy	0	0	0	0
Karnataka	Poor	20	0	83	185
	Wealthy	3	0	29	34
Madhya Pradesh	Poor	22	74	79	183
	Wealthy	2	32	34	52
Maharashtra	Poor	14	75	69	128
	Wealthy	1	12	27	43
Rajasthan	Poor	23	71	84	165
	Wealthy	2	23	38	61
Tamil Nadu	Poor	22	0	0	137
	Wealthy	2	0	0	31

Source: Integrated Watershed Management by Rajesh Rajora.

In Northwest India, the traditional utilisation of grasslands in Saurashtra, Kutch, eastern Gujarat, western Madhya Pradesh and southeastern Rajasthan, a drought prone area, evolved to cater to two basic needs – forage and grazing. Traditionally every village or cluster of villages used to set aside certain areas for livestock grazing, called goucher, and protected other areas from grazing, where the grass was allowed to grow long, to be subsequently harvested and stored for later use. These protected grasslands were called *vidi, veeds, bheeds* or *rakhals*. They were crucial to rural economy as the forage produced was used both during the lean summer months and to tide over drought periods. With the promulgation of Land Ceiling Act, after India's Independence, and population pressures, the land use practices have been altered. Land set aside for grazing (*goucher*) has been encroached upon for agriculture, industrial development and urbanization, as a result of which protected forage producing grasslands came under increasing pressure of livestock grazing. Despite these pressures, there are still some extant grassland, especially in Gujarat and eastern Rajasthan. For instance, in Saurashtra and Kutch there are 137 reserved *vidis* covering a total area of 63,292 ha. The reserved *vidis* are managed by the Forest Department while the non-reserved *vidis* are given to various agencies for protection, *e.g. Gaushalas* and *panjrapoles* (trusts that maintain aged cattle), Maldhris cooperative societies, village *panchayats* and milk cooperatives. There are 471 nonreserved *vidis* in Saurashtra and Kutch, covering an area of 57,602 ha. While the condition of reserved *vidis* is generally good (*e.g.* Rampura grassland in Dahod), the non-reserved *vidis* are in terrible condition due to mismanagement, corruption and neglect.

The CPRs seeks to provide support to the people and their production systems through restoration, protection, regeneration, upkeep and development of grasslands. There is no sound management plan for the development of pasture land and protection of existing grasslands, some of which are unique and harbour rich fauna. We have not even fully documented the value of these grasslands in terms of their biological diversity. The famous Sewan grasslands of Jaisalmer and Bikaner, and the Banni grassland of Kutch have been neglected, resulting in over-grazing, spread of invasive species such as *Prosopis juliflora* and conversion to agricultural crops with dubious results. The highly productive wet terai grasslands of the Gangetic and Brahmaputra floodplains are underrepresented in protected area network of India, except some areas such as Kaziranga, Dudhwa, Jaldapara national parks. The importance of rotational or seasonal grazing, some control on free ranging animals, total protection of selected grassland plots to serve as nucleus for seed bank, secure tenure for pastoralists (both resident and nomadic) over pastures and genetic improvement of livestock (using indigenous breeds, not exotics ones) have not been taken in to consideration in animal husbandry programmes of the country. In our country, only livestock is considered as wealth, not the grasslands on which this livestock does depend nor the traditional knowledge that helps maintain this livestock. Interestingly, protection of forage producing, natural grasslands greatly help in the protection of many endangered species. For example, in Maharashtra, in the late 1970s, a large number of plantation and grassland plots were developed under the Drought-Prone-Areas-Programme (DPAP). The main aim of DPAP was to take conservation measures for the protection of over-used land which was suffering from severe overgrazing and soil erosion. The DPAP not only helped in achieving its aim in certain areas but it also resulted in restoration of wildlife, especially the Great Indian Bustard, Blackbuck and Grey Wolf. Grasslands are the 'common' lands of the community and while there have been robust traditional institutions ensuring their sustainable management in the past, today due to take-over by government or breakdown of traditional institutions they are the responsibility of none. They are the most productive ecosystems in the Indian Subcontinent. Indeed they are often looked at as 'wastelands' on which tree plantations have to be done, or which can be easily diverted for other uses. Such diversions often put even more pressure on adjoining ecosystems for grazing and forage removal, resulting in a cascading chain of degradation. The lack of clear tenure to local communities, confused land records between the Revenue and Forest departments, and other such issues of land rights and responsibilities also compound the problem.

Grassland Protection

Grasslands are not managed by the Forest Department whose interest lies mainly in trees, not by the agriculture department who are interested in agriculture crops, nor the veterinary department who are concerned with livestock, but not the grass on which the livestock is dependent. The grasslands are the 'common' lands of the community and are the responsibility of none. They are the most productive ecosystems in the subcontinent, but they belong to all, are controlled by none, and they have no godfathers. All types of grassland ecosystems are under tremendous grazing pressures. Grassland ecosystems are sensitive to climate change, developmental pressures and invasion by alien invasive plants. Grasslands are the most productive ecosystems in the subcontinent, but they belong to all and controlled by none. Grasslands are the most neglected ecosystems and less attended, under biodiversity conservation in India. Protection, development and sustainable use of grasslands are very important for the livestock economy in the country. There are more than 500 million livestock in the country and more than 50 per cent of the forage for this livestock comes from grazing lands. Many natural types of grassland (such as wet grasslands of *terai, shola* grasslands of the Western Ghats and dry grasslands of *Deccan*) have been converted to plantations, sometimes even in protected areas. Some of the most threatened species of wildlife are found in the grasslands. According to wildlife Institute of India, less than one per cent grasslands come under protected area network. Thus, grassland improvement is important for biodiversity conservation, enhancing productivity of livestock sector and for poverty alleviation (Table 4.7). Considering vital importance of the grassland, there is a need for grassland development and grazing policy to put an institutional mechanism in place for sustainable use of grasslands. For example, in the semi-arid grasslands, the carrying capacity is one Adult Cattle Unit (ACU) per ha (Shankar and Gupta 1992), but the stocking rates are as high as 51 ACU per ha, while in the arid areas, the carrying capacity is 0.2-05 ACU per ha but the stocking rates are 1 to 4 ACU per ha).

Table 4.7 a: Species to Benefit through Better Protection of Grasslands

Habitat/Species	*Schedule WPRA*
Cold Desert Grassland: Black-necked Crane	I
Short Grass Plains: Blackbuck (*Antelope cervicapra*)	I
Wet Grassland: Brow-antlered Deer (*Cervus eldi*); Swamp Deer (*Cervus duvauceli*); Hog Deer (*Axis* porcinus); Hispid Hare (*Caprolagus hispidus*); Pygmy Hog (*Sus salvinius*); One horned Rhinoceros (*Rhinoceros unicornis*); Wild Buffalo (*Bubalus bubalis*); Bengal Florican (*Houbaropsis bengalensis*); Swamp Francolin (*Francolinus gularis*); Lesser Adjutant (*Leptoptilos javanicus*); Greater Adjutant (*Leptoptilos dubius*); Jerdon's Babbler (*Chrysomma altirostre*); Black-breasted Parrotbill (*Paradoxornis flavirostris*); Marsh Babbler (*Pellorneum palustre*); Finn's Baya (*Ploceus megarhynchus*).	I, III, IV

Habitat/Species	*Schedule WPRA*
Hot Desert Grassland: Caracal (*Felis caracal*); Indian Fox (*Vulpes bengalensis*); Grey Wolf (*Canis lupus*); Golden Jackal (*Canis aureus*); Great Indian Bustard (*Ardeotis nigriceps*); Agra Monitor Lizard (*Varanus griseus*); Spiny-tailed Lizard (*Uromastix hardwickii*); Laggar Falcon (*Falco jugger*); Saker Falcon (*Falco cherug*); Peregrine Falcon (*Falco peregrines*); Red headed Falcon (*Falco chicquera*); Lesser Kestrel (*Falco naumanni*); Tawny Eagle (*Aquila rapax*); Steppe Eagle (*Aquila nepalensis*); Imperial Eagle (*Aquila heliacal*); Lesser Spotted Eagle (*Falco pomarina*); Common Buzzard (*Buteo buteo*); Long Legged Buzzard (*Buteo rufinus*); Upland Buzzard (*Buteo hemilasius*); All species of Harriers (*Circus spp.*); Shorttoed Snake Eagle (*Circaetus gallicus*); Red Headed Vulture (*Sarcogyps calvus*); White Backed Vulture (*Gyps bengalensis*); Long Billed Vulture (*Gyps indicus*); Slender Billed Vulture (*Gyps tenuirostrisb*); Eurasian Griffon (*Gyps fulvus*); Egyptian Vulture (*Neophron percnopterus*)	I, II, IV
Grassland: Lesser Florican (*Sypheotides indica*)	I
Shola grassland: Nilgiri Tahr	I

WPA: Wildlife Protected Area.

Table 4.7 b: Ecosystems and Habitats to Benefit from Protection of Grasslands/ Deserts

Habitats/Ecosystems	*States*
Cold Desert	Jammu and Kashmir, Uttrakhand, Himachal, Sikkim, Arunachal Pradesh
Dry Grasslands	Rajasthan, Gujarat, Punjab and Haryana
Hot Desert	Rajasthan, Gujarat, Punjab and Haryana
Tropical short grass plains	Rajasthan, Gujarat, Madhya Pradesh, Maharashtra, Andhra Pradesh, Karnataka
Wet grasslands	Uttar Pradesh, Bihar, West Bengal, Assam, Arunachal Pradesh
Shola Grasslands	Kerala, Tamil Nadu, Karnataka, Maharashtra

Table 4.7 c Protected Areas to get greater benefit from Protection of Grasslands/ Deserts

States	*Name of National Park/Sanctuary*
Rajasthan	Desert NP, Talchapper, Gajner
Gujarat	Desert Wildlife Sanctuary, Narayan Sarovar, Lala and Naliya, Wild Ass Sanctuary, Velavador
Madhya Pradesh	Sailana, Sardarpur, Dahod grasslands
Maharashtra	Bustard Sanctuary, Rehukuri
Karnataka	Rannibennur,

Contd...

Andhra Pradesh	Rollapadu,
Uttar Pradesh	Dudwa, Katerniaghat, Kishenpur, Sohagi-Barwa
Bihar	Valmiki Tiger Reserve
West Bengal	Jaldapara, Gorumara
Assam	Manas, Kaziranga, Pobitora, Laokhowa-Burachapori, Orang, Sonai-Rupai
Arunachal Pradesh	Dirang Memorial Sanctuary

Source: 11th plan Reports on Grasslands and Deserts, Planning Commission, Government of India.

It is considered that in the present scenario, there is an urgent need to have proper policies backed up with institutional support to tackle such an imbalanced situation. Our policies on free range grazing should be less restrictive and rational, based on principles of scientific grazing management.

Legal Protection to Grasslands

The grasslands are the most neglected, abused and least protected ecosystems in India. Grassland should be valued, as grazing resources and appropriate law should restrict their diversion. Such areas can be declared as community reserves. Most of the States have excluded the grasslands and have not identified them as "deemed forest" by the State Expert Committee's pursuant to the landmark order dated 12.12.1996 in the Forest Matter (T. N. Godavarman Thriumalpad V. Union of India and others in W.P. (C) No. 202/95). As per the said order of 12.12.1996, word 'forest' should be given a wide and liberal interpretation. Excluding grasslands and including lands only with tree cover as 'forest' is against the letter and spirit of the said order thereby denying the protection under the Forest (Conservation) Act, 1980. In view of the fact that the grasslands have spontaneous natural vegetative growth, these should also be treated as 'forest land' for the purposes of the Forest Conservation Act and restrictions on diversion of such lands for non-forest use should be applicable to these critical ecosystems as well. The central government should invoke the provisions of the Articles 251 and 254 of the Constitution to direct state governments to instruct Revenue Departments not to divert any grassland identified in the landscape. Such areas can be declared as community reserves. The **National Environment Policy 2006** (NEP) states that while conservation of environmental resources is necessary to secure livelihoods and well-being of all, the most secure basis for conservation is to ensure that people dependent on particular resources obtain better livelihoods from the fact of conservation, than from degradation of the resource. While this can be taken as a guideline for further environmental planning, the NEP totally misses out the Grassland Ecosystems. Many of the grasslands in the country are sensitive to climate change, developmental pressures and invasion by alien invasive plants. By issuing an appropriate notifications under Section 5 of the Environment (Protection) Act,

1986, the Central Government can constitute and declare Ecologically Sensitive Areas or zones for the better protection of the environment and in particular with respect to the grasslands. In case such a notification is issued there would be no need to acquire land or shift people, as has been done in the case of Matheran and Mount Abu areas. This procedure has a clear advantage over declaration of national parks and sanctuaries under the Wild Life (Protection), Act 1972.

A coordinated effort towards conservation and management of Alpine Meadows (Bugyals): Most of the alpine areas fall under one or other category of 'forested land' or Van Panchayat. However, no concerted efforts have been made towards conservation and management of these areas so far. In some of the high altitude protected areas there are conflicts between the local people. Hence there is a need to evolve valley/area specific grazing plans in a participatory manner involving the local and migratory graziers, animal husbandry department, tourism department and Indian army as the case may be so that some of the heavily degraded Bugyals could be brought under a recovery plan.

The recent act passed by the parliament in 2007 – "The Scheduled Tribe and Other Forest Dwellers (Recognition of Forest Rights) Act 2007 (The Forest Rights Act)" – seeks to regularise holdings up to 04 ha per individual in forest areas, including national parks and sanctuaries. This act is expected to take away at least one third of the protected area network (Singh, 2008). Now, bringing more area under reserve forest category and development of areas under various soil and water conservation and afforestation schemes will lead to further reduction of grazing areas (BAIF, 2003, 2006; LIFE, 2007). In villages, grazing has already become very difficult due to reduction in village community lands and other pasture lands. In such a scenario, some policy and institutional support is required to tackle this imbalanced situation. Based on Task Force on Grassland Desert, 11th plan, MOEF, Roy (2009) suggested following broad policy intervention:

1. Instead of formulating more restrictive policies on permitting livestock (including goats) in forest areas, due weight is required on scientific results. There should be a scientific assessment about carrying capacity of forests and based on it, the guidelines for grazing (number of livestock per unit area *etc.*) and/or forage harvests should be developed.
2. Formulation of suitable guidelines for usufruct rights for pastoralists and graziers in various watershed and JFM programmes.
3. Establishment of improved pasture systems with appropriate mixtures of grasses and legumes in open forest lands *etc.* for augmenting forage resources for forest dependent communities. Promotion of silvi/horti-pasture systems for forage and other needs of the communities under watershed or other programmes in a phased manner. The successful examples of good practices on community lands in India (Maarse *et al.*, 2008) may be considered for replication in similar areas through some sort of promotional schemes.

Restrictive forest policy permitting livestock grazing should be developed/ modified in a complementary mode for realizing the forage potential of forest fringe and open forest. The promotional schemes under watershed or other programmes need to be taken up in a phased manner so that enclosures for a minimum possible period and with provisioning of alternate utilization schemes. Such measures are expected to generate more response by the communities for adoption of the technologies.

National Grazing Policy

Despite the fact that India has one of the largest livestock populations in the world, with an estimated 520 million heads, we do not have a grazing or grassland policy on ground. In our country, only livestock is considered as wealth, not the grasslands on which the livestock depends. Currently "The Cattle Trespassers Act' formulated in 1871 is the only Act applicable to regulate grazing in public and forest land. In 1937, cattle conference held in Shimla gave resolution that in all provinces standing forage and grazing committee should be established (Roy, 2009). In the Draft Grazing and Livestock Management Policy, emphasis has been given to develop large blocks of grass reserves away from human habitation for higher production (in arid and semi-arid regions) and as forage banks for drought years. However, we still do not have a grazing or grassland policy on ground. Government of India has formulated 'Draft Grazing and Livestock Management Policy (1994)', and 'Draft National Policy for Common Property Resource Lands (CPRLs)'. These policies have not been implemented effectively in the field. Secure tenure for pastoralists (both resident and nomadic) over pastures, and genetic improvement of livestock (using indigenous breeds, not exotics ones) have not been taken into consideration in animal husbandry programmes of the country. Grasslands are the 'common' lands of the community and while there used to be robust traditional institutions in the past ensuring their sustainable management, today due to take-over by government or breakdown of traditional institutions they are the responsibility of none. Indeed they are often looked as 'wastelands' on which tree plantations have to be done, or which can be easily diverted for other uses, resulting in a cascading chain of degradation. The Forest Policy of 1894 was the most elaborate of all the policies in explaining the modalities of grazing in protected forests. The Forest Policy of 1954 was extremely critical of unrestricted and uncontrolled grazing and refuted it as contrary to scientific management of forests. However, it also admitted that in some forest/grassland types, limited grazing does not do much harm, and may actually improve the grassland/forests. Dhebar Commission (1966) recommended that the Forest Department should promote growth of improved varieties of grasses in forest areas and grazing fees should be regulated. The National Commission on Agriculture (NCA, 1976) also recommended strict control on grazing and regulation on grazing. It also recommended the promulgation of grazing rules by each state specifying the grazing rates and providing for the manner in which grazing should be permitted.

The National Forest Policy (1988) is in consonance with the previous policy on the issue of grazing, except for an important qualifier that grazing in forest areas should be regulated with the involvement of the local community and a committee under the chairmanship of Mr. C. D. Pandya, IGF (Retd.), also recommended that "A National Grazing Policy should come into effect at the earliest."

Recommendations of the Task Force on Grassland and Deserts 11th Plan and Other Policy Advocacies to Ensure the Sustainable Use of Grasslands

- There is an urgent need for a **National Grazing Policy** to ensure the sustainable use of grasslands and biodiversity conservation.
- Necessary modification would be required in the new Environment Impact Analysis **(EIA)** guidelines by including ecologically fragile and environmentally sensitive areas where prior EIAs will have to be made mandatory. Also, presence of representatives from identified institutions and experts should be made mandatory during public hearing whenever an EIA is done in the grassland and desert ecosystems so as to review the identified impacts, prediction and mitigation.
- Detailed and updated GIS based inventory of degraded rangelands in each agro-ecological zone and also measurement of the impact of rehabilitation programmes.
- Higher priority on generation of information on temperate grasslands with emphasis on (i) low input, clover based sheep grazing system; (ii) ideal pasture for mixed grazing systems; (iii) inventory of grazing routes and grazing systems; (iv) designing of suitable production system for migratory graziers of Himalaya and the *Thar* desert.
- To develop a policy of regulated grazing that is managed on scientific principles so that desirable vegetation development could be ensured. As the grazing policy alone can not mitigate the problems of forage availability in the country, a matching approach on forage production, agro-waste-use and fodder trees should be brought under one umbrella in form of a national forage mission.
- Development of Common Property Resources (CPRs) available with village *Panchyats* through improved pasture/silvipasture systems is undertaken by dairy/livestock cooperatives/associations.
- The practice of stall-feeding should be encouraged among livestock owners in order to prevent over grazing and consequent depletion of available forest and forage resources. This should be one of the main issues in the forests being developed under Joint Forest Management programmes (JFMs).
- In arid and semiarid regions of the country, large blocks of lands away from human habitations could be developed as grass reserve and their

production may be preserved in form of hay in fodder banks. The technology of densified bales or enrichment in the form of feed blocks may be practiced for ease in transport and enhancing forage quality.

- In undertaking large-scale range and pasture development programmes, poor availability of quality seeds of range plant species is often a critical problem.
- *Forage* produced from reserved vidis (as in Gujarat) should be given to local people on priority basis, before it is exported to other districts/ regions. Once the local people benefit, they would develop a stake in protection of grasslands.
- Provide a range of incentives to farmers and pastoralists to continue traditional practices that are beneficial for wildlife and help in sustainable use of grasslands and deserts. Encourage and provide appropriate legal backing to community conserved areas containing grasslands and deserts.
- The National Forestry Commission has recently given recommendation for the conservation of grasslands and deserts which are fully relevant in this context. Since India is characterized by tropical monsoon climate and active growths in grazing lands occur only during monsoon months, there is surplus forage during rainy months and deficits of various levels in other months. Thus there is already growing emphasis on animal feed security systems and fodder banks to overcome such problems. The surplus production from grasslands during rainy season is to be carefully preserved in various forms to meet the forage requirements of the lean periods. The post-harvest technologies such as biomass processing, enrichment and densification appear to be the key for better animal husbandry in the deficit zone.
- The prevailing view of looking at grasslands as a single use (forage for domestic animals or wildlife habitat) should be replaced by looking them for multiple uses, incorporating sustainable use, ecosystem functions and biodiversity conservation.
- There is a need of capacity building at various levels for the rangeland development and seed production of range plant species activities with the objective of restoring range health.
- Demonstration of range improvement and management technologies at different locations should receive higher priority and the feed back from the actual beneficiaries and the farmers in the vicinity should be properly accounted for further refinements in the technology.
- There is need to increase grasslands ecosystems in Protected Area, especially in the Thar Desert and the cold desert of Ladakh and Sikkim which are grossly under-represented.
- Considering a wide range of activities and programmes under each Ministry and R&D Institutions, it would be extremely important to identify

some of the cross cutting themes and launching the **Integrated Research and Development Programmes** in the grasslands and deserts.

- ✰ The Ministry of Environment and Forests (MoEF), Government of India should start a division or section to look after the grasslands issues, on the pattern of Wetland Division to be headed by a Joint Secretary.
- ✰ Policy for improvement of forest fringes, permanent pastures and grazing lands (12 million hectares), reclamation of wastelands, and rehabilitation of problematic soils through edible vegetation cover of grasses, hortipasture and silvipasture are of prime importance. Promotion of stall-feeding and feed supplementation to prevent overgrazing is also required.
- ✰ Proper implementation of grazing policy is the need of hour. Some states have made regulations on grazing restrictions. There is a requirement of a rational and holistic grazing policy associating the livelihood support to the disadvantageous communities and the scientific consideration of vegetation recovery and ecological sustenance. Rights of graziers and pastoralist should be properly recognized and their rights be protected in such regulations.
- ✰ Restrictive forest policy needs to be modified in a complementary mode for realizing the forage potential of forest fringe and open forest. The vast tract of unused lands along roadsides, railway line and other wastelands could be utilized effectively through leasing out for a certain period for forage production and regulated grazing to people residing in the vicinity. The policy for increase in forage availability should be based on regional and seasonal (lean period) considerations. We still do not have a grazing or grassland policy on the ground.

Proper implementation of grazing policy is the need of hour. Some states have made some regulations on grazing restrictions. There is a requirement of a rational and holistic grazing policy associating the livelihood support to the disadvantageous communities and the scientific consideration of vegetational recovery and ecological sustenance. Rights of graziers and pastoralist should be properly recognized and their rights be protected in such regulations. The establishment of a nodal agency is required to coordinate production and marketing of quality range seeds, both at regional and national levels, involving commercial seed companies, NGOs and farmers' cooperatives *etc.* There is a need of capacity building at various levels for the rangeland development and seed production of range species activities with the objective of restoring range health. IGFRI and CAZRI exclusively deal with grassland and desert ecosystems, respectively. They have adequate capacities and capabilities to address the respective issues. However, their geographical coverage and overall approach (ecosystem services and intrinsic values) needs to be broadened particularly in a long term ecological and monitoring studies on grasslands in representative zones.

Institutional Mechanisms for Utilization of the CPR and Non-traditional Lands

The area under forage production has remained stagnant. There is ample scope for extending forage production within the available farmland through non-competitive land use techniques and technology, obtaining forage from common property resources (CPR's) and beyond. Institutional mechanisms have to be developed for active community participation and adoption of rotational and deferred rotational grazing systems. Various categories of unutilized lands such as problem soils, along railway tracks and highways, wastelands, along bunds and canals *etc.* could be used for forage resource development with effective support of local institutional mechanism, organisational change, administrative reforms and required policy changes. Planting tree fodder, forage shrubs and other perennials forage grasses under afforestation programmes under social forestry and other schemes can also enhance availability of forage. Such lands could be given on lease to people residing in the vicinity and easy credit may be provided to purchase quality forage seeds and other production inputs. There is need to strengthen the village institutions towards development of common property resources. There are many success stories, where leasing and vesting the ownership rights to the local farmers through self-help groups (SHG's) and various cooperatives has led to the rejuvenation of the degraded pasture and evolving benefit-sharing mechanisms among the stakeholders. The vast tract of unused land along roadside, railways and other wasteland could be utilized productively, if they are leased out to the people residing in the vicinity. In Bangladesh, Grameen Fisheries and Livestock Foundation (Motsho O PashuSampad) has successfully put to use the several stretches of land and water bodies along the highway which has led to substantial increase in livestock and fish production and improved the livelihood of local people in the country.

Managed Stall Feeding and Controlled Grazing of Livestock

The range grazing is the cheapest way to feed the livestock, particularly goat and sheep which provides household security to pastoral as well as a large number of marginal farm families. The government policies have a great influencing factor in livestock farmers' choice towards controlled grazing and or managed stall feeding. Analytical view on present scenario pointed that there is an urgent need for proper enabling policies and guidelines by Government of India as well as state government levels for better utilization of the available natural resources and its further improvement to meet the challenges of livelihood security of the pastoralists.

Utilization of Forage Resources from Forest Margin

There is little interaction among the forestry and the livestock departments. The personnel of forestry department always look with suspicion the livestock in its periphery as destroyer of vegetation. There is need to understand that controlled removal of vegetation/grass in fact spurs its further growth. Hence, a synergistic approach needs to be adopted for the regulated use of grasses in the forest margin, which supply very large quantity of forage to peripheral villages hardly being noticed and accounted. This needed to be regulated and properly accounted on cut and carry system. Further, the requirements of herbivorous wild life in reserve sanctuary or other forest area also need proper attention to develop forage resources because degraded forests do not meet their requirement. The conflict of wild life and livestock has emerged as serious problems in many surroundings of the forests. The forest departments have also started requesting for seeds of various forage grasses and legumes.

Social Forestry and Afforestation

Social forestry and afforestation undertaken by various agencies should invariably take up plantation of trees which has forage value, at least 25 per cent of their target be taken up to mitigate scarcity of forage during drought and floods. This would help to bridge to some extent the gap in forage demand and supply in normal years besides coming to the aid during periods of drought and flood. It may contribute significantly to create live fodder banks.

Utilization of Non-traditional Lands for Forage Resource Development

Vast tract of unused land along roadsides, highways, railway line and other wastelands could also be utilized effectively through leasing out for certain period for forage production and regulated grazing to people residing in the vicinity on lease or users group basis.

Processing, Storage and Transport of Feed and Forage

Since the livestock keeping has emerged as an important livelihood option for a large number of people as well as taken a shape of commercial venture in peri urban areas, they need to be insured against the vagaries of nature affecting the availability of quality forage at reasonable price. For this, a chain of forage banks need to be established in all the regions particularly in vulnerable regions on the lines of Food Corporation of India. In periods of natural calamities, livestock are left to fend off on their own. A fair assessment of the number of livestock in drought and flood prone areas and their feed and forage requirement need to be assessed and a contingency plan developed for creating forage banks based on live fodder trees, grasses and crop residues on the lines of Food Corporation of India to provide succour to the livestock in distress. To facilitate transport and storage, static and mobile type bailing machine and other post-harvest interventions need to be developed. Peri-urban livestock keeping has emerged as an important vocation. The livestock in peri-urban areas often face difficulty in procuring quality forage. In these areas, complete feed blocks to cater the need of animals with different production levels need to be encouraged.

Managing Natural Calamities

During a drought situation, every measure should be taken to save useful cattle. When cattle wealth is seriously depleted, the recovery is very slow. While sheep and goats have a potential for rapid growth while the growth of cows, buffaloes, and camels is much slower. It is necessary to provide support to farmers for forage so that they do not engage in distress selling of cattle. It is also very important they continue to sell milk and other products so that they have an alternative stream of income (NIDM and DAC, 2009). In the event of any disaster, especially drought that prolongs for much larger duration, there is a great concern for forage. Any strategy formulated should ensure adequate forage supplies during the disasters period in order to reduce losses of livestock and enable the poor families to retain the herd. Fodder Bank establishment at village *Panchayat* for seasonal variation in demand, at District level for the year-to-year minor drought situation and at the regional level for major drought situation will be a great help.

Out of the total geographical area of 329 million ha in the country, more than 40 million ha is prone to flood (NDMA, 2008). In the 33.5 million ha area that is flood prone, 19.6 million cattle, 1.0 million buffalo, 6.0 million sheep and 13.1 million goat inhabit. Taking into account these, it has been estimated that annually, 0.43 million cattle, 0.23 million buffalo, 1.4 million sheep and 2.9 million goats are affected by flood for which contingency plan need to be made. Mapping is required for livestock in the flood prone areas and actual number of different categories of livestock affected every year to be estimated. There is a need to have short, intermediate and long-term measures to deal with flood situation. Establishment of a network of fodder banks in a radius of 200-250 km will be helpful where flood occurs frequently, so that relief may be quickly provided to the affected livestock. The fodder bank should ideally be located along major highway and railway line so that it could be quickly transported to the affected regions. Livestock play an important role in the livelihood of people in flood prone areas because most of these areas are under developed and agrarian in character. Most of the studies on flood related problems have overlooked the miseries of the marooned livestock in the flood affected area. In the pandemonium that follows flood, the poor livestock gets the least attention. Restoring livelihood of people after natural calamity is the quickest way to rehabilitate them. Initiating agricultural activities may take some time, but livestock can provide succor right in the midst of such misfortune, provided prior planning and coordination of various stakeholders are in place. However, studies on livestock production system in these niches are lacking elucidating their problems and opportunities. The establishment of Fodder Bank on the lines of Food Corporation of India needs to be addressed. A separate chapter deals on the measures to mitigate drought and flood situations in the country.

Forage Security for Livestock, Forage Mission and Policy on Forage Bank

In India, there is a regular phenomenon of seasonal and regional deficit of forage and often-natural calamities like drought and flood. Policy intervention is required to establish fodder bank in order to regulate deficit and fluctuating requirement and mitigate the natural calamities. Tremendous amount of different varieties of grasses and edible crop residues are available in India. All of these forage resources are highly voluminous and having lower density varying from 40-70 kg/ m^3 due to which there transportation, storage and handling are very cumbersome and expensive. Therefore, forage resource availability round the year needs to be enhanced through promotion of fodder bank which will include feed block, leaf meal and region specific mineral mixture supplementation. Conserving these forage resources scientifically, is the need for livestock sector to strengthen the economic power of the livestock farmers by reducing vulnerability of various kinds resulting in shortage of forage resources. This is possible only by policy interventions to create a network of both types of fodder banks in different zones of the country after assessing the balance sheet of forage production. This will not only provide the timely supply of the feed to the most affected area in the country but it will also serve as community based bank from where a needy farmer can deposit his share, withdraw as per his requirement, sale his share or even he can take loan from the bank with a promise to return with interest in term of forage only and not in cash. The bank would have to maintain its own products in the form of densified blocks, pellets, silage, leaf meal and hay.

Forage and feed availability should be ensured by promoting integrated management of grazing lands, forage production and stall feeding, creation of forage and feed banks and by establishing Livestock Feed Corporations. Second Report of Farmers commission (2005) suggested setting up a Livestock Feed Corporation of India. There is a need to establish Livestock Feed Corporation in every state. The District Consortia should ensure enhanced and sustained production of forage and feed crops and adoption of recommended livestock management technologies.

Fodder bank to be established may be of two types (I) Live fodder bank which could include fodder trees, bamboo plantation along the road, silvipasture on CPR, JFM plantation and other afforestation. (II) Dry forage and edible crop residues and silage based forage bank. Forage bank involves conservation, storage, management and distribution of crop residues, natural grasses, top feed and other feed resources to meet the seasonal and regional deficit as well as requirement during natural calamities like drought and flood *etc.* Thus, it is a cluster of technologies and not just baling and densification and storage.

Sensitization of Policy Makers and All Stakeholders

There is a need for policy on edible crop residues for their diversion for power generation, packaging and other uses and data base on production and productivity of forage crops for policy decisions. Emphasis on forage production is minimal.

More investment in forage resource development is required by eastern states and financial support at national level. There is a need of inclusion of forage line staff in various states up to Block/District level in this region. Considering low seed yield of forage crops as compared to many field crops, pricing should be fixed in a way to encourage farmers to grow forage seed. In fact, seed production need support price.

Oilseed cakes are important ingredients of feed concentrate. Hence, their exports need to be regulated. Oil seed cakes are important contaminant for pesticide residues in milk. Therefore, use of pesticide in groundnut, mustard, cotton; sunflower and soybean need to be regulated. Policy for management of grazing lands including forest fringe and common property resources should be implemented. There is little interaction among forestry and livestock departments. Vast tract of unused land along roadsides, railway line and other wastelands could be utilized effectively through leasing out for certain period for forage production and regulated grazing to people residing in the vicinity. Extending credit for forage production and livestock and developing forage marketing systems will result in economic feeding system and enhance livestock productivity.

Development of a forage marketing system in peri-urban dairies, has commercial potential and extending credit for such venture like forage production, an economically viable project for National Bank for Agriculture and Rural Development (NABARD) and commercial banks.

Linkage between Ministry of Environment and Forest with Ministry of Agriculture (Department of Animal Husbandry, Dairy and Fisheries)

There is little interaction among forestry and livestock departments at present in spite of the fact that both the ministries have schemes related to grassland and forage (Table 4.8). It is primarily on account of the perception that livestock is a destroyer of forest. However, it is important to account the fact that controlled removal of vegetation/grass increases further regeneration of trees and promote tree growth.

Table 4.8: Present Schemes Related to Forage Development Activities

Agency	*Scheme/Role*	*Agency*	*Scheme/Role*
Department of Animal Husbandry, Dairy and Fisheries	Forage block making machines	Ministry of Environment and Forest	JFM Schemes (Fodder)
	Grassland development		Grass reserves around forest fringes
	Forage Seed Production		Afforestation Schemes
	Biotechnology Research		
	7 Regional Stations of Forage Demonstration Farm		
	Central Minikit Testing		

A system may be devised where grasses from forest areas are baled and then supplied for livestock use in view of increased restrictions on grazing animals in these areas in a symbiotic manner to promote better growth of tree plantation. Local level institutional mechanism of users and allowing specific period lease of land available along rail, road and irrigation canal for production of perennial forage crops need to be encouraged.

The forage development programmes in forestry projects need to be linked with livestock in the vicinity for effective utilization, resolving conflict between wildlife and domesticated cattle of forest dwellers. (a) A synergistic approach should be adopted for regulated use of grasses in the forest margins that can supply large quantities of forage at affordable prices. (b) The afforestation drive like Joint Forest Management and Social forestry undertaken by various agencies should invariably take up at least some proportion of fodder trees say 25 per cent of fodder trees and Silvi/hortipasture systems. This will bridge the gap in forage demand-supply in normal years besides combating droughts and floods by creation of live forage bank.

Linkage between Department of Agriculture and Cooperation and Department of Animal Husbandry, Dairy and Fisheries under Ministry of Agriculture

There is a need to develop functional groups between the two Department for monitoring, indents and up-scaling breeder seed requirements in forage crops; formulation of policy on edible crop residues; including data base requirements of DAHDF (NSSO under DAC) and managing Drought and Flood situations in the country.

R&D Linkage and Road Map of Forage Resource Development in the Country

National consultation on "Augmenting efforts on forage resource development and utilization "was held under the chairmanship of Secretary Department of Agricultural Research and Education and Director General, ICAR and Indian Council of Agricultural Research. Based on deliberations, discussion, consultation with all stakeholders of Dairy and animal husbandry sector under various meetings by Indian Council of Agricultural Research, recommendation were made for implementation. Some of the specific recommendations need to be included in various policy decisions, changes in concerned organization to fasten the process of forage resource development.

I. ICAR: IGFRI-Animal Science Institutes-KVK (SAU's) Interface

A consortium of research institutes of Indian Council of Agricultural Research (ICAR) on animal sciences (National Institute of Animal Nutrition and Physiology, Central Institute of Research on Buffalo, Central Institute of Research on Goats, Central Sheep and Wool Research Institute, National Dairy Research Institute, National Research Centre on Yak, National Research Centre on Mithun and National Research Centre on Camel) along with Indian Grassland and Fodder Research Institute (IGFRI) should be made to have interface meeting at least twice in a year to chalk out comprehensive plan for research and development under the joint chairmanship of Deputy Director Generals Crop Science and Animal Sciences. This may be helpful to identify, formulate inter institutional research programmes and monitoring the output and outcome. In turn, an annual interface with state department of animal husbandry and dairy will provide a platform for transfer of technology on forage resource development

1. Need for Forage Production Technology

(a) Cultivated Forages

- ☆ Need to strengthen technology development process of location specific forage varieties, food-forage, forage cropping systems and non-competitive land use strategies for forage production.
- ☆ Economics of forage production *vis a vis* grain production needs to be worked out in peri-urban areas.
- ☆ Varieties/cropping system commensurate with changing climate scenario such as high CO_2, temperature and drought/secondary salinization need to be developed

(b) Forage Production from Non-arable Land

- ☆ Alternate land use system such as hortipasture, silvipasture, and agro-horti-silvipasture should be adopted and upscaled in different agro climatic zones. Opportunities in JFM, Watershed, Rural development programmes need to be harnessed.

(c) Grassland Restoration and Management

- ☆ Degraded forest land and wasteland should be restored by various range restoration techniques such as reseeding, legume seeding and other soil, moisture conservation techniques. Technology is available with IGFRI, SAUs. Need is effective local institutional mechanism for sustainable production and optimum utilization.

(d) Forage Production from Problem Soils

- ☆ Amelioration of problem soils
- ☆ Arid zone for sand dunes and very less rainfall (effective linkage with Centeral Arid Zone Research Institute CAZRI for developing lines in suitable fodder trees, shrubs and grasses such as *Lasiurus, Cenchrus species* etc.)
- ☆ Temperate zone – Suitable forage resources for temperate animals such as Yak, Mithun and migratory sheep and goats.
- ☆ Aquatic and waterlogged zones – Exploitation of aquatic weeds such as Karmi, Para grass, jalkumbhi, Coix *etc.* and forage production after receding of flood

(e) Post-Harvest Technology

- ✰ To mitigate the problem of forage availability during lean period (Nov-Dec and May-June) and the flood and drought situation, effective mechanism should be in place for forage conservation, compaction and transportation (concept of fodder bank)
- ✰ IGFRI and other ICAR institutes have developed several technologies for forage conservation such as bailing, densifying, silage; hay making, complete feed block, leaf meal *etc.* should be utilized by developmental departments which will enable the reduction of the cost of animal feeding, replacement of costly concentrates by complete feed block; silage, leaf meal *etc.*

(f) Seed Production Chain

- ✰ To strengthen the Seed Chain in Forage Crops: Active involvement of State Agricultural University (SAUs), their Krishi Vigyan Kendra (KVKs) and some Animal science institutes is required for strengthening the seed chain of breeder-foundation-truthfully labeled forage seed (TFL). At least 3-5 per cent seed production under Mega seed project is targeted for forage crops by different SAUs. The seed chain can also be strengthened if animal science institutes take up at least 5-10 hectares area under seed production at their research farm depending upon the availability of land. Supply of breeder seed should be ensured.
- ✰ At present, there is no effective chain of breeder-foundation-TFL/certified forage seed production. Hence a huge gap in demand and supply of forage seed exists. It has been noticed that very often breeder seeds are used for forage production. Forage Demonstration Centres may take up foundation seed at least by utilizing there vast land resources. NSC and SSCs should take TFL/certified seed production as a regular activity.
- ✰ There is a huge but irregular demand of forage range grasses and legumes seeds which could not be met. A mechanism of assessing requirement and arrangement for seed production is needed as a prerequisite of range development.

2. Inter-institutional Linkage and Collaboration

- ✰ On the pattern of inter institutional (IGFRI, NRC Yak and CSWRI) R&D projects "IGFRI Outreach Programme on Amelioration of Temperate/ Alpine Pastures for Livelihood Support to Pastoral Communities" as executed in 11th EFC of IGFRI, other research projects may be taken in specific area of forage resource development.
- ✰ A functional linkage among animal science institutes and IGFRI may be established for demonstration of improved varieties of forage crops and production and utilization technologies which will serve to upscale

adoption of forage production and utilization technologies through their regular Kisan Mela and various training programmes already being conducted by each institute. The seed of improved varieties and package of practices will be provided by IGFRI.

(a) Transfer of Technology and Human Resource Development

- Milk collection units can be a single window system for transfer of total livestock feeding and management technologies including forage production and utilization. The network of Krishi Vigyan Kendra can be also a hub to demonstrate the efficacy of technology for forage production and conservation as well as livestock management as a whole. KVK can be a Hub for Up Scaling and Adoption of Forage Technology
- For this a planning and monitoring team under the joint chairmanship of Deputy Director General Crop Science and Agriculture Extension may be constituted with ADG (Agric. Extension), ADG (ANP) and Director, IGFRI. At least 20 per cent of training and demonstration activities of selected KVK (200 KVK in different agro climatic regions) should be on forage resource development and utilization. These KVKs will have functional linkage with three IGFRI Regional Stations and 21 AICRP centres on Forage Crops located all over the country.
- Each KVK needs to be provided with a mobile bailing machine, feed block making machine and leaf meal technology for cost effective balanced feeding of livestock.
- Need for Human Resource Development for various stakeholders such as forestry, animal husbandry, community developers, orchard maintainers, KVKs, pastoralists, nomadic and semi-nomadic tribes.
- The key to success of any programme depend on the quality of human resource. There are few institutions, which run specialized course in forage production and utilization. Revival of Diploma course in *Forage Production and Utilization* at IGFRI with a reduced duration of four months (earlier nine months) needs to be considered. The capacity building of various stakeholders is extremely important to promote forage production and utilization schemes in the country. Specific training needs for various stakeholders are required to be attended.

II. ICAR- DAHDF - NDDB - STATE AHD Interface

- There should be a regular interface meeting between DAHDF and ICAR with the representatives of NDDB and State AHD and Milk cooperatives once/twice in a year on the pattern of DAC-ICAR Interface under the joint chairmanship of Secretary DARE and Secretary DAHDF.
- Seed Chain strengthening for enhanced forage seed availability through a marketing network: (a) There is urgent need of having an effective

functional monitoring team consisting of Joint secretary (Seed) in DAC, ADG (Seed) in ICAR and Joint Secretary (Feed and Fodder) in DAHDF to upscale seed indents of improved varieties from various State AHDs and Milk cooperatives. NSC has to play proactive role at national level not only for production but marketing of forage seed through their network across the Country. The old varieties should be gradually replaced by new improved varieties.

- ☆ Lack of regular and steady demand and non-remunerative price of forage seed was felt by NSC. It requires policy level intervention and joining hand with all stakeholders of forage seed. State AHD Farm; and State Seed Corporations should also focus on production and marketing of forage seed.
- ☆ Revolving Fund Scheme and one time catch-up grant with contractual manpower for the Forage Demonstration Regional Stations (CPSPF and RSFP) of DAHDF need to be taken up to strengthen seed chain and popularization of forage technology.
- ☆ Field level demonstration on forage crops duly funded by DAHDF should be undertaken as currently being done in cereal, oil seeds and pulses crops funded by DAC and executed by various forage demonstration centres of DAHDF, SAUs, ICAR institutes and other Governmental and Non-governmental agencies.
- ☆ The developmental programmes on popularization of chaff cutters, establishment of silage units, mobile bailing machines and feed block making machine and leaf meal technology for cost effective balanced feeding of livestock.
- ☆ The Central Mini-kit Programmes need to be replaced by integrated programmes on Front Line Demonstration on cultivated forage and grasslands on CPR's and individually owned pasture lands.
- ☆ Linkage among Milk cooperatives and central agencies and milk collection units should be developed providing a single window system for transfer of total livestock feeding and management including forage production and utilization technology.

III. Strengthening Forage Production in Eastern Region

Characteristics of Available Forage Resources

- ☆ Excess of edible biomass of low nutritional quality during monsoon and deficit during winter and summer.
- ☆ Crop residues largely paddy straw having high lignin content, high silica and anti-nutritional oxalate, are available.

Approach

- ✰ Improvement of nutritional quality of available monsoon forage resources through protein and mineral supplementation as per mineral map available from National Institute of Animal nutrition and Physiology (ICAR).
- ✰ Conservation of excessive biomass for winter and summer.
- ✰ Encourage forage production through milk cooperatives during rabi and summer.
- ✰ Encourage fodder tree plantation on farm bunds/boundary and under social forestry, JFM and other development programmes.
- ✰ Utilization of by-products of vegetable, fruits and edible aquatic biomass (para grass, jalkumbhi, karmi *etc.*).
- ✰ Post-harvest management
    a. Bailing and enrichment of crop residues particularly paddy straw and other leguminous crop residues for proper storage, balanced feeding with green forage and minimizing wastage and storage loss.
    b. Use of leaf meals to substitute costly concentrate. Leaf meal can be made out of leguminous crop residues (Lentil, gram, Khesari *etc.*) and subabool, stylo, gliricidia *etc.*).

Annexure 4.1: Organizational Structure of Directorate of Animal Husbandry and Veterinary Services

States	Existing Sanctioned Staff Pattern												
	Director	Addl. Director	Jt. Director			Dy. Director			Assist. Director or Farm Manager			FDO	Fodder Supervisor
			Vet	Fodder	Total	Vet	Fodder	Total	Vet	Fodder	Total		
Andhra Pradesh	1	1	22	-	22	-	-	-	279	2	281	12*	35
Assam	1	6	11	-	11	51	1	52	46	1*	47	12	18
Arunachal Pradesh	1	-	2	-	2	7	1	8	-	-	-	1	12
Bihar	1	1	2	-	2	9	7*	16	31	1*	32	1	85
Chhatisgarh	1**	2	5	-	5	22	-	22	595 (VAS)	-	595	-	-
Gujarat	1	1	8	1	9	59	1	60				1*	14
Haryana	1	-	7	-	7	42	-	42	6	1*	7	2*	-
HImachal Pradesh	1	-	3	-	3	12	-	12	12	-	12	-	-
Madhya Pradesh	1**	-	9	1***	10	59	-	59	9	-	9	-	-
Maharastra	1**	2	8	1*	9	64	1*	65	329	-	329	7*	32
Manipur	1	-	10	-	10	?	1	?	?	1*	?	-	4
Meghalaya	1	-	3	-	3	8	-	8	30	3 + 1*	34	3	3
J&K (Jammu)	1	-	3	-	3	2	1	3	8	-	8	-	2
J&K (Srinagar)	1	-	3	-	3	4	-	4	6	-	6	1	-

States	Existing Sanctioned Staff Pattern												
	Director	Addl. Director	Jt. Director			Dy. Director			Assist. Director or Farm Manager			FDO	Fodder Super-visor
			Vet	Fodder	Total	Vet	Fodder	Total	Vet	Fodder	Total		
Kerala	1	2	3	-	3	6	-	6	4	1	5	?	?
Karnataka	1	2	5	-	5	8	-	8	14	1	15	?	?
Mizoram	1	-	4	-	4	17	-	17	-	-	-	1	4
Nagaland	1	1	2	-	2	13	1*	14	14	1*	15	3	4
Orrisa	1	1	3	-	3	18	1	19	12	-	12	40*	8
Punjab	1	-	4	1*	5	23	7*	30	40	-	40	122*	-
Sikkim	1	6	13	1*	14	26	4*	30	8	-	8	4*	10
Tamilnadu	?	?	?	?	?	?	?	?	?	1*	?	21*	?
Tripura	1	1	3	-	3	23	1*	24	29	29*	58	-	-
UP	1	3	9	-	9	70	1*	71	70	8*	78	7*	35
West Bengal	1	5	12	1*	13	67	4*	71	157	25*	182	-	42

*Agriculture/AH background; ** Animal Husbandry Commissioner (Administrative background); *** Milk Commissioner of Dairy Tech background; ? - Not known

Annexure 4.2a: Estimated Breeder Seed Requirement for Selected Forage Crops

Crop	*Area (m/ha)*	*Average Seed Rate (kg/ha)*	*Total Seed Used (Qtls)*	*Average Seed Yield (Qtls/ha)*	*Foundation Seed Required (Qtls)*	*Estimated Certified Seed Required (Qtls) 20 per cent Seed Replacement per Five Years*			*Actual Breeder Seed Demand 20 per cent Seed Replacement per Five Years*		
						2010-15	*2015-20*	*2020-25*	*2010-15*	*2015-20*	*2020-25*
Maize	0.9	20	180000	20	1800	21.6	25.2	28.8	60	70	80
Jowar	2.6	10	260000	10	2600	31.2	36.4	41.6	30	35	40
Bajra	0.9	10	90000	8	1120	16.8	19.6	22.4	3.6	4.2	4.8
Oats	0.25	75	187000	15	9375	561.6	655.2	748.8	210	245	280
Berseem	2.0	20	400000	5	16000	768	896	1024	66	77	88
Lucerne	1.0	15	150000	4	5625	252	294	336	14.4	16.8	19.2
Cowpea	0.3	20	60000	6	2000	80.4	93.8	107.2	24	28	32
Guar	0.2	20	40000	9	890	24	28	32	60	70	80
				Total	39410	1756	2048	2341	468	546	624

Qtls: Qunital.

Annexure 4.2b: Estimated Certified Seed Requirement for Selected Forage Crops

Crop	*Area (m/ha)*	*Average Seed Rate (kg/ha)*	*Total Seed Used (Qtls)*	*Average Seed Yield (Qtls/ha)*	*Estimated Certified Seed Required (Qtls) 20 per cent Seed Replacement per Five Years*			*Actual Breeder Seed Demand 20 per cent Seed Replacement per Five Years*		
					2010-15	*2015-20*	*2020-25*	*2010-15*	*2015-20*	*2020-25*
Maize	0.9	20	180000	20	216000	252000	288000	2160	2520	2880
Jowar	2.6	10	260000	10	312000	364000	416000	3120	3640	4160
Bajra	0.9	10	90000	8	108000	126000	144000	1344	1568	1792
Oats	0.25	75	187000	15	224400	261800	299200	11250	13125	15000
Berseem	2.0	20	400000	5	480000	560000	640000	19200	22400	25600
Lucerne	1.0	15	150000	4	180000	210000	240000	6750	7875	9000
Cowpea	0.3	20	60000	6	72000	84000	96000	2400	2800	3200
Guar	0.2	20	40000	9	48000	56000	64000	1068	1246	1424
				Total	1640400	1913800	2187200	47292	55174	63056

Qtls: Qunital.

Chapter 5

The Impact of Climate Change on Forage Resources

Mechanism, by which global climate change phenomena will influence the performance of earth ecosystems or the likely consequences of these influences in the long term, remains much of it to be unknown. Nevertheless, the change in global climate is putting challenges to agricultural and livestock production systems. Significant changes in physical and biological systems have already occurred in all continents and in the most oceans, and most of these changes are in the direction expected with warming temperature (Rosenzweig *et al.,* 2008). The climate change has also increased frequency of major natural disasters like drought, flood, cyclone and changing pattern of rainfall. Global climate change phenomena would have significant consequences for food, forage and feed production. Climate change has direct effects on livestock productivity as well as indirectly through changes on the availability of forage resources. Climate change is expected to affect livestock at the species level. For example, if the Himalayas turn warmer, the yak could be restricted to higher altitudes where grass and other forage are scarce. Climate changes will also affect nomadic and transhumant livestock keepers. New routes and pastures will have to be found. The impacts of climate change on livestock production systems may be grouped into a) forage crop production and the productivity of pasture and their quality, b) availability of feed grain, c) health, growth and reproduction in livestock and, d) animal disease and their spread. A survey made to factorize perceptions on the impact of climate in India on livestock and interrelated components showed that 79 per cent of the impact of climate change will be on decreasing feed and forage resources and acute deficit of dry forage (95 per cent), drinking water and irrigation water (69 per cent) due to reduced availability of water resources (Singh *et al.,* 2012). Climate change has the potential to impact the quantity and reliability of forage production, forage quality, thermal stress on livestock, water demands for both animal needs and growing forage, and large-scale patterns of forage resource availability (Chaplin Kramer and George, 2013). Major changes can, thus, be anticipated in livestock systems, related to livestock species mixes, crops grown and feed resources and feeding strategies. Elevated atmospheric $CO_{2,}$ shifting high temperature and changing and extreme precipitation patterns, particularly a shorter rainy season, must be considered altogether in order to understand the potential impacts of climate change on forage resources. Shorter growing seasons and increased potential for drought will increase risk. For the emerging climate scenarios livestock farmers will need to strengthen their contingency planning. These vulnerabilities to climate change are not as easily translated to economic impacts as each agro climatic region has a unique set of forage resources and operational conditions requiring niche specific approaches.

Forage Production, Regrowth and the Productivity

a. Response to Elevated CO_2

Limited information is available on the impact of climate change on cultivated forage crops and grasslands. Performance of cultivated forage crops showed that the elevated CO_2 (600 μl^{-1}) increased the leaf size, plant height and fresh and dry matter in berseem and oat (30-35 per cent) than the ambient CO_2 (360 μl^{-1}). In both the forage crops, partitioning of photosynthate improved towards the growth of new branches than the growth of leaves (Madan Pal *et al.,* 1997 and 2004, Bhatt *et al.,* 2010a). In berseem, the growth enhancement effect at the early stage of elevated CO_2 exposure was greater in producing more number of thinner leaves on secondary branches compared to stem (Madan Pal *et al.,* 2004). This was mainly due to less accumulation of starch and faster rate of leaf expansion. The impact of elevated CO_2 concentration also noticed on varietal difference in biomass production in three oat (*Avena sativa* L.) cultivars *viz.* Kent, JHO-822 and JHO-851.

Grassland and pastures support grazing livestock and contribute to the maintenance of overall environmental quality and ecosystem services. Grasslands can be grouped roughly into two categories: pastures that have been improved or created from forest clearing and other human activities for the expressed purpose of raising and feeding domestic livestock, and native grasslands that have evolved due to a particular regime of precipitation and temperature (Lauenroth, 1979). Regarding response to elevated CO_2 on biomass production, Poorter and Navas (2003) found that fast growing herbaceous C_3 species responded more strongly than slow growing C_3 herbs or C_4 plants. CAM species and woody plants showed intermediate responses. When crown in competition, C_4 species showed the lowest responses to elevated CO_2 at high nutrient conditions, although differences were found between slow and fast growing species.

Review on the responses of grasses to CO_2 revealed strong growth enhancements of both C_3 and C_4 grasses due to elevated CO_2, although the response was higher in C_3 grasses (Wand *et al.,* 1999, Bhatt *et al.,* 2008, Bhatt *et al.,* 2010b, Kellog *et al.,* 2001, Morgan *et al.,* 2001, Soussana *et al.,* 2002). The greater sensitivity of C_3 species had been attributed to their photosynthetic metabolism, which is un-saturated at present level of atmospheric CO_2 concentrations. The magnitude of increase in photosynthetic rate of C_3 species was more than the rate of photosynthesis in C_4 plants which could increase due to increases in atmospheric CO_2 level (Le Cain and Morgan, 1998, Bhatt *et al.,* 2010b). Elevated CO_2 stimulated the growth of C_4 grasses from improved water relations and increased leaf temperature in addition to increasing leaf intercellular CO_2 (Ghannoum *et al.,* 2000). The water relations benefit of elevated CO_2 also occurred in C_3 species because stomata's of most plants closed with elevated CO_2, resulting in water conservation and enhanced water use efficiency. Wand *et al.* (1999) reported that both C_4 and C_3 species increased total biomass significantly at elevated CO_2 level, by 33 per cent and 44 per cent, respectively. Differing tendencies between types in shoot structural response were found between C_3 and C_4 species. C_3 species showed a greater increase in tillering, whereas C_4 species showed a greater increase in leaf area at elevated CO_2. In both the C_3 and C_4 species, these structural changes resulted into the higher carbon assimilation rates. At the leaf level, significant stomata closure and increased leaf water use efficiency were confirmed in both types. Tropical grasslands are dominated by C_4 grasses, characterized by their unique biochemistry and anatomy. C_4 plants contribute approximately 20 per cent of global gross primary productivity, and uncertainties regarding their responses to rising atmospheric CO_2 concentrations might limit predictions of future global change impacts on C_4 plant dominated grasslands (Lauenroth, 1979). These uncertainties have not yet been considered rigorously due to expectations of C_4 low responsiveness based on photosynthetic theory and early experiments. In general, the productivity of grasses such as *Bouteloua gracilis* (C_4) *Panicum maximum* (C_4), *Cenchrus ciliaris* (C_4 subtypes), *Pascopyrum smithii* (C_3), two genera of *Setaria* and *Arundinaria* of *Panicoideae aristida* and *Stipa agrostis,* sister genera of subfamily *Aristida oideae* and leguminous *Stylosanthes hamata* had increased at elevated CO_2 (*twice of ambient* CO_2) as compared to ambient CO_2 level (Bhatt *et al.,* 2008, Bhatt *et al.,* 2010b, Morgan *et al.,* 2001, Soussana *et al.,* 2002, Kellog *et al.,* 2001, Ghannoum *et al.,* 2000, Siebke *et al.,* 2002). The reported increase in the productivity showed a wide variation depending upon various production factors and agro climatic conditions in tropical and sub-tropical grassland but in temperate zone, Soussana *et al.* (2002) reported that doubling the CO_2 concentration in the atmosphere would increase the forage production in grassland by approximately 15 to 20 per cent. *Colorado* short grass steppe in semi-arid temperate grassland where vegetation of the site was dominated by the warm-season C_4 grass *i.e. Bouteloua gracilis* (H.B.K.) Lag., but it contained an abundance of cool-season, C_3 grasses, most importantly *Pascopyrum smithii* (Rydb.) *A. love.* and *Stipa comata* Trin. & Rupr and also the varieties of C_3 forbs showed that the C_3 grasses accounted for 61 per cent of the above ground vegetative dry matter, where C_4 grass *B. gracilis* primarily constituted 35 per cent of the total vegetation during the period of study. The remaining four per cent was forbs. Peak above ground phytomass of the grassland occurred in late July (Shoop *et al.*, 1989), with similar production estimated in belowground organs. Aboveground plant productivity had been consistently enhanced in short grass steppe vegetation under a double of ambient CO_2 regime, as indicated by increases in standing aboveground peak phytomass ranging from 20 per cent to 71 per cent. After four years of CO_2 enrichment, no relative differences in growth responses to CO_2 had been detected between C_3 and C_4 grasses, although a trend suggested a slightly higher CO_2- induced production increase for the forb group. Genera, species specific and genotypic response varied considerably in their magnitude that complicates any prediction of response of C_4 plants to future atmospheric change (Wand *et al.*, 1999, Kellog *et al.*, 2001). Comparison of closely related species, however, might lead to intriguing new insight into how regulatory pathways of CO_2 assimilation were modified during evolution. Wilsey *et al.* (1997) reported that elevated CO_2 (700 mL/m^3) caused an increase in total biomass and total productivity (biomass + clippings) only in temperate grassland dominated by C_3 species, and increases in growth occurred primarily in crowns and roots (storage organs). There were no significant CO_2 effects on biomass or productivity in a temperate grassland dominated by a mixture of C_3 and C_4 species, and no CO_2 effects on aboveground biomass or productivity (aboveground biomass + clippings) in species from any of these ecosystems.

Eleven responses were documented due to elevated CO_2, ranging from physiological behaviour at the leaf level to carbon allocation patterns at the whole plant level and also assessed in the context of environmental stress conditions (light, temperature, water and nutrient stress), and the experimental growing conditions. Across the native or semi-native temperate and Mediterranean grasslands and semi-arid systems, tall grass prairie, short grass steppe and calcareous grassland, increased aboveground biomass from CO_2 enrichment was relatively greater in dry years (Morgan *et al.*, 2004). In contrast, CO_2-induced aboveground biomass increased in the C_3/C_4 grassland and the pasture seemed little or only marginally influenced by variation in the soil water, while plant growth in desert was stimulated by CO_2 in a relatively wet year. Vegetative responses to CO_2 were highly varied among species and ecosystems, and did not generally follow any predictable pattern (Parsons *et al.*, 2001, Morgan *et al.*, 2004). The mechanism for this has not been identified, but Ghannoum *et al*, (2000) found increased intercellular CO_2, partial pressure, changes in diurnal CO_2 fixation patterns, improvements in shoot water relations and increased leaf temperature as possible ways in which C_4 species might respond to elevated CO_2. It was noticed that while C_4 species in this system could play an important role during periods of summer moisture deficits, their contribution to total dry matter is small (less than 02 per cent). The reason for the greater stimulation in spring and summer is not clear, but there was evidence to suggest that greater responses might be expected at higher temperatures and, given differences in seasonal growth patterns of plant species, that interactions were likely between this 'seasonal' effect and the CO_2 responsiveness of different species (Newton *et al.*, 1994).

b. Response to Increased Temperature

Mowell *et al.* (2013) reported that increasing global temperatures would alter grassland aboveground production and species composition. Extreme temperatures would likely impact species distributions and abundance through reproduction, competition or survivorship; although species response differences were diverse and difficult to predict (Polley *et al.,* 2000). Global warming had little effect on annual production of grass, either from pure grass or grass-white clover swards in Scotland. On the other hand, white clover as a percentage of total herbage production was estimated to increase from 32 per cent to 46 per cent for a 2°C temperature rise (Topp and Doyle, 1996a and 1996b). At the same time, there were also indications that global warming would increase the percentage of clover in the herbage cut for conservation. The Tibetan Plateau was predicted to undergo climate warming much greater than the global average, as well as shifts in its currently monsoon-dominated precipitation regime. These changes will likely affect the vegetation composition, carbon balance, and nutrient cycling of the alpine, social-ecological system. Winter-grazed meadow sites might be more resilient to climate warming and grazing on vegetative aboveground biomass as compared to the summer-grazed shrub sites (Klein *et al.,* 2000). This resilience might be partially attributed to compensatory biomass growth of different functional groups that stabilises the total biomass responses and partially attributed to the timing of defoliation events in the winter-used meadows. However, it was important to understand the compositional changes in biomass and how they will affect rangeland quality. Hopping *et al.* (2010) reported that Tibet's alpine grasslands (central Tibet's alpine meadow ecosystems, with the turf-forming sedge, *Kobresia pygmaea,* as both the dominant species and preferred forage of yaks at 4870 m altitude) were particularly vulnerable to climate change and soil moisture drying, seems to be the proximate cause of shifts in vegetation composition, net ecosystem productivity, and N availability, with climate warming as the ultimate cause. Some of the negative effects of warming might be mitigated by spring snow storms that provided a pulse of moisture at the start of the growing season. However, in contrast to temperatures that were expected to rise steadily under climate change, snow events on the Tibetan Plateau were much less predictable across both time and space. This mismatch in increased rates of warming and added precipitation to the system would likely affect the landscape patchily. Further, it would leave local herders with the conundrum of how to cope most successfully with the impacts of multiple climate changes, thus illustrating the complexity associated with integrating a fine-scale understanding of ecosystem responses to climate change with coherent strategies for adaptation. Warming and elevated CO_2 might increase biomass and ecosystem respiration in grasslands and an improved understanding of the mechanisms underlying these responses was expected to improve terrestrial carbon cycle predictions. Ryan *et al.* (2013) evaluated the temporal dynamics of ecosystem respiration (Reco) and quantify the potential drivers affecting Reco in the context of the prairie heating and CO_2 enrichment in semi-arid grassland.

Their results suggested that Reco would likely increase with future warming and elevated CO_2, which might be amplified under increased precipitation. Moreover, elevated CO_2 and warming were not only resulting in an increase in Reco, but also in an increase in the temperature sensitivity of Reco. The Tibetan Plateau was predicted to undergo climate warming much greater than the global average, as well as shifts in its currently monsoon-dominated precipitation regime. Relatively large increases in temperature should favour in general warm-season plants (Alward *et al.*, 1999). However, there might be interactions with CO_2 that modify or even cancel temperature effects on vegetation. While future CO_2 levels might favour C_3 plants over C_4 plants, the opposite is expected under associated temperature increases. Changes in evaporation–precipitation ratios modify ecosystem function, particularly in marginal areas, in ways that are not fully understood (IPCC, 2007). For instant while warming might favour warm-season C_4 species, some of the effects of CO_2 enrichment on C3 photosynthesis would tend to counter that response, rendering C_3/C_4 distribution changes relatively insensitive to increases in temperature (Polley *et al.,* 2000). Further, while higher temperatures might enhance the CO_2 production response, differences in plant development might lead to complicated interaction between CO_2 and species response as temperature increases (Newton *et al.,* 1994). The effect of warming on hydrology introduced other uncertainties, by itself a warming trend would increase potential evapotranspiration, leading to desiccation. However, the improved water use efficiency under elevated CO_2 will tend to counter that response. In semi-arid grasslands, significant increases in both temperature and CO_2 might eventually shift the competition more in favour of C_4 grasses because of 1) the overriding benefit of CO_2 to improving plant water relations of most plant species, regardless of photosynthetic pathway, and 2) the adaptation of C_4 species to higher temperatures.

c. Response to Precipitation Variability

Global climate change was predicted to increase precipitation variability, with more extreme rainfall events punctuated by longer dry intervals. Changes in rainfall distribution were especially important in semi-arid environments where precipitation driven spatial and temporal patterns of soil water content were a key control on grassland ecosystem function. The mechanisms and consequences of ecological change in grassland ecosystems were likely to result from these changes in precipitation regimes. Precipitation had been shown to have a strong positive correlation with aboveground net primary productivity in grasslands and had been at the forefront in climate change research in grasslands. Climate models predicted, and empirical evidence confirmed, that more extreme precipitation regimes were occurring in tandem with warmer atmospheric temperatures. These more extreme rainfall patterns were characterized by increased event size separated by longer within season drought periods, and represented novel climatic conditions whose consequences for different ecosystem types were largely unknown. Heisler-White *et al.* (2008) found strong sensitivity to more extreme growing season rainfall regimes, with responses of aboveground net primary productivity, contingent on mean soil water levels for different grassland types along this precipitation productivity gradient. At the mesic end of the gradient (tall grass prairie), longer dry intervals between events led to extended periods of below average soil water content, increased plant water stress and a reduction in aboveground net productivity. The opposite response occurred at the dry end (semi-arid grasslands), where a shift to fewer, but larger, events increased periods of above-average soil water content, reduced seasonal plant water stress and resulted in an increase in aboveground net primary productivity. These results highlighted the inherent complexity in predicting how terrestrial ecosystem would respond to forecast novel climate conditions as well as the difficulties in extending inferences from single site experiments across biomes. Even with no change in annual precipitation amount, aboveground net primary productivity responses in a relatively uniform physiographic region differed in both magnitude and direction in response to within season changes in rainfall event size/frequency. Ross *et al.* (2014) investigated the effects of extended periods of below average soil moisture on the response of grassland's aboveground net production to precipitation included 21 grassland biomes throughout arid-to-humid climates in the United States with continuous surface soil moisture records for 2-13 years during the drought period from 2000-2013. They found that when altered climatic conditions on grasslands led to an increase in the duration of soil water deficit, then the precipitation-to-aboveground net primary productivity relation became non-linear. Non-linearity was associated with extreme grassland dieback and changes in the historic species assemblage. The magnitude of change was related to the precipitation regime, where grasslands in hyper-arid and humid regimes were least likely to be affected by prolonged soil moisture deficit, and semiarid and mesic grasslands were most likely to be impacted, depending on the duration of the deficit. Knap *et al.* (2010) reported that grassland responses to alterations in precipitation inputs might vary depending on the long-term hydrologic regime in mesic grassland. When similar short-term experimental manipulations of precipitation variability were conducted in more arid grasslands, responses in aboveground biomass were opposite to those in mesic grassland. Fay *et al.* (2008) evaluated the aboveground net primary productivity of grassland receiving equal rainfall inputs (from irrigation) and exposed to a continuous CO_2 gradient (250 to 500 μL L^{-1}). At the end of season total aboveground net primary productivity of all species was 35 to 50 per cent greater, showing greater rainfall use efficiency on these soils despite lower water holding capacity. Further, they found that grass dominance of community and ecosystem processes differed strongly among the soils. Individual species showed varying responses among soils to the CO_2 gradient. Interactions between soils and elevated CO_2 had important consequences for the productivity, rainfall/water use efficiency, and species composition of grassland under future atmospheric CO_2 concentrations. Vargas *et al.* (2013) found that the amount and pattern of multiple rain pulses over the growing season were crucial for understanding CO_2 dynamics in burned and unburned water-limited grassland ecosystems. Arid ecosystems, which occupy about 20 per cent of the earth's terrestrial surface area, have been predicted to be one of the most responsive ecosystem types to elevated atmospheric CO_2 and associated global climate change. In an intact Mojave Desert ecosystem, Smith *et al.* (2000) found that new shoot production of a dominant perennial shrub was doubled by a 50 per cent increase in atmospheric CO_2 concentration in a high rainfall year. However, elevated CO_2 does not enhance production in a drought year. Further, above-ground production and seed rain of an invasive annual grass increased more at elevated CO_2 than in several species of native annuals. Consequently, elevated CO_2 might enhance the long-term success and dominance of exotic annual grasses in the region. This shift in species composition in favour of exotic annual grasses, driven by global change, had the potential to accelerate the fire cycle, reduce biodiversity and alter ecosystem function in the deserts of western North America. Collins *et al.* (2009) showed that desert grasslands dominated by *B. eriopoda* were highly sensitive to changes in precipitation regime. More extreme precipitation regimes might increase aboveground productivity, yet because these systems had high potential decomposition rates the overall impact of climate change on carbon dynamics remains uncertain.

Precipitation and temperature drive many aspects of grassland ecosystem function. Chaplin Kramer and George (2013) predicted changes in forage production in response to changes in temperature and precipitation in rangeland in California. They suggested that forage production in the rangelands might be enhanced at least in terms of peak standing crop. However, the timing of production was essentially as important as its peak, and altered precipitation patterns could mean delayed germination, resulting in shorter growing seasons and longer periods of inadequate forage quality. An increase in the frequency of extremely dry years would also increase the uncertainty of forage availability. Many studies had shown that temperature and rainfall changes in the future would modify, and often limit, the direct CO_2 effects on plants (IPCC, 2007). Higher temperatures in lower latitudes might result in more water stress for plants, although higher temperatures in tropical highland areas might increase their suitability for cropping. In terms of impacts on grasslands, sustained increase in mean temperatures resulted in significant changes in rangeland species distribution, composition, patterns and biome distribution (Hanson *et al.,* 1993). Simulation runs showed that predicted climate change would affect plant and animal production for rangelands. Changes in production were more closely related to changes in temperature and precipitation than to enhance CO_2 concentration alone. The key point seemed to be that climate impacts on the plants depended significantly on the precipitation scenario considered. Howden *et al.* (1999) simulated and compared the responses under current tropical climate and CO_2 conditions with four possible future scenarios: (1) doubled CO_2; (2) doubled CO_2 and increased temperature; (3) as in the previous scenario but with a drier climate and (4) as in (2) but with a wetter climate. They found that CO_2 changes alone were likely to have beneficial effects, with increased pasture growth, increased and less variable live weight gain, and increased ground cover. Growth responses to CO_2 were likely to be greater in drier years than in wetter years partly due to nitrogen limitations in the soils. Increases in temperature in combination with CO_2 further increased animal production due to the increased number of growing days in the cooler months. The increased rainfall scenario had few additional positive effects but further increased subsoil drainage. In contrast, the drier scenario had reduced plant and animal production when compared with current conditions even though seasonal transpiration efficiency was increased by 20 per cent due to increased CO_2. Fay *et al.* (2011b) concluded that (1) major ecosystem processes in the grassland might be substantially altered by predicted changes in inter annual climate variability, intra-annual rainfall variability, and temperature, (2) inter-annual climate variation was a larger source of variation in ecosystem function than intra-annual rainfall variability and warming, and (3) effects of increased growing season, rainfall variability and warming were small, but ecologically important. The relative effects of these climate drivers are likely to vary for different ecosystem processes and in wetter or drier grassland ecosystems. Interaction of elevated CO_2 with rising temperature needed to be understood in climate change perspective. Process based ecosystem model after being systematically calibrated against eddy-covariance measurements, meteorological observation, soil inventory data and satellite observed LAI (Leaf Area Index) was used to investigate climate change and rising atmospheric CO_2 concentration driven spatial-temporal changes in vegetation net primary production (NPP) and net ecosystem production (NEP) of Qinghai–Tibetan plateau grasslands from 1961 to 2009 (Tan *et al.*, 2010, Piao *et al.*, 2012). Overall, simulation suggested that the grassland, net primary productivity significantly increased with a rate of 1.9 Tg C yr^{-2} (1 Tg = 1012 g) since 1961. At the regional scale, change in precipitation, temperature, and atmospheric CO_2 concentration accounted for 52 per cent, 34 per cent, 39 per cent of the increase in net primary productivity, respectively, but their relative roles were not constant across the study area. Increase in net primary productivity over the central and southwestern Qinghai-Tibetan Plateau was primarily attributed to precipitation changes, while rising atmospheric CO_2 concentration was the main cause of net primary productivity increase in eastern plateau. Parton *et al.* (2007) in a prairie heating and CO_2 enrichment experiment in southern Wyoming (USA) simulated the projected impact of warming and elevated atmospheric CO_2 on ecosystem dynamics for semiarid grassland ecosystems during the next 10 year. They found that the soil water content, plant production, soil respiration, and nutrient mineralization would increase for the high CO_2 treatment. Soil water content would decrease for all the 10 years, while nitrogen mineralization, soil respiration, and plant production both would decrease and increase under warming depending on the yearly differences in water stress. Net primary production would be greatest under combined warming and elevated CO_2 during wet years. Model results were in consistent with empirical field data suggesting that water and nitrogen would be critical drivers of the semiarid grassland response to global change. In a simulation study, Soussana *et al.* (2002) reported that doubling the CO_2 concentration in the atmosphere would increase the forage production in the temperate zones but if it was combined with a 3°C global warming, no production improvement could be registered, even if length of vegetative cycle increases, considering the summer water deficit. The study on response to temperature of big bluestem (*Andropogon gerardii Vitman*) development, growth and biomass partitioning under low (360 μL L^{-1}) and high (720 μL L^{-1})) CO_2 concentration and five different day/night temperatures of 20/12, 25/17, 30/22, 35/27 and 40/32 °C grown in controlled environmental conditions showed that biomass decreased either above or below the optimum temperature of 30/22 °C (Kakani and Reddy, 2007). The effect of high CO_2 on biomass accumulation (12–30 per cent increase) was visible at less than optimum temperature (30/22 °C) and absent at two high temperatures. With increase in temperature, irrespective of the CO_2, biomass partitioned to leaves increased (35 per cent) whereas that to stems decreased (33 per cent). Water stress did not alter the C_4 grass CO_2 response, except for the loss of a significant positive CO_2 response for above-ground biomass and leaf area. In C_3 species, stimulation of carbon assimilation rate was reduced by overall environmental stress and nutrient stress tended to reduce the mean biomass response to elevated CO_2. In perennial rye grass swards, under elevated (700 μl l^{-1} (CO_2) or ambient atmospheric CO_2 concentration at outdoor and elevated 3°C increase in air temperature, the daily net carbon assimilation increased by 29-36 per cent in elevated CO_2 concentration in temperate climate (Casella and Soussana, 1997). Chen *et al.* (1994) validated a model which predicted that the light saturated leaf photosynthesis would increase by 10 per cent in C_4 grass (*Andropogon gerardii*) with increasing atmospheric CO_2 level from 350 to 700 ppm at 30°C and the optimal temperature would shift from 37 to 38.5°C. Increasing atmospheric CO_2 concentration had significantly reduced standard conductance. Morgan *et al.* (2007) studied the combined effects of warming and elevated CO_2 on a northern mixed-grass prairie and confirmed earlier reports that CO_2 increased productivity in semi-arid grasslands (21 per cent) in seasonal above ground biomass for plants grown under elevated CO_2 compared to non-enriched controls. The response was related to CO_2 induced water savings.

d. Soil Processes and Significance of Symbiotic Nitrogen in CO_2 Enriched Grassland

The longer-term consequences of elevated CO_2 on soil processes also needed to be included. Grassland responses to rising atmospheric CO_2 concentration were regulated by interactions with independent variables including climate and soil type, which set broad bounds on plant and soil water relations and the regional species pool. Morgan *et al.* (2011) demonstrated that elevated CO_2 significantly increased the soil moisture as a result of reductions in plant stomatal conductance, and that warming decreased significantly the soil moisture as a result of desiccation effects. Consequently, warming induced more extreme soil moisture conditions (soil drying after rainfall events), whereas elevated CO_2 dampened soil moisture extremes. The C_3 grasses responded most strongly to elevated CO_2, whereas the C_4 grasses also responded to warming.

These preliminary results indicated potentially strong feedbacks between carbon cycling and warming in rangelands. Growth at elevated CO_2 reduced N concentration and the magnitude of changes were highly species specific (Ross *et al.,* 2000, Rillig *et al.,* 2000). Soil CH_4 uptake was enhanced by elevated CO_2 but reduced by warming, particularly late stages in the year. Soil fluxes of N_2O remained unaffected. The combined warming and CO_2 enrichment indicated soil fluxes of CO_2 increased with elevated CO_2 and warming, but decreased with warming later in the year compared to un-heated controls. Both the soil and leaf factors contributed to the greater growth response of the C_4 grass. In C_{4} - grass *Panicum coloratum* var. *makarikiense* cv. Bambatsi and *P. coloratum,* elevated CO_2 (100 Pa) delayed the changes in carbon metabolism and water relations occurred during drought and that a reduced transpiration rate explained their enhanced growth (Seneweera *et al.,* 2001). Photosynthetic capacity was not reduced in C_{4} - grass *Bouteloua gracilis,* under elevated CO_2, but assimilation rate was nearly saturated at the ambient CO_2 (Le Cain *et al.,* 2003). Current state of knowledge regarding the response of C_4 plants to elevated CO_2 and the likely mechanism was identification of two main routes through which elevated CO_2 could stimulate the growth of both well watered and water stressed C_4 plants studied through enhanced leaf CO_2 assimilation rates and through reduced stomatal conductance and subsequently, leaf transportation rates (Ghannoum *et al.,* 2000). Reduced transpiration rates could stimulate leaf CO_2 assimilation and growth rates by conserving soil water, improving shoot water relations and increasing leaf temperature. Despite mounting evidences that C_4 plants could accumulate more biomass at elevated CO_2, the underlying mechanism of this response is still largely unclear. It was argued that bundle sheath leakiness, direct CO_2 fixation in the bundle sheath or the presence of C_3 like photosynthesis in young C_4 leaves were unlikely explanations for the high CO_2 responsiveness of C_4 photosynthesis. Owensby *et al.* (1999) and Nie *et al.* (1992) reported that peak above-ground biomass in a tall grass prairie ecosystem and canopy photosynthesis of plants in dominant warm-season grasses with the C_4 type of photosynthesis in a rangeland, were greater at elevated CO_2 than ambient CO_2 during years with significant plant water stress, inconsistent with previous findings. The interactions between elevated CO_2, leaf temperature and shoot water relations on the growth and photosynthesis of C_4 plants were identified as key areas needing research.

In C_{3} - grass, *Pascopyrum smithii,* carboxylation efficiency of ribulosebis phosphate carboxylase/oxygenase, and ribulosebis phosphate regeneration capacity were reduced at elevated $CO_2$when soil moisture was adequate in semiarid grassland. (Le Cain *et al.,* 2003). Although photosynthetic capacity was reduced in *P. smithii* (C_3) growing in elevated CO_2, reduced stomatal conductance and transpiration improved soil water content and leaf water potentials at the elevated CO_2, thereby improved assimilation rate of both C_3 and C_4 species during dry periods. These results suggested that photosynthetic responses of C_3 and C_4 grasses in this semi-arid ecosystem would be driven primarily by the effect of elevated CO_2 on plant and soil water relations. Increased infection of roots with vesicular-arbuscular mycorhizae and increased partitioning of phytomass to below ground organs might have also contributed to improved plant water relations of CO_2 enriched steppe grasses through more efficient mining of soil water (Morgan *et al.*, 1994b). Ghannoum *et al.* (2000) and Siebke *et al.* (2002) reported that the reduced transpiration rate stimulated CO_2 assimilation and increased water use efficiency and production in C_4 grasses while Pinto *et al.* (2011) found that the ability of C_3- C_4 intermediate species partially to recycle photo respired CO_2 did not improve water use efficiency or nitrogen use efficiency relative to congeneric C_3 or C_4 species grown under varying CO_2 and temperature conditions.

Local climate, soil moisture, and species composition and change were reciprocally coupled to grassland particularly aboveground net primary productivity (Fay *et al.*, 2011 and 2012). Fay *et al.* (2012) found that CO_2 effects on aboveground productivity might be greater when there were effect of CO_2 on soil moisture, species change and also mechanisms by which the soils would likely cause spatial variation in CO_2 effects on aboveground productivity in mesic, semi-arid and arid ecosystems. On mesic grassland (850 mm yr^{-1} rainfall) aboveground productivity response to CO_2 enrichment was highest on the soils where it increased soil moisture and favoured a productive dominant grass species. In semi-arid short grass steppe (400 mm yr^{-1} rainfall), the CO_2 enrichment to 600 $\mu L\ L^{-1}$ increased soil moisture by 12 per cent and aboveground biomass of C_3 grasses by 34 per cent. However, soil moisture and species change feedbacks were both absent in Mojave Desert and CO_2 enrichment caused no effect on aboveground productivity, species composition, or soil moisture in most of the years.

Although climate change affected terrestrial N cycling with important feedbacks to plant productivity and carbon sequestration, the impacts of climate change on the relative availability of N and P remain highly uncertain. There was an evidence for progressively less N and P being available for plant growth at elevated CO_2 (Allard *et al.*, 2004a). Increased plant production under elevated CO_2 would increase soil carbon (C) which might induce N limitations. Positive relationships were evident between atmospheric CO_2 concentration and net mineral N production around a naturally occurring CO_2 in spring in north New Zealand (Ross *et al.*, 2000). There were many indications that N could limit growth of non-leguminous plant under elevated atmospheric CO_2 under a grass sward (Luscher *et al.*, 2000, Soussan *et al.*, 1996, Zanetti *et al.*, 1997). Owensby *et al.* (1994) also found that biomass production response to elevated CO_2 was suppressed by N limitation with below-normal precipitation. After eight years of elevated CO_2 study, Williams *et al.* (2006) had detected greater amounts of total soil nitrogen, which suggested that rates of ecosystem N flux into or out of tall grass prairie had been altered. In addition to possible changes in denitrification and N_2 fixation, other sources of N such as the re-translocation of N to the surface from deeper soil layers were needed to explain how soil N accrues in surface soils as a consequence of elevated CO_2. They supported the notion that C accrual might promote N accrual, possibly driven by high plant and microbial N demand amplified by soil N limitation. Hungate *et al.* (1996) found that elevated CO_2 stimulated gross N mineralization and plant N uptake in two annual grasslands. Owensby *et al.* (2006) found that N concentration in *Andropogon gerardii* and dicotyledonous herbaceous plants, aboveground biomass was lower and total N higher under elevated CO_2 (twice of ambient CO_2) than in ambient CO_2 plants of tall grass prairie ecosystem dominated by species of *Andropogon gerardii* (C_4), *Sorghastrum nutans* (C_4) and *Poa pratensis* (C_3). However, N concentration of root in growth bag biomass did not differ among treatments but total N was greater in elevated CO_2 plots than in ambient CO_2 plots. Glosser *et al.* (2002) inferred that the amount of N available to plants from soil resources did not generally increase under elevated atmospheric CO_2. Thus, a possible limiting effect of N on primary production could become more stringent under elevated atmospheric CO_2 as the demand of the plant for N increased. Soil mineral nitrogen availability was unaffected by elevated atmospheric $CO_{2.}$ Further, Dijkstra *et al.* (2010) and Carrillo *et al.* (2012) found that soil N availability decreased with elevated CO_2 and increased with warming. In semi-arid grassland, Feike *et al.* (2012) found that elevated CO_2 (600 ppm) increased P availability to plants and microbes relative to that of N, whereas warming 1.5/3.0°C above ambient temperature during the day/night) reduced P availability relative to N. They indicated that soil moisture is important in controlling P supply from inorganic sources, causing reduced P relative to N availability during dry periods. Both wetter soil conditions under elevated CO_2 and drier conditions with warming could further alter N: P. ratio although warming might alleviate N constraints under elevated CO_2, warming and drought could exacerbate P constraints on plant growth and microbial activity in this semi-arid grassland.

In a study with elevated CO_2 and N fertilization over an eight year period, the primary responses to elevated CO_2 were mediated through reduced water use due to reduced stomatal conductance, which improved water use efficiency (Owensby *et al.*, 1993a, b, 1996a, b, 1999; Knapp *et al.*, 1993a, b, 1994, 1995; Ham *et al.*, 1995; Bremer *et al.*, 1996; Hamerlynck *et al.*, 1997, Morgan *et al.*, 2001). Owensby *et al.* (1994) also reported that the increase in above and belowground biomass production of N fertilized plots at elevated CO_2 occurred mainly in a dry year and below ground in relatively wet year. Production increases resulting from growth at elevated CO_2 were often accompanied by decreases in plant N concentration, and there was some evidence that N might limit plant response to rising CO_2 (Ghannoum and Conroy, 1998; Poorter *et al.*, 1996). These results suggested that nutrient availability to plants might change in the longer-term. At high partial CO_2 pressure, species difference was also found to affect photosynthetic nitrogen use efficiency and plant nitrogen use efficiency in C_4 grasses, being higher in *Panicum coloratum* than *Cenchrus ciliaris* (Rudman *et al.*, 2001). At high level of CO_2, plant nitrogen use efficiency (NUE) of *C. ciliaris* was also greater which resulted a 1.6 fold stimulation of dry mass by high *level of* CO_2. Ghannoum and Conroy (1998) also reported that addition of N enhanced the effect of elevated CO_2 by 28 per cent but no effect of elevated CO_2 was found under nitrogen deficiency. In C_3 grass (*P. laxum*), CO_2 enrichment resulted in reductions of CO_2 assimilation rates indicating acclimatization of photosynthesis in plants being more pronounced under N stress. In a high N fertilized grassland, the contribution of fertilizer N to plant growth increased strongly over the years, indicating important net N input into the system

(Hebeisen *et al.*, 1997; Hartwig *et al.*, 1999). Rice *et al.* (1994) reported that added N increased microbial C and N at elevated CO_2 enrichment and stimulated microbial activity under CO_2 enrichment. Increased microbial N with elevated CO_2 enrichment might have indicated that plant production could be limited by N availability. The soil system also could compensate for the limited N by increasing the labile pool to support increased plant production with elevated atmospheric CO_2.

N fertilization was not found economically feasible and therefore not an important issue in short grass steppe rangelands, nor were legumes an important feature of the landscape in some temperate grassland. Nevertheless, the dynamics of soil N cycling plus the seasonality of plant growth could result in a variable available soil N pool, so interactions of N with CO_2 might be important. Further, N was considered the most growth limiting nutrient in grasslands. Productions increased under elevated CO_2 were accompanied by reductions in shoot N concentrations, most notably in the C_3 species (Morgan *et al.*, 2001). This confirmed what had been observed in these species and grassland soils in controlled environment experiments (Hunt *et al.*, 1996; Le Cain and Morgan, 1998; Morgan *et al.*, 1994b; Read and Morgan, 1996; Read *et al.*, 1997). One of the adaptations sometimes observed in plants exposed to elevated CO_2 increased partitioning of biomass to belowground organs (Rogers *et al.*, 1994, 1996), which had been observed in some of the studies with short grass steppe grasses and soils, and appeared to be driven by reductions in plant N concentration (Morgan *et al.*, 1994b, 1998, 2001). It was suspected that much of the reduced plant N concentration under elevated CO_2 was due to enhanced growth in an N limited system resulting in a dilution of plant N (Hunt *et al.*, 1998). These N-deficiency effects of CO_2 enrichment were less evident, or non-existent under conditions of high N fertility (Morgan *et al.*, 2001).

Symbiotic N_2 fixation is one of the main processes that introduces N into terrestrial ecosystems. As such, it might be crucial for the sequestration of the extra C available in a world of continuously increasing atmospheric CO_2 partial pressure (pCO_2). An evaluation of the N sources revealed that all nitrogen that was additionally assimilated in *T. repens* under elevated CO_2, both in mixture and monoculture, derived from symbiotic nitrogen fixation. No additional nitrogen was derived from the soil mineral nitrogen pool. The significance of symbiotic N_2 fixation for grassland ecosystem under elevated atmospheric CO_2 showed fundamental difference in CO_2 response of plant biomass production in grassland ecosystems depending on the presence of *T. repens* in *L. perenne* grassland. *L. perenne* in monoculture showed symptom of N limitation (Zanetti and Hartwig, 1997; Zanetti *et al.*, 1997) in such a way that above ground N yield decreased under elevated CO_2 (Zanetti *et al.*, 1997). This was found in contrast to *L. perenne* with *T. repens* or *T. replens* growing in monoculture, when N nutrition appeared to be adequate (Zanetti and Hartwig, 1997). Total symbiotic nitrogen fixation increased by 66 per cent in the grass legume association under elevated CO_2 (Zanetti *et al.*, 1997; Luscher *et al.*, 2000). Total N yield increased by 60 per cent in grass-legume mixture due to increase in symbiotically fixed N in each individual legume plant.

These led to a simultaneously enhanced apparent transfer of N from clover to grasses (Zanetti *et al.,* 1997). These suggested that increased photosynthate CO_2 fixation under elevated CO_2 although not entirely reflected in production, was counter balanced by an appropriate increase in symbiotic N supply. This would restrict an increase extra carbon sequestration into the grassland ecosystem under elevated CO_2. Symbiotic N fixation was considered to be a crucial driving force which would increase carbon sequestration in a CO_2 rich world. Zaneti *et al.* (1996) reported that the total N yield increased consistently and the percentage of plant N derived from symbiotic N_2 fixation increased significantly in *Trifoilum repens* under elevated CO_2. All additionally assimilated N was derived from symbiotic N_2 fixation, not from the soil. In the mixtures, exposed to elevated CO_2, an increased amount of symbiotically fixed N was introduced into the system. Increased N_2 fixation was a competitive advantage for *T. repens* in mixed swards with pasture grass *Lolium perenne* (a non-fixing reference crop) and might be a crucial factor in maintaining the C: N ratio in the ecosystem as a whole.

Phosphorus (P) is an important determinant of plant productivity, particularly in the tropical grasslands, which contain both C_3 and C_4 species. Few studies had been done to compare the responses of such species to P deficiency. Previous works led to hypothesize that C_3 photosynthesis and the three subtypes of C_4 photosynthesis have different sensitivities to P deficiency. To examine their dynamic response to P deficiency in more detail, four taxonomically related tropical grasses (*Panicum laxum* (C_3) and *Panicum coloratum, Cenchrus ciliaris* and *Panicum maximum*) belonging to the C_4 subtypes NAD-ME, NADP-ME and PCK, respectively, were grown under contrasting P supplies, including P withdrawal from the growing medium (Ghannoum *et al.* (2007). Overall, the study showed that photosynthesis of the three C_4 subtypes behaved similarly. Compared with the C_3 counterpart, photosynthesis of the three C_4 grasses had a higher P use efficiency and lower P requirement, and responded to a narrower range of inorganic phosphate. Although each of the four grass species showed distinctive 1H-NMR fingerprints, there were no differences in responses that could be attributed to the C_4 subtypes. The response of CO_2 assimilation rates to leaf contents of inorganic phosphate was linear in the C_3 grass, but asymptotic for the three C_4 grasses.

Relative growth rate was affected largely by low P in the C_3 species and was correlated with the leaf content of glucose 6-phosphate more than with carbohydrates. To examine photosynthesis of *P. laxum* (C_3) and three C_4 grasses *i.e. Panicum coloratum* (C_4), *P. maximum* (C_4) and *Cenchrus ciliaris* (C_4), were grown. When compared with C_3, three C_4 grasses had a higher P use efficiency and lower P requirement (Ghannoum *et al.,* 2007). Thus, knowledge of CO_2 responses at the community level based on long term field experiments would be a prerequisite to understand, predict or model the response of grasslands to elevated CO_2.

Finally, the relative responses of the C_4 and C_3 photosynthetic types to elevated CO_2 concur only to some extent with expectations based on photosynthetic theory. The significant positive responses of C_4 grass species at both the leaf and the whole plant level demanded a re-evaluation of the assumption of low responsiveness in C_4 plants at both levels, and not only with regard to water relations. The combined shoot structural and water use efficiency responses of these functional types would have consequential implications for the water balance of important catchments and range-lands throughout the world, especially in semiarid subtropical and temperate regions. It might be premature to predict that C_4 grass species would lose their competitive advantage over C_3 grass species at elevated CO_2. Moreover, categorization of plant functional types (PFTs) according to relative growth potential might be more appropriate for predictions of CO_2 responsiveness in C_4 grasses.

Regrowth, Growing Season/Period and Reproduction in Grasses and Legumes

In general, the productivity of grasses and legumes had increased at elevated CO_2 (twice of ambient CO_2) as compared to ambient CO_2 level (Bhatt *et al.*, 2008, Bhatt *et al.*, 2010b; Morgan *et al.*, 2001; Soussana *et al.*, 2002; Kellog *et al.*, 2001; Ghannoum *et al.*, 2000, Siebke *et al.*, 2002). The evaluations of how CO_2 enrichment interacted with defoliation responses were obtained mainly from mechanical defoliation. There was no interaction of CO_2 with defoliation response in the short grass steppe study, although a single mid-summer cutting stimulated production. In the tall grass prairie, the response to CO_2 was greater when defoliation occurred in a dry year, but made little difference in seasonal production in an unusually wet year. Frequent defoliation of L. *perenne* increased root/shoot ratio in the Swiss pasture, but had no impact on the CO_2 response. The only study to utilize grazing animals indicated that sheep grazing enhanced the CO_2 response of legumes. These results were too few and differences between experiments too great to effectively summarize across grasslands, but they clearly indicated a possibility for defoliation of pastures to interact with the CO_2 response through various mechanisms (e. g. water, N, plant reserves). Wilsey *et al.* (1997) reported no interaction between CO_2 and defoliation for any species in the temperate grassland dominated by C_3 species; thus, it appeared that herbivores would not affect how grasses respond to elevated CO_2 (at least under average nutrient conditions). However, Allard *et al.* (2004a and b) reviewed defoliation by cutting and by grazing where they reported significant interactions between CO_2 and management (cutting or grazing) for root growth, soil moisture and plant species composition; consequently significant differences in nutrient cycling through animals fed forage grown at ambient CO_2 or elevated CO_2, emphasizing that it was hazardous to draw conclusions about elevated CO_2 effects on grazed grassland from experiments in which grazing is simulated by cutting. Responses to defoliation were fairly consistent among ecosystems in aboveground productivity, which did not differ between defoliated and undefoliated plants, and in leaf water potentials and percentage of N, both of which increased in response to defoliation. Because the quantity of food was unaffected by the CO_2 treatments and forage N was reduced, grazing mammals in temperate grasslands might be negatively affected. The effects of elevated atmospheric CO_2 concentration (700 µmol mol^{-1}) on defoliated (three clippings at 3-week intervals) and un-defoliated plants for the C_4 grass *Themeda triandra, Forsk* showed that the primary effect of the elevated CO_2 concentration in both defoliated and un-defoliated plants was an increase in individual leaf length and mass of dry matter, linked to higher leaf water content and increased photosynthetic rates at the canopy level (Wand and Midgley, 2004). Photosynthetic down-regulation at the leaf level occurred, but this was compensated by increased assimilation rates and greater canopy leaf area at the elevated CO_2 concentration. Increases in leaf and sheath growth of defoliated plants in the elevated CO_2 concentration treatment were lost following a final 21 days reversion to ambient CO_2 concentration, but it occurred in plants exposed to the elevated CO_2 concentration for the final 21days period only. In conclusion, elevated atmospheric CO_2 concentration increases shoot growth via increased leaf extension, which was directly dependent on stimulation of concurrent photosynthesis. CO_2 responsiveness was sustained following moderate defoliation but it was reduced when plants experienced reduced vigour as a result of maturation or high frequency of defoliation. In a study annual above-ground phytomass accrual was greater on the plots which were defoliated once in mid-summer compared to plots which were not defoliated during the growing season, but there was no interactive effect of defoliation and CO_2 on growth. Increased regrowth and dry matter of C_3 grass species (*Pascopyrum smitti*) and a forage legume (*Medicago sativa*) were obtained when grown under elevated CO_2 in Mediterranean grasslands (Morgan *et al.*, 2001). Owensby *et al.* (1999) reported that above-ground regrowth biomass was greater under elevated CO_2 than under ambient CO_2 in a year with late-season water stress, but did not differ in a wetter year. In temperate climate, the first two year of CO_2 enrichment indicated no significant interactive effect of defoliation on the CO_2 growth enhancement (Shoop *et al.*, 1989). Collectively, these results suggested increased productivity of the short grass steppe in future warmer, CO_2 enriched environments.

Global warming would influence the length of growing season/period of grazing, had not been studied across the agro-climatic regions. In temperate climate, it was projected that global warming would increase the length of the growing season in the range of 12 to 37 days for every 1°C rise in annual mean daily temperature in Scotland (Topp and Doyle, 1996a, b). In pastoral systems, climate change was likely to increase grass production. With increased forage availability, there could be increase in the annual grazing period by a maximum of five weeks for cattle systems, and seven weeks for sheep systems (Topp *et al.*, 2010). In most of the regions, this would allow animals to be kept outdoors for longer period, and could mean a potential reduction in the proportion of the year that animals require housing and/or access to conserved forages. The changes in length of the grass growth season were found generally more substantial in northerly than in southerly regions in Scotland.

Vegetative and reproductive responses to elevated CO_2 were highly varied among species and ecosystems, and did not generally follow any predictable pattern (Morgan *et al.*, 2004). The various mechanisms had been found to drive the changes in species composition at elevated CO_2. In the New Zealand, one important mechanism was increased recruitment from seed (Edwards *et al.*, 2000). The total seed rain of a number of species was greater at elevated CO_2 including grasses (*Anthoxanthum odoratum, Lolium perenne, Poa pratensis*); legumes (*Trifolium repens, T. subterraneum*) and forbs (*Hypochaeris radicata, Leontodon saxatilis*). In the case of *A. odoratum, H. radicata, L. saxatilis* and *T. repens*, this was due to both inflorescences per unit area and more seeds per inflorescence; for the other species it was due to increased numbers of inflorescence. In successive seasons the greater seed rain resulted in greater seedling density of the forbs and legumes and greater abundance of these species. Edwards *et al.* (2000) made the point that recruitment from seed was an important mechanism of species compositional change in this environment and was perhaps an underestimated mechanism in other systems. Whether the effects of elevated CO_2 was nitrogen-dependent in grasses, but independent in N_2-fixing legumes was for reproduction which showed that in both the species of grass and legume, elevated CO_2 had no significant effect on sexual reproduction *i.e.* reproductive growth, flowering phenology, seed development, reproductive success and seed germination in the grass *Lolium perenne* L. and the legume *Trifolium repens L.* under elevated CO_2 at 60 Pa partial pressure as compared to ambient (35 Pa) CO_2 and two levels of nitrogen fertilization (14 and 56 g N m^{-2} a^{-1}) in Switzerland (Wagner *et al.*, 2001). In *Lolium perenne*, reproduction was mainly nitrogen-dependent. There were weak interactions between CO_2 and mineral N supply. Under elevated CO_2, grain maturation was slightly enhanced and grain weight tended to decrease. No influence could be ascertained on the date of anthesis, the temporal pattern of grain growth, the rate of grain abortion and germination. In *Trifolium repens*, grown under CO_2 enrichment at both levels of N fertilization, flowered 10 days earlier, tended to form more inflorescences per ground area and more flowers (8–12 per cent) and seeds (>18 per cent) per inflorescence than at ambient CO_2. The temporal pattern of seed growth was about the same. The embryo development, however, was accelerated under elevated CO_2. The number of aborted seeds per pod, seed size, 1000 seed weight and germination did not show any influence of elevated CO_2. However, under elevated CO_2, plants grown at high N levels were attacked slightly more frequently by seed-eating weevils, which lowered the seed output per pod. In conclusion, the reproductive response of *L. perenne* and *T. repens* to CO_2 enrichment on the flower and inflorescence level was far weaker than expected from the results on vegetative growth. Kakani and Reddy (2007) also studied response of low (360 µL L^{-1}) and high (720 µL L^{-1}) CO_2 to temperate big bluestem (*Andropogon gerardii Vitman*) reproduction and five different day/night temperatures of 20/12, 25/17, 30/22, 35/27 and 40/32 °C grown in controlled environmental conditions. Panicle weight was 6–7 per cent of biomass at 25/17 °C and fell to 1.6 per cent at 40/32 °C.

Persistency, Biodiversity/Species Diversity/Interaction/Genetic Erosion

Climate shifts over this century expected to alter significantly the structure and functioning of plant communities. The analysis for 13 year of grassland community in infertile land showed that responses to temperature and rainfall manipulation portrayed an ecosystem with rapid but relatively minor sensitivity to changes in summer rainfall patterns, and generally unresponsive to winter temperature regime changes in northern England (Philip Grime *et al.,* 2008). However, a long-term climate experiments in natural vegetation were rare and largely confined to systems with the capacity for rapid compositional change. In unproductive, grazed grassland, one of the longest running experimental manipulations of temperature and rainfall revealed that vegetation was highly resistant to climate shifts maintained over 13 year. This resistance in the form might be attributed to (*i*) constant value in the relative abundance of growth forms and maintained dominance by long-lived, slow-growing grasses, sedges, and small forbs; (*ii*) immediate but minor shifts in the abundance of several species that had remained stable over the course of the experiment; (*iii*) no change in productivity occurred in response to climate treatments with the exception of reduction from chronic summer drought; and (*iv*) only minor species losses in response to drought and winter heating. Overall, compositional changes induced by a long 13 years exposure to change in climate regime, were less than short-term fluctuations in species abundance driven by interannual climate fluctuations. The lack of progressive compositional change, coupled with the long-term historical persistence of unproductive grasslands in northern England. This suggested that the community possessed a stabilizing capacity that led to long-term persistence of dominant species. Unproductive ecosystems provided a refuge for many threatened plants and animals and performed a diversity of ecosystem services. Philip Grime *et al.* (2008) supported the view that changing land use and over exploitation rather than climate change *per se* constituted the primary threats to these fragile ecosystems. However, Kakani and Reddy, (2007) found that the decrease in panicle/seed production at two high temperatures (>30/22 °C) in *Andropogon gerardii Vitman* had reduced the species population and dominance in tall grass prairies. The temperature response functions at different CO_2 would be useful to improve the predictive capabilities of dynamic global vegetation models in simulating dynamics of rangelands, where big bluestem is the dominant species.

Global climate and atmospheric changes might interact in their effects on the diversity and composition of natural grassland communities. It became quasi certain that the climatic changes effects would accelerate the biodiversity losses. Climate change is likely to cause nonlinear responses in ecosystem function and threshold changes in species composition. Species composition in the rangelands and some managed grasslands was found to be an important determinant of livestock productivity. As temperature and CO_2 levels changed due to climate change, the optimal growth ranges for different species also changed; species altered their competition dynamics, and the composition of mixed grasslands changed (Hopkins and Del Prado, 2007). In the temperate regions and subtropics, where grasslands often contained C_3 and C_4 species, some species were more prominent than others in the summer, while the balance of the mix reverted in winter. Small changes in temperature could alter this balance significantly and often resulted in changes in livestock productivity. Regarding species losses, IPCC, 2007 cited a study that looked at 1350 European species in terms of their distribution envelopes, and projected that half of these species would be classified as vulnerable or endangered by 2080 because of rising temperatures and precipitation shifts. Fay *et al.* (*2011*) reported that aboveground net primary productivity responses to a continuous CO_2 concentration gradient (250 to 500 μL L^{-1},) in the grassland communities on three soils differing in water holding capacity and other properties. Communities consisting of four C_4 grasses, two C_3 forbs, and one legume were established on a lowland clay (vertisol), an upland clay (*mollisol,*), and an alluvial sand (*alfisol*) and observed for five growing seasons. Aboveground net primary productivity increased with elevated CO_2 almost two-fold more on the upland clay and alluvial sand than on the lowland clay, because of strong linear responses to CO_2 on these soils compared to a saturating response to CO_2 on the lowland clay. On the two more responsive soils, the mesic tall grass *Sorghastrum nutans* replaced the more drought adapted mid-grass *Bouteloua curtipendula* at elevated CO_2, while *B. curtipendula* largely replaced *S. nutans* at low CO_2, especially on the upland clay. Evidence for a similar composition change was not found on the lowland clay. Thus, two soils displayed a threshold change in community composition that accounted for upto 57 per cent of variation in above net primary productivity for these soils. Variation in above net primary productivity and species composition with CO_2 were accompanied by linear increases in soil water content in 0 - 20 cm soil depth, most strongly on the alluvial sand and by weak decreases with CO_2 in soil N. Structural equation models explained 34 to 52 per cent of the variation in above net primary productivity, and indicated that CO_2 effects on above net primary productivity on the upland clay were primarily explained by CO_2 effects on species composition, and on the alluvial sand by CO_2 effects on the soil water content. Responses to elevated CO_2 in soil water content, above net primary productivity, and species composition were explained by reduced stomatal conductance and increased photosynthetic water use efficiency (WUE) in both the grasses. In addition, *S. nutans* gained more in WUE at elevated CO_2 than *B. curtipendula,* while *B. curtipendula* at elevated CO_2 had lower light saturated photosynthetic capacity, quantum use efficiency, and dark respiration than *S. nutans.* Thus, at elevated CO_2, shading by the taller *S. nutans* was likely to lower *B. curtipendula* carbon assimilation and growth. Elevated CO_2 strongly increased above net primary productivity on upland clay and alluvial sandy soils where there were also gains in soil moisture and threshold changes in species composition driven by physiological differences in the two dominant grass species.

In a California grassland, species interactions strongly influenced responses to changing climate, overturning direct climatic effects within a period of five years under simulated seasonality and intensity of rainfall in accordance with projections of leading climate models (Suttle *et al.*, 2007). Changes in seasonal water availability had pronounced effects on individual species, but as precipitation regimes were sustained across years, feedbacks and species interactions overrode the ecological responses to water and reversed community trajectories. Conditions that sharply increased production and diversity through two years caused simplification of the food web and deep reductions in consumer abundance after five years. Changes in these natural grassland communities suggested a prominent role for species interactions in ecosystem response to climate change. Responses of the 10 most common plant species to global changes and to interannual variability of an annual grassland with all possible combinations of experimentally elevated CO_2 (+ 300 μL/L), warming (+ 80 W/m^2, + ~1°C), nitrogen deposition (+ 7 g N m^{-2} yr^{-1}), and precipitation (+ 50 per cent) revealed that the dominant functional groups (annual grasses and forbs) had distinct production and abundance responses to individual global changes (Zavaleta *et al.*, 2003). After three years, N deposition suppressed plant diversity, forb production, and forb abundance in association with enhanced grass production. Elevated precipitation enhanced plant diversity, forb production, and forb abundance but it had affected the grasses little. Warming increased forb production and abundance but did not strongly affect the diversity or grass response. Elevated CO_2 reduced diversity with little effect on relative abundance or production of forbs and grasses. Realistic combinations of global changes had small diversity effects but had more marked effects on the relative dominance of forbs and grasses. The largest change in relative functional group abundance (+ 50 per cent forbs) occurred under the combination of elevated CO_2 + warming + precipitation, which might have larger impact in the future. Strong inter-annual variability in diversity, individual species abundances, and functional group abundances indicated that in the system, (1) responses after three years were not constrained by lags in community response, (2) individual species were more sensitive to inter-annual variability and extremes than to mean changes in environmental and resource conditions, and (3) simulated global changes interacted with inter-annual variability to produce responses of varying magnitude and even direction among years. Relative abundance of forbs, the most species group in the community, ranged after three years from >30 per cent under elevated CO_2 + warming + precipitation to <12 per cent under N deposition. While opposing production responses at the ecosystem level by different functional groups might buffer responses such as net primary production (NPP) change, these shifts in relative dominance could influence ecosystem processes such as nutrient cycling and NPP via differences between grasses and forbs in tissue chemistry, allocation, phenology, and productivity. In addition to predicted productivity increases under elevated CO_2, species composition of plant communities were likely to change, although the exact nature of these changes were difficult to predict (Polley *et al.*, 2000). Owensby *et al.* (1999) reported that the basal cover and relative amount of warm-season perennial grasses (C_4) in the stand changed little during an eight years study period, but basal cover and relative amount of cool-season perennial grasses (C_3) in the stand declined at the elevated CO_2 and in ambient CO_2. In another study, the decrease in panicle/seed production at two high temperatures (>30/22 °C) was found in *Andropogon gerardii Vitman* which reduced this species population and dominance in tall grass prairies (Kakani and Reddy, 2007). The temperature response functions at different CO_2 would be useful to improve the predictive capabilities of dynamic global vegetation models in simulating dynamics of rangelands, where big bluestem was the dominant species.

The spontaneous forage species would be much more affected by biodiversity reduction (Al Faiz *et al.,* 2010; Shaimi *et al.,* 2009). Perennial grasses and legumes which were the forage species most valuable and liked by animals in Morocco, were also the most threatened ones. According to their agro-physiological needs, these species would suffer differently from the climatic changes. For instance, oats, which is normally cultivated in the plains of the center in Morocco having 450-500mm annual rainfall, Mountain regions having 500-600mm rainfall and the West-northern region having 800 mm rainfall. It would probably disappear from the plains of Morocco. Triticale and barley, more rustic and drought tolerant cereals, would be less affected and will replace oats where it could not be grown any more. Some high water demanding forages grown in favorable rainfed area, such as *Trifolium alexandrinum* and Sorghum, would find much difficulty to grow in the same area, without any supplementary or full irrigation. Concerning pasture species in Morocco, some of them should be developed and encouraged to be grown giving the adaptation to low rainfall, such as medics or other perennial forage species with high confirmed summer survival, such as tall fescue and orchard grass. These two latter forages, with their original summer dormancy trait, was well correlated to drought tolerance, provided a promising crop for the future. Some high water demanding species such as *Trifolium fragiferum, Lolium rigidum* would have minimum chance to be grown. It had become quasi certain that the climatic changes effects would accelerate the biodiversity losses (Bounejmate, 1992,1994; Shaimi *et al.,* 2009). However, Sock *et al.* (2005) studied the long-term effects of a CO_2 exhalation, created more than 70 years ago, on a natural C_4 dominated sub-tropical grassland in terms of ecosystem structure and functioning. Long-term effects of elevated CO_2 on plant community composition and system processes in the sub-tropical grassland indicated that very subtle changes in ecosystem functioning and no changes in species composition and dominance which could be ascribed to elevated CO_2 alone. No season length stimulation of plant growth was found even after five years of exposure to elevated CO_2. Species compositional data and soil C^{13} isotopic evidence suggested no detectable effect of CO_2 enrichment on C_3:C_4 plant mixtures and individual species dominance. Contrary to many general predictions C_3 grasses did not become more abundant and C_3 shrubs and trees did not invade the site. IPCC (2007) had reported that substantial genetic erosion of *Lotus., Lupinus., Medicago* species *i.e. M. rigidula, M. rotata, M. secundiflora, M. sauvagei, Vicia spicies., Festuca arundinacea,* and *Phalaris aquatica* which had been found during recent collection trips (Bounejmate, 1992 and 1994). Other recent perennial grasses germplasms collection revealed significant genetic erosion for *Festuca arundinacea* and *Dactylis glomerata* (Shaimi *et al.,* 2009).

Forage Quality

Changes in climate would affect the quality and quantity of forage produced (Valtoria 2002). Few studies had been done on the likely impacts of climate change on forage quality. Moreover, very little results seemed to be relevant to the tropics. Increased temperatures increased lignification of plant tissues and therefore reduced the digestibility and the rates of degradation of plant species (Minson, 1990). This led to reduced nutrient availability for animals and ultimately to a reduction in livestock production. The modeling study of Hanson *et al.* (1993) indicated that mean forage digestibility decreased under all scenarios considered. The models simulated an increase in standing biomass but a considerable reduction in the N concentration of plants during the summer grazing months, large enough to bring about considerable decreases in animal performance. Other studies had shown that an increase in the legume content of swards might partially compensate for the decline in protein content of the non-N fixing plant species under conditions of elevated CO_2 concentrations. At the same time, the decline of C_4 grasses (which are less nutritious than C_3 plants) might compensate for the reduced protein content under elevated CO_2. However, the opposite effect was expected under associated temperature increases (IPCC, 2007). Milchunas *et al.* (2005) reported large negative effects of elevated CO_2 on crude protein concentrations and digestibility of forages during four years study in a semiarid short grass steppe. While the effects were more negative during mid-growing season than autumn, a reduction in already poor quality autumn forage might be more critical to animals. Crude protein concentrations of autumn forage on the elevated CO_2 treatment found to be below critical maintenance requirements during three out of four years. However, elevated CO_2 effects on forage soluble and fibre (celluloses, lignin) constituents were small, even though mid-growing season yield and end of season production increased. Forage digestibility declined by 14 per cent during mid-season and 10 per cent during autumn with elevated CO_2. Further, elevated CO_2 shifted the proportional availability of protein and energy to a species of lower overall quality and the species most negatively affected by drought. Current-year defoliation increased both quality and production of protein and energy compared to non-defoliated plots, but no CO_2 and defoliation treatment interactions were observed. Nitrogen fertilization increased crude protein concentrations and digestibilities, but not in the least nutritious species that increased with elevated CO_2 or in autumn when quality was the lowest. Milchunas *et al.* (2005) also reported that defoliation increased both the quality and production of protein and energy compared to non-defoliated plots during the same year. Studies in other systems also indicated only small effects of CO_2 on cell solubles, hemicellulose–cellulose, lignin fibre fractions of forages (Akin *et al.*, 1994; Akin *et al.*, 1995; Soussana and Loiseau, 1997; Fritschi *et al.*, 1999). Nitrogen fertilization and defoliation generally did not ameliorate the negative impacts of elevated CO_2 on forage quality. Negative effects on quality were most evident in crude protein, to a somewhat lesser degree on digestibility, with only small, mixed effects on quality through altered fibre fractions. The negative effects of elevated CO_2 on forage quality were likely to be greater than the positive effects on quantity, because quality drops to critically low levels that can inhibit utilization of the quantity that was available. The impact of climate change on pastures and rangelands might include deterioration of pasture quality of subtropical C_4 grasses in temperate regions as a result of warmer temperatures and less frost. As a consequence, productivity of grazing livestock could be altered. In common with other experiments there has been some reduction in the protein content within species at elevated CO_2. However, because of increases in yield and in species composition towards greater legume content, the nutrients available for animal growth had actually increased at elevated CO_2. Grazing by animals had the potential to interact with pasture responses to elevated CO_2 (Newton *et al.*, 2000); this might occur as a result of selectivity in the diet of animals, physical damage to plants and canopy, or because of heterogeneity in nutrients return. After three years of experimentation, the differences had started to develop between areas that were grazed by animals and the areas where grazing was simulated by cutting. The most striking difference had involved changes in the species composition with a stronger response to legumes to CO_2 in the grazed areas with the level of CO_2 enrichment which was quite modest 475μL L $^{-1}$), equating the concentration expected in 2030 (Newton *et al.*, 2000). Despite this small increase marked changes in ecosystem processes was observed; in particular changes in species composition to favor dicot species. The increased legume content was beneficial to animal production. However, it would be important to determine whether the greater legume abundance could be sustained or whether the frequently observed oscillations in legume-grass balance would simply had larger amplitude. In a study, acid digestible fiber (ADF) and N values from the ambient and elevated CO_2 diet samples obtained from esophageally-fistulated sheep were used to estimate the growth response of yearling steers grazing tall grass prairie. Estimated gain for steers consuming forage produced under elevated CO_2 was lower than that produced under ambient CO_2 summed over the 150-day growth period, with the greatest reduction in gain coming in the early season. Forage N concentrations were reduced under elevated CO_2 and fibre concentrations increased, both of which should have reduced ruminant productivity. Since ruminant intake declines as forage quality decreases, there could not be a compensatory intake response to maintain productivity levels comparable to current levels. Forage quality of plant community declined under elevated CO_2 because of a decline of N in all tested species as well as shift in species composition towards greater abundance of the low forage quality species (*Stipa comata,* Kandler *et al.*, 2006). Wand *et al.* (1999) carried out literature review (1980-97) and found that leaf carbohydrate status increased and leaf nitrogen concentration decreased significantly in elevated CO_2 only in C_3 species. Sock *et al.* (2005) also reported that total N decreased in the leaf of C_3 grasses but not in C_4 grasses under elevated CO_2 while total non-structural carbohydrate accumulation in the leaf was not affected. Morgan *et al.* (2001) evaluated effect of rising atmospheric CO_2 at twice ambient CO_2 (approximately 720 μ mol mol^{-1}) on important *Pascopyrum smithii* (C_3) and *Bouteloua gracilis* (C_4)

grasses in a grassland. Shoot N concentrations tended to be lower in grasses under elevated CO_2, but only *Stipa comata* (C_3) plants exhibited significant reductions in N under elevated CO_2 compared to ambient CO_2 chambers. In India, Bhatt *et al.* (2010a) reported that accumulation of soluble protein in the oat leaves decreased at elevated CO_2 (600 μl $^{-1}$), which was variety specific (Kent and JHO851). The variety JHO-822 had marginal increase in soluble protein at the elevated CO_2 in a semiarid tropical climate. In berseem crop, leaf nitrogen, soluble proteins, calcium, iron and nitrate reductase activity decreased under elevated CO_2 (600 μl $^{-1}$) as compared to ambient CO_2 (360μl $^{-1}$), thus, adversely affected the nutritional quality of forage but the phosphorus contents increased in the leaf (Madan Pal *et al.*, 1997 and 2004). However, total nutrient per unit area will increase due to increased biomass. This reduction in nitrogen might either be due to the dilution effect as a result of greater carbohydrate accumulation (Rogers *et al.*, 1996) or due to acceleration of plant growth due to increase in leaf area fixing extra carbon by higher photosynthesis (Madan Pal *et al.*, 2004) but not due to increased nitrogen use efficiency (Coleman *et al.*, 1993). Increase in carbon concentration and reduction in nitrogen content due to elevated CO_2 had been reported in various investigations (Gifford *et al.*, 2000; Farage *et al.*, 1998). As a result, Lincoln *et al.* (1993) reported 22 - 80 per cent increase in the rate of consumption by herbivorous insect fed on elevated CO_2 grown foliage. For domestic livestock, diets can be supplemented to compensate for reduced forage quality particularly protein but with wild ruminants, or for ruminants in developing countries, diet supplementation likely would not be an option. This could result in reduced growth and reproduction.

In temperate region of France, perennial ryegrass swards under elevated (700 μl l^{-1} CO_2) or ambient atmospheric CO_2 concentration at outdoor and elevated 3°C increase in air temperature showed that elevated CO_2 resulted in a large increase in the fructose concentration in the pseudo stems and laminas. In elevated CO_2, the hexose and sucrose pool increased by 28 per cent in the laminas but not in pseudo stems. A 3°C temperature increase in elevated CO_2 did not affect significantly the water soluble carbohydrate concentration in pseudo stems and laminas (*Casella and Soussana, 1997*). In perennial ryegrass swards, elevated (700 μl l^{-1}) CO_2 resulted in a large increase in the fructose concentration in the pseudo stems and laminas and increased the hexose and sucrose pool by 28 per cent in the laminas but not in pseudo stems (Casella and Soussana, *1997*). A 3°C temperature increase in elevated CO_2 did not affect significantly the water soluble carbohydrate concentration in pseudo stems and laminas. Field experiments in managed grassland had shown that the response of vegetative growth to elevated CO_2 was nitrogen-dependent in grasses, but independent in N_2 fixing legumes in temperate Switzerland. In Swiss pasture, *Lolium perenne* monoculture at the low N treatment exhibited marked nitrogen deficiency symptoms like increased root growth (Hebeisen *et al.*, 1997) and a highly significant reduction in herbage nitrogen concentration. *L. perenne* growing in association with *T. repens, T. repens* growing in monoculture or mixed culture showed no nitrogen deficiency symptoms (Hartwig *et al.*, 1999); nitrogen yield correlated well with biomass production. Deterioration of *pasture* quality in subtropical C_4 grasses of Argentina grassland was reported as a result of warmer temperatures and less frost (Silvia, 2002). Rebecca *et al.* (2012) reported effect of warming and additional precipitation on forage production and quality of tall fescue. He found that tall fescue pastures might be relatively resilient to future climate change; however, warming induced increases in ergot alkaloid concentrations in endophyte-infected tall fescue suggested that animal issues associated with fescue toxicosis are likely to be exacerbated under future climatic conditions.

Weed Flora and Incursion of Woody Plants into World Grasslands

Interactions between weed and forage species would be affected by any environmental factor that influences growth and nutrient relations, particularly under future climatic conditions. Since stressed ecosystems were also more susceptible to invasion, future drier conditions due to climate change might favour the emergence of weeds. Gilgen *et al.* (2010) found the competition between temperate grassland species and *R. obtusifolius*, a well-known weed in managed grasslands, which showed a competitive advantage to impact forage composition and quality of intensively managed grasslands, depending on site-specific conditions in temperate Switzerland. Above-ground biomass and the resource use in terms of carbon, nitrogen and water showed that *R. obtusifolius* increased its above-ground biomass production in response to drought, by virtue of higher water use efficiency together with a change in N acquisition patterns resulting in a competitive advantage of *R. obtusifolius* over other forage species, favouring the spread of this weed. Their results suggested that a potential increase of weed pressure by *R. obtusifolius* under future climatic conditions, demanding additional management measures to limit its success. Thus, additional adaptive management measures would be needed in the future to avoid an increased spread of *R. obtusifolius* in temperate grasslands. The forbs (C_3) and members of the cyperaceae (C_3) increased in basal cover and relative amount in the stand at elevated CO_2 as compared to under ambient CO_2 (Owensby *et al.*, 1993a, b, 1999; Morgan *et al.*, 2001).

Tropical forests were found expanding into grassy biomes in many regions of North and South America, Australia and Africa. There was growing evidence that elevated CO_2 might also be promoting conversion of savannahs to closed forests (Briggs *et al.*, 2005; Bowman *et al.*, 2011; Favier, *et al.*, 2004; Hoffmann *et al.*, 2009; Lloyd *et al.*, 2008; Loehle *et al.*, 1996; Morgan *et al.*, 2007; Russell-Smith *et al.*, 2004; Silva *et al.*, 2008). Changing atmospheric CO_2 influenced woody plant expansion into both arid grasslands and more humid flammable grasslands of African savannahs (William and Midgley, 2012; Bond and Midgley, 2000; Morgan *et al.*, 2004; Koehler *et al.*, 2010). CO_2 had been implicated in woody increase but attribution to global drivers had been controversial where changes in grazing and fire have also contributed. Changes in arid savannahs seemed less obviously linked to CO_2 effects and might be driven more by overgrazing. Regardless of whether grasses are C_3 or C_4, increasing CO_2 favoured trees more than grasses in flammable (and other disturbed) ecosystems. In more arid savannahs, CO_2 effects might be small relative to the effects of other drivers for large increase in a shrub in short grass steppes under elevated CO_2 (Polley *et al.*, 1997; Sankaran *et al.*, 2005; Van Auken, 2009). CO_2 effects were least likely to happen where edaphic constraints, such as shallow rooting depth and limited tree growth existed. Low soil nutrients might also reduce CO_2 responses in nutrient poor savannahs. There was a clear evidence for CO_2 fertilization contributing to woody thickening over the past few decades in some savannahs. The proportion of browse in rangelands might also increase in the future as a result of increased growth and competition of browse species due to increased CO_2 levels (Morgan *et al.*, 2007). This would have significant impacts on the types of animal species that could graze these rangelands. In sown mixed pastures, elevated CO_2 increased legume development, and this also occurred in temperate semi-natural grasslands. How such results were to be extrapolated is far from clear.

Carbon Sequestration

In terrestrial ecosystems, quantity of carbon in the soil is normally higher than the quantity in living vegetation. Increasing global atmospheric CO_2 concentration has led to concerns regarding its potential effects on the terrestrial environment and carbon balances. There is a need to understand the soil carbon dynamics and its role in terrestrial ecosystem, carbon balance and the global carbon cycle. Soil organic carbon (SOC) includes plant, animal and microbial residues in all stages of decomposition. Carbon input to the soil and its decomposition processes under elevated CO_2 are highly relevant for C sequestering in the soil. Elevated CO_2 generally increases soil C pools. However, greater available C concentrations could potentially stimulate soil organic matter decomposition. The long-term responses of grasslands would be controlled to a large extent by soil processes. Responses of belowground processes to global change might play a major role in terrestrial ecosystem carbon (C) storage and feedbacks to climate, but remained understudied in comparison to aboveground processes. Torbert *et al.* (1997) examined the effects of CO_2 enrichment of cultivated crops on the soil C storage in C_3 soybean (*Glycine max L. Merr.*) and C_4 grain sorghum (*Sorghum bicolor L.*) A significant increase in mineral associated organic C was observed which might result in increased soil C storage over the long-term, however, little change in total soil organic C was observed under either plant species.

Grasslands add greater SOC than under the arable land owing to cultivation. Grasslands are important globally as a reservoir of C that might otherwise add to levels of atmospheric CO_2. Grazing lands are estimated to contain 10–30 per cent of the world's soil organic carbon. Grazing, fire, and fertilization have been shown to affect soil C storage in rangelands. Carbon losses due to soil erosion can influence soil C storage on rangelands, both by reducing soil productivity in source areas and potentially increasing it in depositional areas, and by redistributing the C to areas where soil organic matter mineralization rates are different. Proper grazing management increases soil C storage and new grasslands have been shown to store as much as 0.6 Mg C $ha^{-1}year^{-1}$. Given the size of the C pool in grazing lands we need a better understanding of the current and potential effects of management on soil C storage. Generally, grazing animals convert consumed plants into CO_2 and other greenhouse gases such as methane (CH_4), but research on grazing lands particularly in dry environments, had shown that grazing could often help to lower net ecosystem C emissions and would promote soil C storage through improved grazing management to mitigate climate change (Wilsey *et al.,* 2002; Scurlock and Hall 1998; Allard *et al.,* 2007; Derner and Schuman 2007; Pellant *et al.,* 2005). Natural Grazing lands being used by native herbivores had been largely replaced for livestock grazing by high densities of stocking than the carrying capacity all over the world. Consequently, over grazing decades after decades had degraded these grasslands. The degraded grasslands simultaneously resulted in vegetation degradation, changes and loss of edible plant species, ultimately loss of soil C (Conant *et al.,* 2002; Prins, 1992; Pellant *et al.,* 2005; FAO 2007; Reynolds *et al.,* 2007; Molina Botero *et al.,* 2013). Comparing grazing under managed and natural grasslands under livestock and under native grazers, respectively, need to be understood to realize the full potential of soil C storage in grazing lands (Milchunas *et al.,* 1998). In grasslands, roots comprised about 75 per cent of the biomass, and were responsible for increased inputs of C to soil pools under elevated CO_2. A consequence of CO_2 stimulated plant growth might be an increased demand for below ground resources (water and mineral nutrients). If the demands for below ground resources were not met by increased resource availability or efficiency of resource use or if growth potential was constrained, plant might increase their carbon through root turnover, respiration or exudation. Carbon allocation to these rapid turnover pools would be limiting the quantity of long term carbon storage in response to elevated CO_2 because most of the carbon would quickly return to the atmosphere as CO_2. Many studies had documented changes in the root biomass of plants grown under elevated CO_2 but root turnover, exudation and respiration were difficult to quantify directly for budgets of total carbon portioning to roots (Norby, 1994). Therefore, increased recognition of the importance of the plant rhizosphere in mediating impacts of climate change on ecosystem-scale C cycling had been realized. Future elevated atmospheric CO_2 conditions are likely to alter soil C storage with subsequent potential impacts on atmospheric CO_2. Effects of elevated CO_2 on soil C storage might be direct via plant C inputs to the soil. However, more indirect effects via root associated organisms might also play a role. Root exudation might also be responsible for increased rates of soil organic matter decomposition, or priming, potentially offsetting inputs of new C. Annual budget focusing on below ground carbon cycling for two grassland ecosystems exposed to elevated CO_2 in Mediterranean type climate showed increased carbon uptake in the ecosystem, largely carbon being partitioned to rapidly cycling carbon pools belowground (Hungate *et al.,* 1997; Litvak and Anderson, 2009). This was the reason why numerous elevated CO_2 studies in spite of increased photosynthesis were greater than the increments in ecosystem carbon stocks. The shift in ecosystem partitioning also suggested that elevated CO_2 caused a greater increase in carbon cycling than in carbon storage in grasslands. Thus, increased carbon uptake in one to 10 years experiments did not indicated necessarily a large potential for carbon storage over 50-100 years, thus predicted the timescale when atmospheric CO_2 would be doubled as compared to present level of CO_2. Hungate *et al.* (1997) indicated that elevated CO_2 enhanced carbon partitioning to roots, a prediction that had not received wide support by studies measuring standing root biomass (Stulen *et al.*, 1993; Rogers *et al.*, 1994; Norby, 1994). This study indicated that this shift in partitioning manifested primarily as increased root exudation and respiration, and was thus extremely difficult to detect, providing a possible explanation for difficulty to detect, a possible explanation for the difficulty in accounting for all the additional carbon fixed under elevated CO_2 (Canadell *et al.*, 1996; Hungate *et al.*, 1997). All of the carbon allocated to root respiration was immediately returned to the atmosphere as CO_2. The carbon allocated to exudation was distributed between microbial respiration and labile soil carbon (Parton *et al.*, 1987). Carbon additions to labile pools in the soil could drive substantial but would be difficult to measure sequestration of carbon in the short term (Thompson *et al.*, 1996). The small size and high turnover of the labile pools, however, prevented them from providing quantitatively important long-term carbon storage. Only a small portion of the labile carbon added to soil owing to their chemical nature such as lignin and cellulose, which might persist in soil for many years (Parton *et al.*, 1987; Hungate *et al.*, 1997). Thus, the net carbon balance obtained in short-term CO_2-enrichment experiments tend to overestimate the potential for grasslands to sequester carbon in the soil in the long term. Further, Sock *et al.* (2005) reported long-term CO_2 enrichment also had little effect on soil carbon storage with no detectable changes in the soil organic matter. There were indications that potential soil respiration and N mineralization rates could be higher in the soils close to the CO_2 source. The conservative response of this grassland suggested that many of the reported effects could be short duration experimental artifacts that disappear under long-term elevated CO_2 conditions. More research is needed before the ability of soil to store additional C in a high CO_2 world could be determined. The change in plant community composition at elevated CO_2 resulted in an increase in the rate of litter decomposition by about 10 per cent (Allard *et al.*, 2004a). This additional C was primarily directed belowground resulting in increased root exudation and increased rates of root turnover at the elevated CO_2 (Allard *et al.*, 2004b). Fay *et al.* (2009) found that elevated atmospheric CO_2 would affect carbon cycling processes in mixed species grasslands and monocultures of switch grass differently. Graaf *et al.* (2009) reported that alteration in litter/root decomposition processes directly, via changes in litter chemistry, and indirectly, via changes in plant species compositions in roots from plant species (*Trifolium repens, Lespedeza cuneata* and *Festuca pratense*) grown under long-term elevated CO_2. Elevated CO_2 significantly reduced decomposition of roots of *T. repens* and *L. cuneata*, whereas decomposition of *F. pratense* remained unchanged. Mixing the roots of the three species produced under ambient CO_2 significantly enhanced decomposition rates but when roots produced under elevated CO_2 were mixed, decomposition was not significantly enhanced. These interactions might be particularly important belowground where the roots of different species intermingle and were in direct contact with the soil. These suggested that if elevated CO_2 reduces the quality of the most labile roots in a plant species community, it might decrease overall rates of root decomposition. CO_2 fixation was greater at elevated CO_2 particularly during periods of low soil moisture. Impact of elevated CO_2 showed increased root production (belowground net primary production) of over 60 per cent with elevated CO_2 during years of intermediate levels of precipitation, with smaller effects in a very wet year and no effects in a very dry year in native semiarid grassland (Milchnas *et al.,* 2004). Further, effects of elevated CO_2 on the quantity and quality of belowground biomass showed that elevated CO_2 increased the standing crops of rhizomes (87 per cent), coarse roots (46 per cent), and fibrous roots (40 per cent) but had no effect on root litter (mostly fine root fragments and sloughed cortex material >500µm (Jastrow *et al.,* 2000). C_3 grass grown at elevated CO_2 often increase production and greater belowground C inputs at elevated CO_2 (Donal Zak *et al.,* 1993). William *et al.* (2012) also suggested that decomposition is enhanced under elevated CO_2 compared with ambient CO_2, but that inputs of C are greater than the decomposition rates. Soil C sequestration in tall grass prairie and other drought-prone grassland systems was, therefore, considered plausible as atmospheric CO_2 increases. Soil C sequestration in tall grass prairie and other drought prone grassland systems, were, therefore, considered plausible as atmospheric CO_2 increases (Jastrow *et al.,* 2000).

The C sequestration potential of agroforestry systems was found to be higher than pastures and field crops (Sharrow and Ismail, 2004; Kirby and Povin, 2007). Among forage based production systems, perennial grasses/cultivated perennial forage based systems had high storage of carbon SOC in comparison to annual cultivated forage based systems (Kumar and Faruqui, 2009; Table 5.1). Tillage adversely affected storage of carbon in the soil in subtropical and tropical climates (Rai *et al.,* 2013). In Indian subtropical and tropical climates of semiarid region, silvipasture/hortipasture had storage of 9.5-19.7 tonnes per hectare of total carbon storage within five to six years after establishment followed by northwestern plain and Himalayan region (Table 5.1). The annual storage of carbon per hectare ranged 0.22 - 0.33 g $kg^{-1}yr^{-1}$ in central India. In natural grassland, established pasture, silvipasture and hortipasture, the mean annual storage of soil carbon, ranged 0.74 Mg ha^{-1} $yr^{-1,}$ 9-04 – 18.79 Mg ha^{-1} $yr^{-1,}$ and 15.96 -20.05 Mg ha^{-1} yr^{-1}, respectively, depending upon the composition of grassland and type of tree species grown under various production systems. In aonla based hortipasture in rainfed semiarid climate, it was 6.38 $kg^{-1}yr^{-1}$.

In cultivated forage cropping system, it was 0.19 - 1.95 $Mgha^{-1}yr^{-1}$ depending upon nature of forage crops and cropping systems. When agricultural land is no longer used for cultivation and allowed to revert to natural vegetation or replanted to perennial vegetation, soil organic carbon can accumulate by the processes that essentially reverse some of the effects responsible for soil organic carbon losses when the land was converted from perennial vegetation. Loss of soil organic carbon

Table 5.1: Carbon Storage under Various Forage Production Systems in India

Region	*Forage Production Systems*	*Total C Storage (tonnes C/ha)*	*Remarks*
Semiarid region	*Acacia nilotica* + natural pasture	9.5-17.0	At 5th years age, Rai *et al.*, 2001
	A. nilotica + established pasture	19.7	
	Dalbergia sisoo + natural pasture	12.4	
	D. sisoo + established pasture	17.2	
	Hardwickia binata + natural pasture	16.2	
	H. binata + established pasture	17.0	
North western India	*Acacia/Dalbergia/Prosopis* + *Sporobolus indicus*	1.5-12.3	At 6th years age, sodic soils, Kaur *et al.*, 2002
Northwestern Himalaya	Silvipasture Agri-hortipasture Hortipasture	2.17 1.15 1.08	AICRP Agroforestry, 2006
Central India	*Annona squamosa* + *C. ciliaris* + (*S.hamata and S. scabra*)	0.31g $kg^{-1}yr^{-1}$	Rai *et al.*, 2013 a
	Annonasquamosa + natural grassland + (*S.hamata* and *S. scabra*)	0.22 g $kg^{-1}yr^{-1}$	
Years after Agriculture	*Subtropical Dry Region*	*C Storage*	*Remarks*
Central India CO 5 years	*Cenchrus ciliaris*	14.90 Mg ha^{-1} yr^{-1}	Rai *et al.*, 2013
	C. ciliaris + *Clitoria ternatea*	18.79 Mg ha^{-1} yr^{-1}	
	C. ciliaris + *Macroptilium atropurpureum*	18.26 Mg ha^{-1} yr^{-1}	
	C. ciliaris + *Stylosanthes spp.*	18.1 Mg ha^{-1} yr^{-1}	
	Hardwickia binata + C. ciliaris	15.96 Mg ha^{-1} yr^{-1}	
	H. binata + *C. ciliaris* + *M. atropurpureum*	18.9 Mg ha^{-1} yr^{-1}	
	H. binata + *C. ciliaris* + *Stylosanthes spp.*	19.39 Mg ha^{-1} yr^{-1}	
	H. binata + *C. ciliaris* + *C. ternatea*	20.05 Mg ha^{-1} yr^{-1}	
	Natural vegetation	4.39 gkg^{-1}	Without fertilization, Hazra, 1995
	Improved pasture	9.04 gkg^{-1}	
	Albizia lebbeck + Natural vegetation	6.25 gkg^{-1}	
	Albizia lebbeck + Improved pasture	11.84 gkg^{-1}	
	Natural grassland	0.74 g $kg^{-1}yr^{-1}$	
	Natural grassland + range legumes	0.64-1.83 $kg^{-1}yr^{-1}$	14 range legumes Rai *et al.*, 1980
	Emblica officinalis + *Dichanthium annulatum*	6.38 g $kg^{-1}yr^{-1}$	
	Dichanthium annulatum	5.32 g $kg^{-1}yr^{-1}$	

Contd...

Region	*Forage Production Systems*	*Total C Storage (tonnes C/ha)*	*Remarks*
Years after Agriculture	*Subtropical Dry Region*	*C Storage*	*Remarks*
05 years	Multicut Sorghum – berseem berseem	17.73 $Mgha^{-1}$ (0.19 $Mgha^{-1}yr^{-1}$)	Food + forage, Kumar and Faruqui, 2009
	Groundnut + berseem - Maize + cowpea (Fodder)	19.28 $Mgha^{-1}$ (0.50 $Mgha^{-1}yr^{-1}$)	
	Napier bajra hybrid + cowpea - berseem	29.03 $Mgha^{-1}$ (0.45 $Mgha^{-1}yr^{-1}$)	
09 years	Guinea grass + cowpea - berseem	24.39 $Mgha^{-1}$ (1.95 $Mgha^{-1}yr^{-1}$)	Rai *et al.*, 2013
	Sorghum - berseem	17.33 $Mgha^{-1}$ (1.17 $Mgha^{-1}yr^{-1}$)	
	Cowpea-oat	15.85 $Mgha^{-1}$ (1.00 $Mgha^{-1}yr^{-1}$)	
	Maize + cowpea - oat	15.69 $Mgha^{-1}$ (0.99 $Mgha^{-1}yr^{-1}$)	
Other Countries	*Forage Production Systems*	*C Storage*	*Remarks*
Semi-arid west coast climate Oregon, USA	*Pseudotsuga menjiesii + Trifolium subterraneum*	95.89 MG ha^{-1}	11 years, Sharrow and Ismail, 2004
Tropics, Latin America	*Acacia mangium + Arachis pintoi*	173 MG ha^{-1}	10-16 years, Amezquita *et al.*, 2005
	Cordia alliodora + Guazuma ulmifolia + Brachiaria brizantha	132 MG ha^{-1}	
Florida, USA	*Pinus elliotii + Paspalum notatum*	6.9–24.2 MG ha^{-1}	8-40 years, Haile *et al.*, 2008
Subtropical, Malawi	*Gliricidia sepium + maize*	123 MG ha^{-1}	10 years, Makumba *et al.*, 2007
Conversion to pasture establishment			
Years after Agriculture	*Site History*	*Rate of Change in Carbon Storage (kg $ha^{-1}year^{-1}$)*	*Reference*
Cool temperate steppe			
12	Cultivated to perennial grass	110.0	Gebhart *et al.*, 1994
	Cultivated to improved pasture		White *et al.*, 1976
08	Russian wild rye	6.86	
08	Crested wheat grass	18.87	
5-80	Grass-forbs meadow	4.01	Titlyanova *et al.*, 1988
28-40	Coal mine spoil to dry grassland	28.2	Anderson, 1977
37	Subtropical moist forest	8.11-113.51	Lugo *et al.*, 1986

Contd...

Region	*Forage Production Systems*	*Total C Storage (tonnes C/ha)*	*Remarks*
20	Tropical mature forest cleared to pasture	17.0-74.5	Neil *et al.*, 1997
25	Tropical wet native forest cleared to pasture	(-6.0) – (-87.2)	Veldkamp, 1994
18	Tropical wet native forest cleared to pasture	34.93	Van Dam *et al.*, 1997

Sources: Adapted and Modified from Amita Fanish and Sathyapriya (2013), Post and Kwon, (2000), Nair *et al.*, 2009, Rai *et al.*, 2013.

by conversion of natural vegetation to cultivated use is well known. In recently established grasslands following a period of arable cropping, SOC exceeds the decomposition rate, resulting in a net accumulation of SOC. In ecosystem that are more commonly used for cultivation, accumulation of light fraction organic carbon (LF-OC) can be quite high despite higher decomposition rates than where significant returns of plant litter from forests and permanent grasslands occur. The turnover of LF-OC in cultivated situation is linked to macro aggregate formation and cropping and tillage largely impact its quantity. SOC is transformed by bacterial action and stabilized in clay or silt sized organo-mineral complexes [Fraction High HF-OC]. Time scale of turnover of High Fraction Organic Carbon (HF-OC) is on the order of decades. In permanently vegetated soils such as grasslands, silvipasture and hortipasture, above ground residues are left on the surface to decompose or a portion may be mixed into the soil by animal activity. Substantial gains in SOC are possible with conversion of cropland to grassland particularly with management for high grass productivity. SOC may accumulate at an average rate of 110 g C metre^{-2} year^{-1} in the surface of 3-metre depth in temperate climate, which may decline over time. In subtropical moist situation it was reported to be 33.2 g C metre^{-2} year^{-1}. SOC accumulations are much lower (21 g C metre^{-2}year^{-1}) under arid conditions. SOC is likely to increase when cultivated soil is planted with perennial grasses. Conversion of woody vegetation to grasses and replacement of native tropical savannahs with high producing deep rooted exotic grasses can also result in significant SOC increases for several years (800 - 1300 g C metre^{-2} year^{-1}). At least for tropical moist and wet forest, it is possible for SOC to increase when native mature forest is cleared and converted to pasture at least for 10-20 years. Tropical dry forest showed decreases in SOC when converted to pasture. Some information on carbon storage in forage systems are presented in Table 5.1.

Essential elements of what is known about soil organic matter dynamics that may result in enhanced soil carbon sequestration with changes in land-use and soil management needs to be understood. There will be a large amount of variation in rates and the length of time that carbon may accumulate in soil that are related to the productivity of the recovering vegetation, physical and biological conditions in the soil, and the past history of soil organic carbon inputs and physical disturbance. There will be maximum rates of C accumulation during the early aggregating stage of perennial vegetation growth. Average rates of accumulation are reported to be similar for forest or grassland establishment. However, these observed rates of soil organic C accumulation, when combined with the small amount of land area involved, are insufficient to account for a significant fraction of the missing C in the global carbon cycle that accumulate in the soils.

Long term experiments that address SOC dynamics when land is converted from cultivation with known management histories to perennial vegetation would be valuable in improving our understanding and increase our predictive capability over short and long time scales. There is not enough data available to predict regional SOC but enough information to infer order of magnitude of carbon sequestration rate. We can use the information to investigate some aspects of soil carbon fluxes in the present global carbon cycle. Recently, there has been interest in utilizing grasslands as pathways for terrestrial carbon sequestration. It is entirely reasonable to believe that grasslands can be used in both roles while still maintaining their ability to provide environmental services without degradation of grasslands. It mitigates the effects of CO_2 on global warming. However, some efforts to utilize grassland biomass to produce biofuels to reduce our dependence on fossil fuels appear to be in conflict with each other. The basic questions of how much carbon is sequestrated in the biomass of different grasses, in its roots, in the soil on which these grasses grow and the total aggregates needs to be researched.

Various hypotheses have been proposed concerning the processes involved and regions where various processes may have greater or lesser effects on the rate of net carbon uptake. A linear relationship between the length of time in grass and the amount of SOC sequestered in the surface (60 cm) fit well for time periods from 6 to 60 years. The slope of this function provided an estimate of the carbon sequestration rate, in this case 447 kg C $ha^{-1}year^{-1}$. At this rate, it would require nearly an additional century (98 years) for the 60-year grass site to reach a carbon pool equivalent to that of the prairie.

There are many factors and processes that determine the direction and rate of change in SOC content when vegetation and soil management practices are changed. One of that may be important for increasing SOC storage include (1) increasing the input rates of organic matter, (2) changing the decomposability of organic matter inputs that increase LF-OC in particular, (3) placing organic matter deeper in the soil either directly by increasing belowground inputs or indirectly by enhancing surface mixing by soil organisms, and (4) enhancing physical protection through either intra-aggregate or organo-mineral complexes. Conditions favouring these processes generally occur when soils are converted from cultivated use to permanent perennial vegetation. To obtain a higher predictive capability of detecting changes in SOC, additional empirical studies are needed, combined with a better understanding of the biological and physical processes involved.

The effects of climate warming on C storage can also be positive or negative. There is a high degree of uncertainty on the combined effects of climate warming and atmospheric CO_2 increase on soil organic matter dynamics and its potential feedbacks to climate change. Kakani and Reddy (2007) found that across the temperature biomass partitioned to roots of *Andropogon gerardii Vitman*, decreased by 7 per cent at high CO_2 as compared to ambient CO_2. Pendall *et al.* (2011) found that elevated CO_2 (600 ppm) increased root biomass in mixed C_3/C_4 grassland, a trend that became increasingly significant over the first four years but warming (elevated by 1.5/3°C day/night all year) by itself tended to decrease root biomass in the first two year suggesting a transient negative response of root growth to warming. C_3 grass roots decomposed more rapidly with elevated CO_2 alone, but more slowly when elevated CO_2 was combined with warming, possibly due to soil drying. C_4 grass roots decomposed more slowly than C_3 roots, probably due to higher C/N values, and were less responsive to climate change treatments. C_4 grasses became more competitive with elevated CO_2 and warming, together with their lower decomposition rates and higher root growth under these conditions.

Alpine meadow was the most sensitive region under climate change, and it would become the biggest potential carbon source in Chinese grasslands as climate warming continues to occur. Although recent climate warming benefited vegetation growth, rising temperature did not significantly accelerated net carbon uptake from Qinghai–Tibetan grassland ecosystems due to enhanced soil carbon decomposition accompanying the increase in temperature. The process based ecosystem model simulation in Tibetan grasslands, suggested that rising atmospheric CO_2 concentration and precipitation change enhanced ecosystem production and net carbon source increased from a net – 0.5 Tg C yr^{-1} in the 1960s to a net carbon sink of 21.8 Tg C yr^{-1} in the 2000s, mainly due to the change in the two factors (Tan *et al.*, 2010, Piao *et al.*, 2012). Zhao *et al.* (2012) estimated the sensitivity for soil organic carbon to climate variables in terms of top soil (0-20 cm) and found that a 4°C increase in soil temperature led to a loss of 4.7 per cent of soil organic carbon in the alpine meadow region, but the same temperature increase led to a maximum loss of only 2.3 per cent of soil organic carbon in the temperate steppe region. The effects of precipitation changes on soil organic carbon were varied depending on the moisture level of the local grassland system. The direct effect of CO_2enrichment was to reduce carbon loss throughout northern Chinese grasslands, especially in drought affected regions. Increased atmospheric CO_2concentrations led to net carbon sequestration in all grasslands and tended to diminish the carbon loss driven by precipitation and temperature changes. Li *et al.* (2011) indicated a decrease of 52.39 per cent in soil organic carbon for the increased precipitation and increased temperature + doubling the CO_2 scenario, 49.81 per cent for the unchanged precipitation and increased temperature + doubling the CO_2 scenario and 52.30 per cent for the increased precipitation and unchanged temperature + doubling the CO_2 scenario over the next 50 years. Therefore, soil organic carbon content in the alpine steppe will change because of the changes in precipitation, temperature and atmospheric CO_2 concentrations.

In temperate perennial ryegrass swards under elevated (700 μl l^{-1} CO_2) or ambient atmospheric CO_2 concentration at ambient and elevated 3°C increase in air temperature, the below ground carbon storage increased by 32 and 96 per cent in elevated CO_2 at the low and high N supplies, respectively with no effect of 3°C high temperature than ambient temperature (Casella and Soussana, *1997*).

At ambient and elevated atmospheric CO_2 concentrations and two N levels in *Lolium perenne* L. Ginkel and Gorisen (1996) revealed that after the second growing season, elevated CO_2 had stimulated root growth by 92 per cent and the decomposition rate was affected by the elevated CO_2 and N concentrations at which the grass roots had been grown. The combination of higher root yields at elevated CO_2 combined with a decrease in root decomposition will lead to a longer residence time of C in the soil and probably to a higher C storage.

Although elevation of CO_2 had been reported to impact soil microbial functions, little information is available on the spatial and temporal variation of this effect. Körner *et al.* (1997) found that when fertilizer and CO_2 enrichment were provided jointly, soil microbes were stimulated in alpine grassland at 2470 m altitude. Since responses to elevated CO_2 were absent in both warm and cold growing seasons, they concluded that this late successional plant community was carbon saturated at current atmospheric CO_2 concentrations for reasons not directly related to nutrient supply and climate. Kandler *et al.* (2006) reported that additional belowground carbon input mainly entered the fast cycling carbon pool and contributed little to long-term carbon storage in the semi-arid grassland. Further, Kandler *et al.* (2008) concluded that composition of soil microorganisms during the CO_2 elevation experiment and during the subsequent year after cessation of CO_2 elevation treatment shifted towards a fungal dominated system at elevated CO_2. This might slow down C cycling in the soils and therefore enhanced C sequestration in the short grass steppe in future CO_2 enriched atmospheres. Adair *et al.* (2011) determined the effects of factors on Soil Carbon Flux and elucidated the mechanism(s) behind the effect of elevated CO_2 on Soil Carbon Flux and reported that there was evidence that unlike the response of productivity to elevated CO_2, the response of Soil Carbon Flux was not strongly N limited.

Reid *et al.* (2004) reviewed mitigation options in the pastoral systems of the tropics. In terms of CO_2, protection was already playing a major role for carbon sequestration in pastoral lands, particularly in Africa, where most of the protected areas were located in less productive lands. Better management included conversion of cropland to grassland, reduction in grazing intensity and biomass burning, improving degraded lands and reducing erosion, and changes in species mix of existing protected areas. These kinds of management practices and conversion of crop land to grassland would improve carbon sequestration. This would also help to stop the conversion of rangeland into cropland. Otherwise, this conversion of rangeland to cropland could result in a 95 per cent loss of the above-ground C and 50 per cent loss of below-ground C. Particularly in wetter savannas, payments for maintenance of currently sequestered carbon could be quite effective policy.

While technical options for mitigating emissions did exist, there were problems to be overcome related to incentive systems, institutional linkages, policy reforms, monitoring techniques for carbon stocks, and appropriate verification protocols, for example. For the pastoral lands, mitigation activities had the greatest chance of success if they would build on traditional pastoral institutions and knowledge, while providing pastoralists with food security benefits at the same time. In African smallholder systems, technical options for mitigation also existed - for example, some recent work indicated that modest changes to the systems in Ghana had positive impacts on carbon sequestering (Gonzalez-Estrada *et al.,* 2008). There is also considerable potential for some agroforestry options that could mitigate emissions while at the same time providing opportunities for increasing the resilience of agricultural systems (Verchot *et al.,* 2007).

Optimum livestock density is widely recommended as policy intervention (FAO, 2007) but it has to be confirmed that whether displacement of native herbivore could give equivalent soil C sequestration. Bagchi and Ritchie (2010) found 49 per cent lower soil C in the watersheds converted to grazing lands for livestock than those watersheds grazed by natural herbivores in the Trans-Himalayas. Such difference in soil C might be attributed to indirect effect of reduced vegetation inputs due to change in composition of vegetation and lower biomass production of livestock grazed pasture.

Peat Meadow

Globally, peat lands were considered to be a sink of CO_2, but a source of CO_2 when drained. Additionally, wet peat lands were thought to emit considerable amounts of CH_4 and N_2O. Hitherto, reliable and integrated estimates of emissions and emission factors for this type of area had been lacking and the effects of wetland restoration on methane emissions had been poorly quantified. Hendriks *et al.* (2007) estimated the full greenhouse gasses balance of a restored natural peat in temperate climate, an abandoned agricultural peat meadow by determining the fluxes of CO_2, CH_4 and N_2O through atmosphere and water, while accounting for the different global warming potentials. CO_2 emission had decreased significantly as result of the raised water table, while CH_4 fluxes had increased. In global warming potentials the area was a very small sink of 71 g CO_2 equivalent m^{-2} per square meter over a 100-year period.

Livestock and Climate Change

Primary reason for the present upheaval in anthropogenic sources of greenhouse gases is attributed to the livestock sector. While implicating the livestock sector as the major contributing factor of greenhouse gases in the phenomenon of climate change, it is easily forgotten that the livestock have remained in harmony with the nature for thousands of years, adapting to the vagaries of climate. However, the rearing and feeding systems under intensive livestock production systems need to be evaluated in the context of climate change. Global change in climate might have far reaching consequences for milk, meat and wool production. The direct effect on livestock could be small but the indirect effect brought about by change in feed and foarge supply may severely alter the existing livestock production systems. As the climate changes and become more uncertain, species niches *i.e.* plant and crop substitution will take place. This will modify animal diets and affect the ability of smallholders to manage feed deficits. Depending upon the location, system and species, the productivity and quality of forage crops including grasslands will change. An altered species composition in grassland may affect the carrying capacity and stocking pattern of livestock. In semi-arid grasslands, a reduced growing season may spell doom for the migratory livestock production being practiced for thousands of years. Similarly, the high altitude grassland will move further upward due to rise in snow line throwing haywire the usufruct rights over grazing field and the migratory routes of the nomadic graziers, bringing unrest among the settled farmers and the transhumant graziers. The climate change has also increased frequency of major natural disasters. In coastal areas and adjoining inlands, devastating cyclones wiped away everything along its way. Floods are striking with increased vengeance and engulfing more areas for a longer duration of time. Similarly, the occurrence of drought has also increased, affecting more number of people and livestock.

The unique livestock production system integrated with crop husbandry offers window for reducing the Green House Gases (GHG) in India and elsewhere. It is time to gear ourselves to meet the challenges ahead by putting appropriate policies and adoptable action plan in place so that the millions of livestock farmers and consumers continue to derive income, employment and a healthy nutrition from various livestock and their products.

Methane (CH_4) Emissions from Livestock and its Management

Potentially usable energy was found to be lost in the form of CH_4 during rumen fermentation. The composition of the pasture, the maturity of the forage at the time of harvest and the diet composition significantly influence the emission in the form of CH_4 (Chaves *et al.*, 2006; Molina Boterol *et al.*, 2013). The method used to predict intake could also have a major influence on calculated values of methane emissions by various method to estimate CH_4 using Sulphur hexafluoride (SF_6) tracer technique, calculate with dry matter intake (DMI) estimated by alkane analysis and estimate by the Cornell Net Carbohydrate and Protein System (CNCPS) model.

CH_4 emissions and *invitro* digestibility of the forage were influenced by types of forage. Chaves *et al.* (2006 found 39 per cent higher release of CH_4 by grazing heifers on leguminous pasture of alfalfa than on grass pasture based on dry matter intake (DMI) basis. However, CH_4 loss in rumen fermentation was similar in both types of pastures on the basis of Cornell Net Carbohydrate and Protein System (CNCPS) model. Production of CO_2 was also similar in this situation. In 12 tropical/subtropical forage plant species (legumes: *Arachis hypogea, Trifolium alrxendrinum, Medicago sativa,* and graminaceous: *Avena sativa, Hordeum vulgare, Panicum maximum, Pennisetum purpureum, Saccharum officinarum, Sorghum bicolor*) using the Cornell Net Carbohydrate and Protein System, in a short 24 hours incubation study, Singh *et al.* (2012) found no difference in the loss of energy in the form of CH_4. Release of CH_4 was more in leguminous *Arachis hypogea* and the tree leaves of *Grewia optiva* than in the other forages and tree leaves *Leucaena leucocephala*. Among 16 tropical grasses *in vitro* CH_4 output showed differences among the species (Belete Shenkute Gemeda and Abubeker Hassen, 2014). Based on digestible dry matter (DDM) and digestible organic matter (IVOMD) *Anthephora argentea* and *Stipagrostis ciliata* released the highest volume of CH_4 and the lowest volume of CH_4 was released by *Cenchrus ciliaris, Setaria verticillata* and *Panicum coloratum*. CH_4 production positively correlated with neutral and acid detergent fibre, cellulose and hemicelluloses. It suggested focusing on screening and selecting perennial grass with higher nitrogen content and low CH_4 production to reduce methane production under tropical conditions. Molina Boterol *et al.* (2013) studied CH_4 release by simulating the diets corresponding to an intensive silvipasture systems of (i) *Leucaena leucocephala* with Guinea grass (*Panicum maximum*) and (ii) *Gliricidia sepium* with *Dichanthium aristatum*. They found that CH_4 production out of individual forages and forage mixtures (90:10, 80:20, and 70:30, for grass: forage legume, the mixture composed of *P. maximum and L. leucocephala* (70: 30) had the highest CH_4 production (156 ml) at fourth day, while the lowest CH_4 (121 ml) was released in the case of 100 per cent cattle diet of *L. leucocephala.* The silvipastoral system of *Gliricidia sepium* with *Dichanthium aristatum* showed that CH_4 production at 48 hours by *D. aristatum* and *G.sepium* (70:30) and *D. aristatum* and *G. sepium* (80:20) were higher than the other combinations. No significant differences were found for CH_4 losses between the two silvipasture systems. In general, the legumes had higher degradation rates than the grasses.

Methanogen inhibitors could be used to reduce CH_4. Tavendale *et al.* (2005) found that a polymeric fraction completely inhibited methane production on the common rumen methanogens *Methanobrevibacter ruminantium*, strains YLM-11 and DSM1093, which were extractable condensed tannin found in the two forage legumes *Lotus pedunculatus* (0.10 per cent) and *Medicago sativa* (0.001 per cent). The addition of tannin binding agent increased total volatile fatty acid production more in *Lotus pedunculatus* with more branched chain volatile fatty acid with addition of tannin binding reagent than *Medicago sativa.* The branched chain volatile fatty acid also increased *by 59-67* per cent with the tannin binding reagent addition for *Lotus pedunculatus* but did not affect in case of *M. sativa*. They found that condensed tannin action on methanogenesis could be attributed to indirect effects via reduced hydrogen production (and presumably reduced forage digestibility) and via direct inhibitory effects on methanogens

Leng (2008) reviewed the potential to use nitrates in ruminant feeds to limit enteric methane production and found nitrate as a sole fermentable N source in a diet that could totally inhibit enteric CH_4 production by the ruminants and its reduction to NH_3. The resultant NH_3 could replace reduction of CO_2 to CH_4 as a major sink for disposal of hydrogen in the rumen when it would be included in ruminants diets. The barrier to the use of nitrate was the potential for the accumulation of nitrite in the rumen that is anti-nutritional and can cause reduced productivity and even death when absorbed through the diet of ruminants to reduce enteric methane production. Never the less, nitrate could be used in the diets of ruminants that are fed poor quality diets which in turn produce most of the enteric methane. The use of nitrate in high protein feeds might not be practical. For safe use of nitrate to lower methane production in ruminants, it would be essential that nitrate in the feed is kept within stoichiometric requirements for efficient fermentative digestion, livestock are adjusted to nitrate slowly, and previous diet and the production diet are low in rumen degradable protein. The conditions and precautions for feeding nitrate could be well compared with the safe use of urea in ruminant feeds.

Synthesis

Climate change is expected to bring elevated CO_2, warmer temperature, changes in rainfall pattern and increased frequency of extreme weather (Tim Wheeler & Chris Reynolds, 2013). In general, the effects of climate change on food and commercial crops had been better studied than in forage crops, pasture and rangeland species, in spite of their comparable economic and ecological importance. Expected impacts of rising temperatures, elevated CO_2, precipitation and other associated factors including the potential impacts of increasing variability in precipitation, soil processes *etc.* on rangeland and forage crops showed that rangeland and forage species did not exhibited the differential response (Bhatt *et al.*, 2010a; Izaurralde *et al.*, 2010; Kellog *et al.*, 2001; Morgan *et al.*, 2004; Parson *et al.*, 2001; Soussana *et al.*, 2002). The increasing ozone levels might cause more damage to plants than the other stresses. In addition to the direct impacts of climate there were indirect impacts that resulted from increased weed, insect, and disease pressures.

Vegetative responses to CO_2 were highly varied among species and ecosystems, and did not generally follow any predictable pattern (Parsons *et al.* 2001; Morgan *et al.*, 2004). Increasing the concentration of CO_2 enhanced the productivity of forage due to enhanced rates of photosynthesis (Bhatt *et al.*, 2010a; Drake *et al.*, 2001; Madan Pal *et al.*, 1997, 2004). The reported increase in the productivity showed a wide variation depending upon various growing factors and agro climatic conditions in temperate, tropical and sub-tropical grassland in arid to humid climates. The impact of elevated CO_2 showed varietal difference in biomass production in cultivated forage oat (Bhatt *et al.*, 2010a). Differing tendencies between species and types in shoot structural response were also revealed (Madan Pal *et al.*, 1997, 2004). C_3 grass species showed a greater increase in tillering, whereas C_4 species showed a greater increase in leaf area at elevated CO_2 resulting into the higher carbon assimilation rates in the both types. Fast growing herbaceous C_3 species responded more strongly than slow growing C_3 herbs or C_4 plants (Poorter and

Navas, 2003). This boost to productivity was apparent for forage plants that use C_3 photosynthetic pathway. Although, there was some disagreement about, whether the extent of these benefits could always be found under field conditions (Long *et al.,* 2006). There were, however, a number of important cultivated forage crops like maize, sorghum and millets and many forage grasses and some forbs that use the C_4 photosynthetic pathways. The leaf photosynthetic rates of C_4 plants were not substantially enhanced by elevated concentration of CO_2, hence, yield in these crops grown under elevated CO_2 were much more modest than for C_3 crops and grasses (Kimball *et al.,* 2002). The interactions between elevated CO_2, leaf temperature and shoot water relations on the growth and photosynthesis of C_4 plants were identified as key areas needing research. We could expect increasing CO_2 to benefit the productivity of most of the forage crops and pasture grasses and legumes. In addition, there was also a small improvement in how efficiently both C_3 and C_4 plants use water under elevated CO_2 (Drake *et al.,* 1997). When grown in competition, C_4 species showed lowest responses to elevated CO_2 at high nutrient conditions, whereas differences were also found between slow and fast growing species. Genera, species specific and genotypic responses varied considerably in their magnitude that complicates any prediction of response of C_4 plants to future atmospheric change (Wand *et al.,* 1999, Kellog *et al.,* 2001). Thus, knowledge of CO_2 responses at the community level based on long term field experiments would be a prerequisite to understand, predict or model the responses of grasslands to elevated CO_2. CAM species and woody plants showed intermediate responses to elevated CO_2.

Increasing temperatures would affect the rate at which forage plant species grow, develop, their reproductive performance and potential to survive at the extremes of the temperatures. A prerequisite for increased growth with warmer temperatures was the availability of sufficient water. Generally, grassland needed more water per unit of biomass production than the most arable crops (Ehlers and Goss, 2003). Increasing average daily temperature by 01°C with average moisture conditions would decrease forage production from current levels. Moreover, hotter temperatures would reduce forage production even if moisture conditions were wetter. A 10 per cent change in moisture conditions would mean approximately 09 per cent change in herbaceous production. The effect of warming on hydrology introduced other uncertainties; by itself a warming trend would increase potential evapotranspiration, leading to desiccation. However, the improved water use efficiency under elevated CO_2 would tend to counter that response. In semi-arid grasslands, significant increases in both temperature and CO_2 might eventually shift the competition more in favour of C_4 grasses because of 1) the overriding benefit of CO_2 improving plant water relations of most plant species, regardless of photosynthetic pathway, and 2) the adaptation of C_4 species to warm temperatures. Therefore, future CO_2 levels might favour C_3 plants over C_4 plants; the opposite was expected under associated temperature increases. Changes in evaporation-precipitation ratios modify ecosystem function, particularly in marginal areas, in ways that are not fully understood (IPCC, 2007). For instant while warming might favour warm-season C_4 species, some of the effects of CO_2 enrichment on C_3 photosynthesis would tend to counter that response, rendering C_3/C_4 distribution changes relatively insensitive to increases in temperature (Polley *et al.,* 2000). Besides, above ground production, increasing global temperatures would alter species composition in grasslands (Mowell *et al.,* 2013). Extreme temperatures would impact species distributions and abundance through reproduction, competition or survivorship; although, species response differences were diverse and difficult to predict (Polley *et al.,* 2000). Warming and elevated CO_2 might also increase biomass and ecosystem respiration in grasslands. An improved understanding of the mechanisms underlying these responses was expected to improve terrestrial C cycle predictions (Ryan *et al.,* 2013). In a warmer climate, generally it would be expected that the niche area of existing crops would likely to move northwards in the northern hemisphere. Given the importance of changes in plant development under climate change, the effects of photoperiod and temperature on development rates above the optimum temperature clearly merit further research, and some of the knowledge gaps are identified herein.

Precipitation had a strong positive correlation with aboveground net primary productivity in grasslands and had been at the forefront in climate change research on grasslands. Desert grasslands dominated by *B. eriopoda* were highly sensitive to changes in precipitation regime (Collins *et al.,* 2009). Climate models, predicted, and empirical evidences confirmed that more extreme precipitation regimes were occurring in tandem with warmer atmospheric temperatures. Altered precipitation patterns could mean delayed germination, resulting in shorter growing seasons and longer periods of inadequate forage quality. An increase in the frequency of extremely dry years would also increase the uncertainty of forage availability. Projections of temperature and precipitation patterns during the next 50 years anticipated a 1.5 to 2°C warming trend and a slight increase in precipitation. Many studies had shown that precipitation and temperature changes in the future would modify, and often limit, the direct CO_2 effects on plants which drive many aspects of grassland ecosystem function (IPCC, 2007, Chaplin Kramer and George, 2013)). Simulation runs also showed that changes in production were more closely related to changes in temperature and precipitation than to enhance CO_2 alone. In a simulation study, Soussana *et al.* (2002) found that doubling the CO_2 concentration in the atmosphere would increase the forage production in the temperate zones but if it was combined with a 3°C global warming, no production improvement could be registered, even if length of vegetative cycle increases, considering the summer water deficit. Increased CO_2 concentrations in the atmosphere had a growth-enhancing effect and at the same time led to improved water-use efficiency (Owensby *et al.,* 1993; Soussana and Luscher, 2007); however, this was offset by the rising temperatures which increased the rate of water use. The water limitations would enhance the CO_2 production response partly through improved water use efficiency. However, rangeland plant production would respond little to combined warming and rising CO_2 and might even decline due to increased drought.

The evaluations of how CO_2 enrichment interacted with defoliation responses were obtained mainly from mechanical defoliation which showed that above-ground regrowth biomass was greater at elevated CO_2 in a dry year with late-season water stress, but did not increase in a wetter year (Owensby *et al.,* 1999). The interaction varied across temperate to tropical grasslands and defoliation by cutting and by grazing for regrowth biomass, root growth, soil moisture and plant species composition (Shoop *et al.* 1989; Allard *et al.,* 2004a and b). It is emphasized that it would be hazardous to draw conclusions about elevated CO_2 effects on grazed grassland from experiments in which grazing is simulated by cutting. The study using grazing animals indicated that sheep grazing enhanced the CO_2 response of legumes (Morgan *et al.,* 2001). All these results were too few and differences between experiments too great to effectively summarize across grasslands, but they clearly indicated a possibility for defoliation of pastures to interact with the CO_2 response and warming through various mechanisms (*e.g.* water, N, plant reserves).

Global warming would influence the length of growing season/period of grazing, had not been studied across the agro-climatic regions. In temperate climate, global warming would increase the length of the growing season in the range of 12 to 37 days for every 1°C rise in annual mean daily temperature (Topp and Doyle, 1996a, b). Vegetative and reproductive responses to elevated CO_2 were highly varied among species and ecosystems, and did not generally follow any predictable pattern (Morgan *et al.,* 2004). Increases in atmospheric CO_2 often advanced the time of flowering by a few days, but field experiments suggested that elevated CO_2 had little or no effect on the rate of development other than small advances in development associated with a warmer canopy temperature (Craufurd and Wheeler, 2009). Both pasture and rangeland species might exhibit an acceleration of metabolism and development due to earlier onset of spring and longer growing seasons. One important mechanism was found to be increased selection of plants growing from the seeds of plants under earlier onset of spring and longer growing seasons from seed (Edwards *et al.,* 2000). The total seeds of a number of species falling on the ground was greater at elevated CO_2 including grasses; legumes and forbs. In some cases, this was due to both inflorescences per unit area and more seeds per inflorescence; for the other species it was due to increased numbers of inflorescence. In both the species of grass and legume, elevated CO_2 had no significant effect on sexual reproduction *i.e.* reproductive growth, flowering phenology, seed development, reproductive success and seed germination (Edwards *et al.,* 2000).

Global climate and atmospheric changes might interact in their effects on the diversity and composition of grassland communities. The various mechanisms had been found to drive the changes in species composition at elevated CO_2 (Morgan *et al.,* 2004). However, the ultimate responses of these grasslands to CO_2 and other aspects of global change would involve more complex changes in species dominance and survival along with alterations in soil biogeochemistry, much of which had been still poorly understood. Long-term effects of elevated CO_2 on plant community composition and system processes in the sub-tropical grassland indicated very subtle changes in ecosystem functioning and no changes in species composition and dominance which could be ascribed to elevated CO_2 alone (Sock *et al.*, 2005). Further, it became quasi certain that the climatic changes effects would accelerate the biodiversity losses. Regarding species losses, IPCC (2007) cited a study that looked at 1350 European species in terms of their distribution envelopes, and projected that half of these species would be classified as vulnerable or endangered by 2080 because of rising temperatures and precipitation shifts and had also reported that substantial genetic erosion of *Lotus* spp., *Lupinus* spp., *Medicago* species which had been found during some collection trips (Bounejmate, 1992 and 1994). Other perennial grasses germplasm collection revealed significant genetic erosion for *Festuca arundinacea* and *Dactylis glomerata* (Shaimi *et al.*, 2009). The spontaneous forage species would be much more affected by biodiversity reduction (Al Faiz *et al.*, 2010; Shaimi *et al.*, 2009). Perennial grasses and legumes which were the forage species most appreciated by animals in Morocco, were also the most threatened ones. Some high water demanding forages grown in favourable rain fed area, such as *Trifolium alexandrinum*, would find much difficulty to grow in the same area. However, Philip Grime *et al.* (2008) supported the view that changing land use and over exploitation rather than climate change *per se* constituted the primary threats to the fragile ecosystems.

Realistic combinations of global changes had small diversity effects but had more marked effects on the relative dominance of forbs and grasses. There was a shift in the composition of rangeland species and enhanced population of forbs and shrubs under climate change. Zavaleta *et al.* (2003) reported that the largest change in relative functional group abundance of forbs occurred under the combination of elevated CO_2 + warming + precipitation, which would likely be affected much more in the future. Strong interannual variability in diversity, individual species abundances, and functional group abundances indicated that in the system, individual species were more sensitive to interannual variability and extremes than to mean changes in environmental and resource conditions.

Few studies had been conducted on the likely impacts of climate change on forage quality. Moreover, very little results seemed to be relevant to the tropics. Accumulation of soluble protein in the oat leaves was found variety specific (Kent and JHO851) under elevated CO_2 but in berseem, leaf nitrogen, soluble proteins, calcium, iron and nitrate reductase activity decreased under elevated CO_2 (600 μl^{-1}), adversely affecting the nutritional quality of forage in a semiarid tropical climate (Bhatt *et al.*, 2010a; Madan Pal *et al.*, 1997, 2004) and on crude protein concentrations and digestibility in semiarid short grass steppe (Milchunas *et al.*, 2005). Increased temperatures increased lignification of plant tissues and therefore reduced the digestibility and the rates of degradation of plant species (Minson, 1990). The modeling study of Hanson *et al.* (1993) indicated that mean forage digestibility decreased under all scenarios considered under the climate change during the summer grazing months, large enough to bring about considerable decreases in animal performance. However, some studies in other systems also indicated only small effects of elevated CO_2 on cell solubles, hemicellulose–cellulose, lignin fibre fractions of forages (Akin *et al.,* 1994; Akin *et al.,* 1995; Soussana and Loiseau, 1997; Fritschi *et al.,* 1999). Further, elevated CO_2 shifted the proportional availability of species of lower overall quality and the species most negatively affected by drought. Nitrogen fertilization did not increase crude protein concentrations and digestibility in these least nutritious species. The most striking difference had involved changes in species composition with a stronger response of legumes to CO_2 enrichment (475µL L^{-1}) in the grazed areas (Newton *et al.,* 2000). An increase in the legume content of swards might partially compensate for the decline in protein content of the non-fixing plant species under conditions of elevated CO_2 concentrations and the increased legume content was beneficial to animal production. However, it would be important to determine whether the greater legume abundance could be sustained or whether the frequently observed oscillations in legume-grass balance would simply had larger amplitude. At the same time, the decline of C_4 grasses (which are less nutritious than C_3 plants) might also compensate for the reduced protein content under elevated CO_2. Warming and additional precipitation induced increases in ergot alkaloid concentrations in endophyte-infected tall fescue which suggested that animal issues associated with fescue toxicosis are likely to be exacerbated under future climatic conditions (Rebecca *et al.,* 2012).

Interactions between weed and forage species would be affected by any environmental factor that influences growth and nutrient relations, particularly under future climatic conditions. Since stressed ecosystems were more susceptible to invasion, particularly drier conditions as a consequence of climate change. This might favour the emergence of weeds *viz., R. obtusifolius* in temperate grasslands (Gilgen *et al.,* 2010). The forbs (C_3) and members of the cyperaceae (C_3) increased in basal cover and relative amount in the stand at elevated CO_2 as compared to under ambient CO_2 (Owensby *et al.,* 1993a, b; 1999; Morgan *et al.,* 2001). A potential increase of weed pressure under future climatic conditions might demand additional management measures to limit its success.

Tropical forests were found expanding into grassy biomes in many regions of North and South America, Australia and Africa. There was growing evidence that elevated CO_2 might also be promoting conversion of savannahs to closed forests (Briggs *et al.,* 2005; Bowman *et al.,* 2011;Favier, *et al.,* 2004; Hoffmann *et al.,* 2009; Lloyd *et al.,* 2008; Loehle *et al.,* 1996; Morgan *et al.,* 2007; Russell-Smith *et al.,* 2004; Silva *et al.,* 2008). Changes in arid savannahs seemed less obviously linked to CO_2 effects and might be driven more by overgrazing. Regardless of whether grasses are C_3 or C_4, increasing CO_2 favoured trees more than the grasses in flammable and other disturbed grassland ecosystems. However, CO_2 effects were least likely to happen where edaphic constraints, such as shallow rooting depth and limited tree growth existed. This was a clear evidence for CO_2 fertilization contributing to woody thickening over the past few decades in some savannahs. The proportion of browse in rangelands might also increase in the future as a result of increased growth and competition of browse species due to increased CO_2 levels (Morgan *et al.,* 2007). This would have significant impacts on the types of animal species that could graze these rangelands. How such results were to be extrapolated is not fully understood.

Carbon input to the soil and its decomposition processes under elevated CO_2 were highly relevant for C sequestering in the soil. Elevated CO_2 generally increased soil C pools. This was due to increased carbon uptake, largely carbon being partitioned to rapidly cycling carbon pools belowground (Hungate *et al.,* 1997; Litvak & Anderson, 2009).However, greater available C concentrations could also potentially stimulate soil organic matter decomposition. The shift in ecosystem partitioning also suggested that elevated CO_2 caused a greater increase in carbon cycling than in carbon storage in grasslands. Only a small portion of the labile carbon added to soil owing to their chemical nature such as lignin and cellulose might persist in the soil for many years (Parton *et al.,* 1987; Hungate *et al.,* 1997). The net carbon balance obtained in short-term CO_2-enrichment experiments tended to overestimate the potential for grasslands to sequester carbon in the soil (Sock *et al.,* 2005). Thus, increased carbon uptake did not indicate necessarily a large potential for carbon storage. The long-term responses of grasslands would be controlled to a large extent by soil processes.

Alpine meadow was the most sensitive region under climate change, and it would become the biggest potential carbon source as climate warming, increased CO_2 scenario and precipitation continued to occur (Li *et al.,* 2011). Therefore, soil organic carbon content in the alpine steppe would change due to climate change. In temperate perennial ryegrass swards, the below ground carbon storage increased at elevated $CO_{2,}$ but there was no effect of 3°C high temperature (Casella and Soussana, 1997). The combination of higher root yields at elevated CO_2 combined with a decrease in root decomposition will lead to a longer period sequestration of C in the soil and probably to a higher C storage. In the semi-arid grassland, additional below ground carbon input mainly entered the fast cycling carbon pool and contributed little to long-term carbon storage (Kandler *et al.,* 2006). In African smallholder land use systems, modest changes to the management in the semi-arid grassland had positive impacts on carbon sequestering in Ghana (Gonzalez-Estrada *et al.,* 2008). Globally, peat lands were considered to be a sink of CO_2, but a source of CO_2 when drained (Hendriks *et al.,* 2007). Moreover, wet peat lands were emitter of CH_4 and N_2O. Hitherto, reliable and integrated estimates of emissions and emission factors for this type of area had been lacking and the methane emissions had been poorly quantified. The conservative response of this grassland suggested that many of the reported effects could be short duration experimental artefacts that disappear under long-term elevated CO_2 conditions. More research is needed before the ability of soil to store additional C in a high CO_2 world could be determined.

The impacts of climate change on the relative availability of N and P remained highly uncertain. There was an evidence for progressively less N and P being available for plant growth at elevated CO_2 (Allard *et al.,* 2004a) and addition effects to soil N and P. Growth at elevated CO_2 reduced N concentration and the magnitude of changes were highly species specific (Ross *et al.,* 2000; Rillig *et al.,* 2000)). The soil system also could compensate for the limited N by increasing the labile pool with elevated atmospheric CO_2. Nevertheless, the dynamics of soil N cycling alongwith the seasonality of plant growth could result in a variable available soil N pool, so interactions of N with CO_2 might be important. Williams *et al.* (2006) supported the notion that C accrual might promote N accrual. Whereas warming (1.5/3.0°C) above ambient temperature during the day/night) reduced P availability relative to N. Soil CH_4 uptake was enhanced by elevated CO_2 but reduced by warming. Soil fluxes of N_2O remained unaffected. Further, Dijkstra *et al.* (2010) and Carrillo *et al.* (2012) found that soil N availability decreased with elevated CO_2 and increased with warming. In semi-arid grassland, Feike *et al.* (2012) found that elevated CO_2 increased P availability to plants. Therefore, comprehensive studies were required to be conducted to find the impacts of climate change on the cultivated forage and pasture including various soil and plant processes including grazing regimes, mutualistic relationships (*e.g.,* plant roots effects; N-fixing organisms *etc.*), and on the balance of carbon, essential nutrients, and water.

The responses of pasture and rangeland species to climate change were found to be complex, because, in addition to the major climatic drivers (CO_2 concentration, temperature, and precipitation), other plant and management factors would influence these responses (*e.g.,* community composition, plant competition, perennial growth habit, seasonal productivity, soil processes and management methods). Recent data indicated that grass vegetation could physiologically adjust to elevated atmospheric CO_2 concentrations as an adaptive response to the changing climate (Kohler *et al.,* 2010). One of the adaptations sometimes observed in plants exposed to elevated CO_2 increased partitioning of biomass to below ground organs (Rogers *et al.,* 1994, 1996), which had been observed in some of the studies (Morgan *et al.,* 1994b, 1998, 2001). The legumes would play an important role in the CO_2 responsiveness of grassland. These legumes would be 1) important in supporting CO_2-induced production responses and 2) more competitive in future CO_2 enriched environments.

On the balance, substantial risks could be anticipated due to the variability in production and productivity and quality of forage and volatility in supply chains from climate change. The impact of climate change had been found on productivity; its pattern; regrowth; growing season; the period of growth; forage quality; persistency; species diversity; changing reproduction behaviour; genetic erosion of forage grasses and legumes; weed flora; incursion of woody plants; soil processes; nutrients recycling and finally carbon sequestration into cultivated lands for forage production and grasslands and would also adversely impact peat meadow. These impacts would vary with the level of elevated CO_2, degree of warming and associated changes in rainfall pattern from one location to another but underpinning these impacts would have a number of direct effects on the physiology of cultivated forage crops and grasslands. However, the long-term trajectory of those responses on the whole was poorly understood. Long term global change studies, conducted in field environments, combined with modeling exercise would be required to unravel the complexities or how forage and grassland production would respond to increased CO_2 and climate change. Changes in grazing behaviour due to change in climate would have likely impact on foraging by ruminants. The consequences for animal production were even less understood. High day time air temperature currently would reduce total grazing time for cattle with little or no compensatory night time grazing. For domestic livestock, increased stocking rates might be recommended because of the reduced intake of lower quality forage which might further reduce animal gains. Dietary supplementation would be used to maintain current production levels, but that would be at the increased cost of production. Wild ruminant's diet quality might also be affected, and it was likely that they would have reduced growth and reproduction in them.

Livestock manure hitherto had been given little attention in cutting the GHG, although it had released significant amount of N_2O and CH_4 during storage. Proper storage of manure will emit less GHG.

Adaptation and Policy Issues

Uncertainty surrounding the possible effects of climate change on forage and livestock production demands to put ahead right policies and appropriate action plan in place to meet the challenges of the climate change. Projections of climate impacts on grassland and forage crops showed that there were likely opportunities for increased productivity as well as considerable threats to crop productivity in different parts of the world (Tim Wheeler and Chris Reynolds, 2013). Future food security will depend upon understanding the role of climate change on grassland and forage production and developing adaptive strategies to cope with these changes. Rangelands primary productivity impacts, species distribution and change due to CO_2 and other competitive factors, estimation of carrying capacities, mixed systems localized impacts on primary productivity, harvest indexes, stover production, effective ways to increase livestock water productivity, surface and groundwater supply, and impacts on livestock, needed to be understood in a holistic manner to formulate adaptation strategies. Projected scenarios of the climate change indicated that increased temperatures and CO_2 concentrations have the potential to increase herbage growth and to favour legumes more than grasses, but changes in seasonal precipitation would reduce these benefits particularly in areas with low summer rainfall (Hopkins and Del Prado, 2007). Further implications for grasslands might arise from increased frequency of droughts, storms and other extreme events. Grassland agriculture also contributed to GHG emissions, particularly methane and nitrous oxide, and management of grassland affected net carbon balances and carbon sequestration. Management options were needed to be identified for mitigating grassland's contribution to GHG emissions in a holistic manner including consideration of other pressures. Tropical farming systems were found to be highly complex, and usually involved a mixture of crops of widely differing tolerance to drought and temperature increases. If production levels of human food and livestock feed would decrease or even change relative to each other, the resultant dietary energy deficits had to be met from somewhere else (Jones and Thornton, 2003). Such considerations indicated that while single-enterprise impact assessment of climate change effects was a start, the interactions that existed in most tropical farming system were such that assessment had to be done at the level of the system. Without a systems-orientated assessment of household vulnerability, it would be hard to see how effective adaptation work could be appropriately targeted. From this perspective, the lack of work done on systems' impacts of climate change was a serious void. Adaptation strategies and investment on high quality research at the interface of crop and animal science would be needed. Both would respond to solve the problems arising due to climate change and meet the increasing demand for animal products expected over the coming decades (Tim Wheeler and Chris Reynolds, 2013). There is a great variety of possible adaptive responses available, including technological (such as more drought-tolerant crops), behavioral (such as changes in dietary choice), managerial (such as different farm management practices), and policy options such as planning regulations and infrastructural development. Adaptation options that include the following might prove beneficial for formulating strategy to meet climate change:

- ☆ Micro-level adaptation options, including farm production adjustments such as diversification and intensification of crop and livestock production; changing land use and supplementary irrigation at the appropriate stages of the crop development and altering the timing of operations.
- ☆ Technological developments, such as collecting, conserving and preserving the genetic variability and evaluation of genetic resources of local forage and pasture and other nonconventional plant resources as forage or forage constituents, adopting newly developed and adapted varieties, and promoting new crop varieties and livestock feeds, emphasis on water and soil management, and adoption of improved animal health technology would be a great help for adaptation under the climate change. Breeding tools had been successfully exploited by plants scientists for increased yield, developing stress tolerance, disease resistance and persistency traits in forage crops. On the same pattern efforts were in progress to alter the plant traits for improved ecological efficiency of grasses and forage crops. This approach would be of significance where livestock were largely reared on forage based diet like in India. If the differences in malate contents of forages are genetically governed then there are ample scopes to breed cultivars with high contents of organic acids which will reduce methane production. Alteration in plant composition for reduced methane production from ruminants had been validated through increased digestibility and sugar content of grasses and increased concentration of compounds that alter rumen protein degradation. Modification in plant composition could improve the rumen protein

efficiency through its diversion to assimilation by livestock for production and less for excretion through dung and urine. For example feeding forage with higher water soluble contents enhanced digestibility and reduced excretion in ryegrass. Breeding of forages particularly the legumes with moderate level of tannins would play a significant role in mitigation of greenhouse gases particularly the methane and nitrous oxide. *Lotus corniculatus* and *L. uliginosus* had higher condensed tannins and it could be used particularly in legume forage breeding to meet the dual objective of reduce greenhouse gases emission and bloat prevention. Feeding of red clover and *Lotus coniculatus* to house sheep exhibited reduction in methane production as compared to sheep on rye grass/white clover pasture. Genetic approaches could be also used to breed forage crops with higher nitrogen use efficiency. This would help to grow forage crops with less quantity of fertilizer input resulting in reduced nitrogenous emissions and would slash GHG emissions to increase sustainability of livestock rearing under the climate change. A wide variability had been recorded in the range of plant species for uptake, utilization and retention of nutrients. In many forage crops like maize, barely, *Arabidopsis thaliana* and ryegrass quantitative trait loci (QTL) had been identified by different workers. Preliminary research at IGER (2006) demonstrated that process of root and nodule senescence that contributed to losses of N is genetically controlled and amenable to selection. Genetic changes in forage plant (grasses and legumes) traits for higher N use efficiency in ruminants for reduced GHG emission could be also achieved through increased readily available carbohydrates (energy) and controlled breakdown of protein in rumen for higher microbial protein synthesis (FAO, 2007). Inductions of polyphenol oxidase (PPO) activity from rich red clover to other forage through genetic approaches were found the possible way to reduce the N loss during rumen fermentation. This enzyme helped in conversion of dietary phenols to quinines that would bind the dietary protein, resulting in subsequent reduced rate of protein degradation. Recently a brachialactone chemical compound found in the roots of African grass *Brachairia humidicola,* prevented release of nitrous oxide to the atmosphere. This chemical enhanced nitrogen fixation through microbes and inhibited the biological nitrification.

- Application of feed and forage processing techniques improves the feed efficiency, which in turn results in higher propionate production and reduced methane production in rumen. For example boiling of barely, oat and bajra grains favoured decline in methane production (Tewatia *et al.,* 1997). The decline in methane production due to processing of feeds might be due to gelatinization of starch in cereals thereby altering the proportion of the final end products of fermentation. Reduction in methane production from feeding of alkali treated straws or urea/ ammoniated low quality forages had also been reported in ruminant.

- Defaunation through forage tannins is also possible. Methanogens associated with *ciliate* protozoa accounted 9 and 25 per cent of the methanogens present in the rumen fluid. The hydrogen forming entodiniomorphid protozoa had close physical associations with the methanogenic bacteria. Saponins containing plants had been found to be dafaunating agent. There was decrease in protozoa numbers due to feeding of pure saponins and saponins containing plants. Use of high tanniferous cultivars of forage crops *viz.* trefoil, sulla, vetch, chicory, lotus, *Sericea lespedeza* (*Lespedeza cuneata*) had been found effective in methane reduction when fed to ruminant animals. Addition of forage as propionate enhancers such as organic acids, the intermediates of carbohydrates degradation in rumen stimulated propionate production and reduced methane losses by action as H_2 sink (Newbold *et al.,* 2002). Organic acids *viz.* malate, fumarate, citrate, succinate *etc.* were propionate precursors and their addition in diet reduced methane production. Their use as dietary supplement might be limited due to their cost but they could be found in substantial quantity in forages as intermediate to citric acid cycle. Thus, screening of forage germplasms, utilization of phenols and tannins and other anti-nutritional agents from various plant and chemical resources as potential methane inhibitor or manipulator of rumen fermentation would be helpful. Plant extracts from unconventional sources for both defaunation and anti-methanogenesis properties and comparing digestion characteristics of various forage component and nonconventional feed and forages to develop total mixed ration incorporating properties for reduced methane release and higher livestock production need to be investigated.
- The negative impact of ruminants on greenhouse gases emissions could also be addressed through changes in animal husbandry practices including ruminant's diets and animal stocking ratios to avoid nitrous oxides emissions.
- Herbage quality is influenced by water scarcity and increased CO_2 concentrations (Soussana and Luscher, 2007). Water scarcity leads to quicker plant maturation, which in grassland means lower dry matter digestibility than in physiologically younger swards (Wilson & Ng, 1975). This could be alleviated by earlier cutting. Increasing CO_2 concentrations had been shown to increase nitrogen limitation of grass production (Luscher *et al.,* 2005). This would be beneficial for the competitiveness and growth of legumes in grass-legume mixtures. In pure grass stands, nitrogen fertilizer additions would be more effective in terms of productivity. At the same time, increased temperatures might accelerate soil processes such as mineralization leading to a more rapid supply of nutrients from decomposing organic material if sufficient water is made available, but potentially also causing larger leaching losses of mobile

nutrients (Ineson *et al.*, 1998a, 1998b; Niklaus *et al.*, 2006; Sardans *et al.*, 2006). Increased CO_2 atmospheric concentrations together with nitrogen fertilization had been found to accelerate flowering in forbs, but delay in that of grasses, decreasing phenological complementarity (Cleland *et al.*, 2006).

- ☆ First of all, the use of legumes is one option to supply the swards with sufficient nitrogen if phosphorus is available. It also suggested that legumes might be more competitive in grasslands at elevated CO_2. Introduction of legumes into grasslands might become more important as forage quality declines. N fertilization and the introduction of legumes might be economically viable means by which to respond to rising atmospheric CO_2. In range lands, N fertilization was not economically feasible, and the inter-seeding of legumes is difficult at best with today's technology. The responses of native legumes to elevated CO_2 might be an important factor in how these grasslands would evolve to support grazing by both domesticated and wild animals.
- ☆ Different options are available for dealing with the effects of climate change on grassland production. The botanical composition of swards should be adapted to changing conditions. Deep-rooting plants such as chicory can reach the water that was otherwise unavailable. A combination of species in a diverse sward had been suggested to improve the water-use efficiency of grassland (Caldeira *et al.*, 2001; Chen *et al.*, 2007).
- ☆ Forbs were found to be a small fraction of phytomass in most of the grasslands. Some studies indicated significant growth responses of forbs at elevated $CO_{2,}$ however in the short grass steppe; there was limited proof of a strong forb response to elevated CO_2. But their responsiveness to elevated CO_2 suggested the possibility that they might become more important in future grasslands at elevated CO_2.
- ☆ Adaptation of crop selection and management could counter some negative impacts of climate change, or even turn negative impacts into local opportunities where these were expected (Osborne *et al.*, 2012). Adaptation would certainly produce new areas of potential production. This would require the combined expertise of climate, crop and animal scientists working together to reduce the risks of climate change on grasslands and forage crops. As the animal productivity in India relies heavily on forage crops and other crops residues, it is of utmost urgent to strengthen forage research and feeding management strategies to cope up or abate the further rise in emission of GHG for sustenance of crop-animal production system.
- ☆ Livestock manure hitherto had been given little attention in cutting the GHG, although it had released significant amount of N_2O and CH_4 during

storage. Scientific management and proper storage techniques of manure would emit less GHG in future.

- ☆ Quantitative determination of soil carbon in grasslands on global scale suggested that mitigation of greenhouse emission needed to be handled differently by managing grazing through location specific approach (McSherry and Ritchie, 2013). Grazing not only affect soil carbon storage but grazer had direct bearing on soil moisture and soil nitrogen, needed deeper knowledge to formulate theoretical basis for sustainable soil carbon and grassland management in the world.

Understanding of complex impacts of climate change on grasslands and forages and other feed crops across the body of evidences generated by various experimentation and simulation was found challenging. However, some key aspects of the responses that could be inputs for policy framing are documented. Ruminant livestock (cattle, buffalo, goats and sheep) from forage growing on non-arable land would be more important for food security. There would be considerable spatial variation in the impacts. The expected impacts of climate change would alter over time and vary from one part of the geographical location to another. Uncertainties arise from greenhouse gas emission scenario and from the climate and crop models that were used (Challinor *et al.,* 2009a, b). In spite of systematically varying parameters within climate or crop models, considerable uncertainties to projections would remain. Never the less numerical methods could better define the boundaries of uncertainty (Osborne *et al.,* 2012). Thus, a risk based approach to planning for impacts of climate change would be needed. The impact of climate variability would become increasingly important source of volatility in productivity and quality of forage plants in general, cultivated forage or in grasslands. There would be substantial effects of non-climate factors that could outweigh any climate change signal on forage availability. These might be driven by policy (for example land use and food policies; feedstuff *viz.* edible roughage diversion for power generation and packaging), changing trend in meat consumption in India, competitive return from forage versus food crops. Land use and economic pressures might force livestock production to either remain more reliant on less digestible feeds (crop residues) and less scope to expand area under forages. It might also compel poultry and pig production to be less reliant on grains. Some forage calendar could be adopted specific to the regions. This forage calendar could be complemented by adding minerals (Calcium, Phosphorus and oligo-elements), vitamins A *etc.* Successful conservation strategies must consider various current and future competing demands for the land. The proportion of browse in rangelands might also increase in the future. This will have significant impacts on the types of livestock that could graze these rangelands and might alter the dietary patterns of the community's dependent on them. In sown mixed pastures, elevated CO_2 increased legume component, and this also occurred in temperate semi-natural grasslands. These observations should be sincerely extrapolated and considered for policy formulation.

Epilogue

There is still a great deal that is not well understood concerning the interactions of climate and increasing climate variability with other drivers of change in forage production for livestock systems for and in broader development trends. Multiple and competing pressures are likely on tropical and subtropical livestock systems in the future, to produce food, to feed livestock, and to produce energy crops (Thornton *et al.,* 2009). Livestock and livestock systems are substantial users of natural resources and globally they contribute significantly to global warming, while at the same time they make contributions of critical importance to the livelihoods of at least a billion poor people in rural households, almost all of whom are in developing countries. While recent scientific assessment such as the AR4 represented an accurate reflection of the current state of knowledge but it had been found to represent an unacceptable situation (IPCC, 2007). If future, scientific assessments are to do justice to the topic of livestock keepers in developing countries, then a great deal of urgent work is needed that addresses several broad issues. First, much more clarity is needed concerning the benefits of livestock owners; the negative impacts they can have on greenhouse-gas emissions and the regional and local changes in public goods and disadvantage in terms of benefits associated with livestock before targeting appropriately various technology and policy options for adaptation and mitigation. Lobell *et al.* (2008) indicated that South Asia and Southern Africa were two regions that, without sufficient adaptation measures, would likely to suffer negative impacts on several crops that were important to large food-insecure human populations. Uncertainties varied widely by crop. Second, much more information was needed concerning the nature and extent of the trade-offs possible between different crop and livestock enterprises, and between on and off-farm income sources, in different situations. Key to these broad issues would be the further development and refinement of impact assessment frameworks that could evaluate potential and implemented adaptation and mitigation options at regional and local scales, their effects on livelihoods, and the trade-offs that arise between income, food security and environmental objectives.

Chapter 6

Natural Calamities: Flood and Drought

☆ *K.A. Singh & Anil Kumar*

Natural calamities management endeavour requires basic data on various loss estimates (ground based and remote sensing based observations) based on defined criteria and indicators which may lead to preparedness, monitoring mechanism, early warning systems, contingency plans, mitigation strategies *viz.* loss estimation, immediate and long term measures, mobilizing funds, administrative relief, improved technologies, livelihood sustenance programmes, crop and livestock insurance. It is necessary that there is a shift in public policy from calamities management to mitigation measures. Mitigation implies actions that can be taken before or at the beginning of calamities to help prevention of occurrence of incidence or its impacts. At the national level, there cannot be a standard strategy for calamities mitigation but they need region, intensity and frequency specific mitigation strategies on a time scale *i.e.* short term and long term.

Cattle wealth is the mainstay of the rural economy. Small and marginal farmers constitute about 80 per cent of total farming communities in the country. Their only asset is cattle apart from their small landholdings. It has been found that livestock are paid less attention during natural calamities like drought and flood; hence, they are more vulnerable than the human beings. In such situation, safety to the life of livestock and providing adequate feed, forage and drinking water are the major concern.

I. Flood, Forage and Livestock in India

Rivers act as a lifeline of civilizations. It nourishes the field with periodic silt laden flood waters. People from ancient times have learnt to live with the phenomenon of flood. The normal flood helps crops by spreading fresh silt during rainy season and the moisture content of the soil would be enough to meet the water requirement of the crops in *rabi* season. In the years of abnormal flood, any damage to the monsoon crop would be suitably compensated by a rich harvest of the winter crops. There are two kinds of floods: flash floods (where the capacity of the ground to hold water is exceeded); and riverine floods (where a body of water comes to burst its banks). It's a close call, but riverine floods tend to be the most dangerous. In spite of the periodic flooding, people still found it advantageous to remain in the flood plain zone by protecting themselves from floods by moving to mounds with their cattle and essential belongings, as the river water rose. There has been increased settlement in the flood plains because of construction of a series of embankments giving a sense of protection from floods to the people. Increasing human population and expansion of crop lands also lured people to settle in the flood prone areas. However, the problem of flood has aggravated during the last half century, inundating more land and causing an increasing loss of property and life of human and livestock. The floods have now become more vicious and unpredictable unlike the old days where people had the knowledge of onset and receding of flood waters and they made adjustments accordingly. Sudden breaches of bunds and embankments, an element of surprise has overtaken making people unawares, thus, making the phenomenon of flood as disaster. The magnitude of flood disaster is increasing year after year. During 1975-2010, more than six disastrous floods had occurred in the country (Feasibility study for preparation of Digital Vulnerability, GOI). The impact of flooding is devastating and varied from one location to another. The occurrence of flood creates chaos all around. In the habitation that surrounds flood, people first look for the near and dear ones and the valuables and rush to safety at an elevated place such as highway or railway lines. They also take their animals with them where they spend days together until the flood water recede. But the miseries of livestock are overlooked by all except the people who live with these animals surrounded by water all around. They want forage for their starving animals but it could not be found anywhere. Cattle camps are organized by government agencies at places, but they are so few and far away that they hardly bring any relief to the marooned animals. To deal with various kinds of disasters occurring in the country, National Disaster Management Authority (NDMA) was constituted in the year 2002. NDMA has already inter-alia released guidelines on management of earthquakes, floods, landslides, cyclones and preparation of State Disaster Management Plans but effective plan for livestock in the flood area is still lacking.

More than 45 million hectare of land in the country is prone to floods according to the National Flood Commission. On an average 18.6 million hectare land and 85,599 number of livestock is affected annually (NRSA, 2007). The four major flood zones in the country are:

I. **The Brahmaputra and Barak basins** that comprise the States of Assam, Arunanchal Pradesh, Meghalaya, Mizoram, northern parts of West Bengal, Manipur, Sikkim, Tripura, and Nagaland.

II. **The Ganga basin**, served by numerous tributaries such as Yamuna, Sone, Ghaghra, Gandak, Kosi, and Mahananda, that comprises the States of Uttaranchal, Uttar Pradesh, North Bihar, and South and Central parts of Haryana.

III. **The basins of the northwest rivers** such as the Sutlej, the Ravi, the Beas, the Jhelum, and the Ghaggar that cover the States of Jammu and Kashmir, Punjab, and parts of Haryana and Rajasthan.

IV. **The basins of Central Indian and Deccan rivers** that include Narmada, Tapi, Mahanadi, Godavari, Krishna and Cauvery, and stretch across the entire central and southern India. Of these, the State of Orissa is the most flood-prone. The flood map of India and five major flood prone states are shown in Figures 6.1 and 6.2a-e.

The various districts of five major states suffering from flood have been listed in Table 6.1. The livestock inhabited in major flood prone area, milk production and annual dry matter requirement has been estimated in Table 6.2 and Table 6.3.

Table 6.1: List of Districts Identified as Flood Prone Districts by the Government Flood

Name of States	*No. of Districts*	*Name of Flood Prone Districts*
Assam	09	Barpeta, Dhemaji, Dhubri, Goalpara, Jorhat, Lakhimpur, Marigaon, Nalbari, Sibsagar
Bihar	15	Araria, Bhagalpur, Darbhanga, Purbi Champaran, Gopalganj, Katihar, Madhubani, Muzaffarpur, Purnia, Saharsa, Samastipur, Sheohar, Sitamarhi, Supaul, Vaishali
Orissa	06	Bhadrak, Cuttack, Jagatsinghpur, Jajpur, Kendrapara, Puri
Uttar Pradesh	06	Ballia, Basti, Farrukhabad, Gorakhpur, Mirzapur, Siddharthnagar
West Bengal	03	Birbhum, Murshidabad, Nadia

Table 6.2: Livestock in the Flood Prone States Actually Affected by Flood Every Year and Milk Production

State	*Total Livestock*	*Livestock Density (Number per Sq. km)*	*Annual Dry Matter Requirement (tonnes)*	*Milk Production ('000 tonnes)*
Assam	6534779	176	7817	354
Bihar	19845843	288	26779	2323
Orissa	5017554	150	4922	214
Uttar Pradesh	17831764	243	2466	4857
West Bengal	17660129	469	73	1564

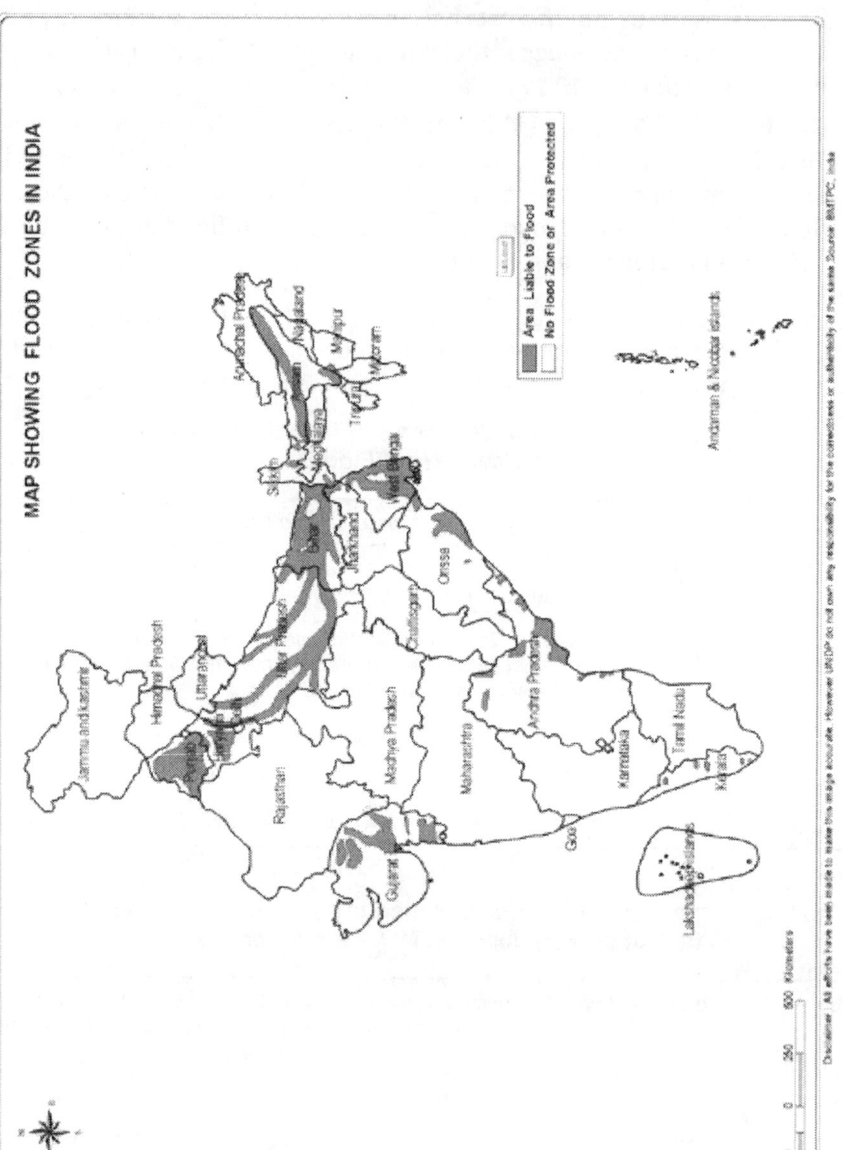

Figure 6.1: Flood Prone Areas in India (National Disaster Management Authority).

***Source*: "Vulnerability Atlas of India" Published by Building Materials and Technology Promotion Council, Ministry of Housing and Urban Poverty Alleviation, Government of India.**

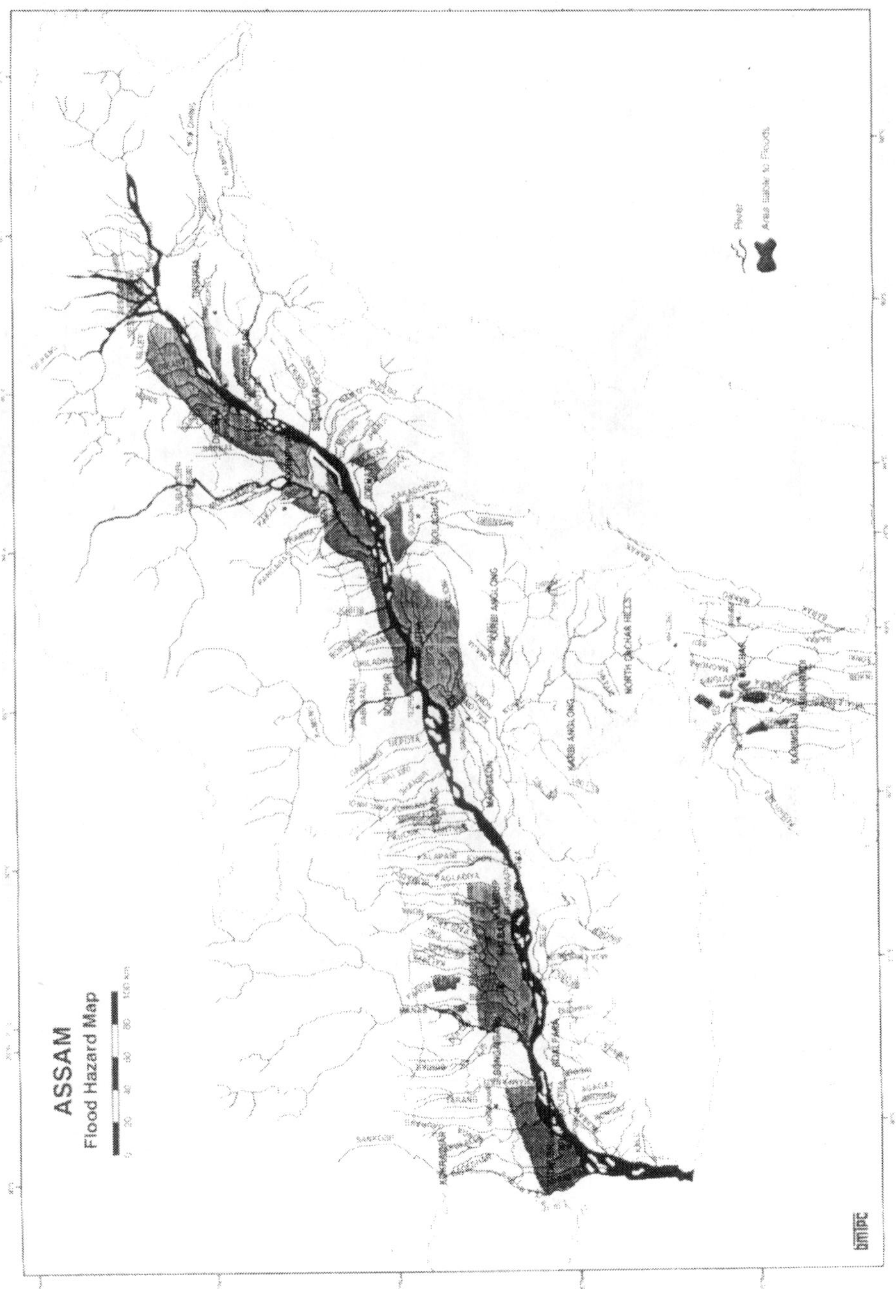

Figure 6.2a: Flood Prone Areas of Assam (UNDP).

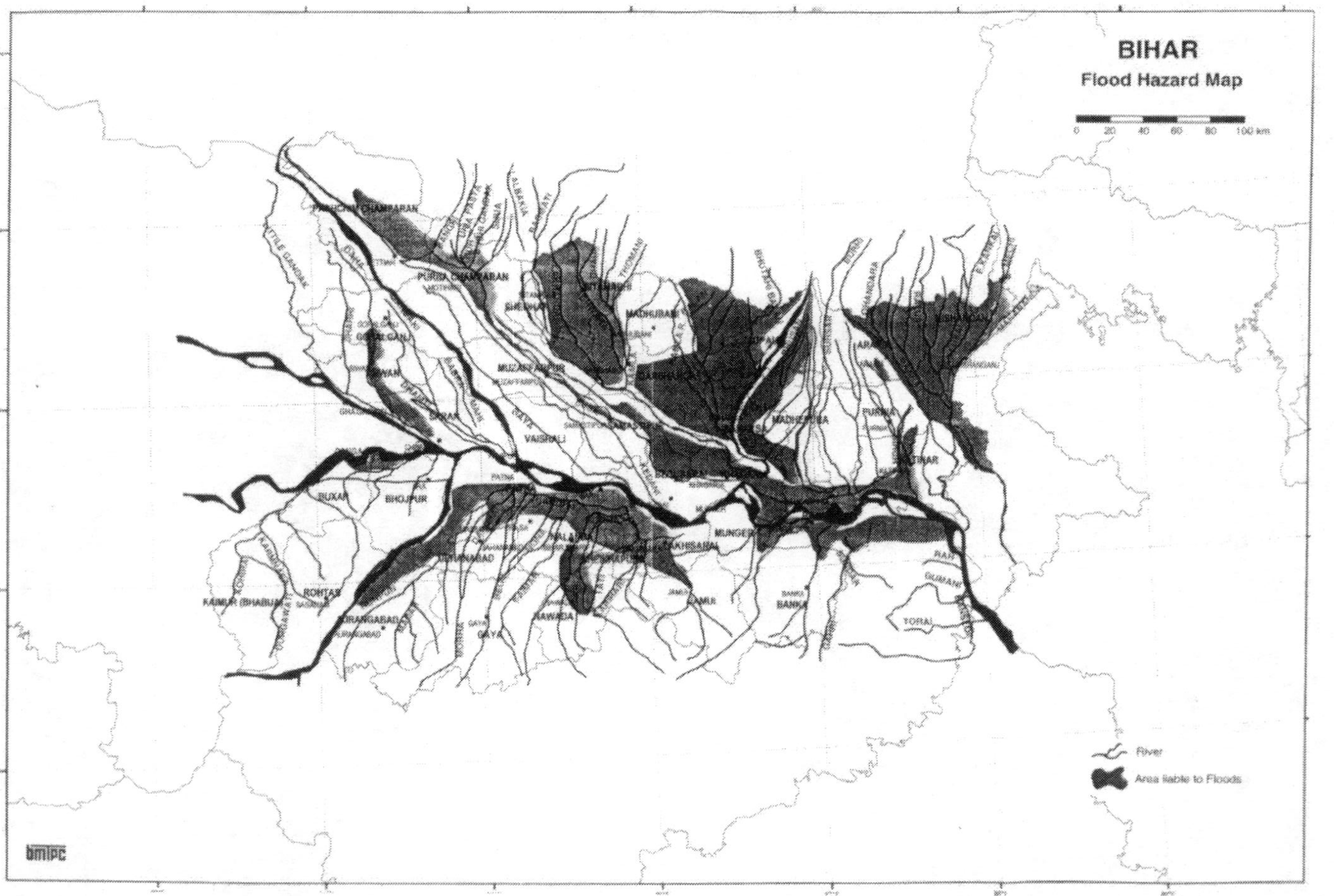

Figure 6.2b: Flood Prone Areas of Bihar (UNDP).

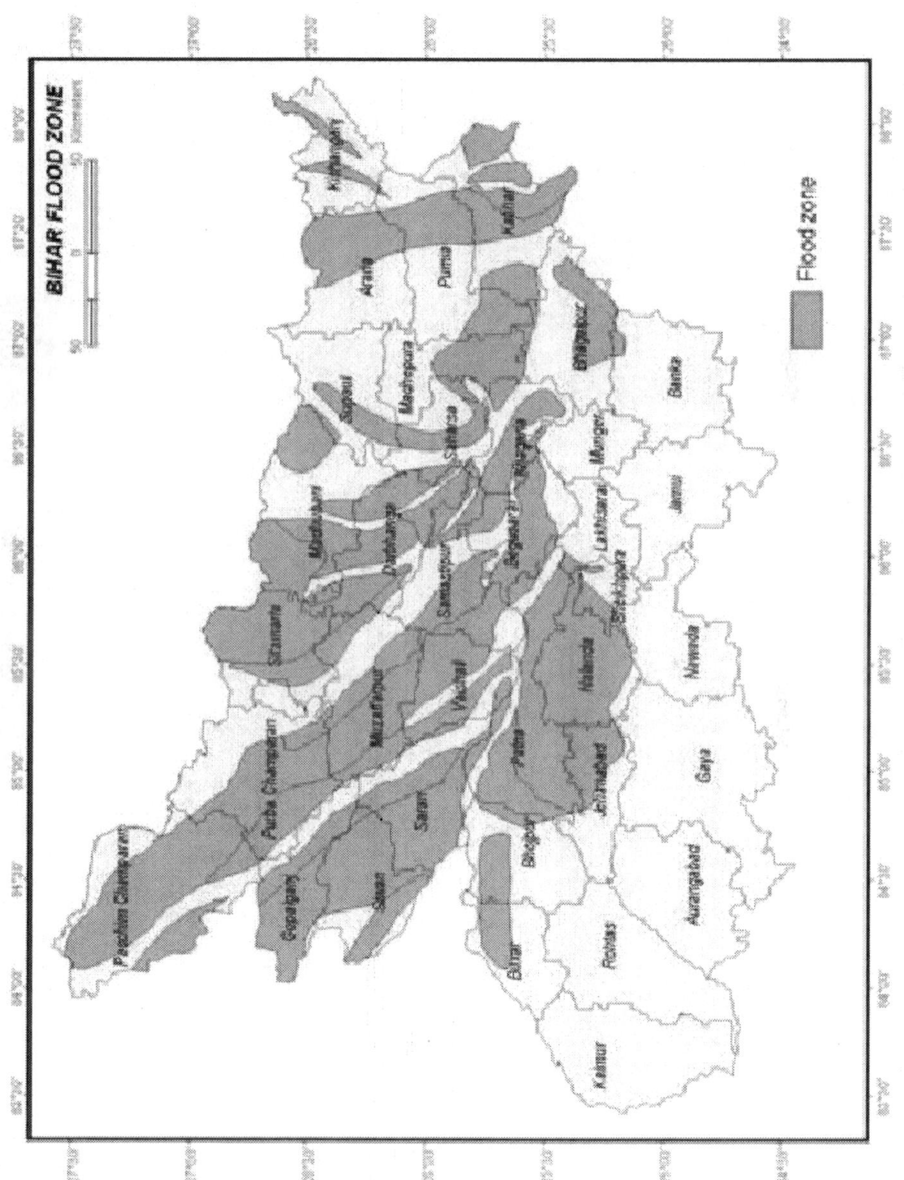

Figure 6.2c: Bihar Flood Zones.

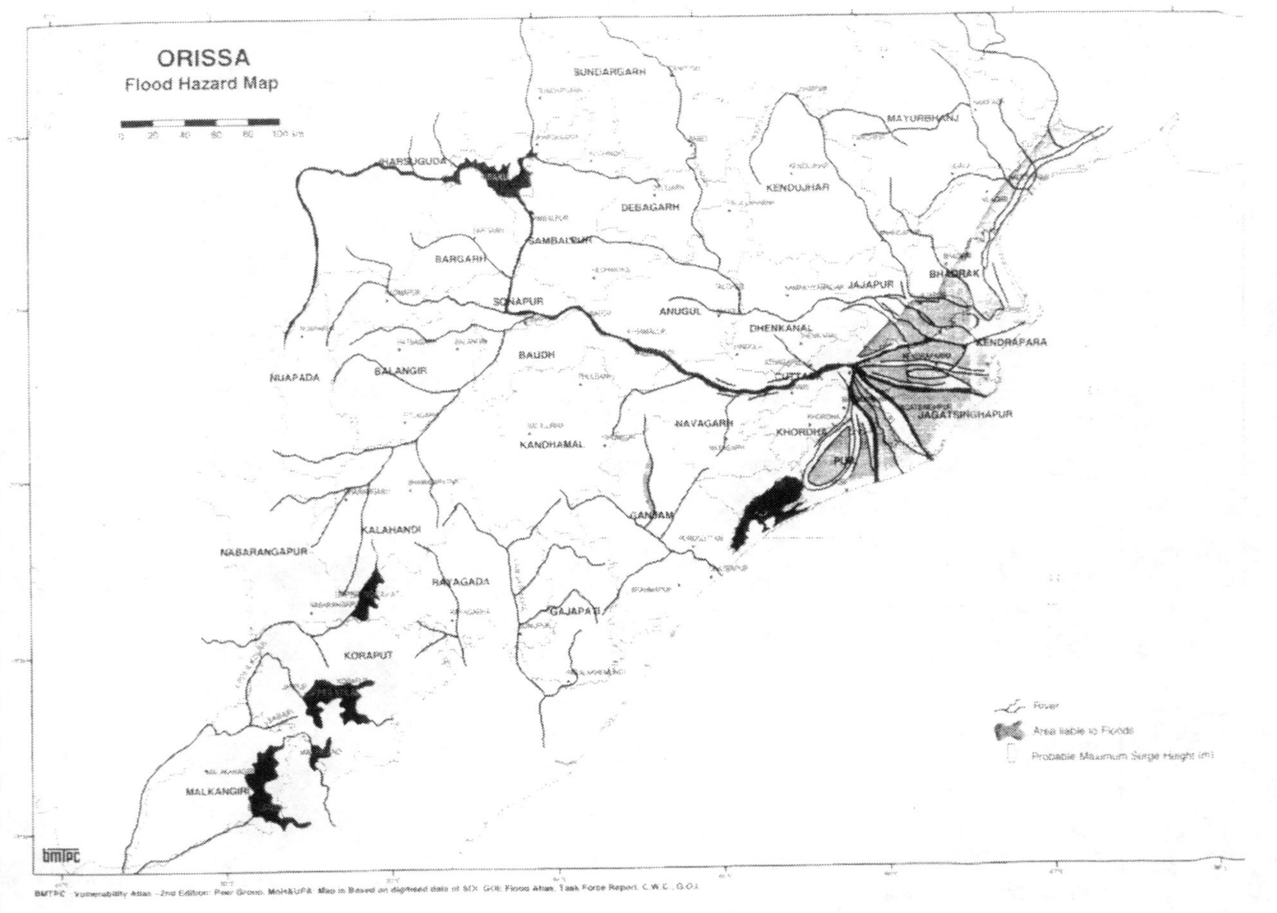

Figure 6.2d: Flood Prone Areas of Orissa (UNDP).

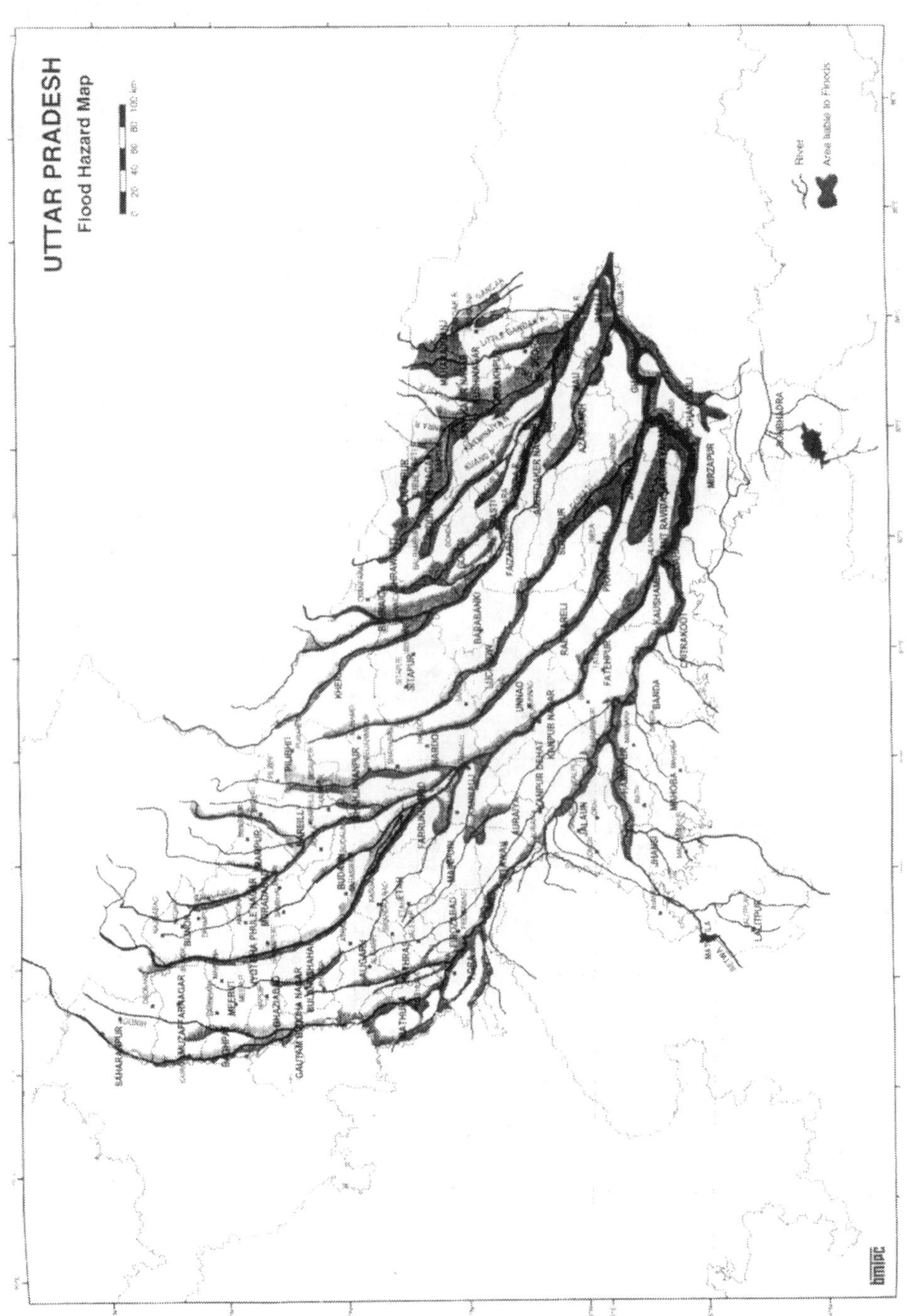

Figure 6.2e: Flood Prone Areas of Uttar Pradesh (UNDP).

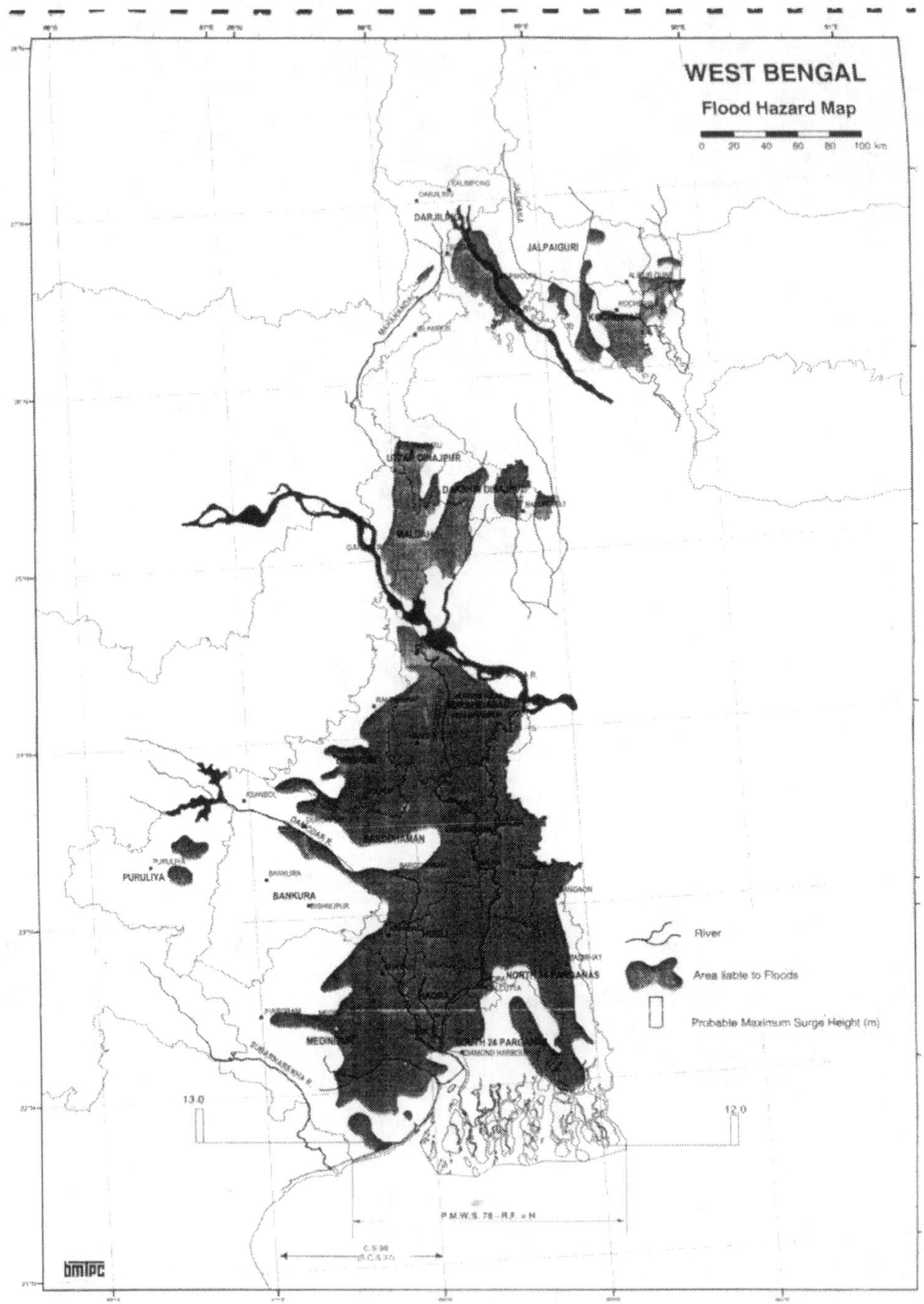

Figure 6.2f: Flood Prone Areas of West Bengal (UNDP).

Table 6.3: Dry Matter Requirement for 3 Months in Flood Prone Areas

Sl.No.	*Flood Prone States*	*Dry Matter Req. 3 months (thousand tonnes)*	*Number of Godown Required*
1	Assam	1954	391
2	Bihar	6695	1339
3	Orrissa	1639	328
4	Uttar Pradesh	7604	1521
5	West Bengal	4215	843
	Total	22106	4421

Size of Godown: 10000 cu m; Density of baled fodder: 500kg/cu m.

Climate Change and Flood

At the global scale, the warming of the atmosphere will increase the capacity of the atmosphere to hold and accelerate the redistribution of water in the atmosphere (Whitfield, 2012). This indicates that flood-generating processes linked to the atmosphere are to increase with all probability. On the other hand, the Intergovernmental Panel on Climate Change projections of future floods involves extremely complex issues that defy simple generalizations. Warming will affect many aspects of the water cycle increasing evaporation, changing precipitation patterns and intensity, and also influencing the processes involved in surface storage of water, including snowpack generation, snowmelt, river ice break-up, and glacial melt. Many of these are active in flood generation, and changes may result in floods to decrease as well as increase. In the possibility, these processes take place not at the global scale but at relatively local scale. It makes generalizations about flooding in future climates difficult and uncertain. At the global scale, the role of land use is generally unresolved, but at a watershed scale, land-use effects can be as important as changes in the meteorological processes. Besides, climate change is expected to have adverse impacts upon sea levels and rainfall intensities, both of which may have significant influence on the flood behaviour at the specific locations. The perceived impact of climate change on the livestock sector has been estimated in percentage. The impact will be 79.1 per cent on feed and forage resources, 95 per cent on shortage of dry forage and decreased water resources.

Flood and Livestock

Livestock play an important role in the livelihood of people in flood prone areas because most of these areas are under-developed and agrarian in character Livestock population is the first to be affected in the precarious situations due to flood. The miseries of livestock are overlooked by all except the people who live with these animals surrounded by water all around. Every year in the flood prone area, 1,02,50,351 number of livestock out of total 6,69,21,044 livestock including 14,60,882 number milch animals out of total 95,37,601 of milch animals are adversely affected. They want forage for their starving animals but it could not be found anywhere. Cattle camps are organized by government agencies at places, but they are so few and far away that they hardly bring any relief to the marooned animals. At that time, large scale transportation of feed and forage becomes more difficult to the affected areas. In that period, main target of rescue, relief and rehabilitation is restricted for the people of affected area. Little or meagre help is extended towards the livestock population. Most of the government and nongovernment organizations generally remain busy for the welfare of human beings. As a matter of fact, due to this negligence, a very difficult situation gradually arises which makes it incapable of controlling the losses due to massive spread of epidemic diseases, particularly during the post disaster period. Most of the studies on flood related problems have overlooked the miseries of the marooned livestock in the flood affected area. In the pandemonium that follows flood, the poor livestock gets the least attention. People would rush away with one or two goats that they own leaving behind many. The chain of Food Corporation of India has helped in minimizing the loss of human life due to starvation but can we vouch that we have taken well care of our livestock during flood. Is there a comprehensive policy in place to bring succour to livestock? Has a fair assessment been made to assess the actual number of livestock lost during flood and the resultant loss in productivity? Do we have an estimate of the number of livestock that get affected? With the increasing dependency of population on livestock for their income and employment, people rue the misery their livestock are subjected during floods. During recent floods in Bihar, the people looted a train carrying forage and found transporting green forge in passenger trains. Bihar Chief Minister sought help from his counterpart in Punjab not for grain for people but forage for livestock.

Therefore, it is high time we take stock of the situation at all levels to define the magnitude of the problem and adopt mitigating strategies. Restoring livelihood of people after a bout of natural calamity is the quickest way to rehabilitate them. Initiating agricultural activities may take some time, but livestock can provide succour right in the midst of such misfortune, provided prior planning and coordination of various stakeholders are in place. However, studies on livestock production system in these niches are lacking elucidating their problems and opportunities. The estimate of flood prone area in the country varies between the different agencies. The total livestock in this area was estimated to be 66.9 million heads. The feed and forage requirement of the livestock in the flood prone areas was worked out to be 92million tonnes. However, the actual area inundated by flood every year is only 6.9 million ha, the actual number of livestock affected by floods is 10.25 million heads requiring 14.1million tonnes of feed and forage. Further, as the flood water remains for a period of four months on an average, the actual requirement of feed and forage for four months is 4.7 million tonnes. Therefore, there is a need to make contingency plans to feed about 10 million heads of livestock affected by floods every year requiring 4.7 million tonnes of feed and forage for a period of three to four months. This calls for creating infrastructure like flood shelters, storage for keeping feed and forage *etc.* to minimize the adverse impact of flood on the livestock.

Disaster Management Plan for Livestock in Flood Prone Area

National Disaster Management Authority (NDMA) has been vested with the responsibility for laying down the policies, plans and guidelines for disaster management. NDMA has already inter-alia released guidelines on management of earthquakes, floods, landslides, cyclones and preparation of State Disaster Management Plans. Flood Management Programme (FMP) amounting to Rs.8000 crores was approved by the cabinet for XI Plan.

Understanding the Problem!

Livestock interventions should always be based on a thorough understanding of the affected system through analysis and identification of vulnerable population. The vulnerability of livestock owners to emergency situations will vary according to their lifestyle, *i.e.* whether this is pastoral (nomadic or semi-nomadic) or incorporates crop production as part of a mixed-farming system. Clearly, conditions change significantly across regions and communities. Feed and forage availability and requirement need to be worked out based on database of different categories of livestock, their production levels, after the flood prone areas have been delineated by the experts.

Consequences

Livestock are unable to find adequate forage and grow weak and die from malnutrition or disease. Availability of supplementary grain and forage on the local market decreases (Photo 6.3). As a result, livestock prices drop too low – and the price of grain climbs too high. Therefore, for pastoralists to earn sufficient profit become impossible.

Photo 6.3: Indigenous Way of Wheat Straw (Bhusa) Storing for Livestock on Highlands (*Source*: Anil Kumar).

What to Do?

Emergency destocking programmes allow for the removal of animals from a region before they die. These programmes involve buying farmers' livestock at a fair price (or offering healthy animals or goods in exchange). It will relieve pressure on natural resources to the benefit of the remaining livestock, livestock numbers are reduced, leaving more grazing or other forage and feed resources for the other breeding animals. It will also provide source of food for crisis-affected families. Destocking is a method of transferring income to maintain the purchasing power of the livestock owner; a transfer of grain instead of cash can be made in exchange for animals, but this removes the power of choice from the livestock owner. Use of multi-nutrient feed blocks (*i.e.* urea molasses) specially formulated to provide energy, nitrogen, and important vitamins and minerals to enable animals to survive until the flood conditions improve and feed and forage availability increase in the area. Most importantly, it reduces environmental degradation by bringing cattle feed from an area outside the normal foraging area and as result provides herders long-term benefits. In most cases, feed security is more important than food security, since keeping animals alive ensures that their families will be able to survive beyond the drought or disaster.

A disaster management plan for livestock in flood prone areas can be divided into three phases: a) Flood preparedness (before flood), b) Relief and rescue during flood period and c) Post flood measures. A comprehensive plan to enhance forage production should be implemented. For various conditions, suitable cropping sequences and forage plant species could be identified. Embankments along water bodies offer a good place to develop the forage resources. There are about 20,511 km of embankments which can be planted with perennial grasses and tree fodder as top feed like *Ficus* species, jack fruit, bamboo plant species *etc.* providing an *in situ* fodder bank.

There is an urgent need for creation of fodder bank in the zone of vulnerability. Suitable sites in the periphery of flood prone area should be identified for creation of fodder bank, from where the supply will be rushed to the affected area.

Individual household level feed storage on the roof of the buildings will play a vital role. The establishment of Fodder Bank on the lines of Food Corporation of India will require many issues to be looked into:

- ☆ The state farms dedicated for the purpose of growing and conserving forage.
- ☆ Bailing of dried forage and their storage.
- ☆ The use of Feed-Block Making Machine: The complete feed block made by Feed Block making machine can produce feed block for different categories of livestock taking into account the varying production levels. Mobile Feed-Block Making Machine can be run as a commercial venture

in normal times, besides coming to help for creation and management of forage reserves in the form of bales/complete feed blocks.

- The cycling of stored forage: The stored forage in the form of bales may be rotated on a three-year cycle, so that a strategic reserve remains in place simultaneously using the older stock gainfully.

IGFRI has done considerable work on bailing, densification, leaf meal, feed pellets and complete feed blocks to suit the requirements of animals with different production levels and developed forage production technologies for various categories of land. For various conditions, suitable cropping sequences and forage plant species could be identified. It has the requisite technologies to assist the central and state agencies in formulating an action plan. The proposed setting up of a National Flood Management Institute should draw experts from livestock and forage background to seek the welfare of livestock which support people during trying times. A comprehensive plan to enhance forage production should be implemented.

Measures to Mitigate Flood Situation

Intermediate Measures

- Sourcing locally available feed and forage in the vicinity of the affected region.
- Use of Urea-Molasses mineral blocks
- Providing baled straw and roughages to overcome hunger of the livestock
- Use of complete-Feed-Block. The complete feed blocks may be made to cater to the needs of different production levels of animals.
- Use of vegetable waste/fruit industry waste *etc.* can provide immediate relief to the affected animals.
- Use of aquatic plants like azolla, duckweed, laminaria *etc.*, which are fast growing under waterlogged conditions, can be grown and fed to animals.

Long Term Measures

- Mapping livestock in the flood prone areas and actual number of different categories of livestock affected every year.
- Working out feed and forage requirement for these animals based on their production levels.
- Establishment of fodder bank in a radius of 200-250 km where flood occurs frequently, so that relief may be quickly provided to the affected livestock. The fodder bank should ideally be located along major highway and railway line so that it could be quickly transported to the affected regions.

The Framework and Response Activity

Flood Preparedness

- ✰ In order to begin the preparedness, there is a need to form the community organization to look after livestock.
- ✰ Organizing information on vulnerable livestock, their dissemination and training for capacity building in livestock care.
- ✰ Planning interface with Government departments particularly Animal husbandry and Veterinary Services.
- ✰ Making provisions for emergency situations to transport livestock.
- ✰ Managing information on livestock, feed and forage requirement and health services for future references. Flood relief activities of the past are, in most cases, criticized. The situation can be better managed if information related to what had been done with what success and failures, collected from previous floods are available to currently concerned people and agencies involved in relief and rehabilitation work. In this way it can be ensured that a community-specific knowledge base is gradually being built and maintained by the local people.

Resource Mobilisation

For this purpose, a community based pooled feed and forage resources may be created. In view of their poverty, the flood affected people may be expected to contribute a small percentage only, say 20 per cent of the pool in cash or in kind depending upon the financial ability of the livestock owners while a major portion should come from state/central government or any NGO active in the field of livestock related activities.

Response to Floods

It is envisaged that these individual actions, if coordinated at the community level and if the community capacity is strengthened through such activities as awareness building, training, and networking, can generate an effective grassroots-based flood management approach for livestock in this area. Key elements of this approach are needed to be identified. Pre-Flood Responses should be quickly assessed, identify the needs and prepare permanent/temporary community cattle shelters on such high lands as those used for weekly village market as well as inner-banks of embankments/highways, feed and forage, veterinary services and quickly prepare *Synthesis Manuals*, plans for evacuation of livestock to a safer place, reallocation within the flooded area looking into accessibility, location, facilities, and capacity *etc.*, and transport ability to bring resources required. Maintain boat/raft transportation facilities. Based on assessment, provision for relief for the most vulnerable livestock particularly arrangement for feed and forage, safe drinking water, sanitation, other requisites be made and day-to-day activities managed in running the flood shelter/camp for livestock.

Post Flood Rehabilitation

Flood affected livestock are keen to get back to normal health and nutrition. After suffering, need to be rehabilitated which in turn need external assistance and help. There is a need to prepare an inventory of losses/damages suffered by each livestock owner based on a questionnaire survey. Prioritizing rehabilitation activities concerning basic infrastructure like shelter, vaccination, arrangement for feed and forage, optimizing livestock number by culling unproductive livestock are required

(a) The animals should be provided with clean water *ad libitum* near their housing place.

(b) In the calamities, there are acute shortage of feed, forage and drinking water for livestock. Saving of animal's life is considered on priority basis. The Feeding strategies can be developed with objectives (i) to feed animals for maintenance that ensures survival of animals. (ii) to feed productive stock, such as pregnant and lactating cows preferentially. In such conditions, livestock are to be fed with the locally available industrial wastes, different tree leaves or improving the coarse roughage which will be able to support the life of the animals. It is useful to produce complete feed for use during flood period and beyond. Biologically the use of complete feeds with an appropriate balance of roughage and concentrates may lead to better utilization of locally available crop residues, agricultural by-products and agro-wastes.

(c) Animals maintained on malnutrition condition for prolonged period are supposed to suffer from different diseases and immunity of animals goes down. So immediately, when the scarcity period is over, animals should be adequately fed taking care of supplementation of different minerals and vitamins. The complete feed blocks may be made to cater to the needs of different production levels of animals.

(d) It has been observed that the livestock populations are also affected during post disaster period due to rapid spread of different epidemic diseases. If suitable measures' are not taken in time, may result huge losses to the farmers.

(e) There is great scarcity of forage in the flooded areas during the monsoons. When the rivers are in overflowing the cultivated areas intermittently and for fairly long stretches at a time, go under water. The cattle populations of entire villages then have to depend for their existence on what little vegetation they can find on the village high lands. The situation becomes still more acute due to the fact that there is no stall feeding which makes the matter more serious and no forage is grown as a crop. To solve this problem it is necessary to have suitable sources of forage throughout the year and this may be arranged (1) by growing the same in suitable season and preserving it either as hay or as silo or (2) by growing suitable tree fodder of which the leaves may be fed to the cattle at times of scarcity of forage. In flood prone area, no provision is made for feeding cattle and no forage crops are grown. There are practical difficulties also in the way, an average cultivator's holding is too small for growing his own food crops, his cash crop and also forage for his cattle. Rice and wheat straw is the only feeding material that the cultivator has for the purpose, but its supply to others after meeting his requirements is not possible.

(f) For the rehabilitation process, an orchestrated strategy should be chalked out to avoid the misuse and to ensure the appropriate use of scarce resources. Livestock being the main stay of livelihood of the people of these areas have been severely lost but still there is enough livestock to fill the gap provided quality of livestock is focused. The game of animals restocking should be strongly discouraged because keeping a few heads of high producing animals is better than keeping more heads of low producing animals.

(g) The existing animals are already a burden on the available forage resources and the area is forage deficient.

Enhancing Forage Availability in Flood Prone Areas

- Cultivated forage: Increasing area under cultivated forager during pre-monsoon and *rabi* seasons and conserving for the period of flooding in the area. Scientific forage production with improved varieties and cultivation practices.
- Tree fodder: There is need of tree fodder species which can grow in the areas subjected to periodic inundation, the species which are extensively lopped for forage and which may be capable of withstanding short inundation on village highlands, the new potential species which might also be worth trying. Forage production on canal bunds and embankments. One of the major areas that can be utilized for forage production is areas available on the embankment (Table 6.4). For example, the length of embankments in Bihar is 3430 km, taking an average of 2.5 m across the embankments slopes available for planting fodder trees/ forage shrubs, an area of 857 ha will be available for plantation. On the embankments fodder tree species can be planted.

Table 6.4: Length of Embankment and Area for Forage Production on Bunds (Considering 2.5 m width) in hectares in major flood prone states

State	*Length of Embankments (kilometer)*	*Area Available on Embankments (hectare)*
Assam	4464.18	1116.0
Bihar	3430.00	857.5
Orissa	6541.00	1635.3
Uttar Pradesh	2097.00	524.3
West Bengal	10539.00	2634.8

The other alternative is to grow some fodder trees on the village land and also on the low cultivated land susceptible to flood. A list is given below of tree/shrub species which are considered to be suitable for lopping for forage and which might be tried for plantations in or near inundated areas:

Acacia arabica, Albizzia lebbek, Albizzia procera, Adina cordifolia, Anogeissus latifolia, Artocrous heterophyllus, Azadiracta indica, Bauhinia purpurea, Bauhinia malabarica, B. racemosa and B. variegata, Bombax malabaricum, Butea frondosa, Carallia integerrima, Celtis tetranda, Ficus benghalensis, Ficus religiosa, Ficus carica, Ficus glomerata, Grewia tiliaefolia, Hardwickia binata, Kydia calycina. Leucaena leucocephala, Madhuca indica, Mytragyna parvifolia, Moringa pterigosperma, Morus alba, Ougeinia dalbergoides, Pterocarpus marsupium, Schleichera trijuga, Sesbama grandiflora, Stereospermum suaveolens. Terminalia arjuna, Terminalia tomentosa, Terminalia belerica, Bamboo plant species (*Arundinaria, Bambusa, Dendrocalamus species*), Banana leaves and pseudo stems and *Opuntia* species (Thornless Cactus).

It is emphasized that many of these species require particular soil and climatic conditions and that only those suitable to the locality in question must be selected, and it is further pointed out that species suggested for growing in the areas subject to periodic inundation are those which, it is believed will stand inundation once they are established. fodder trees grown on the village high lands may be of great help in meeting the situation, as their leaves are nutritious and rich in minerals. There may be many more trees equally or more suitable for the purpose and it is desirable to prepare a comprehensive list of such trees from all over India. Growing of suitable trees for lopping for forage is required which may be fed to cattle at times of scarcity of forage caused by floods. It cannot be ruled out plantations of tree fodders have a comparatively low commercial value. Hence, the tree with timber value whose leaves are also nutritious can be promoted under plantation. It is sometimes possible to establish fodder trees by growing them along with a faster growing thorny or unpalatable species which will protect them during their tender years. As soon as the fodder trees are big enough the thorny nurse trees are cut out. For example, Nim (*Azadirachta indica*) can be sometimes raised successfully by sowing Babul (*Acacia arabica*) or khair (*Acacia catechu*) or *Cassia siamea* along with it. During this period a part of ration may also be obtained from *Dalbergia sisoo* plantation, mango and jamun orchards (leaves fresh or dried fed).

- ☆ Perennial Grasses and others for moist and saturated moisture conditions:
  Some perennial grasses growing on saturated soil moisture condition like para grass (*Brachiaria mutica*), *B. ruziensis, B. decumbens,* German grass, pangola grass are very much usefull to grow. Aquatic water hyacinth leaves (sun dried), azolla, duckweed, ipomeas, laminaria *etc.*, which are fast growing under waterlogged conditions, can also be grown and fed to animals.

- ☆ Agro industrial wastes:
  Availability of different industrial waste as livestock feeds during calamities may be frog meal, poultry byproducts, fish waste, sea weeds,

prawn waste meal, molasses, press mud wastes, bagasse, sugar cane tops, pig processing wastes, blood meal, feather meal, tomato pomace, vegetable wastes, citrus peels, pineapple wastes, rice bran, cashew apple wastes. Use of these industrial edible wastes can provide immediate relief to the affected animals.

Mitigating Drought Effects on the Availability of Forage for Livestock in India

Drought is a temporary aberration unlike aridity, which is a permanent feature of climate. Seasonal aridity, a well-defined dry season also needs to be distinguished from drought. Drought may be defined as a normal, recurrent feature of climate and occurs in all climatic regimes and is usually characterized in terms of its spatial extension, intensity and duration (DAC, Min. Agriculture, GOI, 2009). Drought differs from other natural calamities such as floods, cyclones, earthquakes, volcanic eruptions and tsunamis. Conditions of drought appear when the rainfall is deficient in relation to the statistical multilayer average for a region, over an extended period of a season or the year. Impacts are generally non-structural, cumulative, difficult to quantify and the effects magnify when events continue from one season to the next or one year after another year. Almost all the drought affected states experience enormous fiscal stress in coping with drought. There is shortage of water for human and livestock; shortage of forage for livestock due to crop failure; migration of livestock population and even decrease in animal population. During a drought situation, every measure needs to be taken to save livestock. While sheep and goats have a potential for rapid growth, the growth of large ruminants are rarely one to two percent. It is necessary to provide support to farmers for forage so that they do not engage in distress selling of their livestock.

The Drought Prone Area

The drought prone area was identified by the Central Water Commission (CWC) in the past which have been taken up by various agencies for study and programme implementation. The list as mentioned by the CWC has 99 districts covering an area of 1.08 million sq. km spread over 14 states in India. However, over a period of time there have been several additions. Therefore, an exercise was undertaken to identify the newly carved districts based on 99 drought prone districts. The total number of drought prone districts as of now is 139 including the same area as that of earlier list (Figure 6.4, Table 6.5).

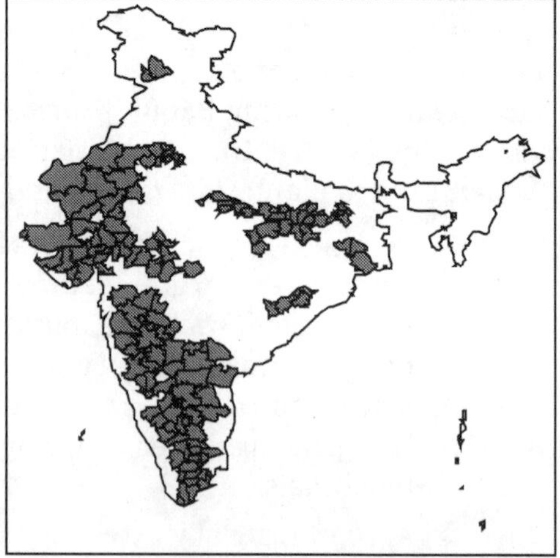

Figure 6.4: Drought Prone Districts in India (Updated).

About 34.78 per cent geographical area of the country is drought prone. In the 139 districts spread over14 states of India, their lives 289.9 million people (28.19 per cent of India). Rajasthan has the highest drought prone area (19.5 per cent) followed by Gujarat (13.8 per cent), Karnataka (13.0 per cent), Madhya Pradesh and Andhra Pradesh (11.4 per cent each) besides others (Table 6.5).

Measures to Mitigate Drought Situations

In the event of any disaster, especially drought that prolongs for much a larger duration, there is a great concern for forage. Our system should ensure adequate forage supplies during the disasters in order to reduce losses of livestock and the poor families are able to retain the herd through a concept of Fodder Bank at village panchayat for seasonal variation in demand, district level for year-to-year minor drought situation and regional level for major drought situation.

Table 6.5: List of Drought Prone Districts (139) in India (updated)

States	*Drought Affected Districts*
Andhra Pradesh	Anantpur, Chittoor, Hyderabad, Cuddapah, Kurnool, Mahbubnag, Nalgonda, Prakasam, Rangareddy
Bihar	Aurangabad, Begusarai, Bhojpur, Buxar, Gaya, Jamui, Jehanabad, Kaimur, Lakhisarai, Munger, Nawada, Rohtas, Sheikhpur
Gujarat	Ahmedabad, Amreli, Anand, Banaskant, Bharuch, Bhavnagar, Dahod, Jamnagar, Kheda, Kutch, Mehsana, Panchmahal, Rajkot, Sabarkantha, Surendranagar
Haryana	Bhiwani, Faridabad, Gurgaon, Jhajjar, Mahendragarh, Rewari, Rohtak
Jammu Kashmir	Doda, Udhampur
Jharkhand	Garhwa, Palamu
Karnataka	Bagalkot, Bangalore, Belgaum, Bellary, Bijapur, Chamarajn, Chikmagal, Chitradurga, Dharwad, Gadag, Gulbarga, Hassan, Haveri, Kolar, Mandya, Mysore, Raichur, Tumkur
Madhya Pradesh	Anuppur, Betul, Datia, Dewas, Dhar, Jhabua, Khandwa, Khargone, Shahdol, Shajapur, Sidhi, Ujjain
Maharastra	Ahmednagar, Aurangabad, Beed, Jalna, Latur, Nashik, Osmanabad, Pune, Sangli, Satara, Solapur
Orissa	Boudh, Kalahandi, Kendujhar, Nawarangpur
Rajasthan	Ajmer, Banswara, Barmer, Bikaner, Churu, Dungarpur, Jaisalmer, Jalore, Jodhpur, Juhnjhunu, Nagaur, Pali, Rajsamand, Udaipur
Tamilnadu	Coimbatore, Dharmapuri, Dindigul, Erode, Kanyakuma, Karur, Madurai, Namakkal, Perambalur, Ramanatha, Salem, Sivaganga, Theni, Thoothukudi, Tiruchira, Tirunelve, Virudhunagar
Uttar Pradesh	Allahabad, Banda, Chandauli, Chitrakut, Hamirpur, Jalaun, Mahoba, Mirzapur, Sant Ravidas Nagar, Sonbhadra, Varanasi
West Bengal	Bankura, Midnapore, Purulia

The plant species of *Cactaceae* are four to five times more efficient in water use efficiency than the most efficient forage grasses (Russell and Felker, 1987). Some of the *Opuntia* species grow rapidly and produce 8-12 tonnes/ha/year fresh fruit and 20-50 tonnes/ha/year vegetative biomass. During favourable monsoon years, these *Opuntia* species get self-protected from grazing due to their spines and during drought year continue to remain succulent and green, with a normal complement of vitamins and carotenoids. During favourable years, they sequester carbohydrates, vitamins, minerals and water in their biomass, which could be made available during drought years as alternative locally available feed for livestock by burning off their spines, when grazing lands have been over-grazed or have become senescent. These *Opuntia* species can be grown as a forage crop on boundary as hedge or on marginal lands where grasses fail to grow optimally. This forage can be of either the spiny or encouraging plantation of spineless *Opuntia* species.

The Short Term and Long term Measures to Combat Drought

Short Term

- ✰ Encourage farmers to undertake forage cultivation on the banks of canals or other areas under irrigations.
- ✰ During *Rabi*, wherever one or two irrigations can be made available, oat and barley should be grown as forage crops.
- ✰ Harvest edible biomass from forests, plantations, hedge, farm boundary *etc.* which can be dried, baled and transported to the needy areas for milk producing cattle. The plant species of *Cactaceae* are four to five times more efficient in water use efficiency than the most efficient forage grasses (Russell and Felker, 1987). Some of the *Opuntia* species grow rapidly and during drought year continue to remain succulent and green, with a normal complement of vitamins and carotenoids. These *Opuntia* species can be grown as a forage crop on boundary as hedge or on marginal lands where grasses fail to grow optimally.
- ✰ Arrange strategic location of feed reserves/fodder bank.
- ✰ Start a special scheme for arranging the supply of nutritious cattle feed to owners of milch cattle with milk cooperatives.
- ✰ Regulate forage and feed price in drought area and create awareness by print and electronic media.
- ✰ Enrichment and baling of available crop residues with protein and mineral supplements. During *Rabi*, Crop residue of *Kharif* crops from irrigated areas can be baled and supplied to drought affected area (Table 6.6).

Table 6.6. Production and utilization of straw and stover in Punjab

Straw	*Annual Production (MT)*	*Annual Requirement/ (MT)*	*Technologies*	*Remarks*
Wheat	14.5	11.0	1. Baling	Such bales/ blocks may be transported to nearby drought and flood prone areas
Rice	10.0	00.0	2. Baling after urea treatment	
Others	01.5	00.0	3. Complete feed blocks	
Total	26.0	11.0		

Source: GADVASU, Ludhiana (2009).

- In late rainfall areas – Grain of sorghum, pearl millet and maize may be used as seed for forage production after germination test.
- Subsidy in the form of seeds should be provided for forage cultivation.
- Arrangement of water tanker based on community drinking water facility for livestock.
- Arrangement for grazing of cattle in army establishments or firing range.
- Set up cattle camp 50-500 units and community cattle feeding centers
- In extreme drought situation, the forage requirement may be arranged through railways from surplus states.
- Explore all possible non-conventional forages and utilize agro-wastes like empty pea pods, cauliflower leaves and stems, cabbage leaves and stems, tomato pomacae, potato chips, potato peels, citrus peels, baby corn husk, sarson crop, cucurbits *etc.* (Table 6.7). Most of the agro-wastes can be solar dried or ensiled for livestock feed.

Table 6.7: Utilization of Agro-by-products Particularly from Fruit and Vegetable Sources

By-products	*Annual Production (MT)*	*Wastes*	*Utilization*
Fruits	170	Empty pea pods, cauliflower leaves and stems, cabbage leaves and stems, tomato pomace, potato chips, potato peels, citrus peels, baby corn husk, sarson crop, cucurbits etc. Only 2.2 per cent of fruits and vegetables are processed by 5293 organized processing units in the country.	1. Most wastes can be used as such 2. Technologies for drying using solar driers and ensiling available for these by-products
Vegetables	159		

Source: National Outlook.

Long Term

- ✰ Strengthen supply chain by enhancing forage seed production and marketing avenues in forage crops
- ✰ Concept of forage bank and baling of crop residues and grass harvest from forest need to be implemented in drought prone area on permanent basis.
- ✰ At least 25 per cent of afforestation should include tree forage plant species as live forage bank.
- ✰ Forage warehouse can be developed at panchayat, block and district levels in the drought area by conserving forager on scientific line.
- ✰ Conservation of forage by individual farmers on roof top may be encouraged.

Chapter 7

Forage Resource Development

In India, the farming systems and livelihood of a large segment of agrarian population is based on the livestock. The livestock also offers livelihood security during natural calamities like floods and droughts. With changing dietary habits and rising income, the demand for livestock products is increasing and projected to increase further (Delgado, 1999). The demand for milk in both rural and urban areas has been reported to be increasing (Down To Earth, 2010). To keep pace with the demand of livestock products, the animal husbandry has to gear itself by increasing the productivity of livestock and providing healthy livestock products. The major constraint identified in the productivity of livestock is non-availability of adequate quantity and quality of forage resources, which account for 70 per cent of the total cost of production of milk and meat. Adequate availability of quality forage is essential for enhancing livestock productivity. Improving forage supply to support livestock production to meet increasing demand for livestock products has remained a formidable challenge in the Indian livestock sector, despite decades of research and extension efforts but meagre allocation by governments and no investments by private sector. In tropical and subtropical climates, a major challenge in feeding livestock is the irregular seasonal rainfall distribution and resultant effect on forage production. In some region, there exist very pronounced single dry season; in others two dry periods are either separated by the long or short rains. The situation becomes critical especially during summer months that soil moisture does not favour the growth of grasses and legumes. These periods required supplementation with concentrates and external forage resources. Livestock production based on green forage is not only economical but also good for the general health of people and wellbeing of livestock. It has also environmental benefits. Dr. M. S. Swaminathan Second Report on "Farmers commission" (2005) suggested that forage and feed availability should be ensured by promoting integrated management of grazing lands, forage production and stall feeding, creation of forage and feed bank and by establishing Livestock Feed Corporation. With proper planning and programme, the forage production, particularly, the cultivated green forage can be doubled. For this, there is a need to target at least 10 per cent of the net cultivable land for growing forage crops. However, forage varieties and technologies have not reached to the target group and the adoptions of forage technologies have remained dismally low. The strategies for increasing forage availability should be based on regional and seasonal (lean period) considerations. Estimation of area and identification of different cultivated forage crops and rangeland forage species grown for livestock at the district level is required for setting regional specific priority for the whole country. There are many issues, which need to be addressed to give momentum to the development of forage resources to ensure forage and feed security of livestock in India (Singh and Sunil Kumar, 2010).

Varieties of Forage crops

Forage crops face unique problems in national perspectives as they are region and season specific (Table 7.1). Multiplicity of plant species constituting basket of forage crops, specificity in their distribution to various growing regions and during different seasons has put a major challenge on forage research system in the country. Forage is always considered as an intermediary crop and hence its direct economic benefit has not been highlighted. Being major input for livestock as feed and forage, make it less prioritized field of agricultural research in the prioritization processes with limited regional spread and scientist per crop. As such, the forage crop is dealt largely as non-commercial at national level. Unique features of forage varieties that further constraint the improvement and availability. The breeding approach to produce high biomass, has inverse relationship with seed yield of improved varieties. Harvesting of forage crop before seed setting for forage unlike other field crops where farmers can use their own seed, has not been in case of forage crops except if farmers are convinced to use oat after first cut and in berseem after third or fourth cut for seed production. Non-synchronous flowering and maturity of spikelet; presence of more sterile glumes; apomictic nature of most of the tropical forage grasses making them difficult for genetic manipulation due to polyploidy nature and complex inheritance. Forage legumes have limitations due to its indeterminate growth habit, uneven pod setting resulting in differential maturity and seed shattering. These characteristics limit the improvement efforts in perennial forage crops.

Most of the Graminaceous forage grasses belong to tribes *Andropogoneae, Paniceae,* and *Eragrosteae.* Among the forage legumes 60 genera with 400 species are edible forage plants. At national level, ICAR-National Plant Genetic Resource Institute in long term storage and ICAR- Indian Grassland and Fodder Research Institute (IGFRI) in medium term storage have conserved forage genetic resources. IGFRI is maintaining about 4,500 forage germplasms in medium term storage module. Besides national collection, introduction from various countries and international CGIAR organizations in the form of new forage plant species like *Stylosanthes* species, from CSIRO, Australia and *Leucaena leucocepahala* from

Table 7.1: Some Suitable Forage Crop Varieties

Forage Crop/Variety	*Recommended Region/State*	*Green Forage Yield (tonnes/ ha)*	*Specific Characteristics*
Pearl millet (*Pennisetum americanum* L.) AVKB-19	North West zone comprising Western Uttar Pradesh, Rajasthan, Haryana and Tarai region of Uttaranchal	35-38	Resistant to downy mildew and blast, suitable for cultivation as dual purpose.
JHPM 05 -2	All zones except South zone *i.e.*, Hill, North West, Central and North East zone	40-45	Superiority for forage yield and crude protein yield
NDFB-2	Saline soils of North East zone	27.0	Superiority for green forage yield as compared to Raj Bajra Chari-2 and It gives higher seed yield than Raj Bajra Chari-2 and Giant Bajra. It is suitable for saline – sodic soil condition
NDFB-3	North East zone comprising eastern Uttar Pradesh, Bihar, Orissa, Jharkhand and West Bengal	31-35	Moderatey resistant to downey mildew, possesses good quality and digestibility.
BAIF-Bajra-1	North West and Central zone	40	Superiority for forage yield and crude protein yield
GFB-1	Gujarat and Western Maharashtra	40-45	Fast growing, profuse tillering high leaf: stem ratio
Napier - Bajra Hybrid Phule Jaywant	Maharashtra	200-225	Foliage dark green and quick regeneration, fairly tolerant to drought and cold
Napier-Bajra Hybrid Suguna	Southern district of Kerala	250-260	Profuse tillering, long and broad lamina. High forage yield and better in quality
Supriya	Southern district of Kerala	270-295	Semi perennial with high tillering and bears dark green leaves, high leaf/stem ratio
RBN-13	Central Zone, Southern part of Uttar Pradesh and Southern Zone	200-225	Grow and provide forage round the year, low oxalic content
APBN-1	Entire Andhra Pradesh	250-300	Stem are less fibrous, more tillers, high leaf stem ratio

Contd...

Forage Crop/Variety	*Recommended Region/State*	*Green Forage Yield (tonnes/ ha)*	*Specific Characteristics*
Sampoorna (DHN-6)	Irrigated areas of Karnatka	120-150	Foliage dark green and quick regenerating ability, no incidence of major diseases and pest
Sorghum (*Sorghum bicolor* L.) Haryana Jowar-513	All zones and Haryana in particular	47	Medium maturity (110-115 days), resistant to shoot fly and stem borer
Maize (*Zea mays*) Pratap Makka Chari (EC-3135)	North West zone covering states of Punjab, Haryana, Rajasthan and Western part of Uttar Pradesh	36	Superiority for forage yield and crude protein yield
Cowpea (*Vigna unguiculata* L.) UPC 607	North West zone comprising states of Punjab, Haryana, Uttar Pradesh, Tarai and Rajasthan	30-35	Suitable for cultivation during summer on irrigated lands
UPC 618	North West, North East and Central zones	30-35	Suitable for summer cultivation under irrigated and rain fed *Kharif* season
UPC-622	North West, North East and Hill zones	26-28	Tolerant to yellow mosaic virus, root rot and anthracnose diseases
UPC -625	North West, Central and North East zones.	35-40	Superiority for forage yield and crude protein yield. Resistant to mosaic, leaf spot, anthracnose, root/collar rot disease and aphids beetle
UPC-628	North East, North West, Tarai region of Uttrakhand and Central zones	28-30	Tolerant to major pest and diseases
CL-367	Punjab state	35-40	Small seeded
Rice bean (*Vigna umbellata* T.) Bidhan Rice bean-2 (KRB-4)	Eastern and North East region	25-30	High crude protein and resistant to major insects-pests
RBL-6	North and Northwest States	30-35	
Guinea grass (*Panicum maximum* Jacq	North West, Hill and Southern zones	45-50	Multicut
PGG-616			

Contd...

Forage Crop/Variety	Recommended Region/State	Green Forage Yield (tonnes/ ha)	Specific Characteristics
Bundel Guinea –1 (JHGG 96-5)	Rain-fed areas of Uttar Pradesh, Madhya Pradesh, Maharashtra, Andhra Pradesh, Gujarat, Rajasthan and Punjab	30-40	Apomictic, perennial, multi-cut, resistant to insect-pests
Bundel Guinea- 2 (JHGG -04-1)	Humid, Arid, and Sub-tropical areas of Himachal Pradesh, Uttar Pradesh, Gujarat, Maharashtra, Karnataka, Tamil Nadu and Kerala	55-60	Drought tolerant, resistant to lodging
Harithasree	Uplands and Homesteads of Kerala	74	Semi perennial, high tillering, dark green foliage
Buffel grass (*Cenchrus ciliaris* L.) Bundel Anjan-3 (*IGFRI-727*)	Arid and Semiarid tracts of Western Uttar Pradesh, Himachal Pradesh, Uttarakhand, Rajasthan. Andhra Pradesh, Punjab, Haryana and Maharashtra.	28-30	High dry matter yield
Sen grass (*Sehima nervosum*) Bundel Sen -1 (IGS-9901)	Central and Western Madhya Pradesh, Southern part of Uttar Pradesh, North West Karnataka, Rayalseema region of Andhra Pradesh, South and Central Maharashtra and Rajasthan	18-20	First variety in this range grass. Higher green forage and Dry matter yield
Black spear grass (*Heteropogon contortous*) Bundel Lampa ghas-1 (IGHCO3-4)	Semi-arid, Tropical and Sub tropical regions of Uttar Pradesh, Gujarat, Himachal Pradesh, Tamil Nadu, Karnataka and Maharashtra	25-27	Resistant to lodging
Setaria grass (*Setaria anceps*) Setaria – 92	Hill zone comprising Himachal Pradesh and Uttaranchal	30-35	Highly tolerant to pests and diseases
Dhraf grass (*Chrysopogon fulvus*) Bundel Phulkara-1 (IGC-9903)	Central and Western Madhya Pradesh, Southern part of Uttar Pradesh, North West Karnataka, Rayalseema region of Andhra Pradesh, South and Central Maharashtra and Rajasthan	25-28	Range grass
(*Coix lachryma* Jobi.) Bidhan Coix-1 (KCA-3)	North East region	30-35	

Contd...

Forage Crop/Variety	Recommended Region/State	Green Forage Yield (tonnes/ ha)	Specific Characteristics
Stylosanthes seabreana Phule Kranti (RSS-2000-95)	Western Maharashtra	45-50	High yielding, good establishment in first year and easy seed collection
Tall Fescue (*Festuca arundinacea*) Hima-4	Temperate Hill zone comprising Himachal Pradesh and Uttaranchal	25-30	Highly tolerant to pests and diseases
EC-178182	Hill Zone of Himachal Pradesh, Uttrakhand and Jammu and Kashmir	30-35	Superior in forage yield and quality parameters
Oats (*Avena sativa* L.) JHO-99-1	Hill regions of the country	40-42	Suitable for single cut in hill zone
Bundel Jai 2001-3 (JHO-2001-3)	Tropical and Subtropical areas in North West and South zones comprising states of Uttar Pradesh, Himachal Pradesh, Uttranchal, Rajasthan, Andhra Pradesh. Punjab, Haryana, Maharashtra, Tamil Nadu and Karnataka	48-53	Suitable for cultivation during *Rabi* season as a single cut crop
Bundel Jai 2000-4 (JHO-2000-4)	All zones except Central zone	35-38	Single cut variety and resistant to root rot, crown rust and powdery mildew
JHO-99-2	North East, North West and Central zones	50-60	Multicut variety (2-4cuts), resistant to major pests and diseases
JO-1	Madhya Pradesh, Uttar Pradesh and Bihar	60-65	Good seed setting ability after first cut for seed
JO-03-91	Central Zone and Southern part of Uttar Pradesh	55-60	Superiority in forage yield and quality
JO-03-93	Central Zone comprising southern part of Uttar Pradesh, Madhya Pradesh, Gujarat and Maharashtra	55-58	Resistant to leaf blight and nematodes
OS-346	Central Zone	50-55	Superiority in forage yield and quality
Phule Harita (RO-19)	All oats growing areas of the country	40-45	Multi-cut variety (2-4 cuts) with broad leaves, high tillering and crude protein

Contd...

Forage Crop/Variety	Recommended Region/State	Green Forage Yield (tonnes/ ha)	Specific Characteristics
SKO-90	Hill zones comprising states of Himachal Pradesh, Uttrakhand and Jammu and Kashmir	35-38	Resistant to leaf blight and sclerotium root rot, moderately resistant to powdery mildew disease.
NDO-1	Saline soils of oat growing areas in the country	31-35	Suitable for saline-sodic soils
Barley (*Hordeum vulgare* L.) RD-2715	Gujarat, Rajasthan, Madhya Pradesh, Chhatishgarh, Kota and Udaipur division of Rajasthan and Jhansi division of Uttar Pradesh	18	Erect with dark green foliage, profuse tillering and resistant to major rust
Berseem (*Trifolium. Alexandrinum*) BL-180	North West and Hill zones	65-75	High green forage and seed yield, profuse tillering
Bundel berseem-3 (JHTB-96-4)	North East zone, comprising Eastern Uttar Pradesh, Bihar, Jharkhand, West Bengal, Orissa and Assam	51-60	Subtropical climatic condition
Hisar Berseem-1	Haryana, Punjab and Rajasthan	90-95	
(HFB-600)			
BL-42	Punjab	85-95	Late maturing
JB-5	Madhya Pradesh, Chhattisgarh	100-105	Moderately resistant to color rot, root rot and good seed setting ability
Lucerne (*Medicago sativa*) AL-3	Central Gujarat	120	Profuse tillering, high regeneration ability, foliage dark green
Red Clover (*Trifolium pratense*) PRC 3	Temperate zone	25-30	Root rot resistant and high yielding

Hawaii, germplasms of guinea grass, *Pennisetum pedicellatum*, cowpea, maize, oat, sorghum, pearl millet, lablab bean, berseem and lucerne have strengthened conventional forage breeding programme. National and international genetic resource enhancement efforts in breeding programme have resulted in the release of 188 varieties in various forage crops (Pandey and Roy, 2011). Further introduction for specific traits are required to broaden the genetic resource base to develop the suitable regional, biotic, abiotic stress and season specific forage plant varieties. Genetic improvement and varietal development need to address biotic stress tolerance as use of agrochemicals has limited option. There is requirement of annual forage legume and range legume to compete with graminaceous forage under mixed culture on cultivated and range land condition to maintain a proper grass legume ratio for production of balanced nutritive forage. The new cultivar responding phosphate will produce high dry matter and augment protein content in dry matter. There is need to upscale present productivity potential and nutritive value as limited lands are allocated for forage production due to high pressure on arable land and reduction in total cultivable lands due to other developmental activities in the country. In this pursuit, lack of genetic variability in general, overlapping of vegetative and reproductive growth phases, uneven pod setting and maturity as well as shattering in forage legumes have been felt as major bottleneck. In the pursuit to maximize the yield of grains in cereal and other food crops, the straw (significant source of roughage for livestock) component of field crops has been reduced to enhance grain yield. Any breeding programme for cereal crops must realize that crop residues are equally important in Indian mixed farming situation for livestock feeding. Therefore, a minimum straw to grain ratio must be ensured before any new crop variety is released. Another possibility is developing dual purpose varieties in various forage crops especially maize, oat, sorghum, Pearl millet and promoting under mixed farming situations

Development of interspecific hybrids, multifoliate and SSR markers in berseem (Roy *et al.*, 2004, Malaviya *et al.*, 2005, Kaushal *et al.*, 2006, and Roy *et al.*, 2011); interspecific hybridization, gene pyramiding for salinity and alkalinity tolerance in oat; *Medicago* species germplasms showing resistance to lucerne weevil; new introduction of *Stylosanthes seabrana;* understanding of apomixes of *Cenchrus ciliaris* and *Panicum maxmum* and identification of sexual plants in *Cenchrus ciliaris;* production of interspecific hybrids of *P. squamulatum* and *P. orientale* with pearl millet, production of trispecific GOS hybrid, production of perennial, thin stemmed, grassy, high tillering, apomictic forage pearl millet is worth noticing. Advancement in hybridization-supplemented apomixes-components partitioning approach by biotech group of IGFRI from sexual x apomictic crosses to obtain apomeiosis (formation of unreduced female gamete), parthenogenesis (fertilization independent embryogenesis) and functional endosperm development (autonomous or pseudogamous). Utilization of recombination of such lines have resulted in the development of ploidy series in *P. maximum* - 3x, 4x, 5x, 6x, 8x, 9x developed from single progenitor through Hybridization-supplemented Apomixis-components

Partitioning Approach (HAPA) (Kaushal *et al.*, 2008). There are also possibility to make efforts to produce edible *chara* vaccine for diseases such as FMD and PPR *etc.*

Specific Scenario of Forage Crops

In last 15 years, area under grain sorghum and pearlmillet is decreasing in favour of other crops. However, it still occupies first position among the forage crops. Its also supply significantly large quantity of stovers from grain sorghum as edible crop residue for livestock. Forage sorghum caters the major forage requirement of commercial dairy farms in northern India. Mostly it is preferred over maize for its drought tolerance and less nutrient need. Traditionally, single cut varieties are preferred over multi cut varieties in *kharif* season to fit it into forage sorghum- wheat crop sequence. However, Multi cut has great potential for intensive forage production under irrigated condition. Now, there is growing market for multi cut forage sorghum hybrids to support the proliferating peri-urban dairy business where private sector is dominating and marketing notified varieties such as MFSH 3, Harasona and non-notified hybrids bred indigenously developed or imported from USA and Australia under new seed category. Attempt made to develop dual type varieties combining high grain as well as stovers yield resulted in the release of dual type varieties namely CSH 10, SPV 462 (Co 26), CSH 13 and CSV 15. It appeared that high stovers yield trait can be combined without a trade off from grain yield in sorghum and pearl millet. In general, both the crops need to be developed as dual type stay green with nutritious stalk for rain fed arid and semiarid region of India. Maize is a succulent nutritious forage crop without any anti-nutritional constituents. It is grown during summer and *kharif.* It is almost free from any anti-nutritional constituents and is most preferred grain for the concentrate feed. Almost 59 per cent of the total maize grain produced is utilized in manufacture of concentrate feed for livestock in the country. As such emerging agriculture scenario is not favourable for forage maize but large expanding area under baby corn, sweet corn and green cob maize will add indirectly one million hectare to provide green maize fodder by 2025. Bringing stay green trait in baby corn, sweet corn and popcorn varieties and increasing demand of the green cobs will increase availability of green stalk of maize for feeding to livestock. Stovers of grain maize are also used as edible crop residue for livestock.

Guinea grass (*Panicum maximum*) and Napier x bajra hybrids are most preferred multi cut high yielding perennial forage crops mainly grown on irrigated lands and in high rainfall areas of south and eastern India. Napier x bajra hybrid and guinea grass are integrated components of most of the round the year forage crop sequences developed. They can grow in acid soils, degraded forests, ravines and coastal saline shaded area under coconut plantation. Under non-competitive land use they can be grown on farm bunds, terrace bunds and terrace risers on undulating sloppy lands. Potential for increasing area under guinea grass can be achieved if it is popularized as under storey crop in various orchards, non-competitive land use and different silvipasture systems.

Photo 7.1: Berseem, a Popular Cultivated Forage.

Egyptian clover (*Trifolium alexandrinum*) is one of the most important nutritious multicut forage legume grown during *rabi* in northern India, occupying about two million hectares (Photo 7.1). Adoption of high yielding varieties is limited. The seeds sold in local market are of poor quality and infested with weeds of *Melilotus* species and *Chicorium intybus.* Usually the farmers do not produce seed and procure poor quality uncertified seed for sowing. They are practicing this to get one extra cut of forage during the lean period of forage availability in April. Failure of imported seed has also been witnessed in last 10 -15 years. Moreover, prevalence of diseases like root rots and stem rot in North West and Central India are major threat to the berseem crop. Now, oat is becoming a competitive crop and farmers are shifting to oats due to less availability of certified seed ensuring quality. The merit of the crop lies in its multi cut nature (4 -8 cuts), long duration of green forage availability (November to April), high green forage (85 tonnes/ha) with high palatability, 20 per cent crude protein in dry matter and digestibility over 65 per cent. Adoption of the new high yielding varieties in even 25 per cent area under cultivation would result in substantial increase in availability of high quality forage. Oat, an important high yielding nutritious forage crop is grown during *rabi* season in Punjab, Haryana, Jammu and Kashmir, Himachal Pradesh, Uttar Pradesh, Madhya Pradesh, Rajasthan, Maharashtra, Bihar and West Bengal. The good quality protein and nutritional composition of forage oat, is suitable for milch cattle and buffaloes. The major problems associated with adoption of oat crop are one time large quantity of forage production in single cut oat varieties traditionally grown and the prevalence of rust in the foothills of Himalayas. Some multi cut oat varieties developed have potential of providing good quality, nutritious forage in two cuts. Lucerne is grown in western parts of the country mainly in the states of Gujarat, Maharashtra and Rajasthan in about one million hectare area. *Medicago* species form a major component of pastures as well as cultivated in vast tract in the high hills of Indian Himalaya. Susceptibility of the crop to lucerne weevil and downy mildew is the major problems. Being a cross pollinated crop, it is difficult to maintain the purity of seed of a specific variety. Forage cowpea occupies 0.3 million hectare out of 0.65 million hectare area under different pulse and vegetable type varieties of cowpea grown in rain fed as well as on irrigated lands. Green forage yield ranges 25-45 tonnes/ha. The cultivated area under cowpea is gradually decreasing because of competition with other crops in changing agricultural scenario and the problems of insects and pests during crop growth and storage. The forage cowpea fits well in intercropping with maize, pearl millet and sorghum and has potential, if dual type varietal development is taken up for producing forage and vegetable. Tolerance to extreme environments such as high temperatures, low rainfall and poor soils makes it suitable for growing during pre-*kharif* season to provide forage for livestock.

Forage Seed

Quality seed availability needs to be produced to match highly fluctuating uncertain demand- supply. Further, limited forage seed marketing agencies and their outlets spread over the country are major concern. Basically, forage crop bred for high biomass is low seed yielder. Low seed yield of most of the forage crops and non-synchronous flowering and resultant seed maturity at different times in range grasses and range legumes result in high cost of seed production. Seed yield of cultivated forage legumes (lucerne, berseem, guar and cowpea ranges 150-1000 kg/ha, cultivated graminaceous forage (pearl millet, sorghum, maize and oat, ranges 600 -2200 kg/ha, perennial forage grasses (*Bracharia mutica, Brachiaria bryzantha, Cenchrus ciliaris, Cenchrus setigerus, Dichanthium annulatum, Panicum maximum, Pennisetum pedicellatum, Pennisetum purpureum* and *Setaria sphacelata)* ranges 25-550kg/ha and perennial legumes (*Clitoria ternatea, Lablab pupureus, Macroptilium atropurpureum* and *Stylosanthes hamata)* ranges 60-1200kg/ha. The low seed yields are the one of the major reasons to limit the adoption and spread of improved forage crop varieties. Forage seed production is largely concentrated in the unorganised sector and effective seed production linking breeder seed to foundation seed to certified seed production chains are lacking. It is mainly because, handling of a large number of forage crops and their suitability to specific regional niches make the process complex and cumbersome. As per current demand of quality seeds, the availability is only 25-30 per cent in case of cultivated forage while it is only 15 to 20 per cent in range grasses and legumes. It is estimated by 2020, 2,77,000 tonnes of certified seed of major forage crops would be required (Anonymous, 1997; 2011).

Considering the specificity and complexity of seed production in forage crops and range grasses and potential specific regions and places should be identified and mapped considering their potential and which is free from specific disease and pests and commiserating package of practices for seed production need to be developed for cost effective seed production. It has to be realized that the weather condition suitable for high forage production will be different than the weather condition required for high seed yield of a forage crop. Hence, weather condition for seed production need to be identified separately for various forage crops. Suitable regions has been identified for some of the forage crops *i.e.* South and South-West regions for sorghum, North-West and South-West regions for cowpea, East and Central regions for Dinanath grass, North-West region for pearl millet, Western and Southern region for buffel grass, Central, South-Central regions for Marvel grass, Central and South-Central regions for siratro and *Stylosanthes* species, Eastern region for rice bean, Central and West regions for berseem, West and South-Central regions for lucerne, North and North-Central regions for oat and South and North-West regions for guinea grass. Seed production protocol has not been developed for all forage crops particularly range grasses and range legumes. The seed production package and practices are available for *Setaria sphacelata, Desmodium tortuosum, Panicum maximum, Stylosanthes hamata, Macroptilium atropurpureum, Cenchrus ciliaris* and *Clitoria ternatea*. Some of the basic aspect to be dealt are role of pollinator, non-synchronous seed maturity and sterile glume as false seed formation in range grasses, techniques and machines for separating false fluff, seed shattering and hard seed coat in range legumes and fodder tree species, The seed chain can be strengthened with the involvement of livestock owners, milk federations, state and national seed producing agencies and forestry sector for seed production range grasses and range legumes, giving due priority to forage seed production and marketing and postharvest baling of forage from forest.

Seed standards for 16 forage crops *i.e.* berseem (*Trifolium alexandrinum* L), buffel grass (*Cenchrus ciliaris* Linn), dinanath grass (*Pennisetum pedicellatum* Trin.), forage sorghum including sudan grass, Guar (*Cyamopsis tetragonoloba*), Guinea grass (*Panicum maximum* Jacq.), Senji (*Melilotus species*), lucerne (*Medicago sativa* L.), marvel grass (*Dichanthium annulatum* Stapf.), napier grass (*Pennisetum purpureum* K. Schum.), oats (*Avena Sativa* L.), Rice bean (*Vigna umbellata* (Thunb.), Setaria grass (*Setaria sphacelata*), *Setaria anceps Stapf.* Ex. Massey, *Stylosanthes species* and Teosinte (*Euchlaena mexicana* Schrad.) have been developed. Seed standards for many range grasses and legumes are yet to be developed.

Among the 158 varieties of 11 crops notified, 70 varieties are in the seed production chain. They are 31 cultivars of berseem (*Trifolium alexandrinum* L), lucerne (*Medicago sativa* L.), oat (*Avena sativa* L. and *A. sterilis* L. var. culta) and gobhi sarson (*Brassica napus* L.) during *rabi* season and 39 cultivars of cowpea (*Vigna unguiculata* L.), maize (*Zea mays* L.), sorghum (*Sorhum bicolour* L.), teosinte (*Euchlaena mexicana* Schrad.), rice bean [*Vigna umbellata* (*Thunb.*)], guar (*Cyamopsis tetragonoloba L.*)and pearl millet (*Pennisetum glaucum* L.) during *kharif* season.

Forage Production on Cultivated Lands

Many annual and some perennial forage crops are cultivated on arable lands for feeding livestock in the country. Graminaceous annuals are sorghum, pearl millet, maize, dinanath grass, oat and perennials are guinea grass, Napier x pearl millet hybrids and Congo signal grass. Cultivated forage legumes are cowpea, guar, berseem and lucerne. Besides, many perennial grasses and legumes are established under silvipasture and hortipasture systems in non-arable and in some cases on the arable lands. Due to pressure on arable lands, still forage requirement is largely met from the various types of non-cultivated naturally growing edible grasses, forbs, legumes, shrubs and trees on common properties resources, forest area and along railway track and roads.

Agro-region Specific Forage Based Cropping Systems

The availability of sufficient forage resource, feed, and their efficient utilization in a particular agro region is required to improve the livestock productivity. Suitable varieties of forage crops identified for an agro region is to be fitted in various forage and forage-food based cropping systems to optimise forage productivity and increase total green forage availability to the livestock in the region (Tables 7.2 and 7.3). Up scaling intensive round the year forage production by incorporating forage legumes and shrubs on the bunds and boundaries has immense potential under irrigated condition all over India. In the periphery of cities, commercial forage production is required to be increased to support adequately the forage requirement of peri-urban dairies. Some of the approaches relevant to forage production is incorporation of forage crops in food based cropping systems (Singh and Pal, 1998); growing short duration forage legume and dual purpose sorghum and pearl millet; integration of suitable forage crops in rice – fallow system by relay cropping of chickling vetch (*Lathyrus sativus*)/cowpea/rice bean in late sown/low land rice cropping in eastern India; priority to grow forage crops on the problem soils *i.e.* acidic, alkaline and saline soils as well as on cultivable wastelands; supplementing migrating grazing with forage production in the hills and central India; in wetlands, aquatic forage plant species grown under low submergence and saturated land situation for example *Brachiaria mutica* and *Zizania species*; sea weeds and other marine plants as forage resource in coastal area of south India, Orissa and West Bengal; forage production in non - competitive land use *viz,* use of sloppy lands, terrace risers, bunds, farm boundaries roadside by planting perennial grasses like golden timothy grass, napier x pearlmillet hybrids, guinea grass, rhodes grass and broom grass augument forage availability (Photo 7.2).

In arid and semiarid area, fodder tree species as shelter belt and in south India and coastal areas, where high raifall occurs (1800-2600 mm) favours the growth of perennial grasses in open and under the plantation crops *viz.* NB hybrid, guinea grass, Congo signal and signal grass as pure crop and forage legumes *viz.* cowpea and horse gram as sole or intercrop and alley cropping under coconut plantations. In the entire north-western region, forage supplementation through sugarcane stalks, potato stems and leaves, vegetable leaves, (cabbage, cauliflower, peas, radish, carrot *etc.*) supply considerable amount of forage during winters. Sugarcane need to be fortified for its mineral deficiencies.

Photo 7.2: Non-Competitive Land Use Technology for Forage Production. (Courtsey: Sunil Kumar)

Table 7.2: Potential Cropping Systems for different States in India

Agro-climate in the State	*Cropping Systems*
Andhra Pradesh- Subtropical to tropical humid, black soil	1. Napier Bajra hybrid (Perennial) + Cowpea 2. Sorghum + Cowpea-Maize + Cowpea-Maize + Cowpea
Andhra Pradesh- Humid tropical red soil	1. Guinea grass/Congo signal grass (under coconut plantation) 2. Napier Bajra hybrid (Perennial)
Assam-Subtropical humid, acidic soils medium fertile soils	1. Perennial Napier Bajra hybrid 2. Maize (F) + Cowpea (F) – Oat – Cowpea (F)
Himachal Pradesh - Sub-tropical sub humid mid hills	1. Maize (F) + Cowpea (F) - Oat + *Sarson* (F) 2. Maize -Wheat (F)
Gujarat- mid Gujarat region, loam to sandy loam and medium soil fertility	1. Perennial Napier Bajra hybrid + Cowpea/Lucerne 2. Pearl millet (F) + Cowpea (F) - Lucerne
Haryana -South western region	1. Green gram (G) –Wheat - Pearl millet (F) 2. Sorghum (Single Cut) - Wheat
Jharkhand Subtropical, Acidic and medium soil fertility	1. Perennial Napier Bajra hybrid + Berseem 2. Deenanath grass - Oat
Karnataka- Tropical black soil	1. Maize (F) + Cowpea (F)-Lucerne 2. Guinea grass/Congo signal grass sole under coconut plantation

Contd...

Agro-climate in the State	*Cropping Systems*
Kerala- Tropical hot humid, red soils	1. Guinea grass/Congo signal grass (sole)/(Under coconut plantation) 2. Napier Bajra hybrid (Perennial) 3. Guinea grass/Congo signal grass (sole)/Under coconut plantation (inter cropped with *Desmanthus virgatus*) 4. Maize (F) + Cowpea (F)-Cowpea (vegetable)-Okra
Madhya Pradesh- Kymore plateau and Satpura hills, clay to clay loam, and medium soil fertility	1. Sorghum (F) – Berseem – Maize (F) + Cowpea (F) 2. Perennial Napier Bajra hybrid + Cowpea - Berseem
Maharashtra- loam to sandy loam slightly alkaline and medium soil fertility	1. Perennial Lucerne 2. Soybean-Berseem-Greengram
Orissa- Coastal humid, Acidic and low soil fertility	1. Pearl millet – Oat – Maize + Cowpea 2. Groundnut – Oat – Maize + Cowpea
Rajasthan - Hyper arid partially irrigated western plain zone	1. Pearl millet (F) + Guar (F) - Oats (F)- Cowpea (F) 2. Pearl millet (F) + Guar (F) - Wheat (G)
Tamil Nadu- Tropical, black soil	1. Napier Bajra hybrid (Perennial) 2. Sorghum (G) – Maize (F) + Cowpea (F)-Maize (G)
Uttarakhand- Tarai region of Shivalik range	1. Rice – Wheat – Maize (F) + Cowpea (F) 2. Rice – Berseem – Maize (F) + Cowpea (F)
Uttar Pradesh (Eastern region) Subtropical pre humid, normal to alkaline and low to medium fertile soils	1. Perennial Napier Bajra hybrid + Berseem 2. Sorghum (F) – Berseem – Maize (F) + Cowpea (F)
Uttar Pradesh (Bundelkhand region) Bundelkhand zone, loam to sandy loam, low to medium soil fertility	1. Groundnut – Berseem – Maize (F) + Cowpea (F) 2. Sorghum (F) -Berseem
West Bengal - Sub humid, acidic to Normal and medium fertile soils	1. Rice – Oat – Sesame 2. Sorghum (F) – Oat – Green gram

F: Forage; G: Grain.

Source: AICRP on Forage Crops IGFRI (2000-2015).

Table 7.3: Production Potential of some of the Forage Production Systems

Region/Cropping System	*Green Forage Yield (tonnes/ha)*
Northern region: White clover + Rye grass	184
Dinanath grass – berseem – maize + cowpea	200
Napier x bajra hybrid – berseem + sarson	211
Napier x bajra hybrid – berseem	176
Central Region: Napier x bajra hybrid + cowpea – berseem	255
Cowpea – sorghum + cowpea – berseem	176
Napier- Bajra + guar – lucerne	252

Contd...

Region/Cropping System	Green Forage Yield (tonnes/ha)
Sorghum + cowpea – berseem + sarson – sorghum + cowpea	169
Southern region: Napier x Bajra hybrid – Hedge lucerne	225
Sorghum + cowpea – maize + cowpea – Maize + cowpea	111
Guinea grass	135
Congo signal grass under coconut plantation	75
Eastern Region: Bajra + cowpea –Maize + cowpea- oat	100
Maize + cowpea – dinanath grass – oat	121
Maize + rice-bean – berseem + sarson	112
Sole Napier x bajra hybrid	106

Source: AICRP on Forage Crops IGFRI (2000-2015).

Water Requirement of Forage Crops

Livestock production in the developing countries must be seen as both essential and an opportunity to increase water use efficiency (Singh, *et al.*, 2004; Amare *et al.*, 2011). Indicative estimate showed that to produce one kilogram of milk, 3751, 2451 and 475 litters of water is required under intensive, semi-intensive and extensive systems of livestock rearing, respectively (Baig *et al.*, 2007). Dairy cows need 1.8 to 2.3 litres of water for each 0.9 litres of milk produced both for maintenance and production (Linton 1965). Based on an estimate water requirement to produce one kilogram of meat is 9,680 litres in buffalo, 8,215 litres in sheep and 9,338 litres in goat (Baig *et al.*, 2007). Buffaloes being vulnerable to heat stress in summer require more water than cows. Water requirement for a high milk producing livestock may be more than a low yielding dairy animal. Thus, it is evident that water not in quantity but quality is going to be critical input for human beings but for livestock production also in times to come. We need to address water productivity issues in a holistic manner.

There is need to factoring productivity of water for livestock and forage production in major planning, management and development (Sonder *et al.*, 2004, Solomon *et al.*, 2009). ILRI (2006) had suggested that integration of livestock production into investments in agricultural water development will be more sustainable and profitable livelihoods for farmers and livestock owner alike. There are challenges towards the goal of enhancing water productivity in forage crops and livestock. Peden *et al.* (2007, 2008, 2009) and ILRI (2006) suggested strategic sourcing of animal feeds, enhancing livestock productivity and reducing herd size, reducing negative environmental impacts as well as provision of drinking water to increase water productivity in animal husbandry. Balancing the four strategies in a site-specific approach would possibly double the livestock water productivity. Regarding strategic sourcing of animal feeds such as grains, crop residues and by-products, pastures and forage crops has been ignored all over the world which accounts for about 90 per cent of the total water requirement for livestock production. Dietary composition (proportion and type of ingredients coming from various field crops and forage crop sources) and livestock rearing systems would highly impact the livestock and forage water productivity. There is the challenge of following different approaches and to continue to work together toward the common goal of increasing water productivity for forage crops and livestock. It's worth noticing that in Africa's pastoral areas achieve only about one-third of their genetic production potential (Peden *et al.,* 2007 and 2008). There is need to improve irrigation and grazing practices that reduce runoff, flooding, degradation and contamination of water resources.

Use of the soil water supply depends upon plant species competitiveness, the rate and extent at which it can make use of water (Haynes 1980). In many crops traits and genes for drought and salt tolerance have been identified. The lessons learnt can be well applied to forage crops for both conventional and molecular-breeding techniques. Setting a goal to increasing forage production without a concomitant increase in evapo-transpiration and stay-green factor is expected to yield results for some forage crops *viz.* maize, sorghum and pearl millet. Forage crop management should make use of supplementary irrigation in rain-fed agriculture and the expansion of drip, trickle and sprinkler irrigation, embrace adoption and adaptation of water productivity enhancing practices in field crops to develop water use efficient forage crop and pasture management. At agro ecological level, the integrated approach to land and water management for the conjunctive management of groundwater and surface water and precise assessment of the gains and losses in water productivity resulting from various interventions have been done. Institutions and policies need to be amended to provide appropriate incentives for livestock farmers. Judicious water management in the forage crops and pasture aims to provide suitable moisture environment to optimise forage yields which commensurate with maximum economy in irrigation water and maintenance of soil productivity. In forage crops, the phasic water requirement are more pronounced in comparison to seed crops as the economic product is edible biomass harvested prior to seed formation stage. Moisture stress become very problematic in forage crops as it results in to increased concentration of toxic/anti-quality compounds *viz.* HCN in sorghum and NO_3 in sorghum and oats *etc.* Water requirement based on various techniques estimated in a semiarid climatic condition and water use efficiency computed for various forage crops and intercropping system presented in Table 7.4 gives an idea of water requirement in forage crops.

Mechanism through which water deficit affects forage yield varies with crop growth stages. The soil moisture stress has a pronounced influence on evapo-transpiration. In Sorghum (MP Chari) peak period of water use is transitional zone of vegetative and reproductive phases (Menhi Lal and Shukla 1985), leaf area index 03 in lucerne (Stegman *et al.,* 1980), in oat it coincides with boot stage (Pradeep Behari and Hazra 1989) and the water utilization of the crop from sowing to jointing stage is only 1/5th to 1/3rd of the total water requirement (Menhi Lal and Shukla 1988 and under maize + cowpea intercropping system, highest crop ET rate (8 mm/

Table 7.4: Water Requirement of Important Forage Crops in a Semi-arid Environment

Forage Crops	*Total Water Requirement (mm)*	*Number of Irrigation*
Berseem	614 - 710	10-18
Lucerne	600	11-13
Oat	340	6-8
Sorghum (PC-6, HC136)	380	8
Sorghum (M.P. Chari)	40-80	1-2
Teosinte	380	8
Cowpea	380	8
Cluster bean	270	5
Hybrid Napier\	380	10
Guinea grass	672.7	5
Triticale	110	2-3
Barley	110	2-3
Turnip	140	3-4
Carrot	380	7-8

Source: Baig *et al.*, 2007.

day) was observed at silking stage (Singh *et al.*, 2002). Conclusively, dry matter is directly proportional to evapotranspiration losses, if soil moisture is not a limiting factor and if soil water is limited, water use is a function of both crop growth stages and atmospheric water need with overriding effect of weather conditions in determining ET losses. Intercropping systems *viz.* MP chari + cowpea, maize + cowpea, teosinte + cowpea enhanced water use efficiency under wide ranges of soil moisture availability and climatic variability and produces highest fresh and dry matter yields (Menhi Lal *et al.*, 1984). For oat + Indian clover/*trigonella*/pea (2:1 planting pattern), the highest green matter production and dry matter accumulation were recorded with irrigation scheduling at IW/CPE ratio of 1.0 followed by 0.8 (Menhi Lal & Shukla, 1987). Integrated effect of soil moisture characteristics, canopy coefficients and atmospheric demand causing a cumulative evapotranspiration of 100 mm provides optimum soil moisture regime for both components in mixed stands of sorghum/maize/teosinte + cowpea. It was concluded that when crops of varying canopy characteristics are grown in association, evaporative demand appeared to be better criterion than degree of soil moisture depletion for irrigation scheduling in mixed/intercropping systems.

Hydroponic Forage

Hydroponics is the most intensive method of crop production that utilizes nutrient-laden water rather than soil for plant nourishment (Bridgewood, 2003; Simply Hydro, 2008). It is an advanced technology to use nutrient solutions *i.e.* water and nutrients with or without any artificial medium to provide mechanical support through gravel, sand, peat, sawdust, rock wool and moss. Liquid hydroponic systems have no any supporting medium for the plant roots. The systems further grouped into open (*i.e.*, once the nutrient solution is delivered to the plant roots, it is not reused) or closed (surplus solution is recovered, replenished, and recycled). Since regulating the aerial and root environment is a major concern in such agricultural systems, production takes place inside enclosures that give control of air and root temperature, light, water, plant nutrition and protect against adverse climatic conditions. Thus, it does not require natural precipitation or fertile land in order to be effective means of agricultural production. Crops are grown with optimal controlled environment with optimal weather conditions. The re-use of nutrient water supplies makes process-induced eutrophication (excessive plant growth due to overabundant nutrients). The chances of general pollution of land and water are unlikely because runoff is weather-independent facilities. The hydroponic system is highly productive and efficient due to the advantage of aeroponic and hydroponic systems as it requires less water and space (can maximize vertical space) as well as almost negligible use or no use of pesticides. However, it's often capital-intensive due to high energy costs as the system requires lighting, pumping, and air moderation systems. Additional costs aside from energy costs include the purchase and purification of fertilizers and water. Hydroponics may not be cost effective for every location.

Hydroponic forage production appears to be exciting approach of forage production duly promoted by various views and news in various ways of communication including web and goggling projecting to make believe (Fazaeli *et al.*, 2012; Naik *et al.*, 2015). The hydroponic forage system is designed to sprout grains which become very nutritious and digestible produced in a temperature and humidity controlled growing room that is specifically designed to sprout grains, put into photo-chemically treated trays without soil and sprayed with nutrient rich water solution at predetermined intervals. Reliable clean water source need to be ensured. Hydro-net and a hydro-cloth are used to ensure higher growth and the best nutritional value possible. Photo-chemically treated trays are used to prevent fungus in which the forage is grown. After seven days, the forage is taken out of the trays to feed the livestock. Hydroponic forage attains plant height of 10 to 20 centimetres. Young tender growth of green shoots are grown from a cereal grain mostly barley or oat. However, other cereals like wheat and sorghum can be used. In India, maize and cowpea seed had been used for hydroponic forage production. It had been reported that 1.5- 3.0 litres of water is used to produce one kilogram of hydroponic maize fresh forage (11-14 per cent dry matter content) in seven days (Rajendra *et al.*, 1998; Naik *et al.*, 2015). The germinated seeds embedded in the root systems are consumed by livestock along with the shoots without any wastage of forage plant parts and nutrients (Pandey & Pathak, 1991). The hydroponic forage was found to be highly palatable with increased digestibility of nutrients which could be attributed to tenderness of the forage (Reddy *et al.*, 1988; Naik *et al.*, 2012; Naik *et al.*, 2014). The digestibility of the nutrients of the hydroponics forage was well comparable with the highly digestible legumes like berseem and other clovers (Pandey & Pathak, 1991). The digestible crude protein and total digestible nutrients contents of the hydroponics barley forage were optimum to meet the production requirement of the lactating cows (Reddy *et al.*, 1988). It had supplied 7.5 per cent digestible crude protein and 4.9 per cent more total digestible nutrients. It's advisable to use hydroponic forage with other roughage (Reddy *et al.*, 1988; Naik *et al.*, 2014). In essence, it replaces grains like dairy meal, pig's feeds and poultry feed concentrates. Globally, hydroponic forage is considered to be the best livestock feed as it has maximum protein, energy and very rich in vitamins such as B-carotene, trace elements, hormones, growth factors and enzymes. The livestock ingest completely roots, shoots and the residual seed. Farmers using this type of forage production can ensure a consistent supply of quality forage 365 days of the year irrespective of rain or sunshine. Naik *et al.* (2015) recommended 5-10 kg hydroponic maize forage in combination with concentrate.

Although hydroponic forage has a great attraction to those livestock farmers who wish to be more self-sufficient in forage supply but the yield, quality, and costs of this system appear not to make it sustainable and an economically viable system of forage production (Reddy *et al.*, 1988; Fazaeli *et al.*, 2011; Putnam *et al.*, 2013; Naik *et al.*, 2014). Moreover, emerging analytical facts, practicability and economic viability do not support its large scale adoption. Obviously, there will be net loss in dry matter yield of 24 to 30 per cent to drive growth of seedlings for six to seven days thus, a loss in feeding value compared to grain as such. Fazaeli *et al.* (2011, 2012) found that true protein decreased, and the non-fibre carbohydrate, metabolic energy, and *in vitro* gas production *decreased* in sprouted barley compared with the raw seed, and there were losses in dry matter yield. Most of the dry matter was found to be in the root/seed mass of hydroponic forage. Its proven fact that the total dry weight of the plant (*i.e.*, seed, root, shoot) decreases during six to seven days because the plant is using up stored carbohydrates from the seed. So such hydroponic systems are likely to lead to a net loss in dry matter and carbon up to about 10-14 days of growth. It may fit for those producers who do not have local sources for hay or forage, or simply want to be more self-sufficient. In situations, where conventional green forage cannot be grown successfully, hydroponic forage can be produced by the farmers for feeding their dairy animals (Naik *et al.*, 2015). Its future growth will depend largely on the development of the systems of production competitive in cost with systems of open field agriculture. For small animal producers (rabbits, *etc.*), this may offer a ready source of palatable feed. Moreover, pastures, alfalfa and field-grown forage crops have other environmental benefits and leguminous forage fix atmospheric nitrogen in the soil.

Tropical and Subtropical Grasslands/Grazing Lands

Natural grasslands are dominated by native species and natural processes that largely determine the type of grasslands along with some management by humanbeings. Man-made or non-natural" grasslands are somewhat inconclusive words used for those grasslands that are often found in forested regions, evolve on the sites that were cleared of trees, and probably planted with native graminoids or simply favoured to be occupied by the natural grassland species. As such, tropical and subtropical grasslands are not climax vegetation, *i.e* truly a natural ecosystem. In fact these tropical grasslands are in the mid successional stage and largely managed by annual or biannual burning and unprotected grasslands subjected to biotic factors including livestock grazing. The development of *seral* community is often controlled by environmental factors and it is retained as sub climax vegetation in semiarid and arid areas. In the areas of high rainfall, forest is the climax vegetation and grasslands evolve after clear felling of forests or edaphically and fluvial factors for example *terai* grasslands of northern India. Considering the ecosystems of these mid successional grasslands and plant species communities in various types of the grasslands, meticulous planning and management are required for sustainable production. In arid and semiarid regions of the country, large blocks of lands away from human habitations could be developed as grass reserve and their production may be stored in the form of hay in forage banks.

Types of Grasslands

Globally, grasslands cover 40 per cent of the area of the earth. Indian grasslands have been grouped into eight types by Whyte (1957) but Champion and Seth (1968) recognized only three broad categories. Indian Council of Agricultural Research conducted grassland surveys during 1954 to 1962 in India (Dabadghao and Shankarnarayan 1973). The five major grasslands are *Sehima –Dichanthium* (1740.000 square km), *Dichanthium- Cenchrus-Lasiurus* type (436,000 sq. km), *Phragmites-Saccharum-Imperata type* (2,800,000 sq.km), *Themeda-Arundinaria* type (230, 000 sq. km) and temperate and alpine cover having large unaccounted area. One separate chapter on forage resource development in mountain and hilly regions deals with alpine and temperate grasslands. In the Western *Ghats* of India, *Shola* grassland confined to the altitude1700 m and above. The *Shola* grassland appears to be climax vegetation which is interspersed with tropical forests in the mountain folds and the valleys. Continuous efforts are necessary for productive conservation in terms of their biological diversity and service functions and utilization for livestock production and sustenance of wild herbivore (Roy and Singh, 2008).

Over last 50-60 years, area under permanent pasture, cultivable wastelands and tree crop and groves reduced from 33.18 to 27.00 million hectares (Roy and Singh, 2008). The forage collection (green and dry forage together) also reduced from 3,04,490 tonnes in 1987 to 45,780 tonnes in 2005 (FAO, 2005).

Range Legumes

Range legumes enrich the protein content to the fresh grazing biomass available in the grasslands, besides improving the soil health of the grazing lands. The legumes in the form of fresh grazing material, hay or leaf meal also reduce the requirement of protein in the concentrate feeds. Skerman *et al.* (1988) listed wide range of legume plant species adapted to various agro-climatic and edaphic conditions for tropical grasslands (Table 7.5). Many of them were found suitable in Indian conditions after adaptive studies.

Table 7.5: Tropical Range Legumes for different Agro-climatic/Edaphic Situation

Rainfall (mm)	*Range Legumes*	*Forage Yields [(Fresh tonnes/ha), (Crude protein in per cent)]*	*Edaphic Condition/Seed Production*
800-2000	*Arachis glabrata*	60-80 (12)	Well drained sandy loam to clayey soils tolerates acidic (pH 4.5) and slight alkalinity.
	Arachis pinttoi	20-35 (13-25)	Sandy loam and river bottom soils, tolerates waterlogging.
250-1000	*Atylosia scaraboides*	8-13 (11-143)	Easy to adapt in wide range of soil, hardy, tolerates salinity, seed production (100-200 kg/ha).
>1500	*Calopogonium mucunoides*	30-40 (15-16)	Establishes easily, covers ground fast, poor tolerance to heavy shade, grows in wide range of soils, suited to acidic soil pH range 4.5-6.5, flood tolerant, become localised weed in humid tropics, low palatability, seed production (600-700 kg/ha).
>1500	*Centrosema pubescens*	15-20 (15-23)	Alluvial soils with medium soil fertility, optimum pH 4.9-5.5, moderately tolerant to poor drainage, stoloniferous root system, grow well in mixture with grasses, seed production (1100-150 kg/ha).
400-1500	*Clitoria ternatea*	50-75 (11-25)	Wide range of soils including highly clayey soil, pH - 4.5-8.7 tolerates salinity and sodicity, susceptible to inundation or water logging, less persistency under grazing situation, palatable and highly nutritious, seed production (500-750 kg/ha).
250-2000	*Desmanthus virgatus*	15-25 (15-22)	Wide range of soils, also suited to calcareous and saline soils, poor tolerance of heavy to medium shade, persistency under grazing, combines well with grasses, seed production (180-350 kg/ha).
900-1300	*Desmodium intortum*	19 (13)	Light to clayey loam soils, pH>5.0, also adapted to poorly drained/water logged lands, shade tolerant, low seedling vigour and low persistency in pasture, susceptible to heavy frost and drought, seed production (100-180 kg/ha).
>900	*Desmodium uncinatum*	18-35 (16)	It is not as palatable as *D. intortum*, low digestibility but robust vigour to compete with other weeds, persistent, tolerant to moderate fire, long growing season, adapted to wide range of soils, good for hay making, seed production (100-200 kg/ha).
900-1700	*Indigofera astragalina*	14—20 (15)	Low fertility sandy loam soils, suited to moderately acidic soil condition,
500-1500	*Lablab purpureus* Varieties: GAUD-1 for central and western India; Bundel Sem-1 (JLP-4) for all over India	25-27(11.7)	Wide range of soils but prefers deep sandy to heavy clayey soils in the range of soil pH 5.0-7.5, a dual purpose short lived legume (forage and green manure), frost susceptible, seed production (1000-2000 kg/ha).

Contd...

Rainfall (mm)	*Range Legumes*	*Forage Yields [(Fresh tonnes/ha), (Crude protein in per cent)]*	*Edaphic Condition/Seed Production*
>900	*Lotonois bainesii*	9-23 (15-20)	Well drained acidic (pH-5.0-6.0) sandy soils but tolerates temporary water logging, frost and heavy grazing, adapted to infertile soil, tolerant to fire, seed production (33-55 kg/ha).
600-1800	*Macroptilium atropurpureum*	15-20 (14-24)	Wide range of light to heavy texture soils and soil pH 4.5-8.0 with good drainage, grows well under full sunlight, drought resistant, susceptible to leaf disease, seed production (100-300 kg/ha).
750-2000	*Macroptilium lathyroides*	15-20 (14-23)	Tolerates acidic, saline or alkaline soils (pH - 5.0-8.0) and poor drainage conditions, widely adapted, poor persistency, susceptible to root knot disease seed production (200-250 kg/ha).
>900	*Macroptilium axillaris*	15-25 (10-12)	Sandy to clayey soils, pH >5.5 with good drainage, tolerates salinity and drought, long growing period, low persistency.
750-1500	*Neonotonia wightii (Glycine wightii)*	12-19 (13-24)	Heavy *black* soil with good drainage, does not tolerate high acidic conditions but moderately tolerant to salinity, drought, susceptible to frost and weevil attack, seed production (500-750 kg/ha).
>1500	*Pueraria phaseolides*	18-24 (14-21)	Sandy to clayey soils with pH 4.0-5.0, shade tolerant, good persistency, susceptible to frost and drought, seed production (300 kg/ha).
>1500	*Pueraria deeringianum*	—	Wide range of soils even on the poor eroded soils, drought resistant, susceptible to frost, highly palatable, good for hay and silage with grasses.
650-2500	*Stizolobium deeringianum*	15-25 (12-16)	Wide range of soils preferably sandy to sandy loam, slightly acidic condition, highly palatable good for soil conservation and green manure, seed production (1000-1500 kg/ha).
>1500	*Stylosanthes guianensis*	15-41 (12-18)	Warm humid climate with temperature range of 15-29°C but withstand up to 43°C, coarse textured to sandy loam soils pH 4.0- 8.3, low P demand, suited for acidic infertile soils, tolerant to Al and Mn, do not tolerate fire, good for cut and carry forage production, less persistent, frost susceptible, seed production (70-200 kg/ha),.
500-1300	*Stylosanthes hamata*	20-30 (11-19)	Wide range of soil textures pH- 5.4-8.0, drought resistant, well suited for seeding in native pasture with low soil fertility, highly persistent under grazing, susceptible to frost and water logging seed production (350-400 kg/ha).
500-1300	*Stylosanthes humilis*	12-18 (12-18)	Prefer sandy to sandy loam textured, well drained low fertility soils, pH - 5.0-6.5 slight acidity and salinity, highly persistent regenerated from falling seed, susceptible to shade and anthracnose disease, hooked seed characteristics make difficulty in sowing, seed production (400-1200 kg/ha) depending upon management level.

Contd...

Rainfall (mm)	Range Legumes	Forage Yields [(Fresh tonnes/ha), (Crude protein in per cent)]	Edaphic Condition/Seed Production
300-1200	*Stylosanthes scabra*	15-25 (9-12)	Wide range of soils with good drainage, tolerates soil salinity, drought tolerant, susceptible to frost and anthracnose, less palatable, seed production (100-250 kg/ha).
400-1200	*Stylosanthes seabrana* Variety- Phule Kranti (RSS-2000- 95) for Western Maharastra	8-21 (14-18)	Wide range of soils, prefer loamy to clayey loam soils, suitable for grazing, hay and leaf meal, seed production (200-300 kg/ha).
>1500	*Vigna luteola*	12—18 (17.4)	Wide range of soils tolerates salinity, highly successful on wet and poorly drained lands, not suitable for grazing, susceptible to frost, drought, insect and virus.

Source: Adapted from Skerman *et al.* (1988), Trivedi (2010), Pandey and Roy (2011).

About 40 forage legume species have been enlisted under genus *Stylosanthes* (Maass and Sawkins, 2004). Some of the *Stylosanthes species* are widely adapted forage legumes, which are already contributing to various forage production systems since last 25-30 years. Particularly, *S. hamata and S. humilis* are emerging as a great promise and becoming popular all over subtropical to tropical belt of India. The new introduction of *Stylosanthes seabrana,* native to Brazil introduced in India through ACIAR-ICAR—Stylo Project in 1998 was found promising. Once the crop established, it can sustain production under mono, mixed culture and silvipasture continuously for subsequent three years due to its good regeneration ability on rangeland conditions (Chandra *et al.,* 2006).

Naturally found *S. fruticosa* in Andhra Pradesh, Karnatka, Kerela, Orissa and Tamil Nadu have also potential but need in depth study for plant improvement. This indigenous species requires detailed studies for realizing their full potential. Field experiences have shown that any forage plant species which is not producing enough seed or easily propagated and established, has a lesser chance of adoption. Poor seed yield and poor establishment have been limiting factor with most of the *Stylosanthes* species, beside, important challenge of the susceptibility to anthracnose disease (Maass & Sawkins, 2004). Tolerance to shade is required for growing under silvipasture and various agroforestry systems. Regrowth of *S. guianensis* is often poor after cutting particularly when cut at maturity stage for seed and leaf meal. Cultivar development in these legumes should aim screening of a large stock of germplasms particularly for shade, anthracnose disease and growing in mixed culture. The comprehensive collections of *Stylosanthes* germplasms have been maintained at CIAT, ILRI, EMBRAPA and CSIRO as well as in India at ICAR-NBPGR and ICAR- IGFRI which can be accessed for the purpose. Seed standards are needed to expand for commercial seed production. Forage legume *Stylosanthes* should be targeted to small holding livestock production systems to reduce the cost and enhance the income of these farmers. Commercially, private sector should be attracted for seed production and leaf meal production. Altogether, these programmes will reduce the cost of livestock products by replacing costly concentrate feed and would make livestock enterprise more remunerative in the country.

Potential for forage production effects on the soil health (nitrogen fixation and addition of organic matter) has been well studied in grazed and mixed forage production systems and environment of *Stylosanthes species* (Changuin *et al.*, 2004; Muhr *et al.*, 1999c; Pengally and Conway, 2000; Obi, 1999). In long run, growing stylo in monoculture or stylo dominated pasture, acidification may be expected in light textured and the soils of low fertility. This happens due to leaching of nitrates down the soil profile as found in Australia, China and Thailand (Noble *et al.*, 2000).

Native range legumes *viz. Atylosia scaraeboides, Calopogonium mucunoides, Clitoria ternatea, Desmanthus virgatus, Lablab purpureus, Macroptilium atropurpureum, Desmodium* and *Indigofera species* are native and naturalized ones with a great promise (GOI, 2002). Suitable range legume species for particular agro climatic and edaphic conditions should be chosen and sown with good quality seed for proper establishment as pure or in mixed culture. The seeds of many range legumes have hard seed coat. To break the dormancy for good germination the seeds require mechanical, acid or hot water treatment. As consequence of poor seed yields in range legumes, availability of adequate quantity of seed are felt the major constraint.

Grassland Management

The grasses are considered to be the most evolved plant species, having a short life cycle as annuals yet a long life as perennials. Unlike trees, when perennial grass plant species are grazed or cut, they regrow instantaneously. Hence, they are most suited for grassland. Besides pasture grasses, range legume species and various fodder tree species are also suitable for incorporation into grassland to provide nutritious protein rich forage to the grazing livestock. As such efforts in India for grassland management and their utilization had been limited. Concept of scientific management of grasslands has not been properly executed due to inherent apathy towards community property resources and policy deficit. Further, grasslands are neither managed by the forest departments, whose interest lies mainly in the trees, nor by the agriculture departments and the animal husbandry department, who are interested only in the field crops and doing veterinary services for the livestock only. None of them is committed to integrate forage production and grassland development with livestock production. Collaboration of village community, non-governmental organizations (NGOs) is must for controlled and restricted grazing practices for sustainable grassland productivity. Village community imposed control such as systematic, restricted and rotational grazing, penalties for rule breakers, selling/conservation of surplus forages, educating villagers about better breeds and supplemental feeding will be helpful in this endeavour. Scientifically planned grazing based on carrying capacity of grasslands would optimise their utilization (Table 7.6).

Table 7.6: Improved Grazing Management of Grasslands

Controlled grazing	Based on the availability of forage from ground vegetation as well as trees/ shrubs; mixed herd grazing may be introduced based on the concept of land's carrying capacity.
Rotational grazing	The grazing area is divided into 4-6 paddocks and livestock species are allowed to graze in rotation for 7-15 days. By the time last paddock is grazed, the first one is ready again.
Deferred rotational grazing	Some of the rotational paddocks are left un-grazed for use in dry season and for proper seed formation. These paddocks should be changed every year to allow good seed bank in the soil in the entire grazing area.
Strip grazing	Grazing animals are confined to smaller areas of pasture paddocks rather than grazing randomly. In this scheme the advantage is that the area is fully utilized and the animal gets access to just right amount of feed requirement.
Cut and carry	In certain situations, livestock are not allowed to graze but the required quantity of pasture is harvested daily and fed to animals in stalls. Although, such a scheme has its own advantage, like trampling losses are less; in most situations it is not economical.

Source: Adapted from Roy, 2009, Patnayak, 1994; Sharma, 2004,

Improvement of forest fringes, degraded grasslands, reclamation of wastelands, and rehabilitation of problematic soils through edible vegetation cover of grasses, hortipasture and silvipasture is of prime importance to meet the forage requirement for livestock population in India. Pastures should be enriched as per the grazing/ browsing habits and forage requirements of livestock. For example nutritionally superior shrubs and tree species for the browser species like goats; proper balances of superior cultivars of grasses and legumes for sheep and cattle. Various range management strategies such as reseeding, protection, controlled burning, removal of unpalatable shrubs *etc.* should be regularly followed under a management plan with the support of grazier community. Promotion of stall-feeding and feed supplementation to prevent overgrazing is also required. There is a requirement of scientific consideration and adoption of technology for recovery of vegetation and its sustainable use for forage production. In natural grasslands and establishing pasture, perennial garss + legume intercropping systems have vast potential to produce nutrient reach forage particularly protein and minerals (Tables 7.7 and 7.8).

Table 7.7: Effect of Introduction of Range Legumes on Forage Yield in Natural Grassland of Central India

Grassland System	*Total Dry Matter Yield (tonnes/ha)*	*Grassland System*	*Total Dry Matter Yield (tonnes/ha)*
Natural grassland (NGLD)	3.3	NGLD + *Macroptilium atropurpureum*	4.1
NGLD + 40 kgN/ha	4.6	NGLD + *Macroptilium lathyroides*	4.9

Contd...

Grassland System	Total Dry Matter Yield (tonnes/ha)	Grassland System	Total Dry Matter Yield (tonnes/ha)
NGLD + *Alysicarpus rugosus*	4.2	NGLD + *Mimosa invisa*	3.7
NGLD + *Atylosia scarabaeoides*	4.1	NGLD + *Stizolobium deeringianum*	4.0
NGLD + *Clitoria ternatea*	4.4	NGLD + *Stylosanthes guianensis*	4.2
NGLD + *Dolichos lablab*	4.7	NGLD + *Stylosanthes humilis*	4.0
NGLD + *Desmodium tortuosum*	4.2	NGLD + *Sesbania sesban*	4.1
NGLD + *Glycine javanica*	3.8	NGLD + *Vigna luteola*	4.2
Cenchrus ciliaris + S. hamata	5.15	*Cenchrus ciliaris + S. hamata + L. leucocephala*	6.05
Cenchrus ciliaris + S. hamata + L. leucocephala + Albizia lebbeck	5.71	*Panicum maximum + S. hamata*	9.38

Source: Rai *et al.* (1980, 1985), Rai (1994), IGFRI Annual Report (2003).

In arid grasslands, introduction of improved perennial grasses, shrubs and fodder trees, improved productivity of grassland which supported more number of livestock (Singh *et al.*, 1983; Singh, 1984; Singh *et al.*, 1984; Singh, 1985; Singh *et al.*, 1985; Singh *et al.*, 1987; Singh *et al.*, 1988; Rai *et al.*, 1988; Roy, 2009). In semiarid regions, re-vegetation of ravine lands with improved management practices increased forage production manifold (Planning Commission, 2012). In arid and semi-arid regions, range grasses (*Cenchrus ciliaris, C. setigerus, Lasiurus sindicus, Sehima nervosum* and *Chrysopogon fulvus*) with or without range legumes (*Stylosanthes hamata/ S. seabrana/ S. humis,* Siratro, *Clitoria ternatea*) can be grown in natural grasslands or in agrisilvi/ silvipasture/ hortipasture/ CPRs/ cultivable waste lands. In this region, tree leaves of khejri, shisham, babul, subabool, wild ber (*Ziziphus* species) account for considerable part of forage resource. Fodder trees are allowed to grow on the arable lands and their field bunds *viz., Prosopis cineraria, A. nilotica, Albizia amara, Azadirachta indica, Ailanthus excelsa, Albizia leebeck.*

Table 7.8: Effect of Grass + Legume Intercropping Systems on Forage and Crude Protein Yields in Central India

Grass + Legume Intercropping	Forage Yield (DM tonnes/ha)	Crude Protein (Kg/ha)	Remarks
Cenchrus ciliaris + *S. hamata* (1:1)	3.57	326.8	Singh *et al.*, 1983
Green panic (*Panicum maximum var. trichoglume*) + *S. hamata* with 26.4 kg P/ha and 49.8 kg K/ha	40.4	531.0	Singh *et al.*, 1983
Blue panic (*Panicum antidotale*) + clusterbean (*Cyampsosis tetragonoloba*)	46.5	—	Singh *et al.*, 1984

Contd...

Grass + Legume Intercropping	*Forage Yield (DM tonnes/ha)*	*Crude Protein (Kg/ha)*	*Remarks*
Blue panic + *Clitoria ternatea*	120.0 (fresh yield)	316.2	Singh *et al.*, 1985
Guinea grass (*Panicum maximum*) + *Macroptilum atropurpureum*	6.68	589.7	Singh *et al.*, 1986
Guinea grass + *Stylosanthes hamata*	19.9		Singh, 1985
Guinea grass + *Stylosanthes hamata*	4.50	427.9	Sunil Kumar and Shukla (2010)
Guinea grass + *Stylosanthes hamata*	6.29	547.0	Ram *et al.*, 2006-2011
Guinea grass + *S. scabra*	6.68	582.0	
Guinea grass + *Clitoria ternatea*	5.41	457.73	
Guinea grass + *M. atropurpureum*	5.60	474.9	
Guinea grass cut at 60 days interval and applied FYM 10t/ha + 60 kgN/ha	8.53	548.0	Ram *et al.* (2006-2010) Ram (2006-2011).

Korangadu is a traditional farming system existing in semiarid tract of Tamil Nadu state of South India *viz.*, Dharapuram, Kangeyam, Palladam, Moolanur and Kallimanthayam areas. Sustainable management of grassland in this tropical region has shown a way out, as they are continuously being managed and utilized in a sustainable way since hundreds of years in a semiarid tropical climate (Vivekanandan and Korangadu, 2007; Kumar *et al.*, 2011). Understanding the inherent potential and limitations of the system, the farmers have organized their grazing lands into paddocks with live fence of *Balsmodendron berryi.* As a result, labour input is reduced to a bare minimum and it provides opportunity for other gainful employment such as handloom *etc.* The grassland dominated with *Cenchrus* species, is sown with legumes like *Phaseolus trilobus* to enrich the pasture nutritionally. Studies under taken on development of Community Property Resources (CPR) and Joint Forest Management (JFM) validated that improved pasture/silvipasture/hortipasture increased forage productivity, and enhanced livestock productivity including livelihood options. The various activities undertaken under soil conservation/reclamation projects, watershed projects, wasteland development projects *etc.* have some component of forage grasses/ forage. A synergistic approach should be adopted for regulated use of such forage resources. The vast tract of unused lands along roadsides, railway line and other wastelands could be utilized effectively through leasing out for certain period for forage production and regulated grazing to people residing in the vicinity. In southern region, cultivable wastelands can be used to be sown with range grasses (*Cenchrus ciliaris, Chrysopogon fulvus, Panicum antidotale, Sehima nervosum, Dicanthium annulatum*) grown as sole or mixed with *Stylosanthes hamata,* and *Centrosema pubescens.*

Livestock particularly goat rearing is considered detrimental to vegetation growth. But various studies had demonstrated that grazing had a role to play in vegetation regeneration, weed control, prevention of forest fires *etc.* (Roy, 2009; Meher-Homji, 1997; Singh, 2002). Grazing and browsing livestock contributed to the soil fertility, and seed dispersal through dung and droppings and control herbaceous growth to promote tree seedling growth (Shankarnarayan *et al.*, 1985; Lu, 1988; Kolars, 1996; Rangnekar, 2006). Thus, grazing livestock have role as regenerators of vegetation. There is need for optimum utilization of grazing resource in and around forest area. A mixed herd grazing of livestock would be helpful for optimal use of such grazing lands. The goat browses the bush and low set tree branches, cattle graze grasses and herbaceous plants whereas sheep confine grazing shorter and close to ground vegetation.

Indian Grasslands, Ecological Services and the Wildlife

In India common property resources (CPRs) are important grazing resources especially in hills, arid and semi-arid region. Small ruminants are largely maintained on CPRs. These CPR, are important sources of livelihood and income for poor people in all the states (Table 7.9). CPRs contribute and allow considerable access to all users, but not all the sections of the rural community are equally attracted by these potentials and opportunities. The proportion of poor households dependent on fuel, forage and food items from CPR's ranged between 84-100 per cent in different villages. In general, the rural poor obtain the bulk of forage from CPR's. CPR's complement private farming by providing crop inputs, sustaining animals and are especially predominant in drought prone area. Any positive or negative change in its position will have impact on several environmental issues.

Grasslands are source of genetic resources in the form of grass and shrub plant species. For example, most of our cereals have originated from wild grasses. They are also habitats of some of the unique wildlife. The semi-arid grasslands, being largest one, spread over 5,48,850 square kilometres *i.e.* 16.60 per cent of India's geographical area in eastern Rajasthan, Gujarat, western Madhya Pradesh, and parts of Uttar Pradesh, Haryana, Punjab and southern parts of Jammu and Kashmir. With growing population of livestock, the grasslands are under tremendous biotic pressure, mainly grazing and conversion to other uses. India having large livestock population, commiserating with high requirement of forage and feed needed various policy interventions and application of ecological principles for conservation and management techniques and technologies for protection, restoration and sustainable use of grasslands in the country. These improved productive conservation and management of grasslands would also serve better way as important catchment for rivers, streams, reservoirs, dams, check-dams and village ponds, greatly helping in the water regime and hydrological cycle including habitat for wildlife. Therefore, it is imperative to recognize the ecological, genetic source, hydrological, economic and sociological functions and the role played by grasslands.

Table 7.9: Grazing Resources in India

Resources	*Area (Million ha.)*	*Percentage*
Forests	69.41	22.70
Permanent pastures, grazing lands	10.90	3.60
Cultivable wasteland	13.66	4.50
Fallow land	24.99	8.10
Fallow land other than current fallows	10.19	3.30
Barren uncultivable wastelands	19.26	6.30
Total common property resources other than forests	54.01	17.70

Source: Roy and Singh, 2008.

Silvipasture

Silvipasture systems optimise land use by enhancing total productivity, diversify outputs and conserve production resources by stabilizing the soil and nutrients recycling (Shelton, 1991). The silvipasture provides opportunity for vertical expansion of forage resource as live fodder bank during drought and flood with efficient nutrient recycling under forage production systems. Desirable traits for adaptation of understory crops are large size thin leaves with thinner cuticles, smaller size cells, lower chlorophyll a/b ratios with higher longevity and horizontal orientation on the stem with lower photosynthetic capacity, compensation point, often reduced production of tiller, few leaves and less roots resulting in lower shoot to root ratio (Shelton *et al.,* 1987). Low midday air and surface soil temperatures and increased relative humidity under trees, reduces evaporation and evapotranspiration by understory forage crops (Wilson & Ludlow 1991). Ultimately, better availability of soil moisture is beneficial for understory pasture species. However, forage production under canopy of various tree species depends upon light interception reaching the under story forage grasses and legumes (Wong *et al.,* 1985). Tree/shrub component in a silvipasture may be grown for forage or for timber and fuel wood. Recently, studies have shown that bamboo based silvipasture can also be developed to get bamboo leaves as forage beside the culms. Shrubs and tree plant species need to meet certain criteria to be qualified as tree fodder for feeding the livestock (Riveros, 1992; FAO, 1994, 1998). It must possess certain requisites characteristics for qualifying which are: i) annual high foliage production ability to obtain significant levels of edible biomass for the consumption of livestock to expect changes in the animal response parameters; ii) agronomic versatility like inherent regenerative capacity to withstand annual lopping management, iii) the palatability and nutritional quality near to most nutritious forage crops with least contents of anti-nutritional factors to make it attractive for animal production; iv) preferably well adapted native species are chosen which can be established easily with simple agronomic practices. Alley cropping in some cases may also provide pruned biomass as forage to be fed to livestock instead of use for soil fertility improvements. It is a form of hedgerow of fast growing nitrogen fixing shrubs and trees intercropping, where hedge is widely spaced to grow crops with benefits of soil erosion control. Tree fodder is mainly obtained in the form of leaf twigs and/or pods, largely fed to small ruminants. This forage is usually rich in crude protein and minerals particularly calcium, phosphorus and micronutrients very much required for dairy animal. However, anti-nutritional factors *viz.* tannins, other phenolic compounds are found in the tree fodders. Tree fodder has added advantage to meet partly lean season (summer) requirement of green forage. Niche area for silvipastures is largely non-arable marginal lands. Many leguminous forage trees play significant role in maintaining soil fertility, conserving the soil, act as windbreaks and provide shade for the grazing livestock during hot summer and sustains productivity of rangelands. Within broad category of silvipasture, several types of system and practices can be identified such as intensively managed cut and carry system, where well laid out fodder trees in combination of improved forage plant species or block or boundaries or live fence posts other designated places of fodder tree plantations is lopped periodically to feed livestock in stalls (Nair, 1993). In extensively managed silvipasture systems, grazing of herbaceous forage plant species and browsing of tender twigs along with leaves, fruits and pods of trees/shrubs are practiced based on livestock carrying capacity. Thus, silvipasture systems have various types of practices ranging from extensive nomadic to intensive cut and carry forage production. Under agro climatic condition of semiarid region, productive silvipasture systems can be developed as per site characteristics. Some of the suitable trees, grasses and legumes are summarized for various region of India in Table 7.10 (Trivedi, 2010).

Protection by locally available fencing resources *viz.* stone walls, trench cum mound, live hedge, barbed wire fencing *etc.* is prerequisite for establishing a silvipasture. During monsoon, planting of seedlings or the grasses and legumes are done in combination with suitable tree species and application of moderate dose of fertilizer (20 kg N/ha and 40 kg P_2O_5) at the time of establishment to accelerate early growth and continuously applied in subsequent years with the onset of monsoon to sustain production (Pathak & Roy 1994). In silvipasture, grazing should be practiced only after three years of establishment, till then forage should be utilized under cut and carry system. Lopping for tree fodder should not exceed 40 to 50 per cent during lean period (Singh & Thompson, 1995). Based on growth parameters of a fodder tree species *viz.* basal diameter, diameter at breast height, crown volume, productivity of tree fodder can be estimated or predicted (Singh *et al.,* 1994).

The tree density of 400-600 trees/ha have been found to be ideal for most of the semi-arid tract of the country for optimum system productivity in short to medium rotation cycles of 10-15 years. By supplementing feeding, mixed herd of livestock (cattle + sheep + goat) could be maintained on such silvipasture. In semiarid parts of the country on degraded grasslands, scrub forest lands, heavily eroded and ravine areas, alkali/sodic soils, mine spoils and water-logged area,

Table 7.10: Range Grasses, Legumes and Tree Species for different Agro-climatic Regions of India

Region	*Grasses*	*Legumes*	*Trees*
1. Western Rajasthan, southeast Punjab, western Gujarat including Kutch and Western Saurashtra	*Cenchrus ciliaris, C. setigerus, Dichanthium annulatum, Lasiurus sindicus, Panicum antidotale*	*Atylosia scarabaeoides, Lablab purpureus, Macroptilium atropurpureum, Rhynchoisa minima*	*Acacia tortilis, Albizia amara, Azardichta indica, Dichrosttachys cinerea, Hardwickia binata, Prosopsis cinerea, P. juliflora, Leucaena leucocephala*
2. Central Punjab, eastern Rajasthan, Western Uttar Pradesh, Northern Gujarat, Maharastra, Western Madhya Pradesh, Andhra Pradesh, Tamilnadu and Karnatka	*Axonopus compressus, C. ciliaris, C. setigerus, D. annulatum, Heteropogon contortus, P. antidotale, Pennisetum pedicellatum, Sehima nervosum*	*A. scarabaeoides, L. purpureus, M. atropurpureum, Mucuna deeringiana, Stylosanthes hamata, S. scabra, S. viscosa*	*A. Tortilis, A. amara, A. lebbeck, A. procera, Azardichta indica, Dalbergia sissoo, H. binata, L. leucocephala*
3. Central and Eastern Uttar Pradesh, Southern Bihar, Southern Gujarat, Coastal Andhra Pradesh and Coastal Tamil Nadu	*Brachiaria mutica, C. ciliaris, Cynodon dactylon, D. annulatum, P. antidotale, P. maximum, Paspalum notatum, Pennisetum pedicellatum, P. polystachyon, Setaria sphacelata*	*M. atropurpureum, Centrosema pubescens, Glycine javanica, L. purpureus, Mucuna deeringiana, Puerraria phaseoloides, Stylosanthes guianensis*	*A. lebbeck, Anogeissus latifolia, Azardichta indica, Erythrina variegata, Gliricidia sepium, Ficus species, L. leucocephala, Morus alba, Moringa oleifera, Tamarix aphylla*
4. West Bengal, Assam, Orissa, Chattisgarh, Konkan region of Maharashtra	*Brachiaria brizantha, B. mutica, Cynodon dactylon, D. annulatum, P. maximum, Paspalum notatum, Pennisetum clandestinum, P. pedicellatum, Setaria sphacelata< Urochloa mosambicensis*	*Calopogonium mucunoides, Centrosema pubescens, Desmodium species, Stylosanthes guianensis, Vigna umbellata*	*A. lebbeck, Bauhinia purpurea, Gliricidia sepium, Morus alba, Moringa oleifera.*
5. Lower hills of Punjab, Himachal Pradesh, Uttarakhand and West Bengal	*Arundinaria species, Chrysopogon fulvus, D. annulatum, P. pedicellatum, Themeda species*	*A. scarabaeoides, L. purpureus, M. atropurpureum, S. hamata, S. humilis, S. scabra*	*A. amara, A. lebbeck, Azardichta indica, Ficus species*

Contd...

Region	Grasses	Legumes	Trees
6. High hills of Uttarakhand, Himachal Pradesh, Punjab, Jammu and Kashmir	*Bromus uniloides, Dactylis glomerata, Festuca arundinacea, Lolium perenne*	*Lespedeza cuneata, Medicago sativa, Trifolium pratense, T. repens*	*Grewia optiva, Quercus incana, Robinia pseudocacia, Salix tetrasperma*
7. Valleys of Uttar Pradesh, Gujarat and Jharkhand	*Axonopus affinis, A. compressus, C. ciliaris, Chrysopogon fulvus, D. annulatum, P. pedicellatum,*	*Lablab purpureus, M. atropurpureum, Mucuna deeringgiana, Pueraria phaseoloides*	*A. tortilis, Dichrosttachys cinerea, Prosopsis juliflora,*
8. Saline and Alkaline areas of Punjab, Haryana, Uttar Pradesh, Rajasthan and Gujarat	*Brachiaria mutica, Bothriochloa pertusa, Chloris gayana, Cynodon dactylon, D. annulatum, Diplachne fusca, Iseilema laxum, P. maximum, S. sphacelata, Sporobolus marginatus*	*Alysicarpus rugosus, Demanthus virgatus, Rhynchoisa minima*	*A. tortilis, Dichrosttachys cinerea, L. leucocephala, Prosopsis juliflora, Sesbania grandiflora*
9. In marshy and wetlands	*B. mutica, B. ruziensis, Iseilema laxum*	*Lotononis bainisii, Sesbania bispinosa*	*Salix species in alpine region*

Source: Adapted from Trivedi (2010).

silvipasture with suitable plant species had proven their potential (Abrol & Dhruva Narayana, 1990). Building microbial biomass is a challenge on the mine spoils. Appropriate tree/shrub, grass and legume coupled with chemical amendments, mulching and top soil replacement *etc.* could help in speeding up of the process of re-vegetation (Jha & Singh 1992).

Studies on silvipasture systems in five agro-ecological regions of India covering semiarid to hilly areas showed that pasture yields stabilized in the third year after establishment (Misri, 2006; Roy 2007). In arid and semiarid region, *Acacia nilotica, Leucaena leucocephala, Acacia planifrons, Acacia catechu, Prosopis cineraria,* date palm and *Hardwickia binata* in combination with natural grass *Sehima nervosum* had been traditionally valued silvipastures for livestock production (Kaushal & Chinnamani 1992). Some newly developed silvipasture in this region are (*a*) *Leucaena leucocephala* (S 24) + *Cenchrus ciliaris* + *Stylosanthes seabrana,* (*b*) *Bauhinia malabarica* + *C. ciliaris* + *S. seabrana,* (*c*) *B. malabarica* + *Chrysopogon fulvus* + *S seabrana,* (*d*) *L. leucocephala* (S 24) + *C. fulvus* + *S seabrana,* (e) *Hardwickia binata* + *C. ciliaris* + *S. hamata,* (f) *H. binata* + *C. fulvus* + *S. hamata,* (g) *H. binata* + *Panicum maximum* + *S. hamata,*(h) *Acacia tortilis* + *C. ciliaris* (Shankarnarayan *et al.,* 1987)showing forage productivity of 4.5 to 6.0 tonnes dry matter per hectare per year in first five years of establishment. Another, hedge row system of *L. leucocephala* (S 24), *B. malabarica and S. sesban* clipped at one metre plant height produced *0.44 to 1.13* tonnes dry matter per hectare per year (Rai & Gupta, 2007-13; and Ram & Shukla, 2008-13). Potential of *Ficus infectoria, Madhuca latifolia, Morus alba* and *Acacia nilotica* trees with understory *P. maximum, C. ternatea, S. seabrana* under silvipasture systems was demonstrated which produced about 38-41 tonnes green forage under tree + grass and 13-19 tonnes/ha under tree + range legumes systems (Anonymous, 2010-14). The carrying capacity of cattle/heifers and sheep per hectare depends on productivity and system of grazing practiced. Continuous grazing is not recommended for silvipasture systems, rather rotational grazing based on compartmentalized paddocks as well as a planned grazing system in which grazing of large ruminants is followed by small ruminants particularly sheep should be practiced. There is potential to diversify a grazing operation and improve economic or environmental benefits on many acres through conversion of pasture to silvipasture. If managed properly, forage production can be maintained while producing fuel and valued timber under silvipasture systems. Silvipasture along with trenching in combination with vegetative barriers as soil conservation measures reduced runoff from 38.9 per cent to 6.8 per cent of the rainfall as compared to natural grazing land (Misri, 2006; Roy 2007). In India, social and economic studies had to be taken to understand various constraints in the adoption of silvipasture at individual farm/community levels and address them to accrue economic and environmental benefits of silvipasture.

Hortipasture

Hortipasture on the degraded lands particularly under rain fed situation has immense potential particularly for land capability class V and VI in a vast tract of rain fed arid to semiarid situations in India. The system would help to reduce the gap between forage demand and supply. Selection of appropriate combination of fruit trees and forage plant species as per site characteristic like soil depth and texture, *in-situ* soil water conservation in a specific agro climatic region and protection from biotic interferences by barbed/woven/chicken wire/stonewall/live hedge fencing or cattle proof trenches are essential for establishing hortipasture. The principle of water conservation should be to create a mini- watershed to collect sheet flow even due to light rain that occurred in the specific surrounding area. Depending on the site slope characteristics, planting could be done on contour, terrace, trenches, gullies or at lower ends of micro-catchments for effective accumulation of run off to recharge the soil moisture and make it available in the root zones of all plant species in the hortipasture. Drip irrigation can be successfully used in dry degraded lands to economise water use under hortipasture. Selection of fruit tree for hortipasture system should be based on complementarity with understory pastures plant species. Singh and Osman (1995) found tamarind (*Tamarindus indica*), wood apple (*Limonia acidissima*), jamun (*Syzygium cuminii*) aonla (*Emblica officinalis*), bael (*Aegle marmelos*) appropriate fruit trees in semi-arid southern region. Other fruit tree species *viz. Annona squamosa, Carissa carandus, Citrus limon, Emblica officinalis, Morus alba, Psidium guajava, Zizyphus mauritiana* are also suitable in ranfed area. In continuous grazing system, custard apple could be one of the ideal options as it is hardy and not browsed by livestock. Ber grows very well in combination with pasture and the foliage is also very nutritious and palatable to livestock. Based on rainfall pattern, Vashishtha (1991) recommended fruit trees for different rainfall zones. Accordingly, ber (*Zizyphus mauritiana*), gonda (*Cordia myxa*), ker (*Capparia decidua*), pilu (*Salvadora oleides*) had been found suitable fruit tree in the less than 350 mm rainfall zone; aonla, khirni (*Manikara hexandra*), jamun (*Syzygium cumunii*) and mulberry (*Morus alba*) in 350-500 rainfall zone; bael (*Aegle marmelos*), mango (*Mangifera indica*), custard apple (*Annona squamosa*), sour lime (*Citrus aurantifolia*) and aonla in 500-700 rainfall zone and for more than 700 rainfall zone, mango, pomegranate (*Punica granatum*), guava (*Psidium guazava*), jackfruit (*Artocarpus heterophyllus*), wood apple (*Limonia acidissima*), tamarind (*Tamarindus indica*) and aonla were found suitable for hortipasture systems in semiarid region. Rao and Rao (2004) also suggested suitable fruit species karonda (*Carissa carandus*), phalsa (*Grewia subinequalis*), sapota (*Achrus zapota*), mango (*Mangifera indica*), fig (*Ficus carica*), passion fruit (*Passiflora edulis*i), sweet orange (*Citrus sinensis*), drum stick (*Moringa oleifera*), wood apple (*Ferronia limonia*), chironji (*Buchanenia lanzan*) and soapnut (*Sapindus mukorosil*) for southern belt of semi-arid conditions. Leaves of fruit trees used as top feed for livestock are ber (*Z. nummularia, Z. mauritian*a), pilu (*S. oleoides*), goonda (*Cordia myxa*), bael (*Aegle marmelos*), tamarind (*Tamarindus indica*), drumstick (*Moringa olefera*), khejri (*P. cineraria*) and ker (*C. decidua*). Horti-pastoral system can supply forage to livestock in two ways. Fruit trees, apart from yielding fruits also give valuable leaf fodder (top feed). Crude protein in the leaves of fruit trees ranges 9.6 to 16.9 per cent on the dry matter basis. These top feed being source of palatable and nutritive forage can be used during lean period and extreme drought condition. The fruit trees such as bordi (*Ziziphus rotundifolia*), pilu (*Salvadora oleoides*), lasoda (*Cordia myxa*), goondi (*Cordia gharaf*), kumat (*Acacia senegal*), khejri, (*Prosopis cineraria*), tamarind (*Tamarindus indica*) are commonly used as top feed for sustaining livestock particularly small ruminants.

Selection of perennial grasses and legumes for understory cropping are done on the basis of compatibility with fruit trees, easy to establish, partially shade tolerant, fast regrowth, amenable to multi-cut specific to agro-climatic region. These are important to optimize forage and fruit production under hortipasture. Aonla (*Emblica officinalis*) had been widely studied as a component of hortipasture in semi-arid region of India and was found, one of the most suitable fruit species (Table 7.11, Photo 7.3). The tree canopy of aonla allows intercepting light and permitting intercropping even after it has attained full growth, Interspaces of old as well as young orchards have a great potential to be put under hortipasture. Sharma (2004), Kumar & Chaubey (2008) and Kumar & Shukla (2011) also reported that understory forage crop combinations did not influenced the fruit yield of *ber* and *aonla* tree species due to different rooting behaviour of both the under storey crops as well as fruit tree.

Grazing cattle in natural stand of coconut without any management and inputs is widely practiced. Now, improved pasture plant species with good management practices are also prevalent in some coconut plantations and cassava crop (Table 7.12). In coconut plantation with native edible plant species many grasses, broad leaved plants and shrubs grows as weeds. These are being grazed but the quality and quantity of forage are at low level (Nair, 1993). Benefits in coconut based silvipasture is associated with a number of management and environmental differences such as plantation density and age, thereby, light transmission; improved or native forage plant species, height of grass; inclusion of forage legumes; soil fertility regimes; weed control; cultivation practices; stocking density and livestock types (small or large ruminants). There have been concerns of plantation owners about likely effects of any intercrops on coconut yields. In existing coconut stands after establishment of improved pastures, slight reduction in coconut yields due to soil/root disturbances and nutrients demands of the sown pasture may be expected. However, ensuring adequate soil moisture and soil nutrient supply to the plantation, the pastures will have no adverse effect on coconut yields. In general, there was increase in yield of coconut reported from Papua New Guinea, Malaysia and Philippines with *Setaria sphacelata* and legumes while in Sri Lanka yield was sustained with inclusion of legumes (Ohler, 2014). In India, Sahasranaman *et al.*, (1983) found increase in coconut yield under mixed farming. The manure and the better nutrient support received from livestock contributed to increase in coconut yield by 28 per cent. Where there is range of choices for intercrops, intercrop of cocoa, coffee, banana gave more return than grazing livestock under coconut. Sometime Cassava (*Mannihot esculentus*) is also intercropped with forage crops.

Photo 7.3a: Aonla + Veranostyl (*Stylosanthes hamata*).

Photo 7.3b: Aonla + Guinea Grass.

Photo 7.3: Aonla Based Hortipasture in Semi Arid Region (*Courtsey*: Sunil Kumar)

Table 7.11: Hotipasture Systems

Hortipasture System	*Fruit Yield (tonnes/ha)*	*Forage Yield (Dry matter (tonnes/ha)*	*Crude Protein (Kg/ha)*	*Remarks*
Aonla + guinea grass + *Stylosanthes hamata*	13.5	5.1	450.7	10 years old orchard, Central India, Sunil Kumar & Roy (2005) Sunil Kumar & Shukla (2010)
Aonla + guinea grass	12.7	4.2	376.4	
Aonla + *S. hamata*	14.8	2.5	317.5	
Aonla + pearl millet single cut + cowpea	13.87	4.25	378.8	13 years old orchard, Central India, Sunil Kumar *et al.* (2013)
Aonla + pearl millet multi cut + cowpea	13.37	4.51	406.1	
Aonla + Guinea grass + *S. hamata*	13.50	5.10	450.7	10 years old ochard, Central India, Sunil Kumar and Shukla (2011)_
Aonla + Guinea grass	12.70	4.20	376.4	
Aonla + *S. hamata*	14.8	2.50	317.5	
Aonla + natural grassland	2.64 in 5th years of planting to 16.63 in 10th year	2.60	—	10 years study since establishment, Sunil Kumar and Choubey (2008) and Sunil Kumar*et al.*, 2009
Aonla + + *Dichanthium annulatum* + 60 kg N/ha		4.55	—	
Aonla + *Chenchrus ciliaris* + *S. seabreana* with out *in situ* soil moisture conservation	0.08-2.07	0.87-7.24 (4.74)		Sunil Kumar *et al.* (2014). early five years
Aonla + *Chenchrus ciliaris* + *S. seabrana* with contour staggered trench method of *in situ* soil moisture conservation	0.48-3.41	0.81-8.95 (5.95)		Sunil Kumar *et al.* (2014), early five years
Aonla + *Chenchrus ciliaris* + *S. hamata*	—	2.33	—	Khan *et al.* (1991), (two cuts per 1000meters length of the bund of orchard)

Hortipasture System	*Fruit Yield (tonnes/ha)*	*Forage Yield (Dry matter (tonnes/ha)*	*Crude Protein (Kg/ha)*	*Remarks*
Ber + *C. ciliaris* + *S. hamata*		4.63		Khan *et al*, 2001
Ber + *Napier* + *S. hamata*		47.10		Khan *et al*., 2001
Ber (var. Gola) + guinea grass + *S. hamata*	6.76	5.87 + 0.92	441.6 + 114.1	Ram *et al.* (2005), 10 years old ber orchard
Ber + Dinanath grass + *S. hamata*	6.65	3.84 + 0.94	330.9 + 116.6	Tree leaves out of pruning of ber
Bael + *C. ciliaris* + *S. seabreana* without *in situ* soil moisture conservation	1.14	4.13		Shukla *et al.* (2014) first five years
Bael + *C. ciliaris* + *S. seabreana* with contour staggered trench method of *in situ* soil moisture conservation	2.41	5.78		Shukla *et al.* (2014) first five years
Guava + *Chenchrus ciliaris* + *S. seabreana* without *in situ* soil moisture conservation	4.14 Lalit 4.71 Shweta	6.34		Sunil Kumar *et al.* (2014) early six years
Guava + *Chenchrusciliaris* + *S. seabrana* with contour staggered trench method of *in situ* soil moisture conservation	5.48 Lalit 4.71Shweta	7.92		
Karonda (*Carrisa carandus* + *Napier* + *S. hamata*		48.10?		Sharma, 2004
Mango + *Stylosanthes hamata* with soil water conservation		10.01		Reddy, 2004
Mango + *Cenchrus ciliaris* with soil water conservation		10.12		

Guava varieties: Lalit and Shweta.

Table 7.12: Coconut and Cassava (*Mannihotes culentus*) Based Hortipasture/Silvipasture in Humid Tropical Situations

System	*Forage Yield (Dry matter tonnes/ha)*	*Tuber Yield (tonnes/ ha)*	*Remark*
Cassava + *Sylosanthes guianensis*	2.22	9.74	Pillai *et al.*, 1994
Cassava + fodder cowpea	10.88 green	22.61	
Coconut + napier bajra hybrid PBN 16	19.64	2142no. (89 trees/ ha)	Manjunath and Singh, 2005, Goa
Coconut + guinea grass	12.24		
Coconut + guinea grass + *Stylosanthes* species/*Centrosema pubescens*	-		Pillai *et al.*, 1994, Kerela
Leucaena leucocephala + *Zea mexicana*	0.7		
Casurina equisitifolia + *B. ruziensis*	0.65		
Casurina equisitifolia + *P. maximum*	1.350		
Ailanthus triphysa + *Panicum purpureum*			Kumar *et al.*, 2001a, Kerala
Acacia tortilis (10m x 10m + *C. ciliaris*	5.58		Shankarnarayan *et al.*, 1987, arid region
Coconut + guinea grass (10 tonnes FYM, 150kg N, 50kg P_2O_5, 50 kg K_2O/ha)	99.53		Sansamma *et al.*, 2000

Source: Adapted and modified from Pillai, *et al.*, 1994, Kumar, 2006.

Several grasses, legumes and fodder trees have been identified as suitable intercrops for the tropical situations of the coconut gardens. The prominent among them are guinea grass (Hamil, Makueni, Riversdale, FR 600, Haritha, PGG 163, PGG 112, PGG 208), Golden timothy grass (Natrok), congosignal grass, *Stylosanthes guianensis, S. hamata, Centrosema pubescens,* fodder trees *Leucaena leucocephala, Sesbania grandiflora* and *Desmanthus virgatus.* On average grasses gave 57 tonnes/ ha and legume 35t/ha fresh forage yield (AICRP 1971 - 2011). Range of nutrient contents in the various sources of forage is shown in Table 7.13. Tree fodders are generally rich in minerals and protein.

Beneficial Association of Microorganism and Insects with Forage Crops

Role played by fauna and microorganism is vital for Forage production as in any agricultural production system (Table 7.14; Saxena *et al.,* 2002). Insects pollinate lucerne, berseem and other clovers embracing *Diptera, Lepidoptera* and *Hymenoptera.* These pollinators are dominated by honeybees, solitary bees and bumble bees belonging to Hymenoptera order and enhance seed production. VAM fungi control parasitic nematodes as a biological control measure. The role of soil

Table 7.13: Range of Nutrients in Various Forage Crops

Forage	*Type*	*CP (per cent)*	*Macro Mineral (per cent)*		*Micro Minerals (ppm)*		
			Ca	*P*	*Zn*	*Cu*	*Fe*
		Ideal >15 per cent	*Ideal >0.5 per cent*	*Ideal >0.22 per cent*	*Ideal >30*	*Ideal 10 or less*	*Ideal 50-100*
Cultivated Forage	Leguminous forage	15-22	0.6-2.0	0.05-0.25	25-35	8-12	200-300
	Non Leguminous forage	8-13	0.3-0.6	0.05-0.2	15-20	4-10	100-200
Tree fodder (Top Feeds)		13-30	1.1-3.2	0.15-0.40	34-118	8-16	128-363
tLeaves of bamboo		9-19	0.9-1.6	0.08-0.15	20-48	7-11	91-750

arthropods become important in the grasslands of arid and semiarid regions where they act as mediators of nutrient cycle. The grasslands have more diversity of soil arthropods than the silvipastures.

Table 7.14: Role of Insect / Microorganism in Forage Crops

Forage Crops	*Name of Insect/Microorganism and its Role*
Lucerne	Important pollinators are - *Apis dorsata, Pithitis smaradula, Megachile rotundata, M. flavipes, M. nana, M. lunata, Certania binghami, A. mellifera, A. cerana, Andrena levilabris, Xylocopa fenestrata, Chalcidoma rubripes* and *Nomia* spp.
Berseem	Important pollinators are - *Apis dorsata, M. flavipes, Pithitis smaradula, Apis mellifera, Lassioglossum cattulum, Halitus spp., Xylocopa spp., Nomia variegata* and *Bomus spp.*
Clovers (*Trifolium pratense, T. repens, T. incarnatum* and *T. hybridum*	Important pollinators are – *Bomus agrorum, B. hortorum* and *B. ruderatus* in *Trifolium pratense* and *Bomus lucorum, B. terrestris, Melitta leporine, Anthidium punctatum* in *Trifolium repens.*
VAM Pratylenchus zeae, Tylenchorhynchus vulgaris	Increased plant growth and reduced nematode population in sorghum and increased plant growth in berseem.
Meloidddogyne hapla	Increased plant growth and reduced nematode population in lucerne.
Meloidddogyne incognita, Heterodera cajani	Improved plant growth in cowpea
Meloidddogyne incognita, M.hapla	Improved plant growth and reduced nematode population in oat
Meloidddogyne incognita	*Meloidddogyne incognita* improved plant growth

Source: Modified from Saxena *et al.*, 2002.

Forage Plant Protection

Forage crops are infested with various pests and diseases causing loss in the yield and quality of forage. Saxena *et al.* (2002), reported the extent of yield losses in various forage crops. The losses were 7-58 per cent in sorghum, 10-45 per cent in pearl millet, 10-30 per cent in maize, 5-55 per cent in oat, 10-47 per cent in cowpea, 7-72 per cent in lucerne and 18-20 per cent in various grasses. Restricted use of agrochemicals limited to seed and soil treatments or only at early stage of the crop growth are advocated. Use of resistant varieties of forage crops and maintain field sanitation should be emphasised in forage production (Pandey & Roy, 2010 and Saxena *et al.,* 2002; Table 7.15).

Non-Government Organisation's Perspectives in Forage Resource Development

I. Anthra (NGO) is the non-governmental organization started by women veterinarians, working predominantly in the field of livestock development since late 90's in different parts of Andhra Pradesh and Maharashtra. The participatory approaches helped to generate a wealth of indigenous knowledge related to

Table 7.15: Resistant Varieties of some of the Important Forage Crops

Forage Crops	*Disease/Pest*	*Cultivars/Varieties/Germplasm*
Sorgum	Shoot fly	CS- 15, CSV – 203, Pusa Chari Hybrid - 106
	Zonate leaf spot	GFS – 203, HC – 308, S – 513, PCH – 109, Gujarat Fodder Sorghum – 5, Haryana Jowar - 513
	Shooty stripe	S – 513, SPV – 1359, SSV- 84, AGS – 306, GK – 908, HD – 19, PCH - 109
	Stem borer	Pusa Chari Hybrid - 106
Pearl millet	Downey mildew	IPFM – 5, CO – 8, Giant Bajra, IP – 36B, IP – 8556, IP – 11890, L – 72, MP – 260, MP – 261, PCB -150, PCB – 151, REB – 1, TNSC – 4, TNSC – 5, TNSC- 6, UUJ – 2, Proagro No. 1, Raj Bajra Chari 2
	Rust	APFM – 2, APFM – 5, APS – 1, CO – 8, Giant bajra
	Shoot borer	APFM – 5, Giant Bajra, IP – 36B, Mallikarjun, MP – 260, MP – 261, PCB – 150, RFB – 1, RFB – 2, TSC – 4, TNSC – 5, TNSC – 6, UUJ – 2, Raj Bajra Chari 2
Sudan grass	Red leaf spot	Meethi Sudan
	Anthracnose	Punjab Sudex Chari 1
Maize	Downey mildew	African tall, APFM – 14, APFM – 22
	Leaf blight	African tall, APFM – 14, APFM – 22
	Aphid	African tall, APFM – 12, APFM – 13, GBM – 84-1, GBM -84-2, GBM – 84-3
	Shoot fly	African tall, APFM – 11, GBM – 84 – 42, GBM – 84-43, GBM – 91-1, RFM- 92, W2, Y 1 x Y2
	Shoot borer	African tall, APFM – 9, APFM – 11, APFM – 12, APFM – 13, GBM – 84-1, GBM -84-2, GBM – 84-3, GBM – 91-1, RFM- 92,
Oat	Leaf spot	DFO – 57, JHO – 94-1, JHO – 94-3, JHO – 94-4, JHO 851, JHO 882, JHO – 893, OL – 240, OL – 633, OL- 793, OS – 201, OS – 242, UPO – 240, UPO – 241, UPO – 242, UPO – 243
	Root rot, crown rust, Leaf blight, powdery mildew	Bundel Jai 2004, Bundel Jai 991 for leaf blight, nematode and grass hoppers, UPO 94 for rust, Kent for rust and blight and Harita (RO -19), for leaf blight
	Root knot nematode	JHO 96-2, JHO 98-4

Contd...

Forage Crops	*Disease/Pest*	*Cultivars/Varieties/Germplasm*
Dinanath grass	Leaf spot (*Helminthosporium*)	Bundel - 2
Berseem	Root rot	Bl -108, BL – 120, JHB – 89-2, JHB -92-3, JHB 93-1, JHB – 94-2, JHTB – 143, JHB– 146
	Root knot nematode	JHTB – 143
	Stem rot	BL – 120, JHB – 90-3, JHB 93-1, Bundel Berseem 3
Shaftal (*Trifolium resupinatum*)	Stem rot	SH 69
Lucerne	Downey mildew	LLC- 3
	Rust	IGFRI – 112, IGFRI – 1212, LLC- 3
	Root knot nematode	RL -88, Sirsa-9
	Aphid	Anand – 2, Anand – 5, CO – 1, LLC – 3, RL -87-2
	Lucerne weevil	Anand – 6, IGFRI – COMP – 193, IGFRI – 112, IGFRI- 1212, LLH – 84, LLC – 3, LLC – 5, Sirsa – 9
Cowpea	Mosaic (CPMV)	CS -90, CS – 91, DFC -1, IFC -9101, IFC – 9103, IFC – 9201, IFC – 9202, UP – 93-1, UP – 9102, RFC – 8903
	Bacterial blight	IFC – 91-2, IFC – 9103, UP – 93-1, UP – 93-3,
	Root rot	Bundel Lobia, CL -334, DFC -1, IFC – 9201, IFC – 9301, IFC – 9303, UPC – 93-1, UPC – 93-2, UPC – 93-3, UPC- 9103, UPC – 9203, RFC - 8903
	Root knot nematode	Bundel Lobia – 2, Bundel Lobia – 3
	Aphid	CL - 235
	Leaf hoppers and defoliators	CL – 324, CL- 331, CS – 55, CS – 98, IFC -9102, IFC – 9201, UPC – 9203, UPC – 9204, UP – 93-1, UP – 93-2, UP 93 – 3, UP -93-4, UP - 9201
Guar	Bacterial blight	GAUG – 42, GL – 536, GL – 546, GL – 562, HFG – 119, IGFRI – 1359-1, IGFRI – 1539-1, IGFRI – 212-1, IGFRI – 23-1, Bundel Guar 1, Bundel Guar 2, Bundel Guar 3
	Powder mildew	Bundel Guar 3
Ricebean	Yellow mosaic virus	RBL 6

Source: Adapted from Pandey and Roy (2010) and Saxena *et al.* (2002).

livestock production, feeding, forage, water resources, health-care practices, grazing systems. These were used in developmental strategies to formulate programmes and their implementation for marginalized goat- and sheep livestock owners to enable them to voice their concerns and negotiate usufruct rights, such as grazing and lopping rights, within mainstream watershed and joint forest programmes (Ghotge and Ramdas, 1998; Anthra, 1999; 2000; Ramdas, *et al.*, 2001, Ramdas, 2003, 2014). The NGO found that forage requirements were met through a combination of crop-residues, and grazing on common lands, private grazing lands and forests, near tanks, fallow agricultural lands and on the harvested agricultural lands. Green forage composed of naturally available grasses, herbs and creepers, cut and fed to the various livestock. Farmers led research based on traditional knowledge, were evaluated to identify the best forage plant species, development of sustainable practices for grazing and lopping of tree fodder by taking into consideration of concerned stakeholders. Innovative experiences on forage and grazing is briefed in three case studies (Ramdas, *et al.*, 2001; Ramdas, 2003; Ramdas, 2014):

Case Study of Anthra I: Building Forage Security in Rural Areas

Anthra conducted a detailed action-research project in different agro-ecological contexts in Andhra Pradesh and Maharashtra to document and validate the traditional knowledge of farming and pastoralist communities with respect to feed and forage situation between 1996 and 2003. In this project, 254 different forage plant varieties were recorded. Out of them 80 tree species, 45 shrubs, 36 climbers, 35 herbaceous non graminaceous, 40 graminaceous plant species of forage value and various crop residues of cereal and pulse crops as feeding resource. Among them, most important tree fodders were *Albizia lebbeck* (dirisena), *Acacia nilotica* (thumma), *Diospyrus chloroxylon* (illintha), *Cleistanthus collinus* (billagodise), *Azadirachta indica* (vepa), *Melia dubia* (pedda vepa), annually lopped by shepherds during summer in Andhra Pradesh. However, their availability remained static since many years. Studies on diversity of forage resource in the different geographic locations in Andhra Pradesh pointed out fast and increasing loss of forage diversity which adversely affected livelihood of poor and marginal livestock keepers. Over 53 per cent of documented forage plant species in the list had been found to be occasional and rare occurrence in every district of Andhra Pradesh. Prominently, they were *Dalbergia latifolia, Hardwickia binata* (yepi), *Melia azaradirachta* and *Pterocarpus marsupium.* In degraded forest, fallow and other common lands in Andhra Pradesh, exotic poisonous/toxic plants like *Parthenium hysterophorus* (congress gaddi), *Chromolaena adorata* (kampudonka) and *Lantana camara* (muriki malle) had invaded the lands and found proliferating all over. It was found that forage availability and its quality were the critical factors impacting livestock productivity. After documentation, five different field trials were conducted to validate/assess traditional ways of utilizing forage in Andhra Pradesh and Maharashtra. Traditional practices of lopping of *Acacia nilotica* trees showed that 'correct lopping does not harm the tree, but enhances its lopped biomass production. Reintegration of traditional practices to augment forage resource was taken up to enhance forage availability in the rural areas. Key elements of the practice adopted were conservation of seeds through community seed banks of traditional forage plant varieties by 600 innovative farmers in nine districts of Andhra Pradesh and four districts of Maharashtra were actively involved in transforming their agricultural practices to mixed cropping with traditional food crops, resulting in improvement in food and forage security. For example, 42 farmers in Medak district, own one or another type of livestock, resulting in a significant increase in the number of buffaloes and goats in the village. The migratory shepherd community showed interest in developing their community land with native fodder trees and forage grasses. Migratory shepherds in Satara district of Maharashtra who returns to their village at the end of the migratory cycle, agreed to cultivate different forage crops such as *Phaseolus aconitifoilus* (moth bean), *Vigna unguiculata* (cowpea), and *Sorghum vulgare* (sorghum) on their lands during the monsoon.

Case Study of Anthra II: Restoring Indigenous Forage

A collaborative study was conducted by Anthra and collaborative partner to find out the availability of forage in the catchment areas of River Papagni, in 10 villages of Chittoor and Ananthapur districts of Andhra Pradesh. In this study, 211 forage plant species/varieties (include trees, shrubs, climbers, herbs and grasses) were identified. In order to prioritize forage plant species for intervention, several interactions and discussions were held with shepherds community to prioritize tree fodder species for plantation. *Hardwickia binata* (Yepi), *Albizia amara* (Chigara), *Ficus religiosa* (Ravi), *Ficus tinctoria* (Jeevi), *Bignonia crispa* (Neeru oddi) and *Azadirachta indica* (Vepa) were listed as priority tree species which are traditionally lopped as tree fodder during summer. Shepherds were involved in collection of seed and helped to establish a central nursery. Raised saplings were planted on the farm bunds of family owned lands, along the road as avenue trees and around the catchment area of Papagni River.

Case Study of Anthra III: Chittoor

Use of Forest Rights Act, 2006 to Defend the Rights to Graze in the Forests

The pastoral communities from Chittoor participated in several state level mobilizations, as part of larger mobilizations across the country to bring pressure on the Government for Forest Rights Act (FRA). Anthra coordinated and mobilised the sheep and goat keepers associations with other NGO under "Fodder and Grazing forum". This mobilization led to the withdrawal of the old forest policy, and setting up a committee to study the issue of grazing and forage to develop a comprehensive forage management policy and plan. In April 2002, the Andhra Pradesh Forest Department announced a revised Forest Policy for Andhra Pradesh through G.O. MS 34. An entire section (Sec 3.1.2) is devoted to the issue of grazing; this states: 'The present policy of free grazing in the forests is detrimental to regeneration and establishment of vegetation. There is need for reasonable restrictions on grazing. Further, the burden of scrub cattle on the already scarce forage resources needs to be reduced by upgrading the cattle stock, and encouraging stall feeding. There would be coordinated efforts of Forest Department with the Animal Husbandry Department of the Government. As a consequence, the act finally got legislated in December, 2006. As per provision of act, the community activists catalysed by Anthra started awareness campaign about their right of grazing through leaflets in local language and meetings with village *samiti* to the members of the shepherd's *sanghams* in KVB Puram Mandal about the legal rights to graze provisions for the forest-dependent communities and other pastoralists in the forests and empowered them by providing each family with a copy of the Act. Women began to carry the copy of the Act with them whenever they entered the forest with their animals, and confidently challenged the forest guards whenever they accosted them and attempted to fine or threaten them. At starting point, the activists accompanied the women into the forests. In the words of one woman grazier, Krishnamma, 'We women are strong, and are no longer afraid, now that we have this Act with us. We just wave it in front of their faces opening the book to the page which talks about our grazing rights, and they are too stunned to oppose us.' Since the past two years, the shepherds, especially the women are pro-actively entering the forests, and staking claims to their rights. They also state that they now have the right to graze and protect the forests from the forest guards who sold out truckloads of trees to smugglers in collusion with the village committee members. *Sangham* activists have also begun to look ahead and think of how to use their next right, which is, to conserve and develop the forests according to communities' livelihood needs, including developing the grazing resources within the forests. In all the districts of Andhra Pradesh, the trend observed was shifted from traditional crops, varieties and cropping systems to dwarf varieties of various food crops and commercial cash crops. This had a negative impact on availability of the edible crop residues. Today, paddy straw constitutes the predominant diet of livestock. Buffalos more efficiently utilized crude fibre rich paddy straw as compared to cows. Further, traditionally grown tall paddy varieties in dry lands had been replaced by high yielding dwarf paddy varieties, resulting in reduced yield of the crop residues. Consequently, many traditional shepherd communities' who never used to migrate earlier had gone on seasonal migration for two to three months searching forage in the districts of Anantapur, Mahabubnagar, Nalgonda and Medakion. Here, shepherds forced to move towards catchment areas of the major irrigated dam such as Nizamsagar dam, Nagarjunasagar after the postharvest *i.e.* from November and they returned in June/July. In Krishna, Guntur, East and West Godavari, Visshakapatnam and Vizianagaram, there was migration during monsoon, where the shepherds migrated into forest regions of the Eastern Ghats, as there were no community and other lands available for grazing around their village. Migratory routes differed from one year to another year based on the type of crops grown during the *kharif* season which was determined by the type of edible crop-residue standing in the cultivated lands for grazing by the small ruminants. Generally, shepherds preferred routes where mixed food crops were grown and also water was available for livestock.

Key concerns emerged from the case studies were:

1. Policies that were drastically changing land use (special economic zones, real-estate, mining, forestry focussed on growing eucalyptus, *Gliricidia*, Australian acacia and teak and bio-fuels, jatropha and karanj plantations that did not need much protection from livestock) made lands completely in-accessible to the small livestock keepers and pastoralists, who needed these lands to graze their stock, collect green forage, cultivate with mixed crops which yield food and crop-residues as dry forage (Anthra, 2008, Ramdas, 2014). Limiting access to lands and thus forage supply was a major reason for the low priced selling of their animals and becoming "non-livestock owners,
2. Investment policies and pricing favouring food production resulted in an un-sustainable competition around land for growing food for people, versus land put to forage for livestock,
3. Intensification of livestock production focussed on high milk and meat yields introduced in dry, arid and semiarid regions through poverty alleviation/watershed/JFM programmes in the name of livelihood enhancement required more resources. Resources required for purchase of improved breeds of cattle, buffaloes, sheep and goat, which increased the requirement of water and needed irrigated lands to produce forage and concentrate feeds, caused a lot of pressure on the forage and water resources in these drought prone regions. The focus, unfortunately, was on breed-replacement and there were minimal comprehensive attempts to address the real issue of forage and water crisis for livestock. Current Land development programs that focus on land, water and biomass development had been unable to respond to the forage/grazing needs of the livestock.
4. Traditional forage plant species like trees, shrubs, herbaceous, creepers, grasses and tall growing food crops which provide forage and edible crop residues to cattle, buffaloes and small ruminants had been systematically and completely neglected and replaced by dwarf varieties of food crops.

Anthra concluded that there is a need for people centred planning while keeping in mind the livestock in the area and understanding of the crops, hills *etc.* in that area, involving women, value their perception and practices to develop strategies and implement project with effective coordination of all stake holders and line departments to augment forage resources and building productive livestock assets and capacity of poor and marginalised livestock farmers (Ramdas, 2003, Ramdas, 2014). In programmes to regenerate lands, the policy of 'zero-grazing' or a complete ban on grazing coupled with a complete ban on goats had regrettably become the pre-dominant formula of "success" in forestry and social forestry plantations being resisted by poor livestock keepers which were forcing them off the land and forcing them out of their livelihood. On the contrary, facts remained that there were negative environmental consequences of "zero-grazing" due to the absence of livestock manure and excessive growth of grasses that catches fire during summer.

II. BAIF Development Research Foundation (formerly Bharatiya Agro Industries Foundation), is a voluntary organization established in 1967 at Urulikanchan, Pune (Maharashtra). It promotes sustainable livelihood in rural India through management of natural resources and livestock development, besides many other associated activities as source of enhancing income of rural families. Its flexible and innovative approach is spread over 5.0 million small and marginal landless families in one lakh villages of 16 states in India.

Case Studies by BAIF

In last two decades, the fact has emerged from various studies that farmers families totally dependent on rain fed agriculture in drought prone semiarid distressed districts of Maharashtra, Karnataka and Andhra Pradesh had more incidences of farmers committing suicide than the rural families partially or fully dependent on livestock for their livelihood, who could mitigate the stress to a greater extent by keeping livestock (Ramdas, 2003, Hegde 2006, Hegde, 2014). Livestock farmers continued to expand their herd and utilize community pastures and forest resources to feed their livestock. Chronic shortage of feed and forage resources during the last few decades indicated that most of the livestock were found to be underfed. In this semiarid and other arid regions of India, community grazing lands had been adversely affected due to increasing population of livestock. Consequently, increasing pressure on forage supply to livestock had forced change in livestock composition from large ruminants to small ruminants particularly goat, to overcome impact of the forage resource scarcity. It was also observed that many goat keepers prefer to breed their goats during April-May to facilitate kidding in September-October when adequate forage was available for the lactating goats just to overcome forage scarcity during pregnancy of goats. Generally, farmers graze livestock on community lands or in the forest areas around their village during the *kharif* season and let out for grazing in agricultural field after the harvest of various field crops in the dairy belt of western Maharashtra (Hegde, (2006). The category of farmers keeping strong breed of bullock for farming had adopted cultivation of green forage. With adoption of high milk yielding buffaloes and crossbred cattle, livestock had become an income generating economic activity. There are potential for cultivating forage to reduce the amount of concentrate feeding there by the cost of feeding by both categories of farmers to optimize the production and profitability of livestock industry but daily harvesting of green forage demands regular engagement of expensive labourers (Hegde, 1991).

In many parts of north India (Rajasthan and Uttar Pradesh), farmers were initially hesitant to make any investment on their non-descript cattle either on breed improvement or on feeding. In the semi-arid tracts of Rajasthan, where BAIF promoted cattle development programme in the early 80's, a large number of small farmers, owning non-descript cattle were motivated to avail the cattle breeding services with the support of the District Rural Development Agency by converging on-going schemes of the Animal Husbandry Department to support the cattle owners participating in this programme. In this collaborative programme, the families living below the poverty line were given opportunity to rear newly born female calves for 28 months by subsidizing concentrate feed cost by 50 per cent. Forage availability was enhanced by improving community pastures jointly and they were provided quality forage seed minikits to cultivate good quality forage on their individually owned arable lands. Timely preventive vaccinations reduced livestock mortality. Finally, milk yield increased over 3000-3500 litres per lactation and improved livestock fetched high price when sold. This integrated approach motivated farmers in the entire region to adopt better dairy husbandry practices and grow forage on their arable lands as well as joined hands to improve and maintain the community pastures. BAIF also executed agri-horti-forestry project in South Gujarat, Rajasthan (Sind taluk of Bhilwara district in Rajasthan) and Karnataka, the bunds and borders of these plots were used for growing forage and fuel tree species like *Leucaena leucocephala, Gliricidium sepium, Acacia and Sesbania species* and interspaces of the fruit tree species were also used to grow forage where food crops were not suitable to be cultivated due to low fertility of the soil. By both ways the agri-horti-forestry sites augmented forage production and became reliable source of green forage. In another project, in Sind taluk of Bhilwara district of Rajasthan, the villagers came together to develop a part of the community grazing land under silvipasture development. The major activities included the establishment of live hedges, gully plugging, contour bunding, sowing of forage seeds such as *Cenchrus setigerus* (Dhaman grass) and *Stylosanthes hamata* (Verano stylo) along with planting of *Acacia species* and *Prosopis cinereria* (*Khejri*) before the onset of monsoon and restricted stray grazing. In the subsequent years, the grass covers increased and the dry forage yield of 3.5–4.0 tonnes per hectare was obtained

In Uttar Pradesh in the 80's, the farmers were not prepared to avail doorstep artificial insemination services even free of cost. The reason was customary practice of not selling milk to ensure nutritional security of the family. Hence, livestock keeping was not being taken as an income generating activity. BAIF started awareness campaign to change the customary practice and increase milk production by promoting cattle improvement and better feeding. This resulted in motivation of the livestock keepers for marketing of the extra milk produced beyond the family requirement. This endeavor also led to the realization of the livestock owners to produce quality forage for feeding of livestock. Over the years, there had been significant changes in the animal husbandry to improve milk production as an economic activity in this region.

Conclusions drawn were that effective coordination among developmental agencies was required to avoid duplication and fill the missing gaps and links for forage resource development. In the absence of good animal breeding and support services, even the available forage resources could not find remunerative price. Thus, forage and feed development should be considered as an integral aspect of the dairy and meat value chain for ensuring success in forage resource development. The economics of milk production has been proved to depend on the feeding of nutritious green forage fed to milch cattle and buffaloes. However, feeding to the livestock of low productivity is not an economically viable option. Moreover, in the absence of market for forage, farmers are not motivated to grow forage on limited land available with them particularly fertile and irrigated agricultural lands. Therefore, in spite of being critical input in animal husbandry, forage resource development was closely linked to the productivity of livestock and the available critical veterinary support services. It's obvious that farmers compare economics of forage cultivation with other field crops. These aspects have to be taken into consideration for developing appropriate strategies for forage resource and livestock development. Moreover, forage research and development could not attract the appropriate attention of animal husbandry departments and policy makers.

Innovations in Forage Resource Development

Many studies highlighted that forage and feed scarcity is the main limiting factor to improving livestock productivity in India (Birthal and Jha 2005). Measures taken to address forage scarcity were often reactive in response to droughts and floods and that provided only short term partial solutions. The challenge is therefore to improve forage resources availability and utilization focussing towards the marginal and small scale livestock owner, contributing major part of milk and other livestock products in India. Despite a number of prominent successes, however, notably in spreading the use of seed-based, high-yielding technologies in Asia, research and extension organisations had been under criticism for their modest impact on both agricultural development and associated impact on socio-economic conditions (Sulaiman and Hall, 2002). Many decades assumption of viewing forage scarcity as technology scarcity had not helped to improve forage resource availability in the country except few limited pockets of success. Forage research led technology transfer had largely failed or had not addressed the forage scarcity issues fully in India or elsewhere (Hall *et al.,* 2007 a, b; Hall *et al.,* 2010). Now, it has been increasingly being felt that forage scarcity is a complex issue and when addressed alone by employing a narrow technology transfer approach will not give tangible results. The review of various experiences (Rajalahati 2008; Puskur *et al.,* 2009; Spielman *et al.,* 2009; Amlaku *et al.,* 2012) had shown that bundling of interventions spanning animal genetics, health and markets were critical to create demand for feed technologies and such packages of interventions had to go beyond technologies and had to include innovative institutional, organizational and policy arrangements. There is need for a change in the approach of forage research and development agencies. Revisiting forage scarcity issues with a new perspective requires implementing a strategy to promote market led intensification of livestock systems and expanding livestock markets while sustainably managing the environment. This was complex and the challenge required innovation in both 'what is done' and 'how it is done' (Abebe Shiferaw *et al.*, 2011; Reddy *et al.*, 2013). Revisiting forage scarcity issues with a new perspective of innovation capacity looked at multidimensional approach, focussing institutional arrangements, institutional innovation and innovation process accompanied by socioeconomic and technical knowhow, pointing a way forward in a holistic manner. Innovation is a process where knowledge is created and used in a new ways in different contexts, to enhance livelihoods of livestock dependent poor which occurs in in a system mediated by the actors and their interactions, facilitated by institutions and policies (Spielman *et al.*, 2009, 2011; Abebe Shiferaw *et al.*, 2011; Puskur, 2011; Reddy *et al.*, 2013). The new concept of innovation systems envisaged that technological change will follow much more easily when policies are set right to create conducive environment and usher more opportunities for marketing development. Among other things, this will entail a shift from narrowly focused sectorial concerns to cooperative, gender-sensitive, integrated approaches, including public– private sector linkages. This requires a collaborative effort among researchers (national, regional and international), extension agencies and policymakers, the private sector and civil societies which constitute the innovation system. The approach embraces an institutional innovation process in which participatory, action-oriented methods drive research for development to solve critical problems. Identification of a comprehensive and integrated set of options are possible through a value chain framework (Kaplinsky & Morris 2001; Rich *et al.*, 2009; Abebe Shiferaw *et al.*, 2011; Puskur, 2011; Vamsidhar Reddy *et al.*, 2012, 2013) which will address the various issues of forage scarcity. It also helps in identifying and linking the necessary actors to form coalitions for joint action and learning to address the challenges (World Bank 2006, 2007). Therefore, objective of the forage related research should shift from mere technology generation and first line demonstration to innovation, which is the process of generation, dissemination, access, adaptation and use of available or new knowledge to finally benefits in socio-economic wellbeing. Institutional and organizational arrangements are also accounted to achieve this uptake of knowhow.

International Livestock Research Institute (ILRI) implemented "Livelihoods of Poor Livestock Keepers through Increasing Use of Fodder" (www.fodderinnovation.org) in India and Nigeria with two distinct phases for six years in mixed crop-livestock systems. In India, three sites selected were for the forage innovation project with the objectives (1) developing a mechanism to negotiate improved forage access in public (wastelands and forest areas) and private grazing areas for poor livestock keepers in Ahmednagar (Maharashtra), (2) developing a mechanism to coordinate complimentary technical (including forage) and institutional support in order to upgrade smallholder dairy systems in Bhilwara (Rajasthan) and (3) establishing an integrated forage production and marketing system based on farmer forage entrepreneurs in Puducherry (South India). Phase 1, looking at scarcity of forage in context of a problem of technology and information scarcity, attempted to solve it by identifying and disseminating new and improved forage and dual-purpose crop varieties to increase forage supply. The technology delivery and use required networks of diverse stakeholders and appropriate institutional environment, to be successful. The learning's in Phase 1 had formed the basis for initiating a different approach for Phase 2 (2007-2009) of the project. The project was very process-driven, building capacities of local rural systems to assess constraints and identify entry points and test innovative interventions to address forage scarcity challenges. The project worked with a range of key partner organizations in three states including government and non-governmental organisations (NGO's) that brokered the formation of innovation networks (Vamsidhar Reddy *et al.,* 2012, 2013). At Ahmednagar, 50 hectares of identified forest area were reseeded with *Stylosanthes hamata* and the leguminous forage was made available to the poor livestock keepers. Plantation of fodder tree species on the private wastelands was also done by the farmers with the support of the Social Forestry Department as a part of the Maharashtra Employment Guarantee Scheme. In Bhilwara, the project objective was to address the forage and water issues by undertaking pasture land development and works on water harvesting under National Rural Employment Guarantee Scheme and National Agricultural Board for Agriculture and Rural Development. It supported watershed development work to augment the ground water and surface water availability and developed 518 ha of pastures in 17 villages. Social systems had been put in place to ensure that benefit sharing in terms of the forage and water with the rightful representation of the poor's in the decision-making processes at the village level. In Puducherry, cultivated green forage was grown and marketed depending upon demand which was determined by the availability of edible crop residue for feeding livestock. Diverse institutional arrangements emerged in the process of implementation to link forage producers and buyers. These types of institutionalization built capacity of livestock keepers (mostly women), farmers and cooperative society and helped to reduce the transaction costs and made forage accessible to the landless livestock keepers, especially women and their Self Help Groups (Puskur, 2011; Vamsidhar Reddy *et al.,* 2013). For example, the forage growers supplied the green forage to Women's Self Help Groups, which in turn supplied them to landless women dairy producers and also channelized the payments. In Ethopia, Abebe Shiferaw *et al.* (2011) found that innovative practices included targeting innovative forage farmers, developing local forages, establishing private forage sources, and forage promotion through diversified capacity building in order to address forage scarcity for increased livestock productivity. Innovation project also had reliable feedback from diverse partners, locations and wider range of farmers as inputs to enable new research and up scaling of improved varieties and production technologies developed. On realizing the merits of a capacity development using the principles of the innovation system approach, effective forage resource development took place because it allowed learning and effective implementation in this new endeavour due to networking, multi-stakeholder implementation, reflection, learning and re-adjustment of activities (Amlaku *et al.,* 2012; Vamsidhar Reddy *et al.,* 2013). Experiences also led to provide holistic approach applied in the broader livestock and forage context and pointed the need to take a broader crop-livestock value chain perspective, to have socio-economic impact. Various research and development organizations need to rethink their performance evaluation and incentive systems to support change in the behaviour of their staff and new ways of working.

Thus, technical changes and institutional/organizational changes cannot be viewed in isolation, they feed on and reinforce each other and lead to livelihood impacts through livestock system upgrading. Appropriate technology introduced through existing stakeholders proved to be a useful catalyst to raise and address broader system constraints (Vamsidhar Reddy *et al.,* 2013). Engaging policy actors from the beginning helps in finding windows for influence and ownership for research results.

Chapter 8

Forage Resource Development in Mountain Regions

Mountains are home to one tenth world's population and occupy one fourth of the world geographical area. About 10 per cent of world's population depends directly on the use of mountain resources for their livelihood and wellbeing (Gerlitz *et al.,* 2014). The mountaineous and hilly regions in India covering western Himalayas (A1), Central Himalayas (A2), Eastern Himalayas (A3) and Assam-Meghalaya hills (A4) give shelter to 41 million people and 29.4 million livestock.

Mountain ecosystems are suitable for forestry, horticulture and animal husbandry. In these hills, livestock rearing is an integral part of the living pattern. These regions have their unique livestock production systems dedicated by the geographical landscape and cultural preferences of the people particularly mithun, yak, sheep and hill goat (Prasad and Singh, 1994; Singh *et al.,* 2008). In the high altitude areas, transhumant livestock production system is practiced wherein animal grazes in the alpine pasture in the summer and over winter in the foot hills and plains. Livestock preference by the people greatly varies between the four regions. There are also 10 hotspot districts in the mountain and hill region where all the three indicators- the human density, livestock density and number of livestock per thousand human beings are on a higher side. They together cover an area of 36151 square kilometres having 4.7 million and 8.0 million livestock. Cattle is important across all the four regions (251 to 398 per thousand human beings, (PTH), buffalo is important for region A1 (145 PTH) and A2 (112 PTH) only; yak and mithun is important for region A3; Sheep for region A1 (269 PTH), goats for all the four regions and pigs for region A3 (235 PTH) and A4 (142 PTH). The threat of land degradation, the forage demand and livestock production need to be carefully assessed and managed accordingly in the each region.

In view of the typical physiographic and fragile ecological condition prevailing in the hills, it is necessary to conserve life-support systems and biodiversity to ensure sustainable use of renewable resources and minimize the depletion of non-renewable land resources to keep within the carrying capacity of the supporting ecosystems (IUCN/UNEP/WWF, 1991). Sustainability can be approached when the farmers learn to manage their natural resources ecologically, to keep intact or improve land, water and biological resources for sustained biological productivity. Therefore, simulation of natural ecosystem based on mixed association of trees with other annuals and perennials including livestock will provide sustainable livelihood and permanent vegetative cover on the hill slopes. In such situation forage production and agroforestry systems have to play a vital role.

The hilly regions favour availability of forage biomass from diverse native annuals, perennials, fodder trees and some introduced forage plant species growing under different types of land uses – arable and non-arable. Forage plant species have also ameliorative and soil conservation role and they are components of nutrient recycling in the prevailing mixed farming situation. However, low arable to non-arable land ratio and practice of free grazing; there is an urgent need to pursue a long term strategy for sustainable forage resource development.

Analytical view of physiographic-climatic-socioeconomic matrix of hilly agro-ecosystems have helped to draw some useful conclusions for identifying potential niches for forage crops and priority areas of forage research and development: a) scopes of annual forage crops are limited due to scarcity of agricultural lands in the hills, b) depleting forage availability in the hills have caused overgrazing and resulted in the degradation of vegetation and land resources, c) potential non-competitive land use for forage crops are bamboo shaded farm lands, thin natural forest near villages, steep un-cultivable hill slopes, fruit orchards, terrace risers/ bunds and top of the hills for integrated farming systems. Intensive dairy/goat farming can be practiced on the gentle hill slopes and the terraced lands, d) about 300 plant species mostly native constitute forage resources. Among them, fodder trees contribute 25 to 60 per cent of the total green forage resources. Fodder trees and forage grasses have conservation as well as ameliorative effects on acidic soils, and enriching the green forage with nutritious forage during monsoon (April to mid-October) and producing green forage during winter months (October to March) should be prime concern for formulating forage resource development strategy in this region.

There is need to understand the migratory routes, the grazing rights, the grazing pressure *etc.* so that overgrazing is avoided and the grasslands get the breathing space to rejuvenate. Poor pasture condition figures highly among constraints. Since there is no baseline data, the evolution of degradation can only be guessed and hence there is a serious need for more monitoring of pasture conditions and trends as a management tool, to measure its environmental impact. There are issues related to land tenure or lack of clarity in grazing rights which often creates a situation of conflict between the nomadic graziers settled people and the forest policy. Another issue of great ecological significance is the camping of migratory buffaloes around high altitude wetlands as practiced by the nomadic *Gujjars* in the western and central Himalayas. What is the total number of such animals that are part of the annual cycle, their migration pattern, the duration of stay up in the hills, the status of grassland surrounding these wetlands, and the degradation over the years; all these questions could be answered only if a scientific study is initiated on this aspect. An integrated approach for pasture improvement and livestock productivity through better grazing management, reseeding with high-yielding forage species and creating awareness among the pastoralists together with addressing the issues of land tenure and grazing rights are needed in alpine and temperate regions.

Forage Resource Development

Carefully planned forage-resource development will preserve bio-diversity, and improve the permanent vegetative cover, leading to conservation of land and water resources, their hydrology favourably and result into a close organic or nutrient recycling on the hill slopes for a self-sustained regenerative agro-ecosystem at micro level and ecosystem development at macro levels.

The Himalayan region have arid to humid and tropical to alpine climates, varied annual rainfall 100 to 2,809 mm with soils rich in organic matter. In these hills, livestock rearing is an integral part of the living pattern. High demand of meat and increasing demand of milk favour livestock production. However, the way ruminants are being reared, is not sustainable for the mountain ecosystem. Considering the fact that the ecosystem is suitable for forestry, horticulture and animal husbandry, high rainfall favouring high forage-biomass production from diverse native annuals, perennial grasses to fodder trees and introduced forage plant species tailored in possible niches in different land uses under arable and non-arable situation. There is an urgent need to pursue a strategy of a long-term approach of research and development for forage-resource development. Diagnostic survey of existing farming practices and analytical view of physiographic-climatic-socio- economic matrix in the hilly area reveal the following:

- ☆ It is clear that the scope of annual forage crops is limited due to scarcity of agricultural lands in the hills.
- ☆ Depleting forage availability in the hills causes overgrazing and results in the degradation of vegetation and land resources on the hill slopes.
- ☆ Potential non-competitive land use for forage crops are bamboo-shaded farm lands, thin natural forest near villages, steep uncultivable hill slopes, orange, guava and avocado orchards, terrace risers or bunds and top portion of the hills for integrated farming systems as well as intensive dairy or goat farming on the gentle hill slopes and terraced lands.
- ☆ Enrichment of the green-forage with nutritious forage during monsoon (April to mid-October) and production of green-forage during winter

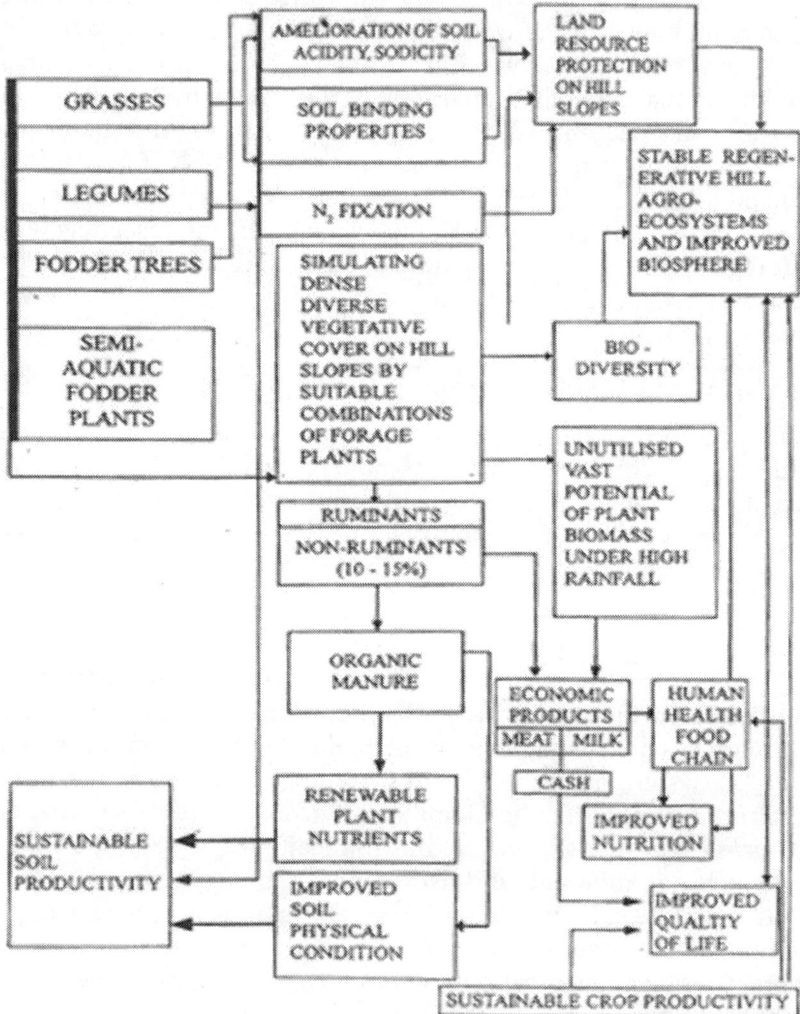

Figure 8.1 a: Interaction of Forage Resource Development (FRD) and their Possible Impact on Agro-ecosystem.

months (October to March) should be the prime concern in formulating strategies for forage-resource development in this region.

- About 300 plant species were enlisted mostly native, that constituted green-forage resources. Among them, fodder trees contributed 25 to 60 per cent of the total green-forage resources. There was paucity of nutritious perennial forage grasses and legumes.
- Fodder trees help conserve soil and forage grasses too may have conservation as well as ameliorative effects on acidic soils.

☆ There is no reliable information on fodder trees productivity and nutritional values of the forage plant resources grown in this region.

Forage Plant Resources at Low to Midhills

Forces that produce great reduction in biodiversity must result into habitat destruction. Therefore, we must develop management strategies for production that incorporates broader diversity and do not interfere with ecological integrity of living things or their ecosystems (Franklin *et al.*, 1986; Singh *et al.*, 1993). A large number of forage plant species having diverse life-span (annual and perennial), plant form (herbaceous, shrubs, climbers and trees), differential exploration systems of nutrients from soil (shallow, deep and fine root systems and nitrogen-fixing plant), row to cover plants having a wide range of adaptability, may be suited to different productive forage-based land-use systems and lead to preserve the native biodiversity (Allan, 1947; Singh *et al.*, 1993; Singh *et al.*, 1996; Singh, 1999). Based on various studies about 300 forage plant species mostly native and few introduced ones constitutes the forage resources in the hills (Balaraman, 1981; Burley, 1985; Peter & Eric Oberholzer, 1985; Shrestha *et al.*, 1990; Singh & Yadav, 1995). Many more can be explored and added to the forage basket. Some of the important productive forage plant species for low to midhills are listed here:

A. Annual Forage Crops

Maize, teosinte, dinanath grass, grassland manava, oat, forage triticale, rice bean, cowpea, soybean, pea, forage sarson, turnip and radish.

B. Perennial Forage Grasses

Bajra (*Pennisetum glaucum*) x napier (*Pennisetum purpureum*) hybrids, thin napier (*Pennisetum polystachyon*), guinea grass (*Panicum maximum*), Congo signal grass (*Brachiaria ruziensis*), para grass (*B. mutica*), Nandi Setaria (*Setaria sphacelata*), green panic (*Panicum maximum var. trichoglume*), dallis grass (*Paspalum dilatatum*), palm grass (*Setaria palmifolia*), broom grass (*Thysanolaena agrotis*), teosinte (*Tripsacum laxum*), *Erianthus longisetosum, E. sikkimensis, Dactylis glomerata, Festuca arundinacea, Loium multiflorum* and *L. perenne (*Singh *et al.*, 1989, Photo 8.1 b).

C. Marshy Lands

Zizinia caudicifolia and *Z. latifolia* (Singh and Pandey, 1995, Photo 8.1 b).

D. Perennial Forage Legumes

Aeschynomene americana var. glandulosa, A. villosa, Alysicarpus molifir, A. rugosus, Calapogonium muconoides, Canavalia ensiformis, Cassia rotundifolia, Centrosema pubescens, Clitoria ternatea, Desmanthus, virgatus, D. distortum, D. heterocarpum, D. podocarpum, D. intortum, D. uncinatum, Glycine javanica, Lablab purpureus, Lotus peduncularis, Macroptilium heterophylum, M. atropurureum,

Congo Signal Grass (*Brachiaria ruziensis*)

Broom Grass (*Thysanolaena agrostis*)

Zizinia caudicflora

Photo 8.1 b: Some of the Important Forage Grasses.

Desmodium hetrocarpum

Desmodium podocarpum

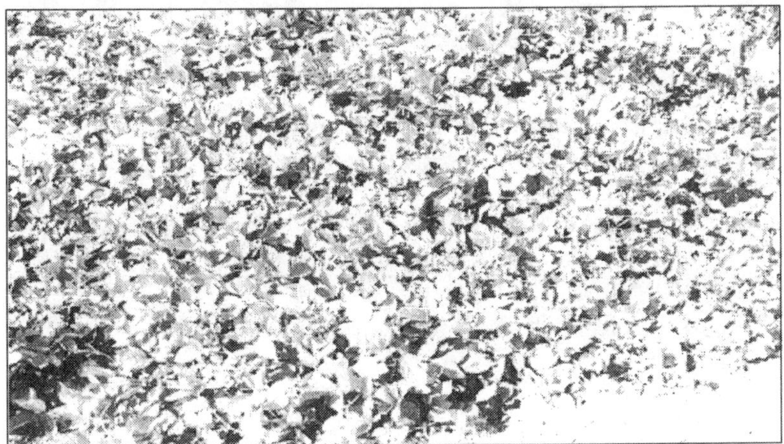

Macroptilium atropureum

Photo 8.2a: Some of the Important Perennial Forage Grasses.

Stylosanthes guianesis

Mimisa invisa

Photo 8.2b: Some of the Important Perennial Forage Grasses.

Milletia auriculata, Mimosa invisa, Mucuna monosperma, Neonotonia whightii, Pueraria javanica, Sylosanthes guianensis, S. scabra, Vicia discarpa and *Vigna parkrii* (Photo 8.2 a, b).

E. Forage Climbers

Bauhinia vahlii, Hedrea helix, Rosa brunii (Singh, 1999).

F. Forage Shrubs

Boehmeria Platyphylla, Buddleia asiatica, B. paniculata, Casia momosoildes, Coriaria nepalensis, Carrisa appaca, Caryopteris foetida, Colquhounia coccinea, Desmodium floribundum, D. microphyllum, D. tiliaefolium, Diplomorpha canesscens,

Ephedra gerardiana, Eurya accuminata, E. japonica, Euonymus hamiltoni, Indigofera cylindrica, Inula cappa, Plectoranthus gerardianus, Polygala arillata, Priotropis cytisoides, Reinwardtia indica, Rhamus persica, R. virgatus, Rhus parviflora, Ribbies acuminata, R. takre, Robinia pseudocacia, Symplocos theaefolia, S. ramosissama, Stachyurus himaaicus, Uraria lagopus and Vitex negunda.

G. Fodder Tree Species

Acer accuminata, A. caesium, A. oblongam, A. pectinatum, Actionodaphne angustifolia, A. reticulata, Adina cordifolia, Aesculus indica, Albizzia julibrissin, A. lebbeck, A. odoratissima, A. laevigatum, A. stipulatum, Ailanthus exceba, A. excels, Ammora wallichii, Anogeissus latifolia, Artocarpus lakoocha, A. integra, A. heterophyllus, Bassica butyracea, Bauhinia purpurea, B. variegata, Benthamidia capitata, Betula alnoides, B. cylindrica, B. hinia, B. longifolia, B. utilis, Brassiopsis glomerata, B. hainla, B. mitis, B. palmata, Bridelia retusa, Butea frondosa (*B. monosperma*), *B. minor, Careya arborea, Castanopsis indica, C. tribuloides, Casearia glomerata, Cedrela toona, C. serrata, Celtis australis, C. orientalis, C. tetranda, Cleyera ochacea, Choerosponddas axillaris, Chukrasia tabularis, Cordia myxa, Cornus oblonga, Dalreglasia velutina, Diospyros dubia, D. embryopteris, D. melanoxylon, Dodecadenia grandiflora, Drymaira cordata, Dubanga grandifolia, Eheria acuminata, E. cerassifolia, E. laevis, E. macrphylla, Elaegnus infundibularis, Eriobotrya dubia, Erythrina arborescens, E. variegata, Evodia fraxinifolia, Ficus auriculata, F. auriculata type II, F. benghalensis, F. benjamina, F. clavata, F. elastica, F. glaberrima, F. glomerata, F. hirta, F. hispida, F. infectoria, F. lacor, F. neriifolia, F. nemoralis, F. racemosa, F. religiosa, F. retusa, F. rumphii, F. semicordata, F. sarmentosa, F. subinicisa, Fraxinus floribunda, Garciana, timitoxia, Garuga pinnata, Gmelina arborea, Gnaphalium luteoalbum, Grewia elastica, G. optiva, G. tilaefolia, Hedera napalensis, Helixanthera liquistrina, Holoptelea integrifolia, Horonia dussis, Hymenodictyon excelsum, Ilex sp. Dipyrena sp., Juglans regia, Kydia calycina, Lannea grandis, Leucaena diversifolia* (K156), *L. leucaena* (S11 and K28), *Ligustrum confusum, L. indicum, Lindera neesiana, Litsea monopetala, L. citrata, L. oblonga, L. polyantha, Macaranga pustulata, Machilus gamblei, M. edulis, M. odorotissima, Madhuca butyracea, Maesa chisia, M. indica, Mallotus phillippensis, Michelia doltospa, Mitragyna parviflora, Moringa oleifera, Morus alba, M. macroura, M. laevigata, Ougenia dabergiodes, Photinia integrifolia, Pieris formosa, Pongamia pinnata, Populnea deltoids, Populus ciliata, P. glauca, Premna barbata, P. integrifolia, P. latifolia, P. mucronata, Prunus ceresoides. P. cornata, P. napaaulensis, Pterocarpus acerifolia, P. marsupium, Pyrus domestica, P. pashia, Quercus dilatata, Q. fenestrata, Q. glauca, Q. incana, Q. lanata, Q. lamellosa, Q. semicarpifolia, Q. spicata, Rhamus nepalensis, Rhus javanica, R. semilata, Rutea* sp., *Romalium mapaulens, Salix babylonica, S. terasperma, Samnea saman, Sapindus mukorosa, Saurauia fasciculata, S. napaulensis, Sauveria facciculata, S. griffithii, S. punduana, S. roxburghii, Schleichera oleosa, S. trijuga, Schima wallichii, Semicarpus anacardium, Strarculia villosa, Streblus asper, Stereospermum personatum, Stravaesia glauccerscens, S. nussia, Symingtonia populnea, Symplocos crategoides, S. paniculata, S. pyrifolia, S. racemosa, S. sumunfia, Talauma hodgesonia, Terminalia*

Brassiopsis hainla

Photo 8.3a: Important Fodder Tree Species.

Litsea monopetala

Sauraula napaulensis

Photo 8.3b: Important Fodder Tree Species.

Bauhinia purpurea

***Ficus auriculata* Type-I Twig**

Photo 8.3c: Important Fodder Tree Species.

Ficus auriculata **Type-I Fruits**

Fisuc auriculata **Type II-Fruits and Twig**

Photo 8.3d: Important Fodder Tree Species.

Ficus hirta

Photo 8.3e: Important Fodder Tree Species.

Photo 8.3f: *Ficus glomerata* Tree Fodder with Cylindrical Canopy and High Forage Yield is a Potential Tree Fodder to be grown under Agroforestry Systems.

alata, T. belerica, T. chebula, T. crenulata, T. myriocarpa, T. orenulata, Toricellia tiliaefolia, Trema orientalis, Trevisia palmata, Trewia nudiflora, Ulmus lancefolia, U. wallichiana, Wrightia tomentosa and *Zanthoxylum acanthopodium* (Photo 8.3).

Ficus elastiea

Ficus laccor

Photo 8.3g: Important Fodder Tree Species.

Ficus hispida

Ficus semicordata

Photo 8.3h: Important Fodder Tree Species.

Ficus subincisa **– Tree**

Ficus subincisa **– Fruit and Twig**

Photo 8.3i: Important Fodder Tree Species.

Ficus rumphii

Grewia optiva

Photo 8.3j: Important Fodder Tree Species.

H. Bamboo Plant Species

Arundinaria hirsuta, A. manii, Bambusa arundinacea, B. balcoa, B. cacharensis, B. multiplex, B. nana, B. nutans, B. pallida, B. polymorpha, B. tulda, B. striata, B. wamin, Chimonobambusa armata, C. callosa, C. griffithiana, C. hookeriana, Dendocalamus giganteus, D. hamiltonii, D. hookerii, D. sikkimensis, Phyllostachys assamica, P. manii, Schizostachys manii, S. pergracile and *S. polymorphum* (Photo 8.4).

Bambusa arundinacea **– Protecting Hill Slopes along the Road.**

Bambusa nutans **on the Bank of Spring Water Course.**

Photo 8.4: Bamboo Plant Species.

Management of Forage Resources

Annual Forage Crops

There is limited scope of growing annual forage crops in the hills. Among annual forage crops, maize, jobs tear, dinanath grass, rice bean, cowpea and soybean can be successfully grown during pre *kharif* and in *kharif* season (Singh *et al.*, 1987). During *rabi* season, oat, pea, forage mustard, Japanese rape mustard, Chinese cabbage and turnip can be grown (Singh and Singh, 1992). Maize can also be grown during *rabi* season. Fresh forage yield of 25-50 tonnes per hectare can be obtained from annual forage crops. Intercropping is a way to combine forage grasses and legumes to obtain high yields of nutritious forage. The forage intercropping systems are well suited for terraced lands or gentle hill slopes up to 30 per cent. Intercropping of forage triticale, March sown maize and orchard grass with rice bean, cowpea, soybean, perennial tall fescue with rice and Napier x bajra hybrid grass with rice bean, cowpea during *kharif* season and oat, pea and turnip during *rabi* season, increases forage yield by 17-42.2 per cent as compared to monocultures (Singh & Singh, 1990; Singh *et al.*, 1996; Singh, 2005). Grassland manava can be well intercropped with rice and finger millet under food + forage intercropping systems. During early phase of establishment of orchards of mandarin, guava, avocado and lemon, annual forage crops like maize, cowpea and rice bean and perennial forage *viz.* fine stylo (*Stylosanthes guianensis*), *Panicum maximum*, Napier x bajra hybrids, *Setaria sphacelata* and *Paspalum notatum* can be grown and the green forage yield of 45 to 138 tonnes per hectare obtained (Borthakur *et al.*, 1979; Singh *et al.*, 1996).

Perennial Forage Grasses and Legumes

Most of the subtropical and tropical perennial grasses grow well at low to midhills and produce forage during April to October (Singh *et al.*, 1995). Among the grasses, *Tripsacum laxum* has extended period of growth during dry winter months, a lean period for green forage availability. Another multipurpose grass, *Thysanolaena agrostis/T. maxima* remains green during the winter months, when it is harvested for livestock. It also produces spike for broom and stem for coarse quality paper and other uses. Congo signal grass is very good for pasture development under medium to high rainfall situation. The yield potential of most productive tropical perennial grasses *viz.*, Napier x bajra hybrids, Congo signal grass, guinea grass (Gatton, PGG1, Acc. 59975), *Setaria sphacelata* and *Tripsacum laxum* ranges 94.0 -22.8 tonnes per hectare fresh yield and dry matter yield of 26.31-58.58 tonnes per hectare (Singh *et al.*, 1996). Most of the temperate grasses are suitable for high hills and alpine region but some of them *viz., Dactylis glomerata, Festuca arundinacea, Loium multiflorum, L. perenne* produce good forage yields with high nutrient contents at mid altitudes of 1000 meter. Theses grasses produces 21.2 -41.5 tonnes per hectare fresh yield and dry matter yield of 5.31-11.17 tonnes per hectare (Singh *et al.*, 1993). The forage yields are low but they are comparatively more nutritious than tropical grasses like *Dactylis glomomerata, Fustuca arundinacea* and *Lolium multifloreum* grown (Table 8.1, Photo 8.5, 8.6), *Medicago falcata* and *M.*

agrostis and *Trifolium species* are grown in Temperate region of Himachal Pradesh and alpine region of Ladakh in Jammu & Kashmir (Photo 8.7, 8.8).

Photo 8.5: Orchard Grass (*Dactylis glomerata*) in temperate region of Himachal Pradesh

Tall Fescue (*Festuca arundinacea*)

Photo 8.6 a: Perennial Grasses in temperate Region of Himachal Pradesh.

Rye Grass (*Lolium multiflorum*)

Photo 8.6 b Perennial Grasses in Temperate Region of Himachal Pradesh.

***Trifolium pratense* in Himachal Pradesh.**

***Trifolium* Species grazed by Sheep**

Photo 8.7: Perennial Legumes in Alpine and Temperate Zones.

Photo 8.8: *Medicago falcata* in Laddakh Region of Jammu & Kashmir.

Table 8.1: Intercropping Systems with Temperate Grasses at 1000 metre Altitude in Sikkim

Intercropping Systems	*Forage Yield (Dry matter tonnes/ha)*
Orchard grass (*Dactylis glomerata*) + maize	8.93
Orchard grass (*Dactylis glomerata*) + rice bean	8.87
Tall fescue (*Festuca arundinacea*) + maize	10.33
Tall fescue (*Festuca arundinacea*) + ricebean	13.14
Italian rye grass (*Lolium multiflorum*) + forage maize	12.84
Italian rye grass (*Lolium multiflorum*) + rice	10.94
Italian rye grass (*Lolium multiflorum*) + finger millet	12.15
Italian rye grass (*Lolium multiflorum*) + soybean	12.38

Source: Singh and Singh (1990), Singh *et al.* (1992).

Among, subtropical and tropical perennial grasses, Bajra x napier hybrids, *Panicum maximum* (Makueni, Acc. No. 59975) and *Brachiaria ruziensis* produced highest amount of forage 45-59 tonnes dry matter per hectare followed by *Panicum maximum* (PGG I) Gatton, Hamil), *Paspalum dilatatum*, *Setaria sphacelata* and *Tripsacum laxum* gave 25-39 tonnes dry matter per hectare during monsoon (Singh *et al.,* 1995,1996). During lean season of winter, *Thysanolaena maxima* is multipurpose grass growing naturally in the hills as well as planted by farmers as hedge row on terrace risers/bunds and block plantation (Singh *et al.,* 1989). The leaves are harvested for forage during winter months, beside spike for broom and stem as sticks for various uses. The leaves in broom grass constitute 27 per cent of the total harvest and the leaf yield ranges 1.59-11.21tonnes/ha/year depending upon growing site. *Zizinia* species are naturally occurring in low hill on the marshy and waterlogged lands. Good potential exists to grow *Zizinia caudiciflora, Z. latifolia* and *Braciaria mutica* on such lands at low hills where other food crops cannot be grown (Singh and Pandey, 1995). Seeds of *Zizinia* species are valuable food for wild fowl during winter. Most of the perennial grasses can be established during June to August from seedlings grown in the nursery during April -May or rooted slips obtained from already established pastures. *Zizinia* species can be propagated from rooted slips, stem cutting or seedlings. Perennial forage legumes can also be established from seed, seedlings and rooted cuttings during monsoon.

Nitrogen is a key nutrient for crop production. Response of annual and perennial non-leguminous forage crops ranged from 60 -200 kg N/ha depending upon inherent potential yield, soil fertility status and growth period (Borgohain, 1983; Dwivedi & Pandey, 1987-2002; Singh, *et al.,* 1995a; Singh, *et al.,* 1995b; Singh, 1996; Singh, 1999a; Singh 1999b; Singh, 1999c). Oat can be fertilized with optimum dose of fertilizer only when minimum two irrigations are available (Singh *et al.,* 1995b). Intercropping of maize and teosinte with rice bean and oat with pea in gave forage yield equivalent to 60 kg N/ha (Singh *et al.,* 1995a; Singh *et al.,* 1995b). Considering the high cost of nitrogen fertilizer, farm yard manure easily available on dairy farm can reduce the requirement of nitrogen (Singh 1996; Singh 1999). Application of 40 tonnes FYM/ha produces 32 tonnes dry matter per hectare per year which is equivalent to 70 kg N/ha applied through fertilizer. Studies on phosphorus and potassium have been limited from acid to neutral and high pH soils. In acid soils, the response was up to 60 kg P_2O_5/ha to no response (Borghain, 1983; Dwivedi & Pandey, 1987-92). Cutting management optimizes the total nutrient and dry matter yields in forage crops. Subtropical and tropical perennial grasses produced high dry matter and crude protein yields when harvested at 40 days interval but napier x bajra hybrids produced maximum dry matter and crude protein yield when harvested at 60 days intervals (Singh & Pradhan, 1995; Singh, 1996). Among temperate grasses, tall fescue cut at 60 days interval, rye grass cut at 45 days and fodder triticale cut at 30 days interval produced maximum forage (Singh *et al.,* 1993). Multicut oat variety JHO 851, first cut at 70 days and second cut at 30 days gave highest total forage yield.

Forage grasses had been proved to have ameliorative effects on the acid soils of hills (Das *et al.,* 1990; Singh, 1996; Singh, 1999b; Singh, 1999c; Singh *et al.,* 2008; Singh *et al.,* 2010). It had been found that there was reduction in soil exchangeable Al, Mn and increase in soil organic matter as well as accelerated its humification. They also increased soil pH, built up available phosphorus, exchangeable K, Zn, Cu and Fe due to growing perennial grasses and natural pastures on the hill slopes.

Seed production in perennial grasses is not successful particularly in eastern hills due to early onset of monsoon. Except *Tripsacum laxum*, all perennial grasses had poor seed yield and very low germination but dinanath grass seed production can be taken up (Singh *et al.,* 1996). However, once procured the vegetative propagation by rooted slip is very successful. Perennial forage legumes can be established from seed, seedlings and rooted cuttings.

The possibility of growing perennial forage legumes particularly at low to mid hills showed suitability of about 25 forage legumes for forage production at low to mid hills (Singh, 1999; Shrestha *et al.,* 1990). Singh (1999) found production potential of introduced perennial legumes *Calapogonium muconoides, Canavalia ensiformis, Centrosema pubescens, Clitoria ternatea, Macroptilium atropurureum, Mimosa invisa, Pueraria javanica, Stylosanthes guianensis, S. scabra.* The green forage yield ranged 20.0 tonnes/ha/year in *Centrosema pubescens* to 55.0 tonnes/ha/year in *Calapogonium muconoides* and the crude protein contents 4.2 to 21.0 per cent in their dry matter. All of these legumes had good nodulation. Perenniality of *S. guianensis* confined to three years in acid soils hence the crop should be sown every fourth year. *S. scabra* had low yield but highly persistent. Some problems of leaf minor and leaf folder insects were found. During winter, cold damage was recorded to the extent of 20-25 per cent in these legumes except *C. ensiformis.* However, the hill region is not found suitable for seed production. The two native forage legumes *D. heterocarpum* and *D. podocarpum* had low yield (11.6 tonnes/ha/year green forage yield) but highly persistent with profuse nodulation.

Fodder Tree Species

Trees are source of green forage during lean period, a forge bank available during lean period as well as when ever need arises for feeding livestock. Growing fodder tree species is an opportunity for vertical expansion in forage production and soil nutrient cycling. Some of the potential single tree species *viz., Ficus glomerata* can feed one dairy cow for 30 days. Forage shrubs and trees provide forage during dry winter months, a lean season starting from October to March at low to mid hills (Singh *et al.,* 1996). The tree fodder constitutes 25-60 per cent of total forage fed to livestock in the hill region. Occurrence of fodder tree species is governed by prevailing temperature, which is further determined by altitude, precipitation and the sunshine hours being controlled by hill slope direction. The greatest diversity in fodder tree species is noticed between 900-1500 meter altitudes. The bio-climatic conditions within the short vertical and horizontal distances enable occurrence of different fodder tree species within a broad ecological range (Pandey, 1982). The occurrence and abundance of a fodder tree species in an ecological range is also influenced by traditional knowledge gained by livestock farmer for its palatability, nutritional value, period of forage availability and providing forage at different times of the year. In addition to this, the distance between natural forest as source of forage and farmlands is also compelling factor to encourage farmers to grow fodder trees on the farm lands. Thus, the diversity of fodder tree species on a farm has a complex interaction between bio-climate at the farm site and traditional knowledge build up on the value of tree fodder. Most probably, a fairly large number of fodder tree species had a wide geographical distribution, without special preferences for soil type, exposure *i.e.* slope direction aspects of the hills and precipitation showing a wide range of adaptability. For example, *Brassiopsis hainla* and *Litsea polyantha* are found between 600 to 1700 metres altitudes and *Garuga pinnata* occurred between 100 to 1100 metres altitudes. *Ficus semicordata* occurs from the *Terai* to 2000 m are small to medium trees which are deciduous for a short time in the year. They are light-demanders and to some extent tolerant to frost but killed by fire. *Saurauia napaulnensis* is found between 750 and 2100 m, being most popular tree fodder on north-facing slopes from 1000 to 1750 m, and on south-facing slopes above 1500 m altitude. It is fairly tolerant to shade, and can be grown on the soils of low fertility. Early growth of *S. napaulnensis* is slow as compared to the growth of *Ficus auriculata* and *F. semicordata* (rai khanyu). After the age of five years, growth is accelerated. Some of the fodder trees were very sensitive to ecological conditions *viz. Artocarpus lakoocha* did not occur naturally in an ecological range but it was individual choice of farmer to plant this tree. Therefore, it was almost difficult to draw distinct grouping the fodder tree species which occurred in different climatic-altitudinal belts. Thus, tree fodder availability was found to be natural one and also based on choice of farmer. Among the introduced fodder tree species, *Leucaena* species, *L. Leucoephala* K156 and *Grewia optiva* are suitable for high rainfall and acidic soil condition of eastern hills. Cultivar K156 is free from major problem of leaf spot disease (*Camptomeris leucaene*) most prevalent in humid climate. Leaves of bamboo plant species are good forage for lean season, need to be encouraged for feeding of livestock. About 26 bamboo plant species analysed showed these as nutritious forage.

In alpine region of Ladakh, *Salix* and *Populous teloides* and other species are the tree fodder grown in and around the farmlands and along various tributary entering rivers (Photo 8.9).

Photo 8.9b: *Populous teldoides* grown under Farming Systems and as Block Plantaiton in Alpine Region of Ladakh.

Studies on time of seed collection, 1000 seed weight, season of seed sowing, method of propagation and age of first lopping are very much limited on fodder tree species. Collection time for seed, its sowing, various methods of propagation and age of first lopping have been given for some important fodder tree species (Table

Photo 8.9a: Block Plantation of *Populous teldoides* In Ladakh.

Photo: 8.9c: *Salix* species and Seabuckthorn (*Hippophae rhamnoides*) Leaves are also Fed to Livestock.

8.2). The most favourable time for planting fodder trees is July to August (Singh *et al.*, 1996). There is feasibility to extend the planting season both before and after the usual planting time for some fodder trees (Pradhan, 1982, Jackson, 1987). *Bauhinia* species, *Ficus auriculata* can be planted right from April to September to avoid conflict with agricultural activities for labourer. Technique of establishing seedlings/cuttings in plastic pots is widely practiced. Bare root planting of some *Ficus* species has been successful. This can be used to avoid movement of soil in plastic pots. Vegetative propagation where possible has added advantage over seedlings to maintain genetic purity of plus tree species identified, reduced time taken to grow into tree for forage, as well difficulties in collecting seed and ensuring seed viability in widely different fodder tree species. Many fodder tree species can be easily propagated by cuttings *viz. Artocarpus lakoocha, Bauhinia species, Butea monosperma, Brassiopsis hainla, Bridelia retusa, Erythrina arborescens, Ficus auriculata, F. lacor, F. neriifolia, Grewia tiliaefolia, Morus alba, Populus ciliata*, and *Salix* species (Carter and Paudyal, 1987, Table 8.2). Seed of *Artocarpus lakoocha* is extracted from fruit to store them for longer periods. *Brassiopsis hainla* hardwood cuttings of 20-30 cm length taken in February, and set within 24 hours of cutting, rooted successfully. In view of the short viability of the seed and the slow growth of seedlings in the nursery the use of cuttings has advantage to recommend it. Seed of *Brassiopsis mitis* should be extracted by soaking the fruit in water until the flesh is soft, and then gently rubbing them together. The germination of fresh seed is 80 per cent but seedlings obtained in nurseries have poor survival. The figs of *Ficus semicordata* are found on the lower part of the leafless branches under leaf litter and humus buried in the surface of the soil. Care is required to collect only seed-containing figs. It regenerates naturally on the eroded sites (Olsson, 1983). Below 700 m altitude, seed is sown in late February to produce planting stock by the monsoon; at higher altitudes seed is sown in early August for planting stock ready in the next monsoon. Germination of seeds is low and time varies from five days to nearly six months.

In *Ficus benghalensis,* the seeds take longer time to germinate. Survival of its seedlings planted in April-June is better. Seed of *Saurauia napaulnens* is separated from the jelly, and carefully dried, can be stored for one year in sealed polythene bags. Germination in warmer localities takes 2-6 weeks and colder area more than 20 weeks (autumn sown). As the seed is very small special care is needed in germinating it. The seedlings show some susceptibility to frost. The seedlings will need shade and protection from heavy rain until they are well established (2-4 weeks). Viability of fresh seed in *Populus ciliata is* 75 to 90 per cent.

Table 8.2: Propagation, Seed, Planting, Age of First Lopping, Period of Availability of Forage in Some Important Fodder Tree Species

Tree Species/ Nature	*Time of Seed Collection*	*1000 Seed Weight (g)*	*Season of Seed Sowing*	*Method of Propagation*	*Age (Years) of First Lopping*
2000 m altitude and above					
Populus ciliata Deciduous	June -July	0.07	Immediately after collection.	Seed, vegetative means	04
Salix species Deciduous	June -July	—	Immediately after collection	Seed, vegetative means Cutting	08
600-2000 m altitude					
Artocarpus lakoocha Deciduous	June - July	—	Extracted seed sown during monsoon	Seed, root cutting	06-08
Bauhinia purpurea Deciduous	May-July	226.0	Spring	Seed, cutting, coppice	—
Bauhinia variegata Deciduous	April-June	278.0	Spring	Seed, cutting, coppice	—
Brassiopsis hainla Evergreen	June-July	28.0	Immediately after collection in May or June	Seed, cutting	04-05
Brassiopsis mitis, Evergreen	May-June	28.0-40.0	Immediately after collection	Seed, cutting	04-05
Castanopsis species Evergreen	October-November	—	Spring	Seed, cutting	07
Celtic australis Deciduous	November -December	—	Spring	Seed	—
Celtis orientalis Deciduous	Novemer-December	—	Spring	seed	07-08
Erythrina arborescens Deciduous	May-June	—	Immediately after collection	Seed, cutting	01-02
Grewia elastica Deciduous	June-July	—	Autumn	seed	—
Ficus auriculata Evergreen	June-July	1.0	Autumn	Seed, cutting, coppice	05-07
Ficus benghalensis Evergreen	May-June	—	Monsoon, seeds take longer time to germinate	Seed, cutting	05-06

Contd...

Tree Species/ Nature	*Time of Seed Collection*	*1000 Seed Weight (g)*	*Season of Seed Sowing*	*Method of Propagation*	*Age (Years) of First Lopping*
Ficus elastica Evergreen	—	—	Monsoon	Cutting	07-08
Ficus glaberrima Evergreen	June-July	—	Monsoon	Seed, cutting, layering	05-06
Ficus hirta Deciduous	June-July	—	Monsoon	Seed	05-08
Ficus infectoria Deciduous	June-July	—	Monsoon	Seed, cutting	5-10
Ficus neriifolia Evergreen	June-July	0.5	Spring/summer	Seed, cutting, layering, coppice	07-10
Ficus semicordata Evergreen	July-October	0.33-0.67	Spring at low and early August at high altitudes	Seed, air layering, hardwood cutting/ coppice	04-05
Ficus subnisa Evergreen	July-August	1.0	Autumn	Seed, layering	04-05
Helixanthera ligustrina Evergreen	—	—	Monsoon	Seed, Cutting	15-16
Litsea polyantha Evergreen	June-July	200.0	Immediately after collection, 60 per cent seed viability	Seed, cutting, coppice	05-06
Litsea citrata Deciduous	June-July	88.0	Immediately after collection	Seed, cutting	05-06
Mallotus philippensis Evergreen	July-August	—	Immediately after collection	Seed	7-8
Prunus ceresoides Deciduous	—	306.0	Immediately after collection. Seeds require scarification, sown during winter, slow to germinate	Seed, cutting, coppice	03-05
Saurauia napaulnensis Evergreen	March-July	0.13-0.25	March to April	Seed, cutting of two year old hardwood, coppice	04-05
Schima wallichii Evergreen	April –June	—	Summer	Seed, coppice	—

Source: Carter and Paudiyal, 1987; Jackson, 1987; Pradhan, 1982; Pandey, 1982.

Not much reseach has been done on diseases and pests of fodder tree species. *Populus ciliata* is affected by the plant parasite *Loranthus elatus;* premature defoliation by *Bipolaris mydis* and *Pseudocercospora salicia* and root rot by *Phorma macrostoma*; ganoderma. *Salix species* planted on the borders of streams are, host to aphids (*Chaitophorus* and other genera) and orange spots rust (*Melampsora species*)on the leaves. *Artocarpus lakoocha* is infected by *Pseudocera spora* fungus. *Ficus auriculata* suffers from leaf spot disease caused by *Cerecosporidium boieanum* and *Coletotrichum* species (Singh *et al.,* 1996). Severe infestation of anthracnose resulted in premature leaf fall. Insect pest psylidae (*Pauropsylla ficicola*) nymph causes gall formation on the leaves in *Ficus auriculata.* The leaves become chlorotic with big and small outgrowth of galls all over upper surface of the leaves. As result, there are leaf deformities and early senescence. It is believed that secretion from the nymphs stimulated gall formation. Adult insects lay eggs on the leaf surface and one nymph remain inside each gall till the moulting takes place inside the gall. Both nymph and adult stage of membracid bug (*Hypsauchem subfusca* Bucket) extract sap from the twig and stems of the tree. Due to sucking of juice by the insects, plant sap oozes out profusely from injured points, making black colouration on the stem. On an average 150 bugs were found in the infested tree. Lopping induces infestation by membracid bug. Tortoise beetle/dark brown beetle (*Cassie pusllula* Boheman) grub and adult feed on the leaves, making holes. After sometime, the insect defoliate the whole tree. The adult insect is found living throughout winter in dry shelter place under bark and come out from hibernation in May to feed on the leaves up to August. Stem borer (*Cerambycidae – Coleoptera*) was found to make bore into trunk and the branches of the tree resulting in to death of tree. More than one adult were found per branch. Severe defoliation was also caused by *Lepidoptera* caterpillars. Hairy as well as nymph caterpillars were found skeletonising the whole canopy of the tree.

Generally, fodder trees are found growing on the terrace risers of upland crop fields, along farm boundaries; water springs passing through farmland and as shade trees in large cardamom plantation (Singh *et al.,* 1989; Singh & Rai, 1990; Singh *et al.,* 1996; Singh, 2005).

Tree fodder is the main source of green forage during dry winter months from October to March, contributing 25 to 60 per cent of total green forage fed to livestock in low to mid hill. Leaves of bamboo are under-utilized forage resource in the hills as well as in the plain area of India. Lopping schedule of fodder trees are mostly based on the requirement of forage on individual farm (Singh and Yadav, 1995, Table 8.3). Mostly tree fodder is harvested during the lean winter months. Lopping commences in October or November and completes in March or April. Generally, lopping of evergreen trees is extended up to April and the lopping of deciduous or semi deciduous fodder tree species are completed before the leaf shed. Some of the fodder trees are lopped during May to August, a peak period for agricultural operation *viz.* sowing and weeding. *Ficus benghalensis, F. lacor, F. glaberrima* and *F. racemosa* are the preferred tree fodder. *Brassiopsis hainla* lopped between the end of October and the beginning of April, but it is preferred mainly for feeding before the monsoon. *Ficus semicordata* trees are often lopped twice in a year; occasionally three times. The lopping is more common during the pre-monsoon (mid-April to mid-June) in most places. *Saurauia napaulnensis* is lopped from November to April. Lopping can take place at almost any time in the year as per requirement of forage but is often extend up to pre-monsoon (mid-April to mid-June) in some places. Productivity of fodder trees are governed by intensity of lopping and the time of lopping during the lean period. Singh (1995) reported significant effect of month of lopping and the intensity on the forager yield of *Brassaiopsis hainla, Grewia optiva, Ficus auriculata, F. benghalensis, F. nemoralis, F. subincisa* and *Litsea polyantha.* The effect of light, moderate and heavy intensities of lopping in a three years old alder (*Alnus glutinosa*) showed that within 24 hours of lopping, transpiration rates decreased resulting in enhanced water use efficiency owing to increased water potential of the trees (Singh & Thompson, 1995). The effects continued up to 48 days of regrowth, till new leaves emerged and grown fully to increase the transpiration rate. Further, the light (30 per cent) to moderately (60 per cent) lopped trees developed larger leaf area than the trees which were not lopped. However, only light lopping of the trees could build up the total biomass in the lopped trees equivalent to the control trees not being lopped. The optimum lopping intensity for fodder tree species should be limited to removal of 30-40 per cent of total canopy leaf area.

Information on nutritional values of tree fodder is available to certain extent but the records of tree forage production not available. Generally, tree fodder yield of a tree species are generalized guess. During a survey of forage yields of different fodder tree species was carried out along with tree growth parameters for formulation of regression models to predict forage yields their yields (Table 8.3; Singh *et al*, 1994, Singh *et al.,* 1996). Prediction of forage yield under a plantation of fodder tree species on a time scale is required for policy decision on tree fodder. It is possible to develop forage yield prediction models based on easily measurable tree growth parameters with limited data (Pradhan, 1982, Wormald *et al.,* 1983, Singh *et al.,* 1994). Accordingly, Singh *et al.* (1994) developed regression models to predict tree forage yield of 18 fodder tree species found at low to mid hills (Table 8.4). To avoid tedious job of determining crown volume, one has to restrict with basal girth or/and girth at breast height may be marginally predicted with less precision. There is need of prediction model taking into account specific growth pattern with age, peak forage yield, site specificity and lopping techniques adopted.

Table 8.3: Forage Yield of Important Fodder Tree Species

Tree Species	*Fresh Forage Yield (kg/tree)*
Populus ciliata	5-7 (8)
Salix species	8-10 (10)
Artocarpus lakoocha	200-240 (15-16)
Bauhinia variegata	50-75 (10-12)
Brassiopsis mitis	50-150 (6-10)
Celtis australis	150 (5)
Celtis orientalis	160 (50)
Ficus auriculata (*Syn. F. roxburghii*)	20-80 (10-20)
Ficus benghalensis	500-700 (30)
Ficus semicordata	270-300 (12-14)
Ficus subincisa (*F. cunia*)	160-200 (7-8)
Ficus glaberrima	60-100 (1-12)
Ficus infectoria	60-80 (15-30)
Litsea polyantha	150 (12-14)
Populus ciliata	5-7 (8)
Helixanthera ligustrina	400 (15-20)
Machilus edulis	60-90 (15)
Mallotus philippensis	160 (50)
Prunus ceresoides	100 (8-10)
Salix species	8-10 (10)
Saurauia napaulensis	30-40 (20-25)

( ) indicates age of fodder tree species.

Source: Singh *et al*., 1996.

Table 8.4: Regression Models for Forage Yields in Fodder Tree Species of Eastern Himalaya

Fodder Tree Species	*Prediction Equation for Lopped Forage Biomass*
Artocarpus lakoocha	Y= 1.54770 + 0.37637 CRV (R - 0.960)
Bauhinia purpurea	Y= – 24.43995 + 0.72309 BG (R - 0.903)
Bauhinia variegata	Y= 0.03095 – 0.06464 BG +0.37638 GBH (R - 0.810)
Brassiopsis hainla	Y= 1.59260 – 0.24886 BG + 0.35045 GBH + 0.37338 CRV (R - 0.943)
Celtis australis	Y= 0.22927 + 0.31486 CRV (R - 0.947)
Ficus auriculata (*Syn. F. roxburghii*)	Y= – 9.03534 + 0.29905 BG + 0.20742 CRV (R - 0.753)
Ficus benghalensis	Y= – 22.65697 + 0.48403 GBH (R- 0.933)
Ficus clavata	Y= – 1.76630 + 0.1333966 CRV (R - 0.978)
Ficus cunia	Y= 0.03187 + 0.13594 CRV (R - 0.999)
Ficus elastica	Y= 0.66650 + 0.21858 BG – 0.32370 GBH (R - 0.892)

Contd...

Fodder Tree Species	*Prediction Equation for Lopped Forage Biomass*
Ficus glomerata	Y= 14.43715 - 0.13676 BG – 0.08749 GBH + 0.73896 CRV (R – 0.951)
Ficus hirta	Y= – 6.41700 + 0.27804 BG + 0.42412 CRV (R - 0.899)
Ficus hispida	Y= 2.05963 + 0.09320 CRV (R - 0.951)
Ficus infectoria	Y= – 0.52109 + 0.28274 CRV (R - 0.987)
Ficus nemoralis	Y= 0.17100 + 0.53135 CRV (R - 0.999)
Grewia tiliaefolia	Y= – 15.63756 + 0.47297 BG + 0.06810 CRV (R - 0.831)
Litsea polyantha	Y= – 18.30264 + 0.18832 BG + 0.16835 CRV (R - 0.960)
Saurauia napaulensis	Y= – 25.55902 + 0.55820 BG – 0.07792 CRV (R - 0.820)

Y – Dry matter yield (kg/tree), BG – Basal girth (cm), GBH – Girth at breast height (cm), CRV – Crown Volume (cubic metre), R- Coefficient of determination.

Source: Singh *et al.* (1994).

Nutrient Contents and Nutritional Evaluation of Forage

Nutritional values of forage crops determine forage quality to meet energy and nutrient requirement of livestock. Nutritional values of some important perennial forage grasses and legumes as found at 1000m altitude in the hills of Sikkim are presented in Table 8.5, 8.6. In perennial grasses, crude protein content was more than 10 per cent in Napier x bajra and bajra x napier hybrids, thin napier, broom grass, Congo signal grass, coloured guinea grass, golden timothy grass, green panic, guinea grass (except Gatton panic), Italian rye grass, orchard grass, tall fescue. In general temperate grasses had more crude protein content than tropical grasses (Singh *et al.,* 1996). In range legumes crude protein content varied from 12.0 in *Stylosanthes hamata* to 21.0 per cent *in Centrosema pubescens.* Calcium content is very important for dairy cattle and buffalo. It ranged 0.40 to 1.56 per cent. In the forage grasses, crude protein increased with increasing level of nitrogen application but under humid high rainfall it decreased due to high yield of dry matter production. Increase in cutting interval from 20 to 60 days decreased crude protein content in the dry matter of Guinea grass, Dallis grass and Golden timothy grass (Singh and Pradhan, 1995).

Growth of cattle calves for beef production is better on established pasture of *Setaria sphacelata* (288 g per day) due to better crude protein content and digestibility of dry matter than native pasture and the pasture of guinea grass variety Hamil (Yadav *et al.,* 1991, Yadav *et al.,* 1992). Growth of Black Bengal goats on established pasture of Congo signal grass was about 42 g per day as compared to 34 gram per day on the native pasture.

Tree fodder and the leaves of bamboo are fodder bank and valuable source of green forage during lean season. Nutritional evaluation of tree fodder shows that they contain higher dry matter of 38 to 46 per cent than other perennial forage crops. Moderate levels of dry matter help in maximum feed intake in the livestock. Proximate analysis of tree leaves shows that they contains good levels of crude protein, rich in minerals *viz.* calcium, phosphorus, zinc and iron and adequate level

Table 8.5: Nutritional Values of Forage Grasses and Legumes (Per cent in dry matter)

Forage	*CP*	*EE*	*CF*	*NFE*	*TA*	*IA*	*Ca*	*P*
Sub tropial and tropical forage grasses								
Napier x bajra hybrid 21	10.2	—	32.5	—	8.76	3.57	0.46	0.37
Bajra x napier hybrid BN82048	10.9	—	33.9	—	8.94	3.89	0.46	0.42
Teosinte	5.6	—	32.5	—	8.02	2.14	0.65	0.28
Brachiaria mutica	5.5	—	30.5	—	6.79	3.83	0.40	0.30
Brachiaria ruziensis	12.6	3.94	24.0	51.0	8.39	—	—	—
Eragrostis curvula	5.5	—	33.9	—	5.55	2.32	0.82	0.27
Erianthus sikkimensis	7.5	—	42.7	—	5.38	1.18	0.60	0.47
E. longisetosum	6.4	—	39.9	—	6.97	2.77	0.66	0.44
Panicum coloratum	14.2	—	26.4	—	5.92	0.60	0.85	0.45
Panicum maximum Acc. No. 59975	12.5	—	27.4	—	9.15	3.42	0.61	0.39
Gatton	8.7	—	32.5	—	9.45	2.90	0.44	0.20
Hamil	11.1	2.61	30.0	46.5	9.86	3.59	0.68	0.54
Makueni	13.2	3.26	29.1	44.7	9.72	2.70	0.86	0.59
PGGI	13.1	—	25.2	—	7.65	3.04	0.68	0.36
Panicum maximum var. trichoglume),	12.0	—	29.9	—	8.07	4.79	0.50	0.40
Paspalum dilatatum,	6.0	—	33.8	—	8.46	3.46	1.56	0.30
Pennisetum pedicellatum	7.4	—	33.4	—	7.65	1.70	0.90	0.53
Setaria palmifolia	8.6	—	37.6	—	7.92	3.38	0.98	0.61
Setaria sphacelata	11.1	3.61	27.4	48.7	9.12	1.60	1.12	0.42
Thysanolaena agrostis/T. maxima	10.2	1.94	30.5	51.6	5.65	3.67	0.82	0.40
Tripsacum laxum	7.7	—	37.2	—	6.08	2.41	0.62	0.30
Zizinia caudiciflora	9.2	2.15	33.2	44.1	11.48	7.89	1.83	0.23
Z. latifolia	9.3	1.85	34.2	44.6	10.14	6.98	1.85	0.22

Contd...

Forage	CP	EE	CF	NFE	TA	IA	Ca	P
Temperate forage grasses								
Dactylis glomerata	18.8	—	33.2	—	8.96	1.76	0.62	0.42
Festuca arundinacea	13.1	—	—	—	—	—	—	—
Lolium perenne	7.7	—	—	—	—	—	—	—
Subtropical and tropical forage legumes								
Centrosema pubescens	21.0	2.18	30.4	37.8	8.66	—	—	—
Desmodium heterocarpum	16.2	—	40.7	—	7.88	0.75	0.48	0.28
Desmodium podocarpum	19.5	—	38.6	—	6.68	0.62	1.42	0.23
Macroptilium atropurpureum	14.2	—	40.5	—	8.18	1.07	1.08	0.40
Stylosanthes guianensis	16.4	2.34	31.4	43.2	6.66	0.14	0.36	0.50
Stylosanthes hamata	12.0	—	40.0	—	4.55	0.45	1.04	0.42
Stylosanthes scabra	14.2	—	37.3	—	5.03	0.57	1.02	0.42
Vigna sinensis	20.8	3.29	21.1	38.9	15.94	—	—	—
Vigna umbellata	19.2	2.01	22.5	49.1	7.21	—	—	—

Table 8.6: Range of Nutrient Contents in Tree Fodder, Leaves of Bamboo and other Forage Crops

Forage		*CP (per cent)*	*Macro Mineral (per cent)*		*Micro Mineral (ppm)*		
			Ca	*P*	*Zn*	*Cu*	*Fe*
		Ideal >15 per cent	*Ideal >0.5 per cent*	*Ideal >0.22 per cent*	*Ideal >30*	*Ideal 10 or less*	*Ideal 50-100*
Tree fodder		13-30	1.1-3.2	0.15-0.40	34-118	8-16	128-363
Leaves of bamboo		9-19	0.9-1.6	0.08-0.15	20-48	7-11	91-750
Cultivated Fodder	Leguminous fodder	15-22	0.6-2.0	0.05-0.25	25-35	8-12	200-300
	Non Legumes	8-13	0.3-0.6	0.05-0.20	15-20	4-10	100-200

of copper being good for dairy animals (Table 8.6; Annexure I and II). The tannins content in the leaves was less than five per cent. Higher crude protein and total ash contents were found in February lopped tree fodder as compared to the tree fodder lopped during November but the tree fodder lopped during November had higher tannins, ether extract and nitrogen free extract than the tree fodder lopped during February (Singh *et al.,* 1996). Generally, upper leaves of the lopped twigs showed higher contents of crude protein, nitrogen free extract and tannin and lower leaves had higher crude fibre, ether extract and total ash. The tannin content of the leaves of *Ficus semicordata* is rather high, with a peak in December, and a subsidiary peak in March (Wood *et al.,* 1992). It is experience of farmers that tree fodder lead to increased milk yield (Shrestha and Pakhrin, 1988). Leaves of bamboo are rich in crude protein and nitrogen free extract, normal to higher in calcium, copper, iron, manganese and zinc contents. However, it has low content of phosphorus.

Type of tree fodder affects the feed intake of ruminants. In sole feeding of six fodder tree species *i.e. Ficus auriculata, F. clavata, F. semicordata, Bauhinia purpurea, Grewia optiva* and *Litsea monopetata*, feed intake of tree fodder ranged from six to ten kilogram (Shreshta and Pakhrin, 1989, *Singh et al.,* 1996). Digestibility of *Ficus semicordata* was higher than *F. auriculata* when the two tree fodders were fed to buffalo. The higher hemicellulose (21 per cent) in *F. auriculata* resulted in 30 per cent digestibility of dry matter as compared to 6.5 per cent hemicellulose in *F. semicordata*, which had 54 per cent digestibility of dry matter. All the tree fodder increased the milk yield except *F. auriculata* in spite of the fact that higher feed intake was recorded with this tree fodder (Allan, 1947, Anderson, 1947, Pandey, 1982 and Shreshta and Pakhrin, 1990). In another study on feeding value six fodder tree species in goat showed that sole feeding of tree fodder of *Brassiopsis hainla* could meet the maintenance requirement of goat. *Ficus auriculata, F. clavata* and F. *infectoria* needed supplementation with phosphorus rich concentrate or bone meal (Gupta *et al.,* 1990). However, Feeding of tree fodder of *Symingtonia populnea* and *Schima wallichii* need both nitrogen and phosphorus supplementation. After establishment, total forage productivity of *Ficus auriculata* + *Thysanolaena agrostis* and *F. auriculata* + Napier x bajra hybrid silvipasture at a mid-hill showed consistent forage yield increase up to 9th year. However, Napier x bajra hybrid need gap filling and spot replacement by rooted slips every fourth year. Strong tufted and tall growing grasses are well suited component to compete with weeds *viz., Eupatorium odoratum, Ageratum conzoides, Bidens pilosa, Gnaphalium affine* and *Spilanthus paniculata*. Silvipasture is productive land use system to utilize wastelands for increasing the availability of green forage round the year particularly tree fodder during the lean period of dry winter months (Singh *et al.,* 1995).

Mixed feeding of tree leaves of *F. auriculata* along with golden timothy grass (*Setaria sphacelata*) to goat kids grown under silvipasture of the two plant species had body weight gain of 69 gram per day per goat (Table 8.7; Balaraman *et al.,* 1988; Singh *et al.,* 1995). Nitrogen, calcium and phosphorus had positive balance indicating usefulness of this silvipasture for goat production.

Table 8.7: Growth Rate and Feed Efficiency of Goat Kids on Nevaro + Golden Timothy Grass Silvipasture System

Attributes	*Values*
Initial body weight (kg)	8.88 ± 4.44
Final body weight (kg)	16.68 ± 3.65
Weight gain (kg)	5.80 ± 1.06
Growth rate (g/day)	69.00 ± 12.41
Dry matter intake (per cent of body weight)	4.09 ± 0.28
Feed to gain ratio	6.74 ± 1.00
Dry matter digestibility	72.32 ± 2.05
Crude fibre	57.98 ± 1.76
Crude protein	64.23 ± 4.76
Ether extractives	57.93 ± 2.86
Nitrogen free extracts	84.30 ± 1.80
Balance of nitrogen and mineral	
Nitrogen	5.91 ± 1.93
Calcium	4.31 ± 1.81
Phosphorus	0.67 ± 0.61

Source: Balaraman *et al.*, 1988, Singh *et al.*, 1995.

Possibility of utilizing forage by mono-gastric livestock has been demonstrated. Maize and bajra forage feeding to younger and finisher pigs have been found beneficial (Yadav *et al.,* 1990 a; Yadav *et al.,* 1990 b; Yadav & Gupta, 1994) When offered, pigs also chew succulent guinea grass (grown up to 50 cm height) and residue fibre is extruded by animal after chewing it (Singh *et al.,* 1996). Rabbit fed on tree leaves of nevaro (*F. auriculata*) and broom grass (*Thysanolaena maxima*) from nevaro + broom grass silvipasture and the growth compared to feeding of the forage rice bean showed marginal loss (13-15 per cent) in body weight gain (Table 8.8; Singh *et al.,* 1995). The added benefit of tree fodder and leaves of broom grass is that they are available round the year unlike rice bean only during the monsoon. Further, establishment silvipasture do not require arable land but can be grown on uncultivable hill slopes. More research is required to reveal the potential of feeding succulent forage to non-ruminants.

Often, alder (*Alnus nepalensis*) is reported as tree fodder but a study by Yadav and Gupta (1994) showed poor palatability and acceptability of this forage in goat (Yadav *et al.,* 1994). Within 20 days of feeding sole alder tree fodder, 75 per cent of experimental goat died due to starvation and increased susceptibility of secondary infection as revealed in post mortem report. The survived goats were also weak and debilitated.

Table 8.8: Body Weight Gain (kg) in Rabbits Fed Green Forage and Concentrates under Nevaro (*Ficus auriculata*) + Broom Grass Silvipasture System

Age of Rabbits (Days)	*Broom Grass Leaves*			*Nevaro Leaves*			*Rice Bean Forage*		
	Mean	*SE*	*CV*	*Mean*	*SE*	*CV*	*Mean*	*SE*	*CV*
42	0.575	0.03	19.36	0.575	0.02	13.69	0.577	0.04	27.1
56	0.770	0.04	19.38	0.808	0.03	13.44	0.875	0.05	20.8
70	0.989	0.06	23.48	1.065	0.04	15.42	1.161	0.05	15.0
84	1.189	0.06	19.94	1.257	0.04	12.78	1.464	0.04	9.1
98	1.400	0.08	21.42	1.498	0.06	15.62	1.691	0.03	6.7
112	1.643	0.06	14.66	1.661	0.05	11.69	1.884	0.04	7.8

Source: Singh *et al.*, 1995.

Potential of Forage Production under Non-competitive Land Use

Possibility of growing forage diverting a part of agricultural lands is not feasible because of limited arable land (less than 20 per cent available for food crops) available in the hills and its priority for annual food crops in the hills. Under the circumstances, terraced lands provides scope for growing forage along with food crops in non-competitive way with added advantage of soil conservation, improved hydrological behaviour and as such enhances the efficacy of terracing on the hill slopes. A non-competitive land use systems can be tailored by growing suitable perennial forage grasses on the terrace risers as well as perennial grasses and fodder trees on terrace bunds and farm boundaries (Photo 8.10–8.13; Table 8.9). These unutilized lands permit growth of weeds and their proliferation into arable lands (Singh, 1998, Singh, 2005).

Photo 8.10 a: *Sesbania aegyptica* with Napier Grass on the Bunds of Terraced Lands in Nepal

Photo 8.10 b: *Leucaena leucocephala* with Napier Grass on the Bunds of Terraced Lands in Nepal

Photo 8.11: Napier Bajra Hybrid on Terrace Bunds in Sikkim.

Photo 8.12: Mustard on the Terraced Land and Guinea Grass on the Terrace Riser in Meghalay.

Photo 8.13: Turmeric on the Terraced Land and Guinea Grass on the Terrace Riser in Meghalaya.

Table 8.9: Forage Yields of Perennial Grasses on Terrace Risers, Terrace Bunds and Bamboo Shaded Area

Terrace risers (tonnes dry matter/ha/year)

Perennial Forage Grasses and Legumes	*Mean*	*Range over 5 Years*	*CV**
Panicum maximum	12.62	6.30-15.74	21.1
Brachiaria ruziensis	28.39	15.95-41.15	56.6
Thysanolaena maxima (Leaves only)	6.26	1.59-11.21	87.2
Macroptilium atropurpureum	3.70	2.40-5.00	
Stylosathes guianensis	5.80	4.70-6.90	

*Coefficient of Variation over age (establishment to 5th year).

Terrace bunds (tonnes dry matter/1000metres/year over 03 years)

Forage grasses		
Napier x bajra hybrids	1.67	0.50 - 2.40
Tripsacum laxum	1.30	0.6 0-1.90
Panicum maximum	1.17	0.70-1.40
Thysanolaena maxima (Leaves only)	2.33	0.30-4.60

Contd...

Terrace bunds/farm boundary			
Fodder Tree Species	*Yield (kg dry matter) per Tree*	*Basal Girth (cm)*	*Crown Volume (m^3)*
Artocarpus lakoocha	10.32-102.32	78-208	26-246
Bauhinia purpurea	4.14-82.61	30-135	14-193
Bauhinia variegata	4.62-54.28	30-218	35-342
Brassaiopsis hainla	6.72-53.77	50-165	14-135
Celtis australis	10.41-104.12	41-213	33-298
Ficus auriculata	1.07-65.95	30-125	02-142
F. benghalensis	3.10-107.16	33-280	11-378
F. elastica	7.99-49.35	37-135	16-87
F. glomerata	13.42-120.77	84-290	63-241
F. hirta	9.30-49.59	40-120	09-46
F. hispida	10.44-63.27	32-160	11-65
F. infectoria	29.57-128.92	55-275	104-447
F. nemoralis	5.34-64.04	48-167	10-121
F. semicordata	4.39-50.99	41-192	33-371
F. subnicisa	3.22-78.25	38-170	26-508
Grewia tilaefolia	3.12-52.04	30-100	11-181
Litsea polyantha	5.92-120.37	43-215	22-318
Saurauia napaulensis	3.26-47.59	46-151	16-278

Forage yield (dry matter tonnes/ha/year) of forage grasses under bamboo shade of (*Bambusa nutans*)

Forage yield (tones/ha.)	
Orchard grass (*Dactylis gloremata)*	11.90
Nandi setaria (*Setaria sphacelata*)	43.00
Dinanath grass (*Pennisetum pedicellatum*)	10.00
Stylo (*Stylosanthes guianensis*)	16.25
Forage production on marshy and waterlogged lands	
Brachiaria mutica, Zizinia caudiciflora and Z. latifolia	8.00-10.00
Forage yield (dry matter tonnes/ha/year) of forage grasses under natural forest	
Setaria palmifolia	14.0-25.0
Pennisetum purpureum	14.0-31.1
Thysanolaena maxima (leaves only)	0.83 – 1.35
Fodder tree species as a shade tree in large cardamom plantation	
Brassiopsis mitis, Ficus hirta, F. auriculata, F. benghalensis, F. cunia/F. benjamina, F. nemoralis, Saurauvia napaulensis, Eurya japonica, Machilus edulis	

Source: Adapted from Singh, 2005, Singh *et al.*, 1996, Singh and Pandey, 1995, Rai *et al.*, 1988, Singh, 1989, Singh *et al.*, 1989, Singh, 1996, Singh *et al.*, 1993).

Micro-watershed based Forage Production Systems

Livestock are valuable component of many sustainable and economically viable farming systems (IUCN/UNEP/WWF, 1991). Forage based feeding is a cheap way of producing livestock products and also converts otherwise waste and not utilizable byproducts from field crops into valuable nutrient rich livestock products. Livestock also provides manures, a cheap source of nutrient for crop production. For developing production systems in the hills, micro-watershed provides a convenient framework to match the need of the people and the capacity of natural resource to meet those needs over a period of time (Singh *et al.,* 2009; Singh *et al.,* 2010). Generally agriculture, forestry, other land uses and soil water conservation and utilization practices are combined to accomplish objectives of micro-watershed management. Accordingly on a demarcated 3-5 hectares piece of hill slope a range of production systems *viz.* agro-pastoral, dairy farming, silvipasture, agri-horti-silvipasture, timber farming and forestry for fuel wood was developed (Singh *et al.,* 2010). Studies on such forage based land use systems had shown their livelihood potential.

Dairy Farming

Dairy farming was considered an option for the farmers located on the periphery of a city (Singh *et al.,* 2010). In the dairy farming forage productivity was increased largely due to replacement of native low producing native forage grasses by high yielding guinea grass and broom grass applying *in situ* generated organic manure obtained from livestock maintained in the system. Round the year green forage availability was ensured by tree leaves obtained from nevaro (*Ficus auriculata*) and broom grass (*Thysanolaena maxima*) during lean period of winter months. Total forage productivity under dairy farming increased from 14.8 tonnes/ha with native grasses without fodder tree to 101.6 tonnes/ha/year under *Ficus auriculata* + *Panicum maximum* + *Thysanolaena maxima* system. The yields of perennial forage plants consistently increased after establishment. Out of total forage produced guinea grass, broom grass and tree leaves contributed 54, 42, and 04 per cent, respectively. Remaining 4 per cent was contributed by annual forage crops grown on few terraced lands on the lower portion of the micro-watershed. Out of the total forage produced, 72 per cent was fed to the livestock reared in the micro watershed. A balance of 21.4 tonnes forage was obtained from guinea grass during monsoon period was conserved for the lean period. This surplus forage was used as buffer stock to avoid any scarcity during the unfavourable year. Forage produced in the system sustained two to three dairy cows along with followers maintained in the system. Mean annual livestock productivity obtained from dairy farming was 5564 litres milk, 278.2 kg gain in body weight of livestock and 18.7 tonnes/ha/year organic manure obtained from cow dung, urine and forage refuse under the system. The cow dung and urine from livestock was collected into a series of manure pits to produce organic manure. The organic manure was annually recycled into the soil for maintaining soil fertility within the micro-watershed. The system could be adopted by farmers in the periurban area.

Agro-pastoral System

A combination of livestock and crop production in a unit land-use management system has been an integral part of the farming pattern in India. Considering the need to improve and develop the traditional land-use system with high livelihood potential as well as to utilize to the maximum capacity, the local man and material resources available with a family, agro-pastoral system was developed on terraced hill slope (Singh *et al.*, 2010). This system was targeted towards marginal or small farmer with assumption that annual requirement of 689 man-days/ha could be largely met from a family for land preparation and sowing (254), weeding, intercultural operation and mulching (127), farmyard manure (FYM) and fertilizer application (66), harvesting and threshing (105) and rearing of a dairy cow unit (137). During the study for 14 years, significant steps were taken to optimize the productivity of agro-pastoral systems (Singh *et al.*, 2005). Terracing of the hill slope improved *in situ* conservation of rain water. This conserved soil moisture, which enabled growing of short duration mustard and radish crops in *rabi* season and enhancing cropping intensity to 175 per cent. Cropping systems included maize, soybean, groundnut, mustard, radish, ginger, turmeric, guinea grass and broom grass. Two years rotation cycle was introduced in the cropping system on the terraced lands by alternating heavy nutrient using crops ginger and turmeric, being rotated with soybean and groundnut – mustard. This helped to maintain soil health of terraced land and also took care of plant protection due to not being continuously cropped with same crop. One dairy cow with follower was maintained in the watershed itself. Composting of livestock refuses, weed and nonedible crop residues provided organic manure 20 tonnes/ha/year for crop production. The manure reduced the nitrogen fertilizer requirement by one third and met the full requirement of phosphorus and potassium.

The terrace risers were utilized for growing forage grasses, lower portion planted with guinea grass and upper portion of the watershed planted with broom grasses. This enabled a non-competitive land use for round the year forage production without application of manure or fertilizers to forage grasses. Moreover, the root systems of guinea grass, being more fibrous, had root-binding factor of 1.677 followed by that of broom grass, with factor 1.294. Both the grasses helped to stabilize the terrace risers and increased the efficacy of terraced land cultivation. Agro-pastoral system developed on a 0.58 ha micro-watershed provided adequate food to a family size of two adults and three children according to a typical rice-based non-vegetarian diet on the standard criterion of one tonne food and one tonne fuel set by Planning Commission of India. The index of farming-system performance kite, formulated by Bimbao *et al.* (1995) indicated that compared with the traditionally practiced shifting cultivation, agro-pastoral systems had 10.32 times higher primary biomass productivity, 0.66 to 0.80 times less crop diversity, three times more numbers of bio-resource flows; and economically more efficient than a typical shifting cultivation practiced on the hill slopes in north-eastern hills of India.

Agri-Horti-Silvipasture

A combined production of annual field crops (maize, groundnut, Indian mustard, radish and ginger) and fruit crops (mandarin, lemon, guava and pine apple), livestock (pig and goat), and fuelwood (a short-rotation silvipasture) in a single land-management unit provided a wholesome strategy to develop a self-sufficient system of land-use management on the hill slope. Agri-horti-silvopasture was ideally a model land use system. In this system, agriculture had 19.4 per cent of the area (Singh *et al.,* 2005). Out of total agricultural lands, 67.3 per cent (0.1341 ha) area was under terraced-land with 173 per cent cropping intensity and 32.7 per cent was under terrace riser planted with guinea grass. The mean productivity of maize, groundnut, Indian mustard, ginger and radish was 2.97, 2.68, 0.48, 14.13 and 32.20 tonnes/ha/year respectively on terraced lands, and 62.50 tonnes/ha/year forage grass on terrace-rises. The organic manure obtained from dung, urine and livestock refuse of goat and pigs, weed biomass and non-edible crop by-product was sufficient for the system. In-situ augmented organic manure was applied to the terraced lands for crop production at the rate of 20-22 tonnes/ha/year and a part of it was applied to fish-pond. It helped to reduce the requirement of nitrogen fertilizer to one third of the recommended dose and completely met the need of phosphorus and potassium for crop production. Under horticulture, *khasi* mandarin, guava and Assam lemon was planted with an intercrop of pine apple. It gave total fruit productivity 15.1 tonnes/ha/year. Moreover, 1.96 tonnes/ha/year tree biomass was built up in fruit trees. Guinea grass inter-planted between fruit trees produced 34.5 tonnes/ha/year green forage. Under silvipasture, the mean forage yield obtained from broom grass was 9.72 tonnes/ha/year for winter-season feeding of goats. Harvest of first cycle of trees for fuel-wood gave sufficient fuel-wood (2.14 tonnes/ha/year) for a standard-size family. In the second cycle of fuel-wood plantation, the annual standing tree biomass build-up was estimated to be 1.73 tonnes/ha/year. Overall, silvipasture gave an estimated fuel-wood yield of 2.01 tonnes/ha/year during the study period. Annual total forage grasses obtained from the micro-watershed was 7.52 tonnes/ha/year. Livestock (10 pigs and 10 goats) was fed on maize grain, radish, guinea grass and broom grass, and other by-products obtained from annual crops, fruits and silvo-pasture within the micro-watershed. Some deficit in individual years was met from purchase out of cash income generated from the sale of livestock and fruits. The annual gain in live weight under this system was 583 kg/ha/year. The goat multiplied in number from 10 to 15 in a year. Keeping stocking rate of 10 in number, 05 adult goats were sold. In a family, the goats might be either sold or used for home consumption. The system could not meet fully the requirement of rice for a family size of 02 adults and 03 children having rice-based non-vegetarian diet. However, from that of balanced items of farm produce after home consumption, *i.e.* ginger, radish and fruits, the deficit in the food requirement can be met by purchase from the market. Farming-system performance kite index of Bimbao *et al.* (1995) indicated that agri-horti-silvopasture system had 6.56 times higher biomass productivity, 1.40 times more species diversity, 05 times more number of bio-resource flows and economically more (Rs 2.94 return per Rupee investment) efficient than shifting cultivation. The agri-horti-silvopasture system compared with agro-pastoral system had less biomass productivity but was richer in species diversity, higher in number of bio-resource flows and also higher in return on per rupee investment. The experience gained in this micro-watershed indicated that various components of agri-horti-silvipasture should be developed on 02 ha land on a moderate hill slope (35 to 45 per cent) for generating complete livelihood potential to sustain a standard-size family.

Symingtonia populnea Based Silvipasture for Goat Production

Farmers having a portion of uncultivable land on his farm may opt for silvipasture for goat production to utilize such land and supplement his farm income. Total plant productivity increased from 08 tonnes/ha to 39 tonnes/ha under silvipasture system after establishment. The edible biomass increased from 3.0 tonnes/ha to 26.5 tonnes/ha in the subsequent years after establishment. The yield of tree leaves increased during the first four years of commencement of annual lopping (seven years after establishment), with a contribution of 32 to 36 per cent to total edible biomass. Due to continuous lopping, yield of tree leaves declined. It contributed just four per cent of total edible biomass in the 14th year. This indicated that the trees cannot sustain annual lopping and alternate year lopping should be imposed for sustained yield of tree leaves. Besides edible biomass, spike (0.21 tonnes/ha/year) from broom grass and fuel wood (2.49 tonnes/ha/year) from annual lopping of trees were obtained as additional output. Livestock productivity was 555 kg/ha/year live weight of goats and livestock refuse (15.4 tonnes/ha/year). The stock multiplied annually by about 1.6 times excluding some mortality of new born kids. If option of retaining younger stock for the next year is practiced, 160-178 kg live weight-comprising adult stock can be sold every year from the goats reared on silvipasture, established on an uncultivable hill slope with intermittent rocky patches. The organic manure obtained was recycled into the soil for improving the productivity of silvipasture. The silvipasture on uncultivable hill slope can be well utilized for goat production to supplement the farm income.

Economics of Forage-Based Livestock Production

Feeding of livestock on forage-based rations is not only a cheap way of producing animal products but is also helpful in converting the otherwise waste edible plant biomass into nutrient-rich food products. In micro-watershed based farming systems on the hill slopes, integration of forage based livestock enhanced the return (Table 8.10). Wherever livestock was integrated with other crop and horticulture or forestry- production systems, net profit increased (Singh *et al., 2002;* Singh, 2008; Singh *et al.,* 2008; Singh *et al.,* 2010). Livestock shared 64 and 70 per cent of the net return under agro-pastoral (dairy cow) and agn-horti-silvipasture (pigs and goats) systems, respectively. Although agro-horti-silvipasture system had a unique feature to combine a number of agro-activities to meet all the requirements of food, fuel and forage and gave a little cash for a farm family that can be produced from land on sustainable basis, where basic infrastructures like market, road *etc.* are yet to be developed.

Impact of Forage Resource Development on Sustained Land Use and Environment

Technical Advisory Committee (TAC) of CGIAR has advised that sustainable agriculture should involve successful management of natural resources for agriculture to satisfy changing human needs while maintaining or enhancing the quality of the environment and conserving the natural resources (TAC/CGIAR, 1989). Current level of technical knowhow the forage production potential under various land use at low to midhills has been shown in Table 8.11 (Singh *et al.,* 1996).

The choice available within the forage plant species is wider than for other crops to structure a land use on the hill slopes with soil conservation as a prime concern. Combinations of fine root systems of perennial forage grasses and deep root systems of fodder trees in a silvipasture systems, stabilizing terrace risers by forage grasses and fodder trees under crop production systems and agri-horti-silvipasture on the hill slopes are few of the various possible interventions that can be structured to achieve complete soil conservation. (Singh *et al.,* 1995; Singh *et al.,* 1996). The trees have the most significant effect on farm-lands (Singh *et at.,* 1989; Singh & Pradhan, 1993). The root system of a single fodder tree *Ficus auriculata* ramified into 6.83 m^3 soil volume, covering 1.54 m^2 surface areas. The fine root system of broom grass (*Thysanolaena agrostis*) well ramified up to 40 cm soil depth. They improve the efficacy of physical soil-conservation measures such as terracing (Singh *et al.,* 2005; Singh, 2002).

Soil Properties and Hydrological Behaviour

Soil properties of soil before and after establishment of forage-based dairy farming, agro-pastoral system, agro-horti-silvipasture and silvipasture structured in hilly micro-watersheds showed that soil pH, organic carbon of the soil, available phosphorus, exchangeable K, Ca and Mg increased. Organic carbon constituents *viz.* humic and fulvic acids increased only in dairy farming. Hydrological behaviour of forage-based systems showed that water yield of these micro-watersheds was similar to natural self-regenerated pine forest (Singh *et al.,* 2005). There was 90 to 95 per cent base flow and only 5 to 10 per cent surface flow. This was possible through knitting of perennial forage grasses *viz.* guinea grass, broom grass and fodder trees such as *Ficus auriculata* on two-third top portion of the watershed and planting of forage grasses, on the terrace riser in the one-third bottom portion for dairy farming and the two forage grasses on the terrace risers of a completely terraced micro-watershed for food production, with a subsidiary cow unit under an agro pastoral system.

Table 8.10: Economics (Rs/ha) of Micro-watershed Based Land Use Systems on the Hill Slopes

Land use	*Gross Return*	*Input Cost*	*Net Return*	*Return on per Rs. Investment*
Green forage based dairy farming	87164	50591	36573	1.72
Goat production on silvipasture (*Bauhinia populnea, Symingtonia populnea, Brachiaria ruziensis, Thysanolaena maxima*) on the uncultivable land	55500	39462	16038	1.41
Agro-pastoral food crops on the terraces and forage on the terrace risers for dairy cow	127701	62866	64835	2.03
Agri-horti-silvipasture (for goat) forage on the top 1/3rd, middle horticulture and bottom 1/3rd food crops on the terraces and forage on the terrace risers	86607	29477	57130	2.94
Silvi-horti system (*Alnus nepalensis*) upper 90 per cent and vegetables on terraced bottom part of the slope)	221312	63052	80713*	
Self-regenerated pine forest	156750	26000	130750 (66042)*	6.03
Mixed block forest (*Alnus nepalensis, Acacia auriculiformis, Symingtonia populnea, Michelia oblonga*)				
Timber farming (*Michelia champaca*)	383610	66254	317356 (161851)	5.79

()* - Real value after 5 per cent discount rate on build up wood.

Source: Adapted from Singh *et al., 2002,* Singh, 2008; Singh *et al.*, 2008, Singh *et al.*, 2010.

Table 8.11: Forage Production Potential under different Land Use Systems at Low to Mid Hills in Himalayan Region

Land Use	*Intervention*	*Green Forage Yield (tonnes/ha)*
Permanent pasture, cultivable wastelands barren and forest lands current fallow	Silvo-pasture	15.6 – 51.6
	Fodder-Trees plantation	18.8 – 70.0
	Forage grasses	42.1– 74.6
	Annual forage crops	25.0 – 54.0
Net area sown	*Terraces*: annual forage crop, perennial grasses and legumes	40.4 – 60.0
	Terrace risers: guinea grass, Congo signal grass, thin napier and golden timothy grass	42.1 – 52.8
	Terrace bunds:	
	Fine stylo	2.8 – 5.0
	Fodder tress	0.03 – 0.7/tree
	Broom grass, Guatemala grass; NB21, BN82048, napier x bajra hybrids and guinea grass	5.7 – 8.3/ 1000m
	Farm boundary: napier x bajra hybrids and broom grass	8.0 –10.0/ 1000m
Horti-pastoral systems	Intercrops of maize ricebean, dallis grass, guinea grass, golden timothy grass and napier x bajra hybrid	12.5 – 15.7
Agro-pastoral system in a micro-watershed	Guinea grass and broom grass on terrace risers-a non-competitive land use	40.0 – 65.0
Dairy farming in a micro watershed	Top two third portion under perennial grasses and fodder trees and bottom terraced land for leguminous forage a and concentrate production	13.7 – 72.0
Bamboo shaded lands	Orchard grass, golden timothy grass dinanath grass and fine stylo	3.3 – 38.3
Waterlogged/Marshy lands	Para grass and Zizania species.	30.0 – 35.0

Ameliorative Effects

The forage grasses, well known for their ameliorative effects on the soil, have a definite role to play in the amelioration of these acidic soils (Singh, 1996; 1999b; 1999c, 2005). Growing of perennial grasses on terraced lands and terrace risers increased the soil pH and organic matter, and reduced the exchangeable Al in 05 years. Application of organic manures at the rate of 20 to 60 tonnes/ha to grasses improved the soil pH and reduced the exchangeable Al by 76 to 93 per cent at 0-20 cm, by 19 to 82 per cent at 20-40 cm, and 67 to 88 per cent at 40-60 cm soil depths. Guinea grass grown on the terrace-risers increased the soil pH and organic C compared with the terraces having maize, rice, ginger and radish crops. Das (1990) studied the soil properties of natural pastures and of the established pastures for 07, 7-14 and more than 14 years on dairy farms at low hills of Assam. She found that the soil pH increased by 0.44-0.75 units, whereas P improved by as high as 19 times due to combined effects of forage grasses and manures applied to produce these grasses compared with those of natural pasture. A concomitant rise in organic C, humic C and fulvic C indicated that addition of farm wastes not only increased the organic-matter status of the soil but also accelerated its humification and the availability of phosphorus in the soil. There was also a substantial build-up of DTPA Zn, Cu and Fe and great reduction in Mn content of the soil.

Fertility build-up in the surface soils was also compared between terrace risers (uncultivated) planted with perennial grasses and terraced lands cultivated for growing various annual crops under dairy farming, agro-pastoral system and agro-horti-silvipasture system. In general, build-up of organic carbon and increase in soil pH were found more on the terrace risers due to perennial grasses than the terraced lands put under various annual crops. Singh (1999) reported ameliorative effects of perennial forage grasses such as guinea grass and broom grass to improve the soil pH by increasing soil organic matter. Build-up of available phosphorus and exchangeable K in the surface soil was found more on the terraced lands than on the terrace risers.

Forage Plants and Feeding Ecology for Mithun (*Bos frontalis*)

This semi wild mithun (*Bos frontalis*) is believed to be a descendent entirely and mainly from the race of *Bos gauras-gauras* of India and Myanmar (Wharton, 1957). It had been listed under vulnerable species in the Red list of International Union for Conservation of Nature and Natural Resources (IUCN, 2002). This semi-domesticated animal is of a special significance for the tribal people of North Eastern Hills region of India (Gee, 1964). The mithun inhabitants of North-Eastern States *i.e.* Arunachal Pradesh, Manipur, Nagaland and Mizoram, have their own indigenous and traditional ways of rearing, basically used for beef purpose besides its use as barter trade, marriage gift *etc.* The livestock has immense potential for meat and milk production under special topographic and climatic conditions.

Photo 8.14: Mithun Feeds on Herbaceous, Shrubs and Tree Leaves in the Forest Area.

Table 8.12: Forge Plant Species Consumed by Mithun in North-eastern Region of India

Herbaceous plants		
Ageratum conyzoides	*Desmodium heterophyllum*	*Nymphia* sp.
Alangium sp.	*Desmodium podocarpum*	*Phragmites* sp.
Arundinaria sp.	*Dimeria* sp.	*Plantago* sp.
Begonia sp.	*Dysophylla* sp.	*Plectranthus* sp.
Borrena hirticulata	*Elusine* sp.	*Polygonum chinensis.*
Brachiaria ruziensis	*Equisetum* sp.	*Potentilla kleiniana*
Bublaja asiatica	*Ethula* sp.	*Pratia* sp.
Capsicicum sp.	*Eugeron* sp.	*Rhynchotechum* sp.
Carex filicina	*Fagopyrum cymosum*	*Saccharum arundinaceum*
Celosia sp.	*Galium* sp.	*Sarcochlamys* sp.
Centella asiatica	*Gnaphalium* sp.	*Setaria palmifolia*
Chenopodium album	*Gynura crepidiodes*	*Sieges beckia*
Coix lachryma	*Hedyotis scanden*	*Smilax zeylanica*
Cuphia sp.	*Impatieus* sp.	*Solanum* sp.
Cuculigo recusvata	*Imperata cylindrica*	*Stellaria* sp.
Curcuma sp.	*Iris* sp.	*Thysanolaena agrostis*
Cynodon dactylon	*Ischaemum barbaratum*	*Tridax species*
Cynolis sp	*Jasminum* sp.	*Viola* sp.
Cyperus rotundus	*Narenga* sp.	*Napier* bajra hybrids

Contd...

Leaves of shrubs, trees and bamboo plant species		
Adina cordifolia	*Ficus nemoralis*	*Osbechia ciliata*
Albizzia sp.	*Ficus prostrata*	*Pinus* sp.
Allophylus sp.	*Ficus racemosa*	*Plecttranthus* sp.
Anarcadium occidentale	*Ficus roxburgii/F. hookerii*	*Pogostemon* sp.
Antidesma sp.	*Gardenia latifolia*	*Polygala arillata*
Aralia sp.	*Globba* sp.	*Prunus nepalensis*
Arundinaria hookeriana	*Glycosmis* sp.	*Quercus polystachya*
Arundodonak Linn	*Gmelina arborea*	*Quercus* sp.
Bauhinia purpurea	*Goniothalamus sesguipedalis*	*Rhododendron* sp.
Bidens bidemata	*Grewia abutifolia*	*Rhus semialata*
Boehmeria platyphylla	*Grewia multiflora*	*Sabia* sp.
Brasiopsis sp.	*Hedychium* sp.	*Sapindus* sp.
Calicarpa arborea	*Helicterus isora*	*Saurauvia napaulensis*
Canthium glabrum	*Jasmimum sambac (L.)*	*Scoparia* sp.
Chromalena odorata	*Lagerstroemia specicosa*	*Sterculia villosa*
Cinamomum sp.	*Lenia indica*	*Stereospermum chelonoedes*
Clerodendrum sp.	*Linaria* sp.	*Strobilanthes vallosa.*
Colguhounia sp.	*Ligustrum robustum*	*Suraria curniculata*
Dubanga grandifolia	*Litsea polyantha*	*Taxus baccata*
Debregeasia longifolia	*Litsea salicifolia*	*Tourmefortia* sp.
Dendrocalamus hamiltonii	*Ludwigia* sp.	*Trema orientalis*
Dendrocalamus sikkimensis	*Macaranga denticulata*	*Vacinium* sp.
Dillenia sp.	*Magnolia pterocarpa*	*Vitex negundo*
Emblica officinalis	*Manihot esculenta*	
Eriamnthus longisetosus	*Micromelum pubescens*	
Erobotrya sp.	*Molinari asp.*	
Eurega acuminata	*Morus* sp.	
Ficus auriculata	*Musa* sp.	
Ficus fistulosa	*Musandra roxburghii,*	
Ficus hirta	*Myristica* sp.	
Ficus insipida	*Neillia* sp.	
Ficus infectoria	*Neolamarckia cadamba*	
Ficus limpongua	*Oreochinida integrifolia*	

Source: Prakash *et al.* (2009), Hazarika *et al.* (2010), Shukla and Khare (1998), Annual Report (1989-90 and 1990-91, 1999-2000, 2007-8), NRC on Mithun, Jharnapani, Nagaland, India.

Mithun entirely feed on the natural vegetation of evergreen forests or semi-evergreen and moist deciduous tropical forests. It is also inhabitants of the peripheral range of deciduous forest. The habitat is characterised by relatively large undisturbed forest tracts with hilly terrain below an altitude of 5,000 to 6,000 feet (1,500 to 1,800 m), being optimal habitat. Mithun prefers cool forested areas with temperature ranging from 20 to 30°C and moderate to heavy rainfall area that are abundant with forage like grasses, reeds, shrubs, trees, bamboos, coarse grasses (Wharton, 1957; Srikosamatara and Suteethorn, 1995; Deng *et al.,* 2007a, b). Their habitat is most significantly characterized by presence of streams, ponds and lakes formed in the villages (Hazarika *et al.,* 2010). Mithun continuous visit to freshwater area makes easier for the owners to trace it in their habitat.

Feeding behaviour of mithun showed that it is herbivorous by nature and both a browser of shrubs, trees and bamboos and a grazer of herbaceous plants (Shukla & Khare, 1998). It graze and browse on a wider variety of forage plant species than any other ungulate species of India, with a preference for the upper portions of plants, such as leaf blades, twigs, seeds and flowers of trees, shrubs, bamboos and grasses of their particular choice (Photo 8.14; Tables 8.12). Forage preference varies by season (Gad & Shyama, 2009). In winter and monsoon, they feed preferably on fine and fresh forage grasses and edible herb species of the leguminosae family, such as *Desmodium triflorum*, but also browse on leaves of shrubs *Strobilanthes callosus, Gardenia latifolia, Grewia obtusifolia, Helicteres* sp. and the *Vitex negundo*. In summer, they also feed on bark of *Tectona grandis*, on fruit of *Cassia fistula*, and on the bark and fruit of *Anacardium occidentale*. Protein rich tree leaves of *Lagerstroemia specicosa* could be incorporated up to 30 per cent in the ration for feeding mithun under semi-intensive/intensive system (Kobu *et al.,* 2007-8). Proximate analysis of forages browsed by mithun showed a range of 10.0 - 17.8 per cent crude protein, 1.0 – 3.7 per cent ether extract, 45.3 – 57.3 per cent neutral detergent fibre, 31.2 – 43.5 per cent acid detergent fibre, 7.4 – 16.9 per cent lignin and 1.2 – 6.8 per cent condensed tannins on dry matter basis (Prakash *et al.,* 2009). The dry matter ranged 23.6 to 39.8 per cent in the browsed feed. Mithun debark trees when preferred feed, forage, minerals and trace elements needed for their nutrition are not available, or for maintaining an optimum fibre/protein ratio for proper digestion of food and better assimilation of nutrients. They may turn to feed on available browse species and fibrous teak bark in summer as green grass and herbaceous resources dry up. High concentrations of calcium (22400 ppm) and phosphorus (400 ppm) had been reported in teak bark, so consumption of teak bark might help animals to satisfy both mineral and other food needs. Long-term survival and conservation of these herbivores depended on the availability of preferred plant species for food. They had been found to cover long distances in search of their selective forage plants, good sheltered areas and a fresh water source. They can climb hilly terrains and move about mountain tracts.

Mithun had been usually found to feed during morning and late evening to avoid hot day temperature and found to spend most of their time on browsing.

Peak grazing activity was observed between 6:30 and 8:30 am and between 5:30 and 6:45 pm. During the hottest hours of the day, 1:30 to 3:30 pm, they take rest in the shade of big trees (Gad and Shyama, 2009). Being selective browsers, mithun feed on specific wild plant species besides those that are consumed by other domesticated ruminating cattle (Baruah *et al.*, 1995). Animals take rest near a cool, shaded area near water sources ruminating themselves when the sun is hot and retire into deep, secure places during evening night hours which had been found to be closer to human settlement. During daytime, when the sun is hot, they take rest under shaded trees and cool areas and drinks water from streams. Besides, feeding on plants they like to lick salt fed by the owners. For salt licking, they visit the owner's house and return to its natural habitat.

Mithun, being wild herbivores had capacity to utilise poor quality forages found in their habitats more efficiently than their domesticated counterparts and these constitute their normal diet (Nelson *et al.*, 2003; Xie *et al.*, 2007). Predominantly, feeding on shrubs, trees, bamboo and reeds plant species, mithun exhibited rumen characteristics different than those of domesticated cattle (Deng *et al.*, 2007a,b). Different ruminant animal species had demonstrated different degradability's of forages particularly under feeding condition of low quality roughages. However, rumen ecology and nutrient strategies of mithun had received not much attention.

Alpine and Temperate Pastures

Indian part of Himalaya constitutes 13 per cent of the geographical area of the country, where pastures and meadows are in 11.4 m ha (Photo 8.15 - 8.18). In Himalayan states, pastures and grasslands are located between 300-4500 meter altitudinal zones spreading over subtropical, temperate and alpine environment (Singh, 1996, Inderdev & Misri, 2003). Alpine meadows exist in the high Himalayan ranges above tree line between 3600 and 5400 meter msl (Khan, 2003). Alpine regions including alpine desert/arid lands, high plateaus, and temperate belt are particularly most vulnerable to any ecological disturbances. In this region, human beings reasonably adapt to pastoralism utilising available natural resources. Pastoral; transhumant; sedentary; and commercial are the four types of livestock production systems prevailing in these region. Livestock farmer practise a nomadic system of animal husbandry, characterised by a continuous search for pasture and the year-round movement of sheep, goats and yak in pastoral system. Traditional pastoralists neither own any land nor do they engage in any farming activities. Livestock graze between the alpine and sub-alpine pastures, typically grazing for a period of four to five months in the alpine pastures and move to lower ranges during remaining period as snowfall set in. Non-local pastoralists graze their livestock by paying to communal land owner for the rights to graze for a specific period. In transhumant system, keeping their principal home atlower altitudes farmers, where they live for approximately seven months of the year and livestock are kept indoors and fed on diet of maize stover, wheat straw and hay during the cold winters.

Photo 8.15: Yak Grazing in Arunachal Pradesh

Photo 8.16: Yak Grazing in Wet Grassland of Ladakh

Photo 8.17: Pashmina Goat Grazing in Alpine Region of Ladakh.

Photo 8.17: Pashima Goat Grazing in Alpinr Region of Ladakh

In the summer, first they graze livestock on the sub-alpine pastures, and finally on the alpine pastures. The movement to and from the mountains typically takes place in a series of well-defined stages in late April or early May, treks to the edge of the conifer forests, stay for three to four weeks, before moving to a third located within the forest itself. Both livestock and people move to the high alpine pastures, where they remain up to two months if favourable climate and pasture conditions further stay of three to four weeks extended. The return journey begins with the first snowfall in late September. There is a variant to this transhumant production system in which the livestock owner remains in the village but hires a shepherd during the summer months to take the flocks to the mountain pastures. Flock consists of 15 to 40 cattle, 100 to 200 goats/sheep, or a mixture of both cattle and sheep/goats (Khan, 2003).

These alpine and temperate grasslands are habitats of several floras, including Central Asian desert species, East Asian temperate, and Himalayan region. Based on edaphic and microclimatic conditions, the pastures situated in high hills (above 3200 m altitude) could be categorized into subalpine and alpine pastures, which remain snow-bound from November to April and are open for grazing to sheep, goats, yaks and horses from May to September (Inderdev and Misri, 2003). Du Guo*et al.* (2000) reported about 2000 forage plant species belonging to the families of compositae, gramineae, leguminosae, rosaceae, cyperaceae and polygonaceae in Tibet. Out of these 86 per cent were found in humid and sub-humid pastures and rest occurs in arid and semiarid pastures. Sub-alpine rangelands are associated mostly with shrubs. Steppe rangelands are associated with *Pinus* and *Juniperus species.* Some of the important forage plant species found in alpine and temperate pastures is listed in Table 8.13. Natural short height *Elymus*-dominated range is normally grazed during winter. *Elymus nutans* is of great importance to the pastoral systems in high elevations. Their potential productivity is not fully realised due to lack of fertiliser use, hence, they get degraded soon. Productivity of such pasture could be increased manifold by introducing comparatively tall growing grasses like *Lolium perenne* L., *Poa annua L., Trifolium repens L.* and *Trifolium pratense L.,* which could also be harvested for hay making (Roder, 2000). On an average typical productivity of pastures ranged 5.8q/ha total digestible nutrient (TDN) in subtropical grassland to 15.4 q/ha. TDN in temperate to alpine grasslands with carrying capacity 0.54 to 1.42 LU/ha. In another study, dry matter yield was 934 kg/ha in alpine pasture to 623 kg/ha in temperate lowland pasture and 1341 kg/ha dry matter in subtropical pasture. Harris & Bedunah (2001) found average standing forage in shrub 115 kg DM/ha in desert shrub to 790 kg DM/ha in desert sub-irrigated pasture.

Bamboos are important forage resources not only for domestic herds, but for wildlife (Taylor, 1987; Yokoyama & Shibata, 1998; Takatsuki, 2009; Wangchuk *et al.,* 2014a). Among bamboo plant species, *Sasa nipponica* in the eastern Himalaya and *Yushania microphylla* in mixed conifer forest of temperate and subalpine region, are the dwarf bamboos subjected to herbivory (Darabant *et al.,*

Table 8.13: Important Forage Plant Species in Alpine and Temperate Pastures

Habitat / Habit / Plant Species
Alpine zone
Graminoid: *Achnatherum splendens, Agropyron caninum, A. dentatum, Agrostis gigantean, Agrostis trachychlaena, Agrostis* sp., *Brachypodium sylvaticum, Bromus* sp., *Chrysopogon* sp., *Cymbopogon* sp., *Dactylis glomerata, Danthonia* sp., *Deschampsia caespitose, Elymus tasiaticus subsp. longe-aristatus, E. nutans, E. tangutorum, Elymus* sp., *Festuca gigantea, Festuca olgae, Festuca ovina, Festuca rubra, Festuca* sp., *Helictotrichon tibeticum, Koeleria litwinowii, Koeleria* sp., *Leymus secalinus, Phleum alpinum, Poa* sp., *Psendophara mites, Roegneria nutans,* Stipa orientalis, *Trisetum* sp. **Semiarid xeric and meso- xeric:** *Agropyron cristatum, Calamagrostis* sp., *Festuca rubra, Poa alpina, P. crymophila, P. dolichachyra, Orinus thoroldii, Oryzopsis* sp., *Roegneria nutans, Roegneria thoroldiana, Roegneria melanthera, Stipa aliena, S. glareosa, S. purpurea, S. subsessiliflora,* ; **Xeric**: *Aristida triseta, Elymus nutans, Orinus thoroldii, Pennisetum flaccidum, P. lanatum, Stipa bungeana* **Desert-** *Leymus paboanus* **Bamboo**: *Yushania microphylla, Sasa nipponica.*
Sedges: *Carex atrofusca, C. infuscata, C. melanantha, C. moorcroftii, C. microglochin, Carex* sp., *Juncus* sp., *Kobresia capillifolia, K. humilis, K. littledalei, K. pygmaea, K. royaleana, K. schoenoides, K. setschwanensis, Kobresia* sp., *Polygonum viviparum* and *P. macrophyllum* **Desert**: *Carex moorcroftii, Carex atrofusca.*
Herbaceous Legumes: *Medicago sativa, Medicago* sp., *Trifolium pratense, T. repens.*
Herbaceous Non-legumes (Forbs): *Alium* sp., *Anaphalis contorta, Anaphalis* sp., *Androsace sarmentosa, Anemone* sp., *Arnebia euchroma, Cicer* sp., *Cousinia thomsonii, Eurotia ceratoides, Eurotia* sp., *Fragaria vesca, Gentian* sp., *Hieracleum thomsonii, Leontopodium alpinum., Lindelophia anchusoides, Nepeta discolour, N. spicata, Pedicularis* sp., *Plantago lanceolata, P. major, P. ovata, Polygonum alpinum, P. aviculare, P. macrophyllum, P. viviparum, Polygonum* sp., *Potentilla* sp., *Primula* sp., *Rheum* sp., *Rumex nepalensis, Saxifraga ciliata, Sibbaldia* sp., *Taraxicum officinalis, Thymus serphyllum.*
***Cushion herbaceous in Semiarid:** *Androsace tapeta, Arenaria musciformia, Oxytropis microphylla.*

Shrub: *Ajana spp., Artemisia wellbyi, Artemisia* sp., *Astragalus grahamiana, Berberis lyceum, Berberis sp., Caragana jubata, C. microphylla, C. pygmaea, Caragana versicolar, Caragana* sp., *Ceratoides compacta, Cotoneaster* sp., *Dasiphora fruticosa, Hippophae tibetica, Juniperus communis, Lonicera* sp., *Myrimanica germanica, Potentilla fruticosa, Quercus* sp., *Rhododendron capitatum, R. przewalskii, R. anthogonoides, Rosa webbiana, Salix petrophila, Sibiraea angustata, Spiraea alpine* **Desert Shrublet:** *Oxytropis aciphylla* **Xeric shrub:** *Artemisia webbiana, Berberis* sp., *Ceratostigma griffithii, Juniperus* sp., *Leptodermis sauranja, Lonicera spinosa, Sophora moorcroftiana, S. viciifolia* **Desert shrub:** *Ceratoides campacta* **Sub-alpine Shrubs:** *Berberis* sp., *Caragana* sp., *Hippophae* sp., *Lonicera* sp., *Potentilla* sp., *Rosa* sp. *and Spiraea* sp.; **Desert Shrub:** *Reaumuria soongarica, Sympegma regelii.*
Temperate zone
Graminoid: *Agropyron caninum, A. dentatum, Agrostis* sp., *Aristida* sp., *Bromus himalaicus, Bromus inermis, Bromus* sp., *Chrysopogon* sp., *Cymbopogon* sp., *Dactylis glomerata, Deyeuxia* sp., *Dichanthium annulatum, Duthiea brachypodium, Elymus* sp., *Eragrostis* sp., *Festuca* sp., *Helictotrichon* sp., *Oryzopsis* sp., *Pennisetum orientale, Phacelurus speciosus, Poa* sp., *Ptilagrostis* sp., *Roegneria* sp. **Bamboo**: *Yushania microphylla, Sasa nipponica*
Herbaceous:*Allium spp., Aconitum* sp., *Astragalus* sp., *Delphinium* sp., *Dracocephalum heterophyllum, Heteropappus altaicus, Iris* sp., *Lathyrus* sp., *Leontopodium* sp., *Ligularia* sp., *Lotus comiculatus, Medicago* sp., *Nepata spicata, Oxytropis* sp., *Pedicularis* sp., *Plantago lanceolata, Polygonum* sp., *Potentilla bifurca, Potentilla fruticosa, Rheum* sp., *Sambucus ebulus, Taraxicum officinalis, Thymus serphyllum, Tulips* sp., *Viola* sp.
Shrubs: *Andropogon* sp., *Artemisia maritime, A. sacrorum, Artemisia* sp., *Calamagrostis* sp., *Caragana versicolar, Caragna* sp., *Ceratostigma griffithii, Cotoneaster* sp., *Daphne oleoides, Deyeuxia* sp., *Ephedra* sp., *Fraxinus xanthoxyloides, Jasminum* sp., *Juniperus macropoda, Lonicera* sp., *Melic* sp., *Parrotia jacquemontiana, Salix* sp., *Sophora moorcroftiana, Sophora sp, Sorbaria tomentosa.* **Desert:** *Ajania fruticulosa, Artemisia* sp., *Calligonum* sp., *Caragana versicolar, Ceratoides latens, Ephedra* sp., *Haloxylon* sp., *Kalidium* sp., *Lonicera* sp., *Nitraria* sp., *Philadelphus* sp., *Prunus* sp., *Reaumuria* sp., *Rhododendron* sp., *Salix* sp., *Salsola* sp., *Spiraea* sp., *Sorbus sp, Sympegma* sp., *Tamarix* sp.

* A cushion plant is a compact, low growing, mat forming plant that is found in alpine, subalpine, arctic, or subarctic environments around the world.

Adapted/Synthesized from: Deng and Gu, (1992); Long *et al.*, 1999; Yu *et al.* (1999); Zhang (1999); Gu (2000); Roder (2000); Holzner *et al.* (2000a); ILRI (2000); Harris and Bedunah (2001); FAO (2003), Inderdeva *et al.* (2003); FAO (Lang, Huang and Wang, 1997; Khan, M. I. (2003); Mishra *et al.* (2004); Miller (2005); Faber-Langendoen and Jose (2010); Wangchuk *et al.* (2014).

2007; Wangchuk *et al.*, 2014a;Wangchuk *et al.*, 2014b). In subalpine regions, *Y. microphylla* is an important source of yak forage in winter. *S. nipponica* differed from *Y. microphylla* in growth habit. While the former is a monopodial bamboo with leptomorph rhizome, the latter is a sympodial bamboo with pachymorph rhizome. The morphological modifications to herbivory in *S. nipponica* had been miniaturization and densification of culms, thinner culms, shorter top leaf, and lower aboveground biomass. Browsed ramets in *Y. microphylla* had higher culm density, shorter and thinner culms, shorter internodes, and shorter top leaf in a Himalayan mixed conifer forest (Wangchuk *et al.,* 2014b). The morphological modifications in root system for the browsed ramets were thinner rhizomes, lower rhizome biomass and dry matter, more nodes, and shorter internodes and the higher root-to-shoot ratio, resulting in overcompensation for aboveground biomass by densification at the expense of belowground biomass as compared to ungrazed/un-browsed. *Y. microphylla.* Thus, high tolerance to frequent grazing was revealed by showing distinctive morphological traits both above ground and belowground. Moreover, grazing herbivory increases the proportion of reserves allocated belowground, even though aboveground overcompensation occurs at the cost of belowground biomass.

Invader non-edible plant species *Pipthanthus nepalensis* encroach the grasslands to make grassland less productive as their population increases year after year (Roder, 2000). Some of the poisonous plant species found in these pasture were *Astragalus malcolmii, Astragalus* sp., *Aconitum* sp., *Eupatorium adenophorum, Euphorbia, Oxytropis, Pteridium, Ranunculus* and *Senecio species.*, which are lethal and very harmful for the livestock (Khan, 2003; Roder, 2000).

Understanding the impacts of environmental factors in relation to grazing is important part of pasture management. Responses to biomass production were dependent on grazing intensity and environmental conditions (Yan *et al.,* 2013). Further, grazing intensity response was dependent on precipitation, which had quadratic relationship with precipitation at low grazing intensity and a linear relationship in the case of moderate to heavy grazing intensity. However, the change in temperature had not influenced grazing effects on the production of aboveground biomass. In general, belowground biomass response to grazing could not be affected by change in the precipitation or temperature. However, water being a major limiting factor in arid and semiarid grassland, precipitation would govern the effects of grazing on the pattern of response for the production of aboveground biomass but it would not affect in the humid grasslands. In general, humid grassland would be more adaptable to grazing disturbances. Therefore, changes in precipitation in future climate change would have consequent influences on grassland ecosystem in response to grazing where water was limiting plant growth (Ruppert *et al.,* 2012; Bai *et al.,* 2012). Plant species composition might change due to grazing, especially the forage with good palatability and rich in nutrient (Liu *et al.,* 2011; Wan *et al.,* 2011). Further, responses of grasslands to grazing are not only affected by heat and humidity conditions, but also by soil moisture and nutrient conditions (Ren *et al.,*

2009; Su *et al.*, 2004; Wang, 2004, Jia *et al.*, 2006). Therefore, more studies should be done to reveal grazing impacts on sustainable use of rangeland. Flexible grazing management based on local growing environment would suit both grassland utilization and conservation together.

Forage availability is very unevenly distributed throughout the year in the areas of extensive grasslands due to restricted growing season. Limited growing period is due to snowfall/limited rainfall or wide changes in temperature. Therefore, managing livestock during the lean season is a major challenge in all types of livestock production systems in the mountain areas. Situations and strategies would vary as do the causes of the deficit and the types of production varied. The availability of forage biomass was found highest in August and got depleted progressively month by month, reaching to a lowest level in the month of May (Mishra *et al.*, 2001; Mishra *et al.*, 2002; Mishra *et al.*, 2003; Mishra *et al.*, 2004). The gross energy of forage grass generally did not change so much between seasons (Xie *et al.*, 1996a).

Yak is largely dependent on natural pastures for its survival (FAO, 2003). Yak was found to graze/browse on 19 plant species, mostly gramineae during summer and other herbaceous plants during winter (Mishra *et al.*, 2001; Mishra *et al.*, 2002; Mishra *et al.*, 2003; Mishra *et al.*, 2004). Graminoids *Carex melanantha* and *Elymus longearistatus* still formed the bulk of the diet. Dzomo was found to graze on 23 species predominantly graminoids during summer as well as winter then herbaceous plants and browse shrubs. *Leymus secalinus* and *E. longearistatus* altogether contributed substantially to dzomo diet during the summer, while during winter, *E. longearistatus* alone formed the bulk of the grazed diet. The importance of *Carex melanantha* in the diet increased during winter grazing. Cattle were found to graze on 26 species during summer and seven species during winter. *Stipa orientalis* and *Carex* sp. contributed significantly to the diet during winter. Graminoid contributed 75 per cent of the cattle diet during summer and 80 per cent of the diet during winter. Horses were found to graze on 29 species during summer, several graminoid species contributing to the diet and 14 species during winter. In winter, yaks and horses grazed similarly. Donkeys fed on 31 species in summer and eight species in winter. They grazed on both graminoids herbs and shrubs in summer (61 per cent graminoids, 30 per cent herbs and 9 per cent shrubs), with several species contributing to the diet. During winter, however, donkeys grazed mostly on graminoid without browsing forage from shrubs. *E. longearistatus* alone contributed more than 40 per cent to the diet of donkey. Small ruminants *i.e.* sheep and goat grazed on 32 and 26 forage plant species during summer and 12 and 14 plant species during winter, respectively. They had a mixed diet of equal number of graminoids and herbaceous plants during summer. Several plant species contributed to the summer grazing, but none of them formed the bulk of the diet. In winter, however, they largely fed on graminoids, with *E. longearistatus* forming the bulk of the diet. Sheep, together with goats, had the widest diet breadth during summer. During summer, their diet was similar that both grazed on graminoids and herbs.

With changing scenario of livestock production, intensively managed pasture introduced exotic forage plant species and their cultivars for harvesting high forage productivity and hay making for lean season. In such newly established pastures, now the farmers do not allow grazing but they cut and store forge for future use. The seeds of exotic forage plant species are also spread through livestock faeces in surrounding forest and alpine pastures to add to forage production. During grazing, they excreta wastes and the exotic grasses grow well and impact on the natural plant communities. In depth research had not been undertaken on the impact of introduced exotic forage plant species on native forage plant communities and its ecological environment.

Nutritive value and nutrient production of native forages varied season wise along with the variation in forage yield. Xie *et al.* (1996a) found that nitrogen (N) in the dry matter of native forages on Tibetan Plateau declined from 2.5 percent in June to 1.0 percent in November, while gross energy (GE) dropped from 17.0 MJ per sq. meter in August to 10 MJ per sq meter in November, resulting in nitrogen and gross energy deficiencies in the diet of grazing yak. Consequently, yak lost much body weight during late winter and spring if appropriate supplementation was not provided. Alpine shrub lands are the habitat and grazing lands for the yak during summer and sometimes during the winter. As such, they played an essential role in yak production (Long *et al.,* 1993; Zhou *et al.,* 1994; Yu *et al.,* 1999; Long 1999; Zhang *et al.,* 1999). Under appropriate grazing strategy, the moderate activity of the livestock would promote the re-growth of the shrubs; therefore, the total phenological biomass and the quality of shrubs were enhanced because of more leaves and tender stems. As a result, the better nutrient supply from shrubs could increase the intake, digestibility and metabolism by the livestock. Thus, livestock performances could be improved. However, inappropriate grazing strategy could destroy the balance of soil–plant–animal ecosystem, resulting implications for the growth of shrub and livestock production. With high intensity of grazing, all of the foliage and tender stems of shrubs were browsed by the livestock, resulting in the disappearance of some shrubs, the destroyed community structure of plant, and the lowered yield of herbage in the swards. Eventually the livestock production would be badly affected and the serious problem of soil erosion would creeps in (Zhou *et al.,* 1994). On the other hand, if the grazing intensity was too low, the re-growth of shrubs could not be stimulated efficiently, and the productive potential of plants was not achieved. Thus, livestock productivity could be considerably affected by the imbalance in both nutrient supply and feed deficiency. Clearly, the harmony between the alpine shrubs and the grazing yak was found to be very important in this alpine ecosystem. Shrub–yak integration must be maintained to optimise the productivity of the yak production system in such situation. With shrubs, the quantity eaten by the yak was relatively small. As compared with edible herbs, the acceptance or palatability of shrubs was much lower, partly due to the physical characteristics (hairiness, bulkiness), and partly because of the presence of some compounds, which might affect the taste and appetite (volatile oils, soluble carbohydrates and anti-nutritional factors). However, the shrubs did considerably contribute to the compositions of daily diets of grazing yak. Pu (1999) reported that on the *Dasiphora fruticosa*, shrubs accounted for 19.2 per cent of the diets of yak. Other researchers have also provided evidences that indicated that some alpine shrubs form major components of diets of grazing yak (Zhang *et al.,* 1999). It was clear from the study that shrubs had a distinct advantage over alpine grasses and sedges due to their superior nutritional values. Some non-legume shrubs such as *Hippophae tibetica and Myrimanica germanica were* comparable to the quality of legume herbages; some legume shrubs such as *Caragana jubata, C. microphylla and C. pygmaea* were excellent. Although shrubs were found to be an ideal protein supplements for the grazing yak as far as nutritive values and economic benefits were concerned, some potentially dangerous compounds (anti-nutritional factors) had been found in some of these forage shrubs. Tannins were the most important ones in alpine shrubs (Long *et al.,* 1999). It is, therefore, necessary to carry out more research to determine appropriate methods of alleviating these deleterious effects in order to upgrade the quality of protein from these shrubs. For optimum utilisation of shrubs, it is essential that details of agronomic characteristics, palatability and nutritive values of the important shrub plant species are determined through both chemical analysis and in Sacco and in vitro method.

Snow storms and pastoral system dynamics showed that where there was sufficient rainfall and the pastoral system appeared to operate in an equilibrium manner with regards to forage production across the Tibetan Steppe. The continental climate and periodic weather perturbations in the form of sudden and brutal snowstorms added to the complexity and dynamic nature of the ecosystem (Goldstein *et al.,* 1990; Miller, 2000). Snowstorms probably served as an important regulatory mechanism in the pastoral system. Serious losses occurred as a result of heavy snowfalls and severe cold weather (Cincotta *et al.,* 1991; Clarke, 1998; Goldstein *et al.,* 1990; Miller, 1998a; Schaller, 1998).

Rangeland degradation covers large number of factors emerging issues of sustainability due to over stoking leading to overgrazing, wildlife – livestock competetion for forage, unregulated grazing in forest area, cutting of peats and turf from grazing lands, invasion by non-edible plant species, changing composition of plant communities due to changes in climate and social change in pastoral communities. Increasing the gap between forage supply and demand had caused rangeland degradation (FAO, 2003). A good understanding of yak nutrition under various grazing conditions, which could help to alleviate some of the problems, was found still rather inadequate. Alpine and temperate grassland resources support a wide range of wild animals of which the takin (*Budorcus taxicolor*), the blue sheep (*Pseudovis nayaur*), the sambar (*Cervus unicolor*) and the musk deer (*Moschus leucogaster*), Kiang (*Equus kiang*), bharal (*Pseudovis nayaur*) (Roder, 2000; Bhatnagar *et al.,* 2006; Mishra *et al.,* 2004). In Changthang Plateau of Laddak,

Trans-Himalayan pastoral communities do face serious forage competition in key areas such as moist sedge meadows by livestock herder and wildlife *Kiang* (*Equus kiang*; Bhatnagar *et al.,* 2006). Changthang's rangelands are overstocked with kiang, resulting in forage competition with livestock. A study of conflict between seven species of livestock (yak, cow, horse, dzomo (yak x cattle hybrids), donkey, sheep, goat and wild herbivore, mountain ungulate *bharal* (*Pseudovis nayaur*) for their competition for forage showed that livestock grazing caused significant reduction in forage availability particularly graminoid availability to bharal in the high altitude of Spiti Valley in Indian Trans-Himalaya (Mishra *et al.* 2001; Mishra *et al.* 2002;Mishra *et al.,* 2003; Mishra *et al.,* 2004). The issue of competition between livestock and wild herbivores had been contentious over centuries. Considerable diet overlap was noticed among the herbivore species. In summer, *bharal*, yak, horse, cow, and *dzomo* (yak–cow hybrids) used to graze predominantly on graminoids, while donkey, sheep, and goat grazed both graminoids and edible herbaceous plants. During winter, less availability of graminoid caused *bharal*, yak and horse to graze more on other herbaceous plants, while the remaining livestock species fed predominantly on graminoids, diet overlap being less in winter. In both the seasons, all important forage plant species were consumed in substantial amounts by one or more species of livestock as well as *bharal.* Herder centred participatory action would only ensure sustainable use of rangelands by livestock herder and wildlife such as *Kiang* and *Bharal.*

Further, there was popular perception that forest grazing damages trees and inhibits regeneration, leading to degradation. In a study of forest dynamics, active regeneration of tree seedlings was found in yak-grazed plots of *Abies spectabilis* woodland, which supported that warm-slope vegetation as a diverse and enduring shrub land did not degraded forest due to grazing (Brower & Denis, 2000). These findings challenged much of the received wisdom about the dynamics of Himalayan forests in general and suggested that yak and yak–cattle hybrids here have been unfairly targeted as agents of forest destruction. The cutting of peat and turf was one of the major factors responsible for pasture degradation (Holzner *et al.,* 2000). Two main types of vegetation in Tibet that accumulated peat were Giant Kobresia (*Kobresia schoenoides*) bog and pygmy Kobresia (*K. pygmaea*) turf. Giant Kobresia made good pastures, very conspicuous due to their dark green colour. Due to overgrazing, the tussocks are partly decomposing and slowly and gradually becoming flatter and flatter, a wide spread elements in the grazing lands. The decomposing tussocks were often found to be covered by dense micro-sedges (*Carex microglochin*) vegetation, also good pasture plant species. *Carex microglochin* was found to be less productive than the *Kobresia schoenoides.* Cutting of such flattened peat and turf for household use is degrading alpine pasture. Higher density of burrow of small mammals also degrades pasture as well as found to be a carrier of pathogen for disease (Zhang *et al.,* 1998).

Principles to sustainbilities of pastoral systems would be in one hand adaptation strategies to a stressful environment, conservation of ecosystem diversity and also mobility that make pastoralism sustainable and in other hand issues of empowerment of pastoralists to engage in decision and policy-making processes as well as tackle marginalization alleviate pastoral poverty (Gliessman, 2005). In order to reverse the degradation of rangeland, traditional local institutions should be revived for better use of traditional knowledge to ensure that pastoralism continues to be practised and managed effectively and autonomously, without external pressures. Multi forage plant species grazing maximizes the use of forage but requires complex management. Multiple forage plant species minimizes the risk of total loss from disease or winter storms. McIntire & Powell (1993) found that the central characteristics of traditional pastoralism - low productivity, high variability in forage and livestock production, low production density and high market transaction costs - mean that conventional markets in land, labour and capital have not become well developed in Africa. Never the less, Tibetans often developed quite sophisticated arrangements for meeting their labour requirements, for managing grassland without exclusive private property rights and for allocating their livestock as capital in the absence of financial markets.

Pasture Condition in the Hindu-Kush-Himalaya (HKH)

Rangeland covers 60 per cent 4.3 million km^2 of Hindu Kush Himalaya (HKH) covering Afghanistan, Bhutan, China, India, Nepal and Pakistan. It provides ecosystem service to 1.3 billion people living downstream. It also regulates the climate of the Tibetan plateau which is linked to global climate circulation. Rangelands in HKH contain diverse geographical and cultural landscapes which have been shaped by the forces of nature and the communities that inhabit them. However, the important role that pastoralists play as the custodians of rangelands ecosystems has not been sufficiently valued by decision makers. Rangelands remain a neglected sector in terms of research, legislation, government planning and particularly of spending for sustainable development. The most important drivers of change in HKH region are globalization, land use transformation and climate change. The rate of contemporary climate change makes the already variable rangeland environment even more unpredictable it starts the adaptive capacity (resilience) of pastoralists and alters their socioeconomic condition, with ramification that ultimately impacts the ecological security of HKH region and beyond.

Most of the case studies from the Hindu-Kush-Himalaya state that pastures are generally in poor condition; this may not apply to all the pastures of the zone but most are said to be seriously overstocked (FAO, 2003). The causes of pasture degradation are the usual ones associated with poor pasture management *i.e.* overstocking, putting stock to graze too early in spring before there is adequate herbage growth and keeping them on pasture too late in the season (where there is no snow cover); grazing every available piece of pasture all the time without sparing for rest may also cause damage; the stock of sedentary owners may be more culpable than those of transhumant, since the latter move seasonally. Further degradation of the vegetation is caused by excessive felling of trees, and uprooting of shrubs for fuelwood and harvesting of medicinal herbs. Many of the middle-altitude hay lands in the Himalayas are closed to grazing throughout their growing season and only mown at the end of the monsoon. This probably explains why they generally maintain a good cover, generally of a few grasses (*Chrysopogon fulvus* is often prominent), with very few herbaceous legumes since these are shaded out by the management regime. Some of the studies from Nepal have revealed very high stocking rates indeed. Many countries and localities have grazing rules and regulations, often old and societal but quite logically formulated and designed to maintain the natural vegetation while allowing livestock production to continue along traditional lines, especially in the forest areas. It is clear, however, that these have little effect in controlling stock numbers or in improving grazing management. Rules are not a solution in themselves, they require the means for their application and a political will, at all levels, to see that they are respected. Fortunately, it appears that most of the pastures are quite resilient and respond to resting, so management changes could be used as a means of improving their vegetative cover. The effect of short-term protection has been shown in many of the enclosed versus open cutting trials. Another demonstration of rapid growth of rested pasture was that of Afghanistan during the troubles, when most of the livestock had left the country. Manipulation of grazing pressure and management is only likely to affect the pasture vegetation – improvement of the tree layer may be more difficult.

Rangeland Issues in HKH Region

Present day pastoralist and range management is challenged more than ever before by population pressures over the last 50 years, the number of people doubled and the livestock population has quadrupled. Besides, pastoral production is not competitive in a globalised market place and potential of climate change further aggravate the situation by variability in rainfall, prolonged drought, more severe and frequent snow storms. The major issues that are relevant to rangelands ecosystem sustainability and pastoral developments are emerging changes in the role of rangelands calls for an increased understanding of their multiple values and the trade-offs that can arise from livelihood alternatives that will allow reaping maximum benefit from these options while conserving the rangeland resources. Quality of Life and Economic Viability of Mountain Pastoralist and their marginalization – low productivity, a long production cycle, access to markets, products are undervalued due to remoteness. Hence, new adaptation strategies are needed which are potentially being complex and encompasses elements beyond the purview of local pastoralists. Pastoralists are vulnerable to new polices and developments in this 21st century. Maintenance of rangeland ecosystem services is we neither understood nor appreciated by decision makers/GO's. Legislation that restricts mobile livestock from grazing across administrative units is conflicts between conservation and livelihood related interests restricting the access by and benefits to pastoralists. In last 10 years protected area increased from 12 to 39 per cent in HKH. Hence, creative compromises need to be worked out to accommodate people and biodiversity in the Himalayan ecosystems. Linkage and coordination among governments with administrative boundary and institutions who presently manage rangeland resources across administrative boundary are weak or no effective coordination exists. As a result, capability is deminished and the responsibility blurred. Rangeland degradation and desertification in HKH is also affecting rangeland ecosystem service, diminishing sustainability and livelihoods. There is a need of adopting a holistic approach and to acquire long term monitoring data to assess the impacts that the changes arising from various driving forces for better inputs to develop sustainable adaptation strategies and ultimately adjust policies accordingly.

The Future of Herding as a Lifestyle

In much of the Himalayan rangeland area, herding is under stress from incursions by agriculture and the stock of settled groups, loss of grazing land, and interference with traditional migration routes and protective forest law. There are also serious problems associated with increasing human and livestock populations. Modern transport and means of communication have raised the awareness of many of the herding community to other ways of life and to the attractions of urban or settled life. In some areas, such as the Northern areas of Pakistan, improved education has encouraged the young to seek employment in "more skilled" jobs; but then other ethnic groups have taken over the herding tasks, so the pastures are still used. Nomadic herding is anachronistic and that it will have disappeared by the middle of this century, however, it is difficult to imagine herding's demise. While many would prefer a more settled life, and many governments would like to settle nomads, alternative employment would have to be found for them; in the Himalayan context this would mean finding livelihoods in a labour market that is already oversupplied. If one group settles, then it is likely that other will arrive to scrape a living from the pastures. In the cold semi-arid zone, extensive herding seems to be the only practical way of earning a living from the land. Wildlife and tourism may have a niche but it is so cold and isolated an area that such prospects are probably limited. It is likely that Mongolia pastures will be exploited by livestock, under extensive grazing systems, for a long time to come. Transhumance need not exclude herders from access to services and education. In semi-arid and arid areas where the climate is harsh and herding risks high, mobility is very necessary to allow skilled herders room for risk-avoidance strategies; this is even more important in situations where they have no access to agricultural products for use as emergency feed. Herding is a low external input, labour intensive system that provides employment in lands where other work is scarce; it provides both subsistence and income. Ranching is the "modern" capitalist system of using extensive grazing land for livestock production; it is far less labour intensive than traditional systems. Subsistence herders are the only secondary users of vegetation who depend on milk much more than meat. All others' from carnivores to ranchers and capitalist herders depend on meat. No commercial dairy producer would choose arid and semi-arid grazing as the basis for their production. The logic for relying on milk is that it is available daily, but meat only sporadically, and the system can provide subsistence for far more people per unit area than any other arid zone production method. No data are available for temperate Asia. For many authors, the need for continuing herd mobility and the need for planners and policy-makers to consider the special conditions of the extensive grassland environments remains. Merkle (2002) notes that Scholz (1995) proposed a modern form of mobile livestock keeping, but suggests that policy-makers and planners need to give priority to subsistence rather than market-oriented husbandry, to job security rather than to increased productivity, and to resource conservation rather than increased yields. Richard (2002), when reviewing the potential for rangeland development in yak-rearing areas of the Tibetan Plateau, notes that planners must recognize that animal husbandry and livelihoods on the Plateau are still subsistence based, and that the environment upon which these livelihoods depend is marginal, with limited potential for intensification. Conclusions from the Himalayan studies briefed below:

- Promoting livestock mobility to prevent environmental degradation
- Developing legal mechanisms to protect both individual and communal rights to resource access,
- Building on the strengths of local communities, and increasing social cohesion through collaborative management of rangeland resources. Overuse is a common feature and range condition is generally poor.
- Forage production is low, partly as a consequence of a short growing season.
- Alpine pasture contributes only a small part of annual feed (30 per cent) in goat/sheep and cattle migratory systems, but in the *Chauri* (yak hybrid) systems subalpine pastures are used for year-long grazing.
- Seasonal forage shortages, mainly in late winter and the early part of the grazing season, are common and critical.
- Changes are occurring- *e.g.* reductions in herd size-and pastoralists are becoming sedentary.
- On migration-more detailed knowledge of livestock and household movement has been collected.
- Grazing rights-information has been recorded on traditional grazing rights, but the studies have also exposed uncertainties about these traditional grazing rights.
- There is need for more information on growth in enclosure cages to measure response to specific periods of defoliation and rest for the determination of stocking rates and formulation of site-specific improved range management practices.
- Better identification is needed of opportunities for forage intervention over sowing for fodder banks or pasture improvement and integration of sown forage on arable land.
- There is a lack of cost-effective methods for controlling unwanted plants.

- ☆ Clarity is needed in grazing rights and land ownership so that communities and households have the appropriate rights to manage and benefit from their range resources. Herders would welcome "improved" seeds for over sowing, provided that the government provides them. None are willing to pay higher grazing fees. All would like to see reduction in weeds and more veterinary attention.

The Himalayan studies overall have shown that most of the usual technical "grassland" suggestions for improving pasture management and herding productivity are impracticable. Some useful suggestions are: better grazing management; reseeding with high-yielding species; and providing herder training. However, any management improvement would require major changes in the present vague allocation of grazing rights (especially in some of the cases where two or three very different groups use the same land at different seasons). Reseeding could only be done if all the management faults that caused the degradation had been resolved and suitable ecotypes of desirable forage plant species identified, deciding on economic sustainability of the activity and the operational problems of reseeding in very inaccessible regions on steep slopes. As to imparting training, what can a technician tell a herder who has no real secure right to land or protection from trespass? Herders probably know fully well the glossier faults of their practices: grazing too early in spring, overstocking, and so on, but usually they have no choice but to feed their stock as they can today since they have not the finance to plan for the longer term. In XIth Plan IGFRI had executed a network project on "*IGFRI outreach programme on amelioration of temperate/alpine pastures for livelihood support to pastoral communities*" involving NRC on Yak and CSWRI Regional Station, Garsa and IGFRI Regional Station Palampur which understood the current status of pasture productivity and the socio-economic scenario of pastoralists; suggested technology refinement and on farm trials/demonstrations for enhancing livelihood options; to enhance capacities of the beneficiaries for development, management, utilization and conservation of pasture resources, and mapped for monitoring forage biodiversity in temperate grasslands. The outcome of the project may be used for developing sustainable pasture production and utilization technology. Indian Remote Sensing Satellite (IRS) P6 LISS-3 data of year 2009 were selected for the identification, characterization and mapping of pasture lands of Jammu and Kashmir, Arunachal Pradesh and Himachal pradesh. Total 22 scenes of IRS P6 L-3 B2345 dated June-Sep., 2009 were selected for geometric and radiometric correction; image analysis and classification work using Erdas imagine image processor. GPS based GT (Ground Truthing/Survey) was conducted for Leh and Laddak region. During this field survey pasture lands were found between 10850 to 17500 feet (between Pong ground-Nakila-Leh-Khardungla). These pastures were assessed sampled and GPS locations were recorded for the supervised image classification of satellite images. For the pasture improvement total 08 field sites located at Jammu and Kashmir, Himachal Pradesh, Sikkim, and Arunanchal Pradesh were selected. The testing of suitability of different grasses/ legumes had been completed.

ICIMOD Project - Regional New Rangeland Programme in Himalayas

Current Rangeland Issues in HKH

- Quality of Life and Economic viability of mountain Pastoralist and their marginalization – low productivity, a long production cycle, access to markets, products are undervalued due to remoteness. Hence, new adaptation strategies are needed which are potentially being complex and encompasses elements beyond the purview of local pastoralists.
- Pastoralists are vulnerable to new policies and developments in this 21st century.
- Maintenance of rangeland ecosystem services is neither well understood nor appreciated by decision makers/GOs.
- Legislation that restrict mobile livestock from grazing across administrative units.
- Conflicts between conservation and livelihood related interests restricting the access by and benefits to pastoralists. In last 10 years protected area increased from 12 to 39 per cent in HKH. Hence, creative compromises need to be worked out to accommodate people and biodiversity in HKH.
- Linkage between Go's with administrative boundary and institutions are weak reducing the capability and blurring the responsibility.
- Rangeland degradation and Desertification in HKH affecting rangeland ecosystem service, diminishing sustainability and livelihoods in China where 90 per cent rangelands are degraded.
- Wetland ecosystem is linked to rangeland ecosystem which needs a holistic approach.
- Issue of long term monitoring data for better inputs to sustainable adaptation strategies to be put in place.

Objectives of Expert Consultation

- Consultative expert discussed the key issues pertinent to rangelands.
- This helped to formulate some consensus on how best to develop a regional programme on rangeland management.

Key Questions for the Expert Consultations

- Aspects of knowledge generations (by identifying gap areas)?
- Institutional innovations (by considering both formal and informal institutions)?
- Livelihood improvement/diversification (by using the concept of multiple uses of rangelands; and livestock development)?

- Capacity building (by considering skill of all the stakeholders)
- Policy inputs/innovation (by visualizing country-specific requirements and regional commonality)?

Following Expert Groups were constituted

A. Rangeland Degradation and Desertification

What are the strategic elements that can be prioritized for addressing the issues of rangeland degradation and desertification in the HKH region?

B. Rangelands and Wetlands Interface

What are the priority research areas in developing and understanding on the relationship between wetlands and rangelands in a holistic manner.

C. Long-term Monitoring and Assessment

How best to approach long-term monitoring and how to assess the impacts caused by climate and socio-economic changes in the HKH region?

D. Vulnerability and Adaption

What are the risks involved? And what are the vulnerabilities and adaption possibilities for pastoralists in the HKH region?

E. Payments for Ecosystem Services

How can the ecosystem services arising from rangelands be evaluated in economic terms and how can a reward system for pastoralists are developed?

F. Carbon Storage and Financing

How can new potentials be integrated in terms of both adaptation and mitigation in rangeland management of the HKH?

Finally, it was concluded that healthy grassland, healthy livestock and associated livelihood constituted a win-win option for addressing a climate change.

ICIMOD's Regional Rangeland Programme (RRP) was developed with the aim of helping to alleviate poverty among rangeland dependent mountain people and improving the productivity of rangeland ecosystems in the Hindu Kush-Himalaya-Tibet Plateau region. The Programme 1999-2009 has completed. Overall, the programme focused on the promotion of ecologically appropriate, socially equitable, and gender sensitive innovations for enhancing the livelihoods of herders and the ecological health of the rangeland ecosystem on which they depend, paying special attention to women and other vulnerable groups. The first phase focused on developing baseline information, and the second on testing innovations for overcoming the social, economic and physical vulnerabilities faced by rangeland dependent mountain communities. The focus in Phase III ending December, 2009, had shifted from policy dialogue to policy formulation and amendment and scaling up of innovative pastoral development and rangeland management practices, building on the experience and achievements of the previous phases. Co-management of rangeland resources by all the stakeholdsers involved was promoted as a strategic approach in all the phases. The core of the co-management approach was negotiated and the essence was 'learning by doing'. RRP was also facilitating institutional arrangements in all its pilot project countries by applying a co-management approach, using both formal and informal mechanisms. Regional New Rangeland Programme Development had been executed. Potential of climate change to further aggravate the situation by variability in rainfall, prolonged drought, more severe and frequent snow storms are expected.

Annexure 8.1
Details of Nutrient Contents in the Leaves of Fodder Tree Species as Influenced by Season in Eastern Hills of India

Fodder Tree Species	*Per cent in Dry Matter*										*ppm in Dry Matter*					*Total Tannins (per cent)*	*Calorific Value (k Cal/kg)*
	CP	*CF*	*EE*	*NFE*	*TA*	*IA*	*P*	*K*	*Ca*	*Mg*	*Na*	*Cu*	*Fe*	*Mn*	*Zn*		
	Important fodder tree species																
Artocarpus lakoocha, November	12.25	15.25	2.24	58.29	11.73	3.48	0.150	1.42	1.80	1.13	667	08	128	109	58	1.76	3.64
February	17.86	12.46	1.03	56.75	11.91	4.28	0.275	1.28	1.33	0.49	622	07	363	100	74	1.95	3.58
Bauhinia purpurea November	16.08	25.57	1.67	48.81	7.88	1.29	0.190	1.33	2.60	0.41	475	09	114	28	34	2.62	3.77
February	16.08	22.75	1.34	54.17	5.88	1.06	0.198	0.79	1.70	0.45	338	10	274	64	44	2.74	3.84
B. variegata November	17.73	22.20	1.55	49.97	8.56	0.70	0.222	0.82	1.86	0.43	678	10	139	33	92	2.97	3.74
February	14.59	21.06	1.77	57.31	5.33	0.90	0.180	0.77	2.08	0.43	363	11	282	61	47	3.31	3.87
Brassiopsis hainla November	12.75	23.79	2.02	51.70	9.73	1.25	0.252	1.74	1.14	0.44	622	10	207	64	64	2.65	3.71
February	12.25	18.99	3.83	55.62	9.32	1.44	0.341	1.14	1.82	0.33	522	05	192	89	135	2.52	3.82
Celtis australis November	23.33	18.42	2.50	36.49	19.28	6.82	0.323	0.87	3.02	1.45	424	09	188	39	41	1.26	3.36
Ficus auriculata November	12.57	18.36	1.00	55.59	12.49	2.19	0.128	0.96	1.35	0.59	740	06	150	125	106	1.85	3.55
February	15.99	18.28	1.81	50.96	12.97	3.73	0.136	1.43	1.37	0.48	600	11	168	220	89	2.27	3.58

Contd...

Fodder Tree Species	Per cent in Dry Matter										ppm in Dry Matter					Total Tannins (per cent)	Calorific Value (k Cal/ kg)
	CP	CF	EE	NFE	TA	IA	P	K	Ca	Mg	Na	Cu	Fe	Mn	Zn		
F. benghalensis November	12.43	16.13	3.67	57.11	10.67	1.18	0.135	1.38	1.89	0.30	689	12	132	32	42	2.03	3.76
F. elastica November	8.93	21.98	4.77	53.92	10.42	0.75	0.184	1.55	1.52	0.14	460	08	128	27	84	1.32	3.82
February	10.03	20.57	4.01	54.09	11.31	2.27	0.217	1.69	1.74	0.23	490	07	177	31	86	2.35	3.75
Ficus glaberrima November	9.17	17.93	0.93	61.92	10.07	3.83	0.181	1.45	1.74	0.35	470	06	130	24	118	3.33	3.64
February	10.48	15.48	0.69	61.34	12.02	3.74	0.142	1.65	0.70	0.39	510	04	80	20	130	3.43	3.56
F. hirta November	18.68	9.51	2.95	50.10	19.67	4.94	0.132	0.74	4.39	1.34	678	09	1315	34	72	0.71	3.32
F. hispida November	14.39	11.91	2.62	54.62	16.47	7.20	0.183	1.37	1.58	0.32	500	07	152	64	80	2.29	3.48
February	18.33	10.65	0.53	48.28	22.30	10.40	0.219	1.43	3.17	0.98	548	07	311	104	67	1.06	3.15
F. lacor November	11.99	25.37	1.86	53.32	7.62	1.68	0.166	1.01	0.76	0.55	511	08	77	37	47	1.61	3.80
February	14.24	22.23	1.61	49.10	12.93	5.32	0.163	1.09	1.67	0.69	511	08	202	38	60	2.40	3.56
F. neriifolia November	12.57	11.24	5.19	62.64	8.38	2.15	0.170	1.20	1.44	0.42	440	08	209	193	95	2.10	3.93
February	12.24	9.67	7.20	58.47	12.43	4.21	0.192	1.59	1.90	0.54	500	06	350	205	84	1.21	3.87
F. racemosa November	12.31	25.81	3.08	45.50	13.31	8.48	0.122	0.96	1.67	0.42	767	22	96	14	67	4.09	3.63
February	13.91	28.03	2.62	41.29	14.17	9.61	0.119	0.52	1.94	0.59	545	08	155	42	47	3.47	3.57

Contd...

Fodder Tree Species	*Per cent in Dry Matter*										*ppm in Dry Matter*					*Total Tannins (per cent)*	*Calorific Value (k Cal/kg)*
	CP	*CF*	*EE*	*NFE*	*TA*	*IA*	*P*	*K*	*Ca*	*Mg*	*Na*	*Cu*	*Fe*	*Mn*	*Zn*		
F. semicordata November	13.47	20.72	1.25	54.63	9.82	3.07	0.233	1.20	1.35	0.34	530	07	155	37	78	4.71	3.68
February	11.85	12.61	0.95	55.03	19.57	11.40	0.190	0.98	0.95	0.24	410	06	144	23	84	2.24	3.27
F. subinicisa November	15.69	10.70	1.00	62.64	9.98	1.63	0.164	0.98	1.16	0.42	420	08	104	80	98	4.69	3.65
February	13.67	8.69	0.70	61.09	15.86	7.96	0.142	1.22	1.80	0.42	450	08	242	59	85	2.64	3.41
Grewia optiva November	26.94	22.03	0.92	45.14	4.98	0.74	0.171	1.66	1.53	3.18	450	16	162	97	61	3.64	3.85
February	26.88	15.87	1.47	48.26	7.29	1.18	0.162	1.49	1.84	2.21	450	17	383	140	62	3.78	3.78
Leucaena diversifolia K156 November	31.37	19.03	1.96	41.91	5.73	0.93	0.146	1.25	1.84	0.46	375	11	100	25	39	6.36	3.87
February	33.68	16.26	1.37	42.76	5.93	1.22	0.146	1.43	1.92	1.08	450	13	103	24	24	5.15	3.83
L. leucaena S11 November	25.00	14.10	1.49	55.26	4.15	0.60	0.125	1.03	1.32	1.37	375	13	175	11	25	6.27	3.91
February	30.00	14.59	1.34	43.98	10.09	2.00	0.120	0.73	1.12	1.80	275	11	237	21	32	5.53	3.66
L. leucaena K28 November	31.25	19.50	1.62	40.81	6.82	1.71	0.156	1.53	1.56	0.60	375	19	168	41	36	4.55	3.81
February	32.19	18.80	1.26	40.60	7.15	2.17	0.151	1.30	1.28	1.34	375	12	231	49	44	4.21	3.78
Litsea monopetala November	18.60	16.72	1.44	58.62	4.63	0.19	0.193	0.85	1.62	1.19	325	11	124	277	37	5.41	3.89

Contd...

Fodder Tree Species	*Per cent in Dry Matter*										*ppm in Dry Matter*					*Total Tannins (per cent)*	*Calorific Value (k Cal/ kg)*
	CP	*CF*	*EE*	*NFE*	*TA*	*IA*	*P*	*K*	*Ca*	*Mg*	*Na*	*Cu*	*Fe*	*Mn*	*Zn*		
Saurauia napaulensis November	13.22	14.35	2.38	60.84	9.22	0.89	0.107	0.79	1.29	0.54	544	09	290	19	76	6.32	3.75
February	16.08	13.68	2.31	57.72	10.23	0.79	0.176	1.40	2.53	0.64	539	12	139	14	61	1.94	3.71
	Fodder tree species of secondary importance																
Anogeissus latifolia	8.25	28.92	1.32	50.06	11.45	0.53	0.167	0.93	1.88	3.14	375	13	342	80	50	3.45	3.61
Betula alnoides	15.00	25.41	1.40	52.39	5.80	0.65	0.182	1.25	1.16	1.32	375	15	222	159	73	1.50	3.84
Bridelia retusa	12.58	17.76	1.17	59.60	8.89	0.95	0.260	1.33	1.96	0.86	425	10	139	39	47	3.40	3.70
Castanopsis indica	20.43	22.13	1.65	52.62	3.17	0.18	0.146	0.88	0.64	1.13	275	07	104	180	33	6.26	3.96
C. tribuloides	21.75	16.13	0.44	56.27	5.25	1.46	0.136	0.93	0.76	1.51	350	57	130	42	41	3.80	3.81
Elaegnus infundibularis	21.69	13.47	2.37	57.51	4.96	0.06	0.182	0.98	1.24	2.28	300	13	158	63	32	2.55	3.92
Erythrina arborescens	31.25	25.55	1.38	34.84	6.98	0.90	0.115	1.15	1.60	0.58	400	10	241	319	26	2.43	3.79
Evodia fraxinifolia	18.87	13.04	0.41	59.11	8.57	0.71	0.156	1.70	1.24	1.75	375	10	225	125	21	3.41	3.68
Ficus auriculata type II	14.44	18.95	1.34	55.03	10.24	2.22	0.125	1.45	1.68	0.67	500	10	114	154	37	1.38	3.66
Lindera neesiana	18.62	16.22	1.64	59.02	4.50	0.35	0.162	1.08	0.52	1.46	325	16	200	53	28	1.46	3.90
Macaranga pustulata	22.50	17.96	2.24	51.30	6.00	2.31	0.135	4.40	1.08	1.66	500	11	116	179	52	9.25	3.87

Contd...

Fodder Tree Species	*Per cent in Dry Matter*										*ppm in Dry Matter*					*Total Tannins (per cent)*	*Calorific Value (k Cal/ kg)*
	CP	*CF*	*EE*	*NFE*	*TA*	*IA*	*P*	*K*	*Ca*	*Mg*	*Na*	*Cu*	*Fe*	*Mn*	*Zn*		
Maesa chisia	18.50	17.34	1.21	54.10	8.85	0.70	0.281	1.88	1.88	0.43	575	10	146	32	40	2.83	3.71
Mallotus phillippensis	18.38	24.07	1.04	47.08	9.43	4.26	0.151	0.70	1.80	1.32	350	08	222	202	38	7.38	3.68
Morus macroura	20.81	08.03	0.47	55.97	14.72	5.19	0.250	2.30	1.72	1.34	450	11	214	37	25	1.62	3.44
Premna mucronata	20.75	23.85	2.32	42.75	10.33	1.89	0.271	2.18	1.28	0.74	525	10	115	35	20	1.95	3.70
Prunus ceresoides	15.12	9.73	1.25	65.40	8.50	1.12	0.292	1.43	0.68	1.90	375	08	184	26	34	2.68	3.72
Rhus javanica	14.25	12.02	2.03	65.75	5.95	0.21	0.099	1.63	1.08	0.96	400	07	79	14	29	7.21	3.86
Stereospermum personatum	15.13	19.83	0.50	53.22	11.32	6.67	0.162	0.60	1.60	0.96	375	07	109	26	28	4.15	3.57
Schima wallichii	10.94	21.89	1.22	60.60	5.35	0.67	0.073	0.68	1.16	0.43	325	09	174	480	28	6.53	3.85

Source; Singh *et al.,* 1994, *Balaraman, 1981.

Annexure 8.2
Details of Nutrient Contents in the Leaves of Bamboo Plant Species in Eastern Hills of India

Fodder Tree Species	*Per cent in Dry Matter*										*Part per million in Dry Matter*					*Calorific Value (kcal/kg)*
	CP	*CF*	*EE*	*NFE*	*TA*	*IA*	*P*	*K*	*Ca*	*Mg*	*Na*	*Cu*	*Fe*	*Mn*	*Zn*	
Arundinaria hirsuta	15.75	21.20	2.06	52.15	8.80	5.62	0.115	1.08	0.72	3.46	375	7	177	91	28	3.75
A. manii	8.56	25.82	3.24	57.32	5.06	1.49	0.151	1.08	0.96	2.16	400	8	258	48	160	3.96
Bambusa arundinacea	15.75	24.47	0.85	51.47	7.46	4.52	0.125	1.13	0.96	0.96	425	17	119	221	27	3.74
B. balcoa	16.33	23.81	0.78	50.56	8.52	5.60	0.135	0.88	1.00	0.82	475	9	251	410	34	3.70
B. cacharensis	14.80	27.32	2.76	41.95	13.17	10.23	0.099	0.65	1.64	0.58	400	8	164	1270	38	3.61
B. multiplex	15.31	23.14	3.22	47.23	11.10	8.55	0.104	0.83	1.28	0.43	425	10	162	550	33	3.72
B. nana	8.55	26.09	2.22	53.26	9.88	7.16	0.109	0.78	1.24	0.72	375	10	211	319	29	3.72
B. nutans	15.31	25.71	2.02	48.14	8.82	5.07	0.115	1.13	1.36	0.55	450	8	219	430	39	3.75
B. pallida	14.50	23.99	1.65	51.87	7.99	5.26	0.130	0.95	1.20	0.27	450	9	164	680	25	3.76
B. polymorpha	12.76	24.27	1.15	53.31	8.51	7.06	0.115	0.90	1.40	0.22	425	7	224	325	55	3.72
B. tulda	12.76	27.46	2.29	49.43	8.06	5.57	0.120	0.93	1.28	0.55	425	8	323	450	39	3.79
B. striata	13.78	24.80	3.42	50.24	7.76	5.08	0.130	1.03	1.08	0.98	475	10	159	470	37	3.86
B. wamin	15.82	25.24	1.64	49.44	7.86	4.83	0.115	1.03	1.16	1.44	450	9	170	370	20	3.77
Chimonobambusa armata	15.31	21.66	2.43	51.60	8.77	4.94	0.130	1.48	1.40	0.84	600	11	137	370	51	3.77
C. callosa	12.76	17.89	1.10	51.11	16.20	14.26	0.083	1.00	0.92	0.22	350	6	206	740	37	3.37
C. griffithiana	12.25	25.30	2.45	52.73	7.27	5.38	0.099	0.85	1.56	0.36	425	7	126	331	31	3.83

Contd...

Fodder Tree Species	Per cent in Dry Matter										Part per million in Dry Matter					Calorific Value (kcal/kg)
	CP	CF	EE	NFE	TA	IA	P	K	Ca	Mg	Na	Cu	Fe	Mn	Zn	
C. hookeriana	16.07	22.05	1.20	52.26	8.42	5.35	0.110	0.90	1.12	0.82	325	10	269	128	40	3.72
D. giganteus	12.76	24.81	1.81	49.70	10.92	6.99	0.089	1.05	0.96	0.41	400	8	222	147	48	3.65
D. hamiltonii	7.88	22.53	0.76	57.33	11.50	8.58	0.146	0.75	0.72	0.84	375	7	157	77	27	3.58
D. hookerii	14.29	34.42	1.47	40.58	9.33	5.74	0.115	1.08	1.72	0.58	450	9	110	550	29	3.70
D. sikkimensis	18.88	25.78	2.57	45.43	7.34	4.65	0.120	1.08	1.16	0.51	500	7	123	275	46	3.83
Phyllostachys assamica	13.27	23.32	3.88	48.43	11.10	8.53	0.115	0.80	1.28	0.36	400	9	236	750	47	3.75
P. manii	14.80	20.17	1.73	54.20	9.10	6.71	0.109	0.85	1.48	0.24	400	9	137	400	47	3.72
Scizostachys manii	16.19	26.33	0.98	45.22	11.28	8.34	0.115	1.03	1.24	0.87	500	8	120	550	29	3.60
S. pergracile	12.76	16.44	0.48	60.94	9.38	7.01	0.104	1.23	1.40	0.29	475	11	198	650	13	3.65
S. polymorphum	12.25	25.71	1.79	50.79	9.46	6.56	0.115	1.18	1.16	1.10	475	7	196	375	112	3.71

Chapter 9

Conservation, Preservation and Utilization of Forage Resources

Livestock farming provides natural resource based livelihood for vast majority of the rural population in India. The Indian animal husbandry can be most productive when adequate supply of forage and feed is ensured all-round the year. The supply of adequate quantity of green forage in all through 365 days has become a challenge due to limitation of cultivated land available for forage production. The most of the regions face the problem of acute shortage of forage resources. Forage crop production and crop residues availability is seasonal and found not produced in adequate quantity. In most of the regions of India, monsoon period (July - October), forage availability exceeds the forage requirement in grassland and the sometimes the cultivated forage, then, scarcity commence and felt during post monsoon, a lean period particularly during summer months. Similarly, regional imbalances also occur in forage availability due to varied agro climatic plain and the hills condition. Moreover, availability fluctuates from one year to another year due to occurrence of natural calamities like drought, flood and ever changing agro climatic conditions. Therefore, it is necessary to conserve surplus green forage and preserve its nutrient content for use during the lean periods. Under the circumstances, conservation and preservation of forage resources *i.e.* green forage produced on cultivated lands and grasses, legumes collected from forest margins, community grazing lands and edible crop residues from field crop production are the way to ensure continuous forage supply to livestock and also balancing nutrient contents in the livestock feed. Conservation and preservation has to commence with right stage of harvesting and collection, transporting to the processing place, processing in various feed products, balancing nutrient content in these products by various combination and additives to meet the nutritional requirement of livestock, its proper storage and transporting the processed products for marketing or use by individual livestock farmer. Proper storage is very important to stop wastage of available forage resources.

Crop Residue

In India majority of livestock rearing is under mixed crop livestock production systems integrating crops and livestock on the same piece of land. Limited area available under cultivated forage crops, diversion of grazing lands to other land use in recent past has increased the dependency on crop residues in most of the agro regions. Under the circumstances, conservation, preservation and fortification of crop residues are essential as potentially useful forage resource. Crop residues are the single largest bulk feed material available at the door steps of the farmers. In the northern part of the country, wheat straw (*bhusa*) is more intensively utilized, while feeding of paddy straw is common in eastern and southern regions and a part of the western region, particularly in coastal areas. Sorghum stovers are fed in the central and western regions and in parts of the southern region. Feeding millet and pulse straw is also observed in certain localities. In sugarcane growing areas of the country (part of Uttar Pradesh, Maharashtra and Gujarat), sugar-cane tops are extensively fed to dairy animals during the harvesting season from October to May. In recent past the attempts were made to assess the requirements and availability of feed and forage resources relying upon secondary data in different agro climatic regions (Ramamchandra *et al.*, 2007). IGFRI, Jhansi attempted to provide estimate based on actual field survey and on the availability of major crop residues as well as their nutritional variability in the country (IGFRI Annual Report, 2007-11, Misra *et al.*, 2007) These variability in nutritional parameters among varieties offers the scope for plant breeders to develop the varieties with superior quality straw quality. India produces about 500 million tonnes (Mt) of crop residues annually (Table 9.1, MNRE 2009). Traditionally crop residues from various agricultural crops (cereals, pulses, oilseed, sugarcane, fruits, vegetables *etc.*), have numerous competing uses such as animal feed, forage, biofuel for industrial use, in brick kiln, packaging and composting, domestic fuel for rural homes, composting and roof thatching, (FAO, 2012). About 25 per cent of nitrogen, 25 per cent phosphorus, 50 per cent of sulphur and 75 per cent of potassium uptake by cereal crops are retained in the crop residues. However, a large portion of the cereal crop residues, about 140 Million tonnes, is burned in the field primarily to clear the field from straw and stubble after the harvest of the preceding crop under irrigated agriculture of rice-wheat cropping system in north-west India particularly in Punjab and Haryana. Among different crops, cereals generate 352 million tonnes (Mt) residues, oilseed (29 Mt), and pulses (13 Mt). Sugarcane residues comprising tops and leaves generates 12 Mt is either used for feeding of dairy animals or burned in the field for *ratoon* crop.

Farmers use crop residues either themselves or sell it to other landless livestock owner or marketing intermediaries, who in turn sell the residues to industries. Processing of agricultural produce through milling and packaging also produces substantial amount of edible crop residues that are used for livestock feeding. Crop residues are natural forage resources with tremendous value to the livestock farmers.

Table 9.1: Crop Residues Produced in Different States in India

States	*Residue Generation (million tonnes)*	*States*	*Residue Generation (million tonnes)*
Andhra Pradesh	43.89	Maharashtra	46.45
Arunachal Pradesh	00.40	Manipur	00.90
Assam	11.43	Meghalaya	00.51
Bihar	25.29	Mizoram	00.06
Chhattishgarh	11.25	Orissa	20.07
Goa	00.57	Punjab	50.75
Gujarat	28.73	Rajasthan	29.32
Haryana	27.83	Sikkim	00.15
Himachal Pradesh	02.85	Tamilnadu	19.93
Jammu & Kashmir	01.59	Tripura	00.04
Jharkhand	03.61	Uttarakhand	02.86
Karnataka	33.94	Uttar Pradesh	59.97
Kerala	09.74	West Begal	35.93
Madhya Pradesh	33.18	India	501.76

Source: Ministry of New and Renewable Energy Resources (2009) www.mnre.gov.in/relatedlinks/ biomass resources.

Forage Material Handling and Transport

Traditional transport of cereal straw and harvested grasses in loose form (40-70 kg/m^3 low density) by trucks, tractor trolleys and by bullock carts violates the traffic rules due to bulging and causes accidents and inconvenience on the road. Densifying (density 350-400 kg/m^3) of these forage resources into bales and other forage products could help in cutting cost of transport by increasing quantity of transport, comply with traffic rules and reduce the chances of accidents. Transporting densified bales and other products of different forage grasses, legumes, and edible straw of the field crops by densifying could help in transporting more quantity, following traffic rules with less chances of accident. Pathak *et al.* (2008) reported that densified paddy straw can be transported five times more in comparison to loose paddy straw.

Densification of Crop Residues/Grasses

Now days, mobile type bailing machines are available in the market which are becoming popular to make densified bales of crop residues and dried forage grasses for safe transport to peri-urban area and other marketing places and for better storage to be used during scarcity period. Several, manual and bullock-operated presses were designed earlier. Now, the baling machines are more efficient at moderate cost. A mobile type densification machine operated by a 35 horse power tractor is however found more suitable for making bales of whole plants of mixed natural grasses and crop residues or of those materials having higher chop length (>30cm) like sorghum, bajra maize *etc.* Forage grasses/crop residues are also densified with the help of stationary type high density forage baling machine developed by IGFRI, Jhansi operated by an 18.5 KW-3 phase, electric motor with V-belt and pinion-gear power transmission system and another by IARI, New Delhi based on the principle of hydraulic pressing for feed blocks (400 kg/m^3) weighing one to four kilograms. The capacity of the machine is 100 kg per hour. IGFRI baling machine can be adjusted with the help of a spring tension mechanism provided at the outlet point of the machine. Using these machines compound herbage feed mixtures with ingredients of grass, straw and tree leaves may be densified up to 350-400 kg/m^3. For baling wheat *bhusa* 20 per cent hot molasses and 2 per cent urea can be mixed thoroughly with wheat *bhusa* and the admixture is densified into 200 kg/m^3 complete feed blocks and tied by 1.25 cm wide plastic tape and the steel wire across the bale. The complete feed blocks may also be blended to enhance nutritive value by blending with leguminous forage, concentrate mixture, minerals and vitamin additives. Densified bales are 6-7 times less in volume and are easy and economical to transport and store.

Densified Total Mixed Ration Blocks (DTMRBs) and Densified Complete Feed Blocks (DCFBs)

Innovative feed products like DTMRBs and DCFBs will also help in removing regional disparities of forage resource availability and both can be effectively utilized to manage emergency situations arising due to natural calamities *viz.* floods, droughts and man-made deficits by creation of fodder and feed banks (Blümmel, *et al.*, 2009; Walli, 2008 a; 2008b; Walli, 2009a; 2009b; 2010; 2011a; 2011b; FAO, 2012). These processes are different than green forage or silage based total mixed ration for feeding livestock. In total mixed ration concentrate is mechanically mixed with chopped green forage. Substantial research efforts in the country has resulted into standardize the optimum conditions for block making by designing different static and mobile prototypes of densifying machines, taking momentum of commercialization of manufacture of the machines and use had been incentivised up to 50 per cent in different states through dairy cooperatives and state livestock boards. Efficacy of Feed block based on non-conventional forage resource *viz.* grasses and tree leaves from forest in the North–Eastern region of India (Das, Haque & Rajkhowa, 2009) and mixing sugarcane bagasse up to 30 per cent with crop residues had been successfully demonstrated (Singhal, 2009). The strategic and catalytic supplements such as micronutrients and catalytic supplements and feed additives *i.e.* vitamins, minerals, bentonite (binder), probiotics, enzymes, antioxidants, immune-protective agents, antitoxins and herbal extracts, among others can be appropriately added to increase the productive and reproductive efficiency of the livestock and decrease ruminal methanogenesis. Considering the environmental sustainability and economic viability, value addition to feed blocks through R&D efforts on feed additives has immense potential. The lead organizations which had taken the real initiative for the development of the prototype machines for block making, especially the mixing and pressing of the feed ingredients are Indian Grassland Research Institute (IGFRI), Jhansi (UP), Indian Agricultural Research Institute, New Delhi and the Poshak Agrivet Pvt. Ltd, Karnal (Gupta *et al.*, 1998, Singh & Jha, 2009; Zombade, 2009). The evaluation of straw based DTMRBs had been evaluated on different livestock i.e, growing buffalo (Singh *et al.*, 1998; Verma *et al.*, 1996; Diwedi *et al.*, 2003; Lailer *et al.*, 2006); milch buffalo (Lailer *et al.*, 2009; Sehgal & Jha, 2008); growing cattle (Pachauri *et al.*, 1999; Singh *et al.*, 2004; Das *et al.*, 2004); milch cattle (Berwal *et al.*, 1997); growing sheep, goat and camel (Jakhmola, 2005) and growing mithun (Das, *et al.*, 2009). Addition of anaerobic rumen fungi (*Orpinomyces* and *Piromyces* species) isolated from wild blue bull in wheat straw feed blocks and feeding to buffaloes resulted in better growth of livestock and increased milk yield (Sehgal & Jha, 2008). The acceptability of densified feed blocks has been more in green forage deficient hilly terrains and arid/semi-arid regions and during the scarcity period due to natural calamities. Slowly, it is penetrating in various regions of the country. Densified total mixed ration can also be manufactured as densified straw based feed pellets from straw which are highly lignified with hard fibre (Walli, 2009a; Blümmel *et al.*, 2009). Hard straw of soybean, mustard, sugarcane bagasse, groundnut shells, cotton balls and straw, maize and millets can be processed into pellets. The benefit of this pellet is that it does not require wrappers or high density polythene bags. India has installed Straw-based pellet making plants with the technical support of NDDB.

Forage Storage

Storage of crop residues with lower density (70-120kg straw/m^3) is different than grains. The physical properties vary with type of crop residues, their shape and size, location of material and ultimate use. Several temporary and permanent structures have been traditionally used for storing these edible crop residues, forage grasses and legumes. They may be permanent and temporary structures depending upon the construction material used, underground or above the ground pace of storage depending upon ultimate use of these materials. However, most farmers prefer to construct their own storage structures depending on the availability of local construction material and the cost of construction (Walli *et al.*, 1995). Loss of straw during storage period was mainly due to earth evaporated gas, rat, termites, *anjona* (a reptile pest), poultry birds and excessive rainfall (Photo 9.1). It was therefore necessary to advice farmers to improve their storage systems. The method of storing crop residues and natural grasses consists of making stacks on the ground in open area by putting layers of straw or grasses or edible crop residues and making the shape like the dome of a mosque. Polythene, corrugated iron or coconut tree leaves can also be used to improve storage condition. Stacks of rice straw and grasses are built under trees to give some protection from rain and put on raised wooden platforms, covered by polythene, corrugated iron or coconut tree leaves also used as cover to improve the storage conditions (Doyle *et al.*, 1986). A cover and plinth storage structure for storing bales/blocks of five different products of forage grasses, wheat stubbles and paddy straw was made at

IGFRI, Jhansi, which includes three platforms of size 9 m x 3 m with 0.76 m height from ground level and covered with polythene sheet (of 1000 gauge) for the stored bales. It costs about Rs. 35,630 with capacity of each platform storing around 1100 to 1600 bales up to a height of three metres (13-32 tonnes) (Photo 9.2).

In temperate zones of India, bundles of dry forage grass from the grassland is collected and carried by both men and women on their back (Photo 9.3) and is stored in *toli* which is pyramid shaped and on the roof of house and branches joining main trunk of the trees (Photos 9.4, 9.5) (Das *et al.*, 2002, Das *et al.*, 2003). Hay or forage is stored in open fields from September to October for meeting needs in winter. These are piled up in a circular shape of pyramid or *telang*. A cloth made of yak hair, called *thobi*is used for covering these structures. Stones of heavy weight are used for pressing it and to keep the hay/forage in its place. Natural grasses from grasslands, paddy straw and stovers of maize and sorghum are stored on the trees for feeding the animals during the lean period (Photo 9.4).

Photo 9.1: Stacking Stovers for Livestock in Plain Area is a Common Practice.

Photo 9.2: Scientific Stacking of Baled Grasses on Raised Plateform. (Courtsey: P. K. Pathak)

Photo 9.3: Collection and Storage of Dry Forage Grassing and Residues in Himachal Pradesh for Feeding during Winter Months.

Photo 9.4: Storing Crop Residues and Dry Forage Grasses on Trees is a Common Practice in the Hills of Himachal Pradesh.

Photo 9.5: Storage of Leguminous Hay of *Medicago falcata* and *Medicago glutinosa* on Roof of Houses in Alpine Region of Leh in Jammu and Kashmir.

Green Forage

Harvesting and Collection

Adhering to optimized harvesting schedule of green forage resource enable accumulation of high drymatter and achieve maximum nutrient content in the forage. Early harvest than the optimum stage reduce dry matter and total nutrient yield while late harvesting stage makes the forage more fibrous with less digestible nutrients. For example, *Stylosanthes seabrana* harvested at 55 days where as *S. scabra* at 60 days and *S. guianensis* at 80 days, *Leucaena leucocephala* 4-6 cuts and lucerne 8-9 cuts taken per annum depending upon their growth, produced highest quantity of leaf meal with maximum nutrient contents (Guodao *et al,*. 2004; Leng, 1997; Singh, 1988). Lopped biomass of leguminous tree fodder *Leucaena* and *Sesbania* species and herbage from *Stylosanthes* species and lucerne can be used for making leaf meal. Capacity of manual harvesting of annual crops by sickle is 0.0108 ha per hour. China has developed harvest machine for *Stylosanthes guianensis* for making leaf meal production (Guodao *et al.,* 2004). Fodder trees are lopped by axe, saw and kata. Now, spinning discs and pod plucker or twig cutter used for lopping the twigs up to 12 mm diameter of fodder trees having height up to 6.6m are available in the market which could be utilized for lopping tree fodder. For forage grasses rotary grass mower, having the high tip speed of revolving knives, being collected in a window by using a 45 horse power tractor. It has effective field capacity of 0.3 ha/hour. Flail type forage harvesters with effective field capacity of 0.2 ha/hour is suitable for harvesting all types of forage crops and grasses for silage making and fresh feeding. This put harvested forage in a side by side moving tractor trailer and a walking type self-propelled vertical conveyor reaper is also used for harvesting oats, berseem, barley and wheat crops. The field efficiency of machine for *Stylosanthes,* oats, berseem, barley and wheat were found to be 55 to 71 per cent.

Processing and Product Formation

Forage resources are processed for better digestibility and balanced nutrition of livestock. Various methods of processing used are physical *i.e.* chaffing, soaking, grinding, pelleting, boiling, steaming under pressure and gamma radiation; physiochemical methods *i.e.* NaOH/pelleting, urea/pelleting, lime/pelleting, chemicals/steaming and NaOH/temperature; chemical treatment methods *i.e.* sodium hydroxide, calcium hydroxide, urea-ammonisation, ammonium hydroxide, anhydrous ammonia, sodium carbonate, sodium chloride, chlorine gas and sulphur dioxide and biological methods *i.e.* addition of enzymes, white rot fungi and mushrooms. During treatments of the crop residues deficient in some nutrients can be supplemented. Chopping green forage makes more homogenous meal before drying as leaves could be mixed well. In a study for developing leaf meal production technology for ruminants at IGFRI showed the benefit of using the thresher for leaf meal production. The main problem of biological upgrading of low quality roughages into feed is to find suitable microorganism with metabolic patterns different from those of the rumen flora and fauna for use in the process of delignification and upgrading. "Ideal microorganisms" for upgrading lignocellulose crop residues into animal feed, should have a strong lignin metabolism with poor degradation of cellulose and hemicellulose so that in the treated material, carbohydrates are preserved for their utilization by rumen microbes as an energy source in the ruminants (Zadrazil *et al.,* 1999). Complete feed making require chemical as well as biological treatments for improving palatability to enhance intake and increasing digestibility of nutrients. Most common methods are chopping and urea ammonisation, recently composite feed block being popular. Complete feed blocks cater the need of animals with different productivity levels. It deserves to be popularized on a large scale.

Drying of harvested forage grasses and legumes is required before making hay, leaf meal and baling (Das *et al.,* 2009). Traditional practice of sun drying is influenced by sunshine hours, prevailing temperature and humidity as well as chances of sudden rainfall. Quick drying preserves the nutrients in the herbage. For processing green forage into hay, crop should be harvested after moisture settled on the leaves get evaporated and harvested materials left for some times in scattered or in small heaps in the field to get it partially wilted. After partial wilting, it should be collected with the help of hand rake and placed on the structure by tying into loose bundles. Most common technique of hay making is done by using A – frame, tripod, zigzag fences and single pole structures (IGFRI Annual Report 1990). For proper draining of water if there is any rain, upper most layers may be placed in the form of a thatch roof. Normally, two quintal forage spread (6-8 cm thick) uniformly is placed on the frames in zigzag fence at a time. It takes four to seven days depending on weather in the month of September to reduce its moisture to a level of 18-20 per cent for safe storage. During three to seven days, dry matter loss in *Stylosanthes* was reported to be 15-40 per cent (Amodu, 2004; Guodao, *et al.,* 2004). Alternatively, A-Frame using local material built *i.e.* bamboo or wood or iron rod or angle; two tier drying structure of subabool logs made; three/two tier hay drying shed; mechanical hay drier, a forced batch type hay dryer; hot air drying and a solar drier can be used for drying processes. Hot air drying improves the quality of dried herbage and requires less time for drying for commercial manufacture of leaf meal (Samul & Anap, 1980; Chaisang *et al,* 2004; Leng, 1997; Guodao *et al.,* 2004). The high temperature drying was reported to be costly to dehydrate leaf meals due to expensive machinery used and the operational cost. The additional cost of drying is however, reported to be compensated for by the quality product marketed at the high price (Guodao *et al.,* 2004). A design approach to a five tonnes capacity portable hay drier was found useful and cost effective by Samual & Anap (1980) to dry hay from 40 per cent moisture content to a limit of 20 per cent within three days. A solar drier has been installed at IGFRI, Jhansi for drying green forage up to the drying temperature of 65°C can also be used for drying forage. Two stage drying by first drying in sun of freshly cut forage having 45-60 per cent moisture content to a lower level of 20-25 per cent then finally to 9-12 per cent by using quick drying process of hot air (Guodao *et al.,* 2004).

Stylo leaf meal can supply protein and other nutrients as feed for ruminants and non-ruminants. Many studies had demonstrated usefulness of stylo leaf meal in livestock ruminants as well as non-ruminants. In china, pig feed was successfully incorporated 10 per cent stylo leaf meal 3-5 per cent in poultry feed and one per cent in fish compound feed (Changjun *et al.*, 2004). Indian studies at ANGARU, Hyderabad and TNAU, Coimbatore, showed benefit of 3-5 per cent stylo leaf meal in poultry feed prepared from *S. scabra, S. guianensis* and *S. seabreana.* Subabool (*Leucaena leucocephala*) leaf meal suitably replaced mustard cake up to 45 per cent in lamb diet (Mahantha *et al.*, 1999) in terms of reducing the feed cost. Table 9.2 showed the recovery of leaf meal from three *Stylosanthes species* in India. Stylo leaf meal can be made into pellet or powder for mixing in feed rations. The standard as compared to International Standard of leaf meal is slightly inferior.

Table 9.2: Recovery of Leaf Meal in *Stylosanthes species* in India

Species	*Fresh Yield (tonnes/ha)*	*Recovery of Leaf Meal (per cent)*
Stylosanthes seabrana	83-86	40
Stylosanthes scabra	80-86	32
Stylosanthes guianensis	90-95	28

Source: Guodao *et al.* (2004).

Pelleting

Pelleting is a way to pulverise unpalatable and difficult to handle feeding material, sometime powdered form feed material and process by application of heat, moisture and pressure into larger particle and palatable feed material for improved feeding of various types of livestock, which can be easily handled (Rekhate *et al.*, 2007; Suresh *et al.*, 2009). Latest type of pellet mills use a ring die turning about two fixed rollers mounted in a vertical plane and consist of feeder, conditioning, pelleting device, speed reduction device, prime mover and the base. The feeder is generally fitted with a variable drive for speed control. A power operated, auger type feed pelleting machine developed by Indian Grassland and Fodder Research Institute, Jhansi, is very useful for making feed pellets (Photo 9.5). It is operated by a 5 hp electric motor. The capacity of the machine ranged between 65 and 85 kg/hour with die size of 28 mm and 38 mm. The length of feed pellets varied from 10-15cm.

Photo 9.6: Pelleting Machine (Courtsey: P. K. Pathak)

Silage

Silage is a fermented feed produced under anaerobic conditions from succulent chopped forage. The silage making process and the quality of silage depends upon proper harvesting stage of the crops which maintain its succulence at 30-35 per cent dry matter at the time of ensiling (IGFRI 1990; IGFRI, 1991; Kamra, 1983). In general, forage grasses should be harvested just before flowering. In case of high moisture content; the harvested forage is allowed to wilt to 30-35 per cent dry matter content. Proper stage of harvesting for silage is pre-flowering stage chaffed to about 2.5 cm size especially maize, sorghum, and Napier x Bajra hybrids with about 75 per cent moisture content. Irrespective of the amount of silage to be made, the forage must have a high nutritive value, ensure chopping no longer than 2.5 cm for proper compaction to expel air within the chopped forage, the forage should not be contaminated with soil and ensure quickest way of sealing the silage pit for good silage. The size and type of silo depends on the herd size, quantity of available feed and the period of feeding. The most commonly used small type and size of silo pits are of 1.75 x 1.7 x 1.75 metres dug up in the ground. Boundaries should be raised so that rain water cannot enter silo pit. One cubic meter silo require seven quintal forage and to feed a unit of 10 ACU for a month nine to 10 tonnes silage prepared in total volume of 16 cubic meter silo pits. The filling should be rapid with proper pressing by use of tractor after each filling to remove air and completed within 4-7 days after thorough pressing, and top should be covered with polythene followed by soil layer of 6 inches depth. Silage undergoes anaerobic fermentation just after 48 hours of covering the silo and the silage become ready within a period of 15-20 days. The controlled fermentationof forage by microbes present in fresh herbage put in an airtight silo results in production of lactic acid, acetic acid and formic acid, which prevent decomposition and kill the growth of the unwanted microorganisms which otherwise spoil the silage. Eventually, silage can be preserved for the period as long as 10 to 15 year under anaerobic condition. The primary factors affecting the success of silage fermentation are water-soluble carbohydrate content, buffering capacity of the forage, moisture content of the forage, type of bacteria which predominate during fermentation. If ensiling procedures are such that lactic acid producing bacteria aren't favoured, *clostridia* type microorganisms will grow which will adversely affect the quality of silage. Then fermentation process can also be initiated by additives to maintain anaerobic conditions and discourage *clostridia* growth that produce butyric acid. Molasses (3-4 per cent) to low sugar grasses; urea (0.5-1.0 per cent) to graminaceous forage to increase protein content and lime (0.5-1.0 per cent) to maize to increase acid production is added as per requirement. The silage preserve about 85 per cent of nutritive value of forage crops, conserve excess forage to ensure supply during lean period when green forage are not available. It requires less space and gives better nutrition than hay. Silage also helps to eliminate anti-nutritional metabolites (*e.g.* tannins and alkaloids) in certain forages (*e.g.* herbaceous and woody legumes and cassava leaves). Quality of silage is judged by green, brown or golden colour (black

Photo 7.1: Berseem, a Popular Cultivated Forage.
P (184)

Photo 7.2: Non-Competitive Land Use Technology for Forage Production.
(Courtsey: Sunil Kumar)
P (188)

Photo 7.3a: Aonla + Veranostyl (*Stylosanthes hamata*).

Photo 7.3b: Aonla + Guinea Grass.

Photo 7.3: Aonla Based Hortipasture in Semi Arid Region (*Courtsey*: Sunil Kumar)

P (211)

Congo Signal Grass (*Brachiaria ruziensis*)

P (234)

Broom Grass (*Thysanolaena agrostis*)

P (234)

Zizinia caudicflora

Photo 8.1 b: Some of the Important Forage Grasses.

P (234)

Desmodium hetrocarpum
P (235)

Desmodium podocarpum
P (235)

Macroptilium atropureum
Photo 8.2a: Some of the Important Perennial Forage Grasses.
P (235)

Stylosanthes guianesis
P (236)

Mimisa invisa
Photo 8.2b: Some of the Important Perennial Forage Grasses.
P (236)

P (238)

Brassiopsis hainla
Photo 8.3a: Important Fodder Tree Species.
P (238)

Litsea monopetaia
P (239)

Sauraula napaulensis
Photo 8.3b: Important Fodder Tree Species. P (239)

Bauhinia purpurea
P (240)

***Ficus auriculata* Type-I Twig**
Photo 8.3c: Important Fodder Tree Species. P (240)

***Ficus auriculata* Type-I Fruits**
P (241)

Fisuc auriculata **Type II-Fruits and Twig**
Photo 8.3d: Important Fodder Tree Species. P (241)

P (242)

Ficus hirta
Photo 8.3e: Important Fodder Tree Species. P (242)

Photo 8.3f: *Ficus glomerata* Tree Fodder with Cylindrical Canopy and High Forage Yield is a Potential Tree Fodder to be grown under Agroforestry Systems.
P (243)

Ficus elastiea
P (244)

Ficus laccor
Photo 8.3g: Important Fodder Tree Species. P (244)

Ficus hispida
P (245)

Ficus subincisa – **Tree**
P (246)

Ficus semicordata
Photo 8.3h: Important Fodder Tree Species. P (245)

Ficus subincisa **– Fruit and Twig**
Photo 8.3i: Important Fodder Tree Species. P (246)

Ficus rumphii
P (247)

Grewia optiva
Photo 8.3j: Important Fodder Tree Species. P (247)

Bambusa arundinacea **– Protecting Hill Slopes along the Road.**
P (248)

Bambusa nutans **on the Bank of Spring Water Course.**
Photo 8.4: Bamboo Plant Species. P (248)

Photo 8.5: Orchard Grass (*Dactylis glomerata*) in temperate region of Himachal Pradesh
P (250)

Tall Fescue (*Festuca arundinacea*)

Photo 8.6 a: Perennial Grasses in temperate Region of Himachal Pradesh. P (250)

Rye Grass (*Lolium multiflorum*)

Photo 8.6 b Perennial Grasses in Temperate Region of Himachal Pradesh. P (251)

***Trifolium pratense* in Himachal Pradesh. P (251)**

***Trifolium* Species grazed by Sheep**

Photo 8.7: Perennial Legumes in Alpine and Temperate Zones. P (252)

Photo 8.8: *Medicago falcata* in Laddakh Region of Jammu & Kashmir. **P (252)**

P (256)

Photo 8.9b: *Populous teldoides* grown under Farming Systems and as Block Plantaiton in Alpine Region of Ladakh. P (256)

Photo 8.9a: Block Plantation of *Populous teldoides* In Ladakh. P (257)

P (258)

Photo: 8.9c: *Salix* species and Seabuckthorn (*Hippophae rhamnoides*) Leaves are also Fed to Livestock. P (258)

Photo 8.10 a: *Sesbania aegyptica* with Napier Grass on the Bunds of Terraced Lands in Nepal. P (271)

Photo 8.10 b: *Leucaena leucocephala* with Napier Grass on the Bunds of Terraced Lands in Nepal. **P (271)**

Photo 8.11: Napier Bajra Hybrid on Terrace Bunds in Sikkim. P (272)

Photo 8.12: Mustard on the Terraced Land and Guinea Grass on the Terrace Riser in Meghalay. P (272)

Photo 8.13: Turmeric on the Terraced Land and Guinea Grass on the Terrace Riser in Meghalaya. (273)

Photo 8.14: Mithun Feeds on Herbaceous, Shrubs and Tree Leaves in the Forest Area. P (283)

Photo 8.15: Yak Grazing in Arunachal Pradesh P (287)

Photo 8.16: Yak Grazing in Wet Grassland of Ladakh P (287)

Photo 8.17: Pashmina Goat Grazing in Alpine Region of Ladakh. P (288)

Photo 8.17: Pashima Goat Grazing in Alpinr Region of Ladakh. P (288)

Photo 9.1: Stacking Stovers for Livestock in Plain Area is a Common Practice. P (318)

Photo 9.2: Scientific Stacking of Baled Grasses on Raised Plateform. (Courtsey: P. K. Pathak) P (318)

Photo 9.3: Collection and Storage of Dry Forage Grassing and Residues in Himachal Pradesh for Feeding during Winter Months. P (319)

Photo 9.4: Storing Crop Residues and Dry Forage Grasses on Trees is a Common Practice in the Hills of Himachal Pradesh. P (319)

P (320)

Photo 9.5: Storage of Leguminous Hay of *Medicago falcata* and *Medicago glutinosa* on Roof of Houses in Alpine Region of Leh in Jammu and Kashmir. P (320)

Photo 9.6: Pelleting Machine (Courtsey: P. K. Pathak) P (323)

Table 9.3: Concentration Limit for Pesticide Residues

Pesticide	*Maximum Concentration Limit (mg/litre)*
Aldrin	0.001
DDT	0.05
Dieldrin	0.001
Chlordane	0.003
Endrin	0.0002
Hephtachlorepoxide	0.0001
Hephtachlor	0.0001
Lindane	0.004
Methoxychlor	0.1
Toxaphene	0.005
2, 4D	0.1
2, 4, 5-T	0.01

colour indicate poor silage), free from mould, gives good smell of lactic and acetic acid like vinegar. Silage of a pit should be completely fed after opening because over exposure to air spoil the silage.

Silage is sound technology and beneficial at lower cost which has been well reported favourably by research and extension worker in India. Special programmes on silage was implemented and subsidies were given by governments to construct silo pits and silo towers and purchase chaff cutter. It was expected to at least ensiling surplus obtained during rainy season for lean periods. Still use of silage in India is very much limited in spread over various states and even at state government's farm except Punjab. A study by Rangnekar (1999) identified various reasons for silage not making ground in animal husbandry in central and western India. It showed lack of in depth knowledge and practical experience of technical knowhow of silage making among the extension worker had negatively impacted faith in silage technology; being reported labour intensive for digging pits and non-familiar with use of plastic sheets, being felt costly not commiserating with benefits by farmers and farm women keeping low milk yielding animals. Further, chopping forage found cumbersome over traditional practice in the area *i.e.* convenience factor; no surplus forage for ensiling in rainfed area; preference of growing two to three forage crops in irrigated area than silage making and women feeding the livestock did not like smell of silage are limitations to promote and expand silage making across the country. Unless high milk producing animals are reared for high returns, concentrate versus silage benefits cannot convince the livestock farmers. With low producing livestock, supplement with dry forage and low quality roughage is found more convenient, hence, spread of silage is becoming difficult. Now, with spread of high productive livestock all over the country, silage technology can benefit farmers. Constraints should be further analysed, technology modified and promoted through experienced extension worker in silage making for better adoption.

Fodder Bank

In India, there is a regular phenomenon of seasonal and regional deficit of forage. Often-natural calamities like drought and flood aggravate the situation to the level of acute scarcity. Considering the situations of perennial deficit of various magnitudes, policy intervention is required to establish forage banks across the country in order to regulate deficit, meet fluctuating demand and mitigate deficit during natural calamities. Conserving the forage resources scientifically, is therefore a need for livestock sector to strengthen the economic power of the livestock farmers by reducing vulnerability of various kinds resulting in shortage of forage resources.

Forage banks may be established under two categories (1) Live forage bank - fodder trees and bamboo plantation along the road, silvi/horti pastures as common property resources, inclusion of fodder trees and grasses in joint forest management plantation and other afforestation programmes and (2) construction of fodder banks for proper storage of dry forage and edible crop residues and silo for silage. Bailing and enrichment of crop residues particularly straw of paddy, wheat and other leguminous crop residues along with green forage help in balanced feeding of livestock and minimize wastage and storage loss. Use of leaf meals from leguminous forage and crop residues from pulses (Lentil, gram, grass pea (*Khesari*) *etc.* and leaves of subabool, stylo, *Gliricidia etc.*) will help to substitute costly concentrates. Plenty of forage grasses, legumes and tree fodder in the forest margins during monsoon and edible crop residues from field crops are available in the country for storing in forage banks. All of these forage resources can be densified and stored in the forage banks in the form of various products *viz.* feed block, leaf meal and mandatory region specific mineral mixture supplementation to create a network of both the types of forage banks in different zones of the country after assessing the balance sheet of forage production.

Forage resource availability round the year can be well enhanced through promotion of fodder banks. Fodder banks will not only provide the timely supply of the feed to the most affected area in the country but it will also serve as community based fodder bank from where a needy farmer can deposit his share, withdraw as per his requirement, sale his share or even he can take loan from the bank with a promise to return with interest in term of forage only and not in cash.

The fodder bank would have to maintain its own products in the form of densified blocks, pellets, silage, leaf meal and densified hay. It is estimated that for storing bales of crop residues required for 1000 animals for five months of lean period needs a space of 12000m^3. These bales could be stacked on the pattern of Food Corporation of India under CAP (cover and plinth storage structure) system. The stored bales could be used in phased manner after keeping for about five months for creating space for next season storage. Live fodder bank of tree fodders are important source of feed during natural calamities as well as for small ruminant production. Nutritionally they are rich in minerals like calcium and phosphorus, very important for milk production. Establishment of fodder tree plantation, a minimum of 25 per cent under social forestry and joint forest management as well as their regulated use will serve the purpose of a live fodder bank.

Mineral Nutrition for Improved Livestock Production

In most of the tropical and sub-tropical countries, deficiencies and imbalances of mineral nutrients reduce the production performance of livestock. Forage and feed grown on the soil deficient in essential minerals do not meet the mineral requirement of livestock. Further, the mineral contents in the soil keeps on changing due to imbalanced fertilization in the soil under various forage based cropping systems, which may also alter their contents in feeds and forage, affecting the mineral nutrition of livestock. Macro-elements required for livestock are calcium, phosphorus, magnesium, sulphur, potassium, sodium and chlorine; trace or microelements are cobalt, copper, iron, manganese, iodine, molybdenum, selenium and zinc. Devising effective nutritional management strategies for livestock requires an understanding of a broad range of nutrient contents of various forage crops and grazing lands. Although protein and energy are the two key nutrients commonly deficient for at least part of the year but potential mineral deficiencies are also frequently reported. State and zone wise survey conducted to identify mineral deficiency scenario in India by National Institute of Animal Nutrition and Physiology, Bangalore, revealed that phosphorus is the most critical mineral and its deficiency is spread over about 72 per cent agro-climatic zones followed in descending order by zinc (68 per cent), calcium (66 per cent) and copper (48 per cent), Mn (25 per cent), Mg (18 per cent), Fe (14 per cent), Co (7 per cent), Fe (7 per cent) and Co (3 per cent). Deficiencies of these mineral nutrients are fast emerging in various agro-climatic zones. Minerals, particularly excess availability of trace elements also causes toxicity in livestock. Among various minerals, toxicity of Mo, As, Se, Ni, Al, Cr and Fe are being reported in many parts of the country.

Mineral composition of feeds and forages revealed that most of the livestock requirement can be met from feed and forage feeding practices except for Ca, P, Na, Cl, Cu and Zn. Forage legumes and top feeds contain over twice the amount of calcium required as compared to cereal forages and grains fed to the livestock. Concentration of magnesium and sulphur are just sufficient except in the crop residues and non-leguminous green forage to meet the requirements of growth and gestation but additional requirement for milch livestock need to be supplemented. P, Na, Cu and Zn are the minerals that are largely deficient in forages in most of the agro-climatic zones of India. However, the concentration of iron is many folds higher than their required levels of 50-100 ppm. The shrubs and tree leaves have higher contents of iron (Ramana *et al.,* 2000). High iron content interferes with the copper absorption and metabolism leading to low blood copper status. The molybdenum is within the safe limit in dry and green forages, shrubs and tree leaves.

Forage mineral composition is dependent on many factors including soil characteristics, stages of plant growth, climatic condition and management practices (Brady, 1974; Mulliniks *et al.,* 2010). In most of the forage crops and pasture the content of mineral like Fe is extremely high due to soil contamination. The high Fe intake may interfere with the absorption and utilization of Cu in the livestock. The crop management and climatic conditions also influence the mineral levels of forages. Fertilizer application results in faster pasture growth which dilute the contents of trace minerals. The application of molybdenum fertilizer to improve the growth of forage crops may reduce copper absorption and the application of phosphorus and nitrogen fertilizers to the soil will reduce selenium, cobalt and copper uptake by forage crops. Mineral deficiency, imbalances and toxicity both can be avoided by adopting various nutrient management strategies for forage and grazing lands as well as mineral supplementation strategies. A complete mineral mixture usually contains all the essential minerals such as calcium, phosphorus, magnesium, sulphur, potassium, iron, copper, cobalt, manganese, and zinc. But the supplementation of all the above mineral elements may not be needed always, because many of them such as magnesium, sulphur, potassium, iodine, cobalt, iron and manganese are already present in the required quantities in the feeds and forages to satisfy the need of dairy livestock producing 8-10 kg of milk daily. Area specific mineral mixture is the best approach to optimize animal production and reproduction to be considered in isolation but as a part of whole nutrient management system, which includes protein, energy and vitamins.

Forage Resources and Feeding Practices

In Indian agriculture, dairy farming plays a vital role for livelihood of farmers under crop-livestock mixed traditional agriculture particularly for resource poor farmer's *i.e.* landless, marginal and small farmers, contributing to major share in total milk production in the country. Small ruminants *viz.* camel, goat and sheep largely depend on grazing. In all the agro-ecosystems, livestock are grazed during monsoon due to large quantity of green biomass available in the grasslands, along the rail, roads and in the vicinity of forest as compared to summer season. Resourceful large and medium holding farmers offer forage produced on cultivated lands. Local forage grasses and edible weeds are also utilized as forage resource by all categories. In arid and semi-arid region milch cow and buffalo are grazed for more hours as compare to grazing in sub humid region. In arid ecosystem, seasonal forage grasses along with weeds were the most common green forage offered to cows and buffalo. Dry livestock offered more roughage as compared to milch livestock.

Contamination of Livestock Products with Pesticide Residues and Heavy Metals

Safe limit of pesticide residues and heavy metals in feed and forage is the pre-requisite for producing safe and healthy livestock products (Chhabra, 2004). Forage and feed are the major possible sources of contamination with pesticide residues, heavy metals and many other contaminants. Therefore, care must be taken at all the stages of production (sowing of the crops, their management and harvesting) of feed and forage, their processing and storage to prevent contamination originating due to pesticides residues and various sources of heavy metals in forage or ingredients of feed. Various contaminants through dumping of industrial wastes on the dump site reaches to cultivated lands and in turn from the soil taken up by forages and grasses which is the main source in contaminated feed and forage offered to livestock for consumption. Feeding of such feed and forage may act as a direct source of contamination in livestock products. Drinking water offered to livestock may also sometimes be a source of contamination (Choubisa *et al.,* 2012). The agreement on the application of sanitary and phytosanitory measures of World Trade Organisation has provided guidelines to produce safe food for international trade and recognizes Codex standards of reference points related to food safety issues (Table 9.3).

Dairy milk and its products get contaminated with residues of pesticides, particularly the persistent organochlorides (Kalra *et al.*, 1980; Khandekar *et al.*, 1981; Kaphalia, 1982; Visalakshi *et al.,* 1993; Awasthi & Ahuja, 1995; Chhabra, 2004; Battu *et al.,* 2004; Kan & Meijer, 2007; Nag *et al.,* 2008). The reasons for existence of pesticide residues in milk or its products is not due to any direct application but they originate mainly from contaminated feed and forage (*e.g.* pasture, edible crop residue, cultivated forage and ingredients of concentrate) consumed by the dairy livestock. The significance of pesticide contamination in milk has been noticed since 1960s. Most of the studies are being conducted on organochlorides and organophosphate for obvious reasons and the data revealed the presence of DDTs and HCHs, often in very alarming concentrations (Kaphalia, 1982; Jadav, 1986). Organophosphates are less persistent and easily metabolized in the plant system but they are highly toxic to mammalians. Many of them particularly the organochlorides and synthetic pyrethroids are lipophilic in nature and have a tendency to accumulate in the adipose tissues. They do not get converted into water soluble metabolites. Hence, the concentration in the fatty tissues exceeds a threshold level, the excess residues are translocated into the lactating glands of the animals from where they are secreted in the milk. The presence of polychlorinated biphenyls (PCB) in milk has been traced back to cattle feed and other sources such as silos coated with sealant or paint, oils and grease containing PCBs, and sewage and sludge application to agricultural pasture. In few cases the presence of aldrin and dieldrin were also recorded in the livestock products. Generally, butter and ghee have been found to be contaminated with organochlorides and organophosphates in India (Kalra, *et al.,* 1979; Kalra, *et al.,* 1980; Takroo *et al.,* 1985). The equilibrium between uptake and release with the milk fat was reached between 40-80 days after the beginning of a continuous exposure to dioxins/furans.

Heavy metal contaminants lead (Pb), cadmium (Cd), arsenic (As), mercury (Hg) *etc.* enter into livestock system through pollution of air, drinking and irrigation water, and soil into the feed and forage (Dawd *et al.,* 2012). The contaminated soil and heavy metal loaded ground water used for irrigation act actually as a reservoir of the toxic metals in the feed and forage. Combustion of fossil fuel and emission of industrial smoke pollute the atmosphere and surrounding environment and release heavy metals, which ultimately are taken up by nearby growing forage crops. They also get deposited in the soil for subsequent uptake by the various crops. From tissues of animals, they may enter lactating glands through physiological processes and may be secreted through milk. Normally, very low concentrations of these metals are secreted into milk because the carry over rate of metals from forage to milk is approximately 1:500 (Bluthgen *et al.,* 1997; Bluthgen, 2000). The uptake of Cd by forage plants especially at lower pH is the main route of uptake into the food chain. According to a survey conducted in two villages (declared to be arsenic contaminated) in two districts of West Bengal, feed, roughages given to animals contained highly toxic levels of arsenic (Datta *et al.,* 2012). The average excretion of arsenic in the milk was found to be 77ppb (Singh *et al.,* 2002). Ozdemir *et al.* (2009) found that the Pb content (3.24 – 80.69μg/kg) above the regulatory limits was found in the milk of cows, grazing at the road side pastures due to pollution from vehicle emissions in Turkey. Toxicity of fluoride has also been noticed in cattle (Choubisa *et al.,* 2012). Studies on pesticide residues and heavy metals in meat products are limited. Occurrence of DDT and BHC had been reported in various meat sources (Battu *et al.,* 1984; Vijayan *et al.,* 2006).

Chapter 10

Forage for Non-ruminants

Non ruminant's livestock *viz.* pig; poultry and rabbit's play an important role in livestock production and the livelihood of the farmers. In spite of differences in the digestive systems of non-ruminant livestock's *viz.* pigs, rabbit and the poultry birds, they have similar general feed digestion patterns, having implications for identifying suitable forage plant species and processing technologies. Generally, feed of non-ruminant livestock requires higher nutrient densities which are made of grains and various by-products of cereal, pulses, oil seeds *etc.* These ingredients largely compete with human food. Therefore, it has become a challenge to reduce human-livestock food-feed competition by way of substituting the feed partly by succulent and less fibrous green forages which will make non- ruminant production more efficient and cost effective. Forages being a versatile resource have potential for supplemental and *adlibitum* feeding in non-ruminants. They are good sources of nutrients particularly the energy, fat, protein, amino acids and minerals, easily digestible, palatable and slightly laxative in nature.

The humid and sub-humid tropical agro-climates in India provide year round growing of forage under seasonal water deficits and excesses, constraints that are usually manageable (Ly, 1990). Chosen forage basket for non-ruminant may be comprised of diverse plant species including aquatic ones, each varying in growing requirement, chemical composition, supplying important carotenoids, xanthophyll's pigments and other neutraceutical *etc.* In this context, screening the suitable forages for non-ruminant, harvest them at the most suitable stage for them having different digestive physiology, will be a starting step. The harvesting stage of forage for the non-ruminants will be generally earlier than the stage it is harvested for ruminants. Further, processed forage products *viz.* leaf meal, hay and silage have potential to enhance year-round animal productivity and reduce cost under various feeding systems. Over past 30 years, there has been a growing interest in the use of forage and the processed forage as sources of protein and other nutrients for non-ruminant animals (D'Mello, 1992). The diversity of about 650 genera of legumes with some

18,000 plant species exists in the tropics and subtropics (NAS, 1979). Many of them have been collected for evaluation as tropical forages with gene banks at the Centro Internacional de Agricultura Tropical (CIAT, over 20,000 accessions, Marcel Dekker, 2011), the International Livestock Research Institute (ILRI, 20,000 African forage accessions (Karaimu, 2011), together with the Australian Tropical Crops and Forages Collection and the collection of Cenargen-Embrapa, at ICAR-Indian Grassland and Fodder Research Institute in India. Forage legume as source of protein with a desirable amino acid profile, especially lysine, methionine and other sulphur containing amino acid and high levels of vitamins and minerals adequately meet the requirements of non-ruminants unlike graminaceous forage plant species and existing concentrate based feeding systems (Lopez and Tapia, 2003; Lajide *et al.,* 2008; Aarti *et al.,* 2005; Mosha *et al.,* 1995; Imungi and Potter, 1983; Garcia *et al.,* 1996; Leterme *et al.,* 2006). Constraints imposed by chemical composition particularly anti-nutritional factors can be solved by processing and additives. The major groups of biochemical to be dealt are polyphenols, cyanogenic glycosides, alkaloids, saponins, steroids, toxic proteins and amino acids, non-protein amino acids, phytohema glutinins, triterpenes, and oxalic acid (Kumar, 1992; Liener, 1980). They act as anti-nutritional factors or toxic to livestock. Polyphenols are a major group often related to taste, odour, and colour. Many forage plants contain condensed tannins, a complex heat-stable phenolic compound. They are found in many forage plants, especially shrub legumes such as *Gliricidia sepium, Acacia species, Leucaena leucocephala and Albizia falcataria* (Jeroch *et al.,* 1993). Saponins depress growth in chicks (Jeroch *et al.,* 1993), but there are conflicting results in pigs, with a positive effect on both sows and piglets (Hauptli & Lovatto, 2006), compared with fewer stillborn piglets but reduced growth performance (Ilsley & Miller, 2005). Moreover, phytate significantly reduces the availability of minerals such as zinc, calcium, and magnesium (Rimbach *et al.,* 2008). Oxalates limit especially the availability of calcium, magnesium, and iron (Weiss, 2009; Rahman *et al.,* 2011).

Nutritional Potential of Forage for Non-ruminants

Nutritional potential of cultivated forage, perennial forage grasses and legumes, tree fodders and the aquatic forage resources to supplement feeding of non-ruminant livestock's (pig, rabbit and poultry birds) have been well proven by experimentation and examples of traditional feeding in some pockets across the Asian countries with proven economic advantages. Use of forage resources in non-ruminants revealed use of 11 forage legumes, 04 forb plant species, 09 fodder tree species and 06 aquatic forage plant species in poultry (Table 10.1), 07 forage legumes, 03 grasses, 11 forbs, 07 tree fodder and 06 aquatic forage plant species in pig (Table 10.2) and 29 forage legumes, 22 grasses, 13 forbs, 15 tree fodder and 07 aquatic forage plant species in rabbit (Table 10.3). Characteristics of the forage traditionally used as local feed resources for non-ruminants showed certain merits of quality and the quantum availability (Foulkes *et al.,* 1977). The roots and vines of sweet potato and the leaves of cassava are rich in crude protein

(Hang, 1998). Beside rich in crude protein (20-30 per cent), the leaves of cassava are also rich in biologically available amino acids i.e valine, methionine, isoleucine and serine continuously (Eggum, 1970). The crude protein content in the leaves of cocoyam is more than cassava or sweet potato but it contains low amount of histidine, lysine, leucine, and tryptophane and methionine amino acids. The leaves of cocoyam also contain sufficient Ca, P, Fe, vitamin C, thiamine, riboflavin and niacin. *Trichantera gigantea* called Nacedero in Vietnam is a multipurpose forage tree having18 to 22 per cent crude protein content in dry matter, very rich in true protein with a good amino acid balance and calcium in the foliage twigs used for all types of livestock (Ha & Phan, 1995; Nhan *et al.*, 1996; Galindo *et al.*, 1989; Toung *et al.*, 1994). There is potential usefulness of a wide variety of protein rich plant materials which could replace soybean meal in feed formulations (D'mello, 1992). Various forage legumes and grain legumes *viz.*, berseem (*Trifolium alexandrium*), lucerne (*Mediago sativa*), peanut (*Arachis hypogaea*), *Stylosanthes* species, pigeon pea (*Cajanus cajan*), sword bean (*Canavalia gladiata*), soybean (*Glycine max*), hyacinth bean (*Lablab purpureus*), moth bean (Vigna *aconitifolia*), mung bean (*Vigna radiata*), rice bean (*Vigna umbellata*), cowpea (*Vigna unguiculata*), broad bean (*Vicia faba*), green gram (*V. radiata*)vetches, and lupins (*Lupinus albus, Lupinus mutabilis*), peas (*Pisum sativum*), velvet bean (*Mucuna pruriens*), horse gram (*Macrotyloma uniflorum*) harvested at 45 to 60 days of growth can be partly supplement or substitute protein in the form of fresh or hay in non-ruminant's. *Stylosanthes* is grown for fresh feed, hay and commercial leaf meal production, being fed as plant protein source for pigs, ducks and chicken in China (Gupta & Singh, 1983). Tree fodders like *Leucaena leucocephala, Gliricidia sepium, Caliandra* species are rich in protein and containing the essential amino acids, carotenoids and good sources of minerals. *Gliricidia sepium* can yields 5-16 tonnes/ha of leaf dry matter, or up to 43 tonnes/ha fresh leaves annually (Andrew Speedy, 1994). Aquatic plant species are succulent, high in crude protein and low in crude fibre with reasonable forage digestion coefficient to supplement feeding for non-ruminants (Linn *et al.*, 1973). Among tropical and warm aquatic fern, *Azolla caroliniana Willd.* ; *A. filiculoides Lam.*; *A. mexicana C.Presl*; *A. microphylla Kaulf.* ; *A. nilotica Decne.* Ex. Mett. and *A. pinnata R. Br.* are important. In India, yields of 37.8 fresh weight/ha (2.78 tonne DM/ha) with 19-30 per cent crude protein in dry matter and lysine for nutrition have been reported for *Azolla pinnata*, being promoted as livestock feed (Hasan *et al.*, 2009). The *Azolla* is considered a good feed for pigs and ducks in Southeast Asia; poultry in Vietnam and for pigs in Singapore and Taiwan (FAO, 1994). Environmental constraints such as very high temperatures, low humidity, limited water availability and poor quality of water can restrict the adoption of *azolla* production (Giridhar *et al.*, 2013). Other aquatic plant species Duckweed (*Lemna minor*) and water spinach (*Ipomoea aquatica*) are also good sources of protein (25-30 per cent), fat (5-10 per cent) and metabolizable energy (2.1-2.5 Kcal/g) for pigs and ducks (Hang, 1998; Amornsak *et al.*, 2004; Becerra *et al*, 1995; Bui *et al.*, 1995, 1996). However, Crouch (1964) found that feeding aquatic forage at high level would lead to problem of palatability.

Processing technologies can help to enhance palatability, intake, and digestibility, to conserve, detoxify the anti-nutritional factors as mentioned earlier in the text, or concentrate nutrients (Akande *et al.,* 2010; Close, 1993). Forage plants can be processed by soaking, heat treatments, grinding/milling, pelleting, extraction/chemical treatment, use of enzymes, fermentation (ensiling). Leaf meal on leguminous forage resource is one of the popular choices comparison to cereal fed to non-ruminants. The leaf meal is largely fed to meet the requirements of protein and other nutrients to pigs, poultry and ducks (Gadzirayi and Mupangwa, 2014). The leaf meal is relatively balanced amino acid composition and rich in certain essential amino acids. In many instances, fibre content in the leaf meal may equal or exceed crude protein content. With limitation of fibre content in the leaf meal, digestibility of the crude protein fraction of many leaf meals is low which tends to depress overall crude protein digestibility when leaf meals constitute a significant proportion of the diet (Tangendjaja *et al.,* 1990). Therefore, leaf meals cannot achieve fully to replace high quality ingredients in diets for monogastric animals. Besides, limitation of metabolizable energy and sulphur amino acids also prevails due to increased bulkiness of leaf meal and other processed forage products. However, the diets based on cottonseed cake and coconut meal, substitution by leaf meal provide better option. It would not be advisable to exceed for example *Leucaena* leaf meal more than 05 per cent for broiler chicken and 10 per cent in for laying hens, growing pigs and rabbits of the total diet. Pelleting increases the digestibility in chicks of protein and starch and apparent metabolisable energy values of *Vicia faba* (Lacassagne *et al.,* 1988). The feasibility of small scale ensiling of forages in plastic bags or containers need to be worked out. Ensiling with or without ammonisation may be a way out to detoxify tannins, mimosine and other toxins, and provide storage of feed for the dry season (Raharjo *et al.,* 1986). Dietary additives such as ferric sulphate and polyethylene glycol (PEG) are effective in overcoming growth inhibitory properties of *Leucaena* (D'Mello & Acanovic, 1989).

At every individual farm level, ecological, economical and technical requirements can be well matched based on already available feed ingredients for achieving good economic gain. Creative approach and systematic research are also required to establish the actual value of underutilized forage plant species for different non-ruminants. However, there exist large knowledge gaps of the digestibility of total plant or crude protein, and even the direct effects of feeding some of the forage plant species to different non-ruminants. It is important to identify the optimum inclusion level for the available forage species as well as the best administration form in mixed rations to formulate optimal diets for non-ruminant livestock to improve sustainability of livestock production in different types of livestock farming systems (Schultze-Kraft & Peters, 1997; Savon, 2005). Under the prevailing circumstances and compelling requirement, partly substitution of potential green forage including tree leaves in the feeding system of non-ruminants would play a greater role in developing economically viable feeding system. Tropical legumes by and large could serve as the forage base for non-ruminant production while more research is needed on the individual forage plant species. In the end, the gap need to be filled by systematic research taking into account agronomic, economic, and sociological studies to harness the full potential of forage plants in pig, poultry and rabbit feeding system for eco-efficient and market-oriented livestock production. The need is to expand and utilize the techniques and technology at commercial scale. Adaptation research is required for certain modification to upscale and make it viable at commercial scale. Future research should focus on enrichment for amino acids, vitamins, minerals; harness carotenoids, xanthophyll's, anti-oxidant and other neutraceutical properties of forages for quality products in poultry, pig and rabbit.

Poultry

Poultry is comprised of domestic chicken (*Gallus domesticus*), turkey (*Meleagris gallopavo*), rock pigeons (*Columba livea*), red wattle guinea fowls (*Numidame leagris*), muscovy duck (*Cairina moschata*), wild mallard (*Anas boches*), turtle dove (*Streptopelia turtus*), grey leg goose (*Anser anser*), white footed goose (*Anser albiformis*), ring neck pheasant (*Phasianus colchicus*), north African ostrich (*Struthio comelus*) and south African ostrich (*Truthio australis*) which are raised for their meat and eggs (Chiba, 2009). Forages have positive effects when included in the diets of poultry. Generally, leaf meals and in some cases fresh forge are being used to partly substitute or supplement feeding in poultry. Leaf meals of about 12 forage legumes, *Arachis glabrata, Cajanus cajan, Centrosema pubescens, Lablab purpureus, Medicago sativa, Mimosa invisa, Pueraria phaseoloides, Stylosanthes guianensis, S. scabra, S. seabreana, Trifolium alexandrium* and *Vigna unguiculata;* grass meal; seven leguminous forage tree species, *Albizia jullibrissin, Calliandra calothyrsus, Gliricidia sepium, Leucaena leucocephala, Moringa oleifera* (Lam.), *Sesbania aegyptica* and *Sesbania grandiflora;* five forage forb plant species, *Cnidoscolus aconitifolius (Mill.) Johnston, Ipomoea batatas, Manihot esculenta* and *Tithonia diversifolia and Vernonia amygdalina* and six aquatic forage plant species, *Alternanthera sessilis, Azolla pinnata R. Brown, Eichhornia crassipes, Ipomoea aquatica, Lemna minor and Salvania molesta* are being used in poultry (Table 10.1). Primary benefit of forage consumption as a nutrient source, is that they are high in both vitamins and minerals, contain components such as fibre, protein, energy (calories), and other compounds like carotenoids and omega-3 fatty acids that are important for metabolic functions. Replacement of leaf meals of various types depend upon its inherent specificity of a particular type of forage plant species having different biochemical composition in context of affecting palatability and containing anti nutritional factors. In forage legumes, substitution of leaf meal/hay ranged ranged 2-20 per cent (Bhattacharya *et al.*, 2008; Fapohunda, 2002; Gorski, 2000; Hendricksen & Minson, 1985; Karsten *et al.*, 2010; Krishna *et al.*, 2008; Lopez-Bote *et al.*, 1998; Muhamed Arif *et al.*, 2000; Nworgu *et al.*, 2007; Nworgu and Fasogbon, 2015; Nworgu and Egbunike, 2013; Nworgu & Egbunike, 2013; Odunsi, 2003; Rybina & Reshetova, 1981; Karsten *et al*, 2010; Teguia *et al.*, 1997; Udedible & Igwe, 1989; grass meal 14 per cent (Davtyan & Manukyan, 1987; Gorski, 2000; Karsten *et al.*, 2010; Lopez-Bote *et al.*, 1998; Rybina & Reshetova, 1981), forage forbs 5-30

Table 10.1: Effects of Forage Incorporation in Poultry/ Duck Feeding Systems

Forage Plant Species	*Form*	*Acceptable Level in the Total Diet*	*Substituted Mainly to*	*Effects Growth/ Maintenance/ Other Effects on Poultry/ Duck*	*Reference*
Forage Legumes					
Arachis glabrata	Leaf meal	≤ 20 per cent	—	(+/–) Broilers	Teguia *et al.*, 1997
Cajanus cajan	Leaf meal	7.5-10 per cent of the diet	White maize-based layers diets	(+) Give a yolk colour score of 8 on the Roche colour fan	Udedible & Igwe, 1989
Centrosema pubescens	Leaf meal	2-6 per cent pullets 12 per cent broiler diets	Basal diet (Groundnut cake and soybean in the diets	(=) Growing pullets/ broiler	Nworgu *et al.*, 2007; Nworgu and Fasogbon, 2015
C. pubescens	Leaf meal	2.0 per cent	Leaf meal of *Mimosa invisa; Mimosa invisa*	(+) Growth	Nworgu and Egbunike, 2013
Lablab purpureus	Leaf meal	10-15 per cent	—	(–) Growth of layers but deep yolk colour, strong shells and high Haugh unit	Odunsi, 2003
Medicago sativa (Lucerne)	Lucerne meal	14 per cent	Basal diet	(+) Decreased cholesterol in their eggs, increased vitamin A and carotene content	Rybina and Reshetova, 1981; Karsten *et al.*, 2010; Gorski, 2000;. Lopez-Bote *et al.;* 1998
Mimosa invisa	Leaf meal	2.0 per cent	Basal diet	(=) Growth of broiler chicken	Nworgu & Fapohunda, 2002; Nworgu and Egbunike, 2013
Pueraria phaseoloides	Leaf meal	3.0 per cent	Basal diet	(=) Growth of broiler chicken	Nworgu & Egbunike, 2013
Stylosanthes scabra, S. guianensis and S. seabrana	Leaf meals	3.0-9.0 per cent	Maize Soya diets in isocaloric and isonitrogenous rations	(=) Growth of broiler chicken	Krishna *et al.*, 2008
Trifolium alexandrium (berseem)	Fresh berseem	Choice feeding	Basal diet	(+) Turkey	Bhattacharya *et al.*, 2008

Forage Plant Species	*Form*	*Acceptable Level in the Total Diet*	*Substituted Mainly to*	*Effects Growth/ Maintenance/ Other Effects on Poultry/ Duck*	*Reference*
T. alexandrium	Sun dried	6 per cent	Basal diet	(+) Broiler chicken	Muhamed Arif *et al.*, 2000
Vigna unguiculata (cowpea)	Leaf meal	10-15 per cent	Basal diet	(+) Improved egg quality (deep yolk colour, strong shells and high Haugh unit)	Hendricksen & Minson, 1985
Forage Grasses					
Grass meal	Grass meal	14 per cent	Basal diet	(+) Decreased cholesterol in their eggs, increased vitamin A and carotene content	Davtyan & Manukyan, 1987; Rybina & Reshetova, 1981; Karsten *et al.*, 2010, Gorski, 2000;. Lopez-Bote *et al.*; 1998
Forage Forbs/ Others					
Cnidoscolus aconitifolius (*Mill.*) *Johnston* (Chaya)	Leaf meal	25 per cent in the diet	Basal diet	(+) Broilers, high carcass dressing, Increased red blood cells, and decreased serum cholesterol	Armstrong *et al.*, 1999
C. aconitifolius	Leaf meal	(a) 15 per cent (b) 25 per cent mixed with maize	(a) Maize-soyabean diet (b) Maize only	(+) Growth	Sarmiento-Franco *et al.*, 2002, Sarmiento-Franco *et al.*, 2003
Ipomoea batatas (Sweet potato)	Dried leaf	20-30 per cent	Replacing 300 or 200 g/ kg maize	(–) Broilers	Teguia *et al.*, 1997, Inthapanya & Preston, 2009
Manihot esculenta (Cassava)	Dried leaf	50 per cent	Basal diet	(=) Broiler chicken	Eruvbetine *et al.*, 2003
M. esculenta	Leaf meal	(1) 30 per cent Soybean protein (2)50:50 leaf meal of cassava and leucaena	14 per cent Soybean meal based ration	(=) Growth of broiler chickens	Onibi *et al.*, 2008

Contd...

Forage Plant Species	*Form*	*Acceptable Level in the Total Diet*	*Substituted Mainly to*	*Effects Growth/ Maintenance/ Other Effects on Poultry/ Duck*	*Reference*
M. esculenta	Leaf meal	15 per cent	Coconut oil meal	(+) Growth of broiler chick	Ravindran *et al.*, 1986
M. esculenta	Leaf meal	5-20 per cent	Basal diet	(-) Growth	Ross & Enriquez, 1969
Tithonia diversifolia Hemsl. A. Gray (wild sunflower)	Leaf meal	10 per cent to substitute maize	Commercial broiler mash as a base line diet	(=) Growth and body weight without any adverse effect	Togun *et al.*,2006
Vernonia amygdalina (bitter leaf meal)	Air dried leaf meal	10 per cent	Basal diet	(+) Growth of broiler chicken	Durunna *et al.*, 2011
Tree Fodder (Legume and Non-legume)					
Albizia jullibrissin	Leaf meal	2 per cent of total diet	Maize and soybean based diet	(+) Yolk colour of Laying hens, cost effective	Dei *et al.*, 2012
A. jullibrissin	Leaf meal	5 per cent of total diet	Maize and soybean based broiler diet	(+) Growth of broiler chicken	Dei *et al.*, 2012
Calliandra calothyrsus (Calliandra/ Red calliandra)	Leaf meal	<5 per cent	Basal diet	(+) layers, yolks colour increased within 3 days	Paterson *et al.*, 2000
Gliricidia sepium	Leaf meals	—	White maize-based layers' diets	(+) Give desirable yolk colour	Udedible & Igwe, 1989
G. sepium	Hay	4.5 per cent	White maize-based layers' diets	(+) Increased egg number with high egg weight and better yellow colour of the yolk	Odunsi 2002; FAO, 1994
G. sepium	Leaf meals	—	White maize-based layers' diets	(+) Give desirable yolk colour	Udedible & Igwe, 1989
G. sepium	Hay	4.5 per cent	Basal diet	(+) increased egg number with high egg weight and better yellow colour of the yolk	Odunsi 2002; FAO, 1994

Forage Plant Species	*Form*	*Acceptable Level in the Total Diet*	*Substituted Mainly to*	*Effects Growth/ Maintenance/ Other Effects on Poultry/ Duck*	*Reference*
Leucaena leucocephala	Leaf meals	—	White maize-based layers' diets	(+) Yolk colour	Udedibie & Igwe, 1989
L. leucocephala	Leaf meals	—	Basal diet	(+) Pigmentation of egg yolks and broiler carcasses	D'Mello & Acamovic, 1989
L. leucocephala	Leaf meals	21 per cent	Groundnut cake meal in the diets of growing chicken	(=) Equally good substitute for broiler chicken	Ayssiwede *et al.*, 2011b
L. leucocephala	Leaf meal	5-10 per cent	Basal diet	(=) Satisfactory laying performance	Vohra *et al.*, 1972
L. leucocephala	Leaf meal	5 per cent CP of fish meal	12 per cent CP from fish meal in the basal diet	(=) Equally good substitute for fish meal for broiler chicken	Khan *et al.*, 2009
L. leucocephala	Leaf meal	10 per cent	Basal diet	(+) Quadratic effect on the egg laying rate, egg colour and mass (g/hen/day) and feed conversion	Abou-Elezz *et al.*, 2011
L. leucocephala	Leaf meal	50 per cent	Soybean meal	(+) Broilers chicken weight	Eichiel *et al.*, 2015
L. leucocephala	Leaf meal	5-15 per cent	Basal starter and finisher diet	(-) Broilers chicken weight	Zanu *et al.*, 2012
Moringa oleifera (Lam.)	Leaf meal	24 per cent	Basal diets	(=) Broiler chicken body weight and feed conversion ratio,	Ayssiwede *et al.*, 2011a
M. oleifera	Leaf meal	8-16 per cent.	Basal concentrate diet	(=) Indigenous chicken	Bello *et al.*, 2010
M. oleifera (Lam.)	Leaf meal	5 per cent	Cassava based diets	(+) Broiler chicken	Onwudike, 1995; Olugbeni *et al.*, 2010
M. oleifera (Lam.)	Leaves	1-2 per cent of the basal feed	Basal feed	(+) Broiler chicken	Teteh *et al.*, 2013

Contd...

Forage Plant Species	*Form*	*Acceptable Level in the Total Diet*	*Substituted Mainly to*	*Effects Growth/ Maintenance/ Other Effects on Poultry/ Duck*	*Reference*
M. oleifera (Lam.)	Leaf meal	0.3 per cent of basal diet	Corn-soybean meal diet	(+) Broiler weight and enhanced the ability to resist heat stress	Hassan *et al.*, 2016
M. oleifera (Lam.)	Leaf meal	10 per cent	Basal diet	(+) Egg yolk colour	Abou-Elezz *et al.*, 2011
Morus indica	Shade dried	6 per cent	Basal diet	(+) Egg production and yolk colour in layers	Narayana & Shetty, 1977
Sesbania aegyptica	Leaf meal	5 per cent CP of fish meal	12 per cent CP from fish meal in the basal diet	(=) Equally good substitute for fish meal in broiler chicken	Khan *et al.*, 2009
Sesbania grandiflora	Leaf meal	5-10 per cent	Basal diet	(-) Reduced feed intakes and weight of broiler	Ash & Peteia, 1992
Aquatic Forage					
Alternanthera sessilis	Leaf meal	5 per cent	Soybean meal and wheat bran based diet.	(+) Weight of broiler	Albert Nii Armah Tagoe, 2012
Azolla pinnata R. Brown	*Azolla* meal	10 per cent	Soybean meal and maize bran equalized crude protein content of chick diets	(+) Growth of pullet chicks	Alalade & Iyayi, 2006
A. pinnata	*Azolla* powder	5-15 per cent	Maize -soybean based diet	(+) Improved daily weight gain and feed conversion ratio in broiler chicks.	Hozhabr Naghshi *et al.*, 2014
A. pinnata	Azolla meal	5 per cent	Sesame meal by Azolla meal.	(+) Growth broiler chicken	Basak, *et al.*, 2009
A. pinnata	Sundried azolla meal	4.5 per cent	Standard broiler ration	(+) Growth broiler chicken	Balaji *et al.*, 2009
Azolla	Azolla meal/ sundried	5 per cent	Soybean-maize Basal diet	(+) Final weight of broiler chicks	Dhumal *et al.*, 2009; Shoukat *et al.*, 2015

Forage Plant Species	*Form*	*Acceptable Level in the Total Diet*	*Substituted Mainly to*	*Effects Growth/ Maintenance/ Other Effects on Poultry/ Duck*	*Reference*
A. pinnata	Dried raw azolla,	200 g per chick per day	In addition to Commercial feed (basal diet)	(+) Final body weight and egg production but feed intake reduced in Nicobari foul	Sujatha *et al.*, 2013
A. pinnata	Dried	5 per cent	Isonitrogenous and isocaloric starter-cum-grower rations	(=) Growth Japanese quails (*Coturnix coturnix japonica*)	Shamna *et al.*, 2013
Azolla species	fresh azolla	15-45 per cent	Basal diets isonitrogenous, and isocaloric	(+) 15 per cent for growth and 45 per cent for economic efficiency of Fayoumi growing chicks	Namara *et al.*, 2010
Azolla species	Dried	10 per cent	Basal diets	(+) Weight of broiler and increased the number of eggs.	Mahadevappa *et al.*, 2012
Eichhornia crassipes (water hyacinth)	Fresh	*adlibitum*	With concentrate feed	(+) weight of duck and increased the number of eggs	Lu *et al.*, 2008
Ipomoea aquatica (Water spinach) *Lemna minor* (Duckweed)	Mixed with basal diet	—	Basal diet	(+) Growth improve color of skin and egg yolk of chickens and duck poultry, quails	Toung *et al.*, 1994;, Bui Xuan Men *et al.*, 1995 and 1996, Giang & Preston, 1997, Hang, 1998,
Salvinia species	Dried	10 per cent	Basal diets	(+) Weight of broiler and increased the number of eggs.	Mahadevappa *et al.*, 2012
Salvinia molesta	Dried meal.	18 per cent	Basal diet	(+) Produced better muscle protein mass and higher body weight gain	Binti Ma'rifah *et al.*, 2013

(+) positive effects; (–) negative effects; (+/–) both positive and negative effects; (=) equal effects compared to control.

Source: Siriwan *et al.*, 2012 and others as in reference.

per cent (Armstrong *et al.*, 1999; Durunna *et al.*, 2011; Eruvbetine *et al.*, 2003; Onibi *et al.*, 2008; Ravindran *et al.*, 1986; Ross & Enriquez, 1969; Sarmiento-Franco *et al.*, 2002; Sarmiento-Franco *et al.*, 2003; Teguia *et al.*, 1997; Togun *et al.*, 2006), tree fodder 2-50 per cent (Abou-Elezz *et al.*, 2011; Ash & Peteia, 1992; Ayssiwede *et al.*, 2011a; Ayssiwede *et al.*, 2011b; Bello *et al.*, 2010; Dei *et al.*, 2012; D'Mello & Acamovic, 1989; Eichiel *et al.*, 2015; FAO, 1994; Hassan *et al.*, 2016; Khan *et al.*, 2009; Narayana & Shetty 1977; Odunsi 2002; Olugbeni *et al.*, 2010; Onwudike, 1995; Paterson *et al.*, 2000; Teteh *et al.*, 2013; Udedible & Igwe, 1989; Vohra *et al.*, 1972; Zanu *et al.*, 2012) and aquatic plant 05- 45 per cent (Alalade & Iyayi, 2006; Albert Nii Armah Tagoe, 2012; Amornsak *et al.*, 2004; Balaji *et al.*, 2009; Basak, *et al.*, 2009; Becerra *et al*, 1995; Binti Ma'rifah *et al.*, 2013; Bui Xuan Men *et al.*, 1995, 1996, Dhumal *et al.*, 2009; Hang, 1998; Giang & Preston, 1997; Hozhabr Naghshi *et al.*, 2014; Lu *et al.*, 2008; Mahadevappa *et al* ., 2012; Namara *et al.*, 2010; Shamna *et al.*, 2013; Shoukat *et al.*, 2015; Sujatha *et al.*, 2013; Toung *et al.*, 1994). Leaf meal of cassava induces greater reductions in growth rate of chickens than dehydrated alfalfa meal at all inclusion levels from 5 to 20 per cent in the basal diet (Ross & Enriquez, 1969). However, if substitutions of cassava leaf meal are accomplished at the expense of poor quality ingredients such as coconut oil meal in the control diet, then dietary inclusion rates of cassava leaf meal of upto 15 per cent may be incorporated without reducing growth of broiler chicks (Ravindran *et al.*, 1986). In most of the cases it was cost effective to partly substitute the basal concentrated diet without adversely affecting the performance. In leaf meal/dry forage of three forage plant species out of 30 forage plant species (*Lablab purpureus, Ipomoea batatas and Sesbania grandiflora*), the performance were reported neagative. In the case of *Sesbania grandiflora*, there may be signs of acute toxicity with inclusion rates as low as 100g/kg diet, with 100 per cent mortality in chickens fed leaf meal at 300g/kg diet (Brown *et al.*, 1987). In case of *Leucaena leucocephala,* it increased or gave equal performance in seven research reports on the leguminous fodder tree but the reseach report also found negative performance as compared to basal concentrate diets (Ayssiwede *et al.*, 2011b; Zanu *et al.*, 2012). The depressed performance of birds fed leaf meal based diets was reported due to adverse effect of amino acid imbalance in the *Leucaena leucocephala* based diets (Agbade & Olator, 2003). This may also be due to high fibre in the diet of leaf meal based diets and anti nutritional factors (Nwokolo *et al.*, 2000). Performance is measured in terms of cost effectiveness (Bello *et al.*, 2010; Ho Thanh Tham 2015; Krishna *et al.*, 2008;, Namra *et al.*, 2010; Nworgu *et al.*, 2007; Nworgu & Fasogbon, 2015; Onibi *et al.*, 2008; Shamna *et al.*, 2013;Shoukat *et al.*, 2015; Sujatha *et al.*, 2013), increase feed conversion efficiency (Abou-Elezz *et al.*, 2011; Basak *et al.*, 2009; Eruvbetine *et al.*, 2003; Hozhabr Naghshi *et al.*, 2014; Nworgu *et al.*, 2007; Nworgu & Fasogbon, 2015; Sujatha *et al.*, 2013), final live weight and quality of meat (Alalade & Iyayi, 2006; Albert Nii Armah Tagoe, 2012; Balaji *et al.*, 2009; Basak *et al.*, 2009; Dhumal *et al.*, 2009; D'Mello & Acamovic, 1989, Durunna *et al.*, 2011, Gorski, 2000; Hassan *et al.*, 2016; Hozhabr Naghshi *et al.*, 2014; Khan *et al.*, 2009; Mahadevappa *et al.*, 2012; Olugbeni *et al.*, 2010; Ponte *et al.*, 2008a; Ponte *et al.*, 2008b; Shoukat *et al.*, 2015;

Sun *et al.*, 2012a; Sun *et al.*, 2012b), increased layers performance and egg quality *viz.* quality increased vitamin A, omega-3 content and decreased cholesterol in their eggs (Abou-Elezz *et al.*, 2011; Amornsak *et al.*, 2004; Armstrong *et al.*, 1999; Becerra *et al*, 1995; Bui Xuan Men *et al.*, 1995, 1996, D'Mello & Acamovic, 1989; FAO, 1994; Giang & Preston, 1997; Gorski, 200;, Hang, 1998; Ho Thanh Tham 2015; Ige *et al.*, 2006; Karsten *et al.*, 2010; Long & Alterman, 2007; Lopez-Bote *et al.*, 1998; Mahadevappa *et al.*, 2012; Narayana & Shetty, 1977; Onibi *et al.*, 2008; Odunsi *et al.*, 2002; Odunsi, 2003; Paterson *et al.*, 2000; Rybina & Reshetova, 1981; Udedible & Igwe, 1989), increased fertility (Davtyan & Manukyan, 1987). Incresed yellow pigmentation (due to -carotene, precursor of vitamin A) was found in egg (Dei *et al.*, 2012) and meat (chickens are able to incorporate dietary components into their muscles) of poultry (Onibi *et al.*, 1999; Agbede & Aletor, 2003)

Pig

Pigs are monogastric and omnivores fed on plant and non-vegetative feed resources. Forage are fed as fresh, wilted, chopped and dried forage, leaf meal, and silage or grazed on the pasture. Most of the works on pasture and forage for feeding pig has been done in the western countries. It is yet to take root in developing Asian countries, where limited research and recommendations are available. Forages selected for pigs need to be succulent, short and sweet, capable of high biomass production, highly palatable, the pasture containing a high percentage of legumes (high in protein and vitamins), which has sustained production over a long period during a year. In western countries, forage resources used for pig feeding were alfalfa, red clover, white clover, lespedeza, soybean, orchard grass, rye grass, timothy, tall fescue, brome grass, barley, brassica rape, sorghum and Sudan grass. Orchard grass has ability to withstand trampling by pigs. Pig preferred white clover and alfalfa (*Medicago sativa*), and rooting and eating white clover (*Trifolium repens*) compared with buffalo grass (*Buchloe dactyloides*) or tall fescue (*Festuca arundinacea*) (Rachuonyo *et al.*, 2005). Rates of ground cover loss were less for less preferred tall fescue and buffalo grass than for the more preferred forage alfalfa and white clover. Brassica rape seed crop can be utilized for pasturing pigs provided overgrazing is avoided (Le, 1999). It contains 28 per cent crude protein and able to provide abundant, palatable forage. Less preferred forages should be promoted as pasture for pig when the primary objective is ground cover maintenance rather than nutrient supply. In fact, forage legumes are preferred over forage grasses due to its higher crude protein contental though both the forage group has similar energy contents. Therefore, most of the research on the use of forage in the diets of pig has been with alfalfa (Kenneth, 1998) in western countries. It is a most practical forage crop for the pig because it can be used for both pasture and silage but not more than 65 per cent alfalfa should be used in the pig diets at commercial scale. Annual forage Sudan grass is also highly palatable to pigs. When it is seeded thickly, it yields large quantity of forage during the hottest period in the summer. Sudan grass contains a cyanogen compound at the early stage of growth. However, the non-ruminants are less sensitive to the cyanogen than the ruminants due to

Table 10.2: Effects of Forages Incorporation in Pig Feeding Systems

Forage Plant Species	*Form*	*Acceptable Level in the Total Diet*	*Substituted Mainly to*	*Effects Growth/ Maintenance/ Other Effects on Poultry/ Duck*	*Reference*
Forage Legumes					
Arachis pintoi (Forage peanut)	Dried forage	+ 20 per cent of the diet CP		(+) Growth	Posada *et al.*, 2006
Centrosema pubescens	Fresh forage	——	Substituted as protein source in the basal diet	(+) Growth of growing pigs	Ugwu and Chukwuka, 2001
Desmanthus virgatus	Leaf meal	18 per cent	Basal diet	(–) Growth	Ly *et al.*, 2001
Medicago sativa (alfalfa)	Grazing	—	—	(+) Growth of growing pigs	Rachuonyo *et al.*, 2005
Stylosanthes guianensis	Fresh forage		Adlibitum	(+) Growth	Phengsavanh & Stür, 2006
Trifolium repens (white clover)	grazing	—	—	(+) Growth of growing pigs	Rachuonyo *et al.*, 2005
Canavalia brasiliensis, Centrosema brasilianum, Clitoria ternatea, Cratylia argentea, Desmodium velutinum, Fleminigia macrophylla, Lablab purpureus, Stylosanthes guianensis Vigna unguiculata (Cowpea)	Leaf meal	30 per cent	Basal diet	(+) Growth	Sarria *et al.*, 2010, Heinritz *et al.* (2012)
Forage Grasses					
Buchloe dactyloides (buffalo grass)	Grazing	—	—	(+) Growth of growing pigs	Rachuonyo *et al.*, 2005
Festuca arundinacea (tall fescue)	Grazing	—	—	(+) Growth of growing pigs	Rachuonyo *et al.*, 2005

Forage Plant Species	*Form*	*Acceptable Level in the Total Diet*	*Substituted Mainly to*	*Effects Growth/ Maintenance/ Other Effects on Poultry/ Duck*	*Reference*
Panicum maximum	Young succulent	—	Fresh forage for feeding pigs	—	Obua *et al.*, 2012
Forage Forbs/Others					
Chenopodium quinoa	Lef meal	5 per cent	Basal diet	(+) Growth	Diaz *et al.*, 1995
Colocasia esculenta	—	—	As per availability	(+) Growth	Rameswar Deka *et al.*, 2007
Colocasia esculenta	Colocasia ensiled with water spinach	—	Basal diet	Digestibility of CP (>80 per cent) and the N retention (70.5 per cent). No adverse effect on reproduction	Giang & Preston, 2011 and Malavanh *et al.*, 2008
C. esculenta)	Foliage mixed with banana stem	——	Rice bran based diet	High palatibility dietary nutritional value	Chhay *et al.*, 2014
Hibiscus rosasinens	Fresh leaves	—	—	(+) high digestibility	Ly *et al.*, 2001
Ipomoea batatas (sweet potato)	Wilted leaves	60 g per kg	Fish meal/ groundnut cake as the protein source	(=) Growth	An, 2004; Mora *et al.*, 1992
I. batata	ensiled sweet potato leaves with lysine	100 per cent of CP	Fish meal/ groundnut cake as the protein source	(+) Growth	An *et al.*, 2005
I. batata	Fresh foliage	*ad libitum*	The animals received 1.6 kg/day of a feed mix	(=) Growth	González *et al.*, 2003
I. batata	Chopped leaves	02 kg supplement	Commercial diet	(=) Growth	Gabino (2012)
Manihot esculenta	Fresh leaf	50 per cent	Basal diet	(+) Growth	Ty *et al.*, 2011
M. esculenta	Leaf meal	26.7 per cent	Coconut oil meal	(+) Growth	Ravindran *et al.*, 1987
M. esculenta	Ensiled leaves	Replace	sweet potato tops	(+) Growth	Hang, 1998

Contd...

Forage Plant Species	*Form*	*Acceptable Level in the Total Diet*	*Substituted Mainly to*	*Effects Growth/ Maintenance/ Other Effects on Poultry/ Duck*	*Reference*
Musa paradiosica (banana)	Leaves, pseudo-stem	—	—	—	Rameswar Deka *et al.*, 2007
M. paradiosica	Leaf meal	15 per cent	Basal diet	(=) Growth	Garcia *et al.*, 1991b
M. paradiosica	Fresh pseudo stem	02 kg supplement	Commercial feed	(=) Growth	Gabino (2012)
Asystasia gangetica, Amaranthus hybridus, Alternanthera bettzickiana, Ipomoea batatas, Manihot esculenta, Telfairia occidentalis, Syndrella nodiflora	Fresh forage	—	Fresh forage for feeding pigs	—	Obua *et al.*, 2012
Tree Fodder (Legume and Non-legume)					
Gliricidia sepium	Fresh leaves	—	—	(+) High digestibility	Ly *et al.*, 2001
Gmelina arborea	Leaf meal.	——	Part of basal concentrate diet	(+) Growth	Nkwocha *et al.*, 2008
Leucaena diversifolia	alternative protein supplement		-		Heinritz *et al.*, 2012
Leucaena leucocephala	Leaf meal.	10 per cent	Basal diet	(+) Growth	Onwudike, 1995).
L. leucocephala	Leaf meal	10 per cent	Commercial ration	(+) Growth	Malynicz, 1974
L. leucocephala	Leaf meal	16 per cent	Basal diet	(=) Growth	Chen *et al.*, 1981
Microdesmis puberula	Sundried leaf meal	—	Basal diet	(=) Growth	Obua & Davidson (2011)
Moringa oleifera	Fresh forage	—	—	(+) High digestibility	Ly *et al.*, 2001
Morus alba	Fresh leaves	21 - 28 per cent of the total DM intake	Basal diet	(+) Growth/high digestibility	Ty *et al.*, 2007; Ly *et al.*, 2001

Forage Plant Species	*Form*	*Acceptable Level in the Total Diet*	*Substituted Mainly to*	*Effects Growth/ Maintenance/ Other Effects on Poultry/ Duck*	*Reference*
Morus alba	Leaf	15 per cent	Commercial concentrate	(+) Growth	FAO, 2002
Trichantera gigantea	Leaf meal	20-25 per cent of diet CP	Basal diet	(-)Growth	Sarria *et al.*, 1991, Le, 1999
Aquatic Forage					
Azolla pinnata	Dried	20 per cent	Conventional concentrate ration	(+) Growth	Cherry *et al.*, 2014
Azolla filiculoides	Dried	15 per cent	Basal diet	(+) Growth	Letermeet *al.*, 2010, FAO, 1994
Eichhornia crassipes (water hyacinth)	Water hyacinth supplemented diet	—	Farmers rice based and common diet for pig	(=) final body weight but back fat thickness was minimum	Men *et al.*, 2006; Rameswar Deka *et al.*, 2007
E. crassipes	Water hyacinth meal	6 per cent	Concentrates of pig diets	(=) final body weight	Son & Trung, 2002, Ho Thanh Tham, 2015
Ipomoea aquatica	Fresh whole plant	15 per cent	Basal diet	(+) Growth	Nguyen & Preston, 2011
I. aquatica	Fresh forage with added 0.5 per cent DL-methionine.	*Adlibitum*	Basal diet of broken rice	(+) Growth	Le *et al.*, 2002
Lemna minor (duckweed)	Fresh duckweed	5 per cent of dietary dry matter as CP supplement	Ensiled cassava root, brewery by-product (brewers' grains), rice bran and sweet potato vines	(+) Growth 35 per cent gain than control diet	Hang, 1998, Le, *et al.*, 1997
Salvinia molesta		15 per cent	Basal diet	(+) Growth	Leterme *et al.*, 2010

(+) positive effects; (–) negative effects; (+/–) both positive and negative effects; (=) equal effects compared to control.

Source: Siriwan *et al.*, 2012 and others as in reference.

the near-neutral pH in the rumen. When screening potential of tropical forages for the nutrition of non-ruminants livestock, low ADF concentration should also be taken into the criteria of selection as it is correlated to enzymatic digestibility (Heinritz *et al.*, 2012). Among various 49 tropical forage plant species, 14 forage legumes, *Arachis pintoi, Canavalia brasiliensis, Centrosema brasilianum, Centrosema pubescens, Clitoria ternatea, Cratylia argentea, Desmanthus virgatus, Desmodium velutinum, Fleminigia macrophylla, Lablab purpureus, Medicago sativa, Stylosanthes guianensis, Trifolium alexandrium* and *Vigna unguiculata;* three grasses, *Buchloe dactyloides, Festuca arundinacea* and *Panicum maximum*; nine leguminous forage tree species, *Acacia auriculiformis, Gliricidia sepium, Gmelina arborea, Leucaena diversifolia, Leucaena leucocephala, Microdesmis puberula, Morus alba, Moringa oleifera* (Lam.) and *Trichantera gigantea*; five non-leguminous forage tree species *Artocarpus heterophyllus, Borassus flabellifer, Cocos nucifera, Eucalyptus* spp., *Hibiscus rosasinensis,* 12 forage forb plant species, *Amaranthus hybridus, Asystasia gangetica, Amaranthus hybridus, Alternanthera bettzickiana, Chenopodium quinoa, Colocasia esculenta, Hibiscus rosasinens, Ipomoea batatas, Manihot esculenta, Musa paradiosica, Telfairia occidentalis* and *Syndrella nodifloraand* six aquatic forage plant species, *Azolla filiculoides, Azolla pinnata, R. Brown, Eichhornia crassipes, Ipomoea aquatica, Lemna minor* and *Salvania molesta* are being used in feeding pig as per survey of the literature (Table 10.2). Among 49 forage plant species studied, feeding of 32 forage plant species had enhanced growth and reproductive performance or was equal to concentrate based diet, as a part substituted in the total basal diet but feeding of *Desmanthus virgatus* and *Trichantera gigantea* had negative effects on the performance. The highest *in vitro* N digestibility coefficients were found in *Cratylia argentea, Vigna unguiculata, Moringa oleifera* (79.2 per cent), *Hibiscus rosasinensis* (74.2 per cent), *Gliricidia sepium* (69.4 per cent) and *Morus alba* (47.9 per cent), and the lowest value of 9.4 per cent was recorded in *Cocos nucifera* and 5.1 per cent in *Borassus flabellife*r (Ly *et al.*, 2001). 15 per cent foliage of mulberry substituted in the basal commercial concentrate diet of growing pigs was better utilized than inclusion of vine of sweet potato (FAO, 2002; Ty *et al.,* 2007). In Vietnam, fresh or cooked water hyacinth was found to replace up to 6 per cent of the concentrates in pig diets (Ho Thanh Tham, 2015; Son & Trung, 200; Heinritz *et al.*, 2012). High production potential of cassava crop (fresh roots 24.0-36.0 tonnes/ha and foliage of 4.3-6.0 tonnes/ha) can be used as a potential feed for pig with only limitation of presence of cyanogenic glycosides as linamarin and lotaustralin. Le (1999) reported that sweet potato foliage has higher crude protein (15-20 per cent) than the roots (4-7 per cent) but deficient in total sulphur containing amino acids and lysine in terms of ideal protein for pigs. Fresh sweet potato vines can be included up to 27 per cent of dry matter and 40 per cent of the total dietary protein for feeding of growing/finishing pigs. The vines are high in protein content compared to other protein-rich forages (An, 2004; An, 2005; Mora *et al.,* 1992; González *et al.,* 2003). González *et al.* (2003) reported that it is possible to obtain good performance traits in pigs fed fresh sweet potato foliage *adlibitum*, provided a feed supplement is given for protein to meet the different requirement of protein during the growing and finishing periods, respectively. The leaves of cassava and sweet potato are used fresh, dried or as silage, and they replace fish meal and groundnut cake as a protein source for growing pigs under small farm conditions in central Vietnam. In many tropical countries, leaves of banana, *Trichantera gigantea* called Nacedero (used in Vietnam), leaves and roots of cassava, forage *Brassica*, aquatic forage plant species *viz.*, water spinach (*Ipomea aquatica*), duckweed (*Lemna minor*) and azolla can be used for pig feeding (Hang, 1998; Le, 1999; Le, *et al.*, 1997; Ly *et al.*, 2002). Gabino (2012) recommended two kilogram chopped banana trunk or two kilogram of chopped sweet potato leaves and the vines along with commercial feed under backyard scale piggery to utilize the byproducts as valued pig feed. Among aquatic forage, tropical and warm aquatic fern (*A. filiculoides, A. pinnata*) contains *Anabaena azollae* within cavities in their leaves, capable of fixing nitrogen and highly productive to yield nine tonnes of protein per hectare per year in impounded condition. Azolla has been used as a feed for pigs and ducks in Southeast Asia; poultry in Vietnam and for pigs in Singapore and Taiwan (FAO, 1994). It is described as an excellent substitute and may replace up to 50 per cent of the rice bran used as feed for pigs in Vietnam. Being very low in dry matter, it contains 24 per cent crude protein in dry matter and the amino acid composition compared well with reference protein sources. It has excellent amino acid balance even better than soybean meal. *Azolla filiculoides* and another fern *Salvinia molesta* as feed ingredients for pig revealed that *Azolla filiculoides* was more palatable but the inclusion level in rations should be limited to 150 g per kilogram of the total diet considering its low digestibility, energy density and the negative impact on the digestibility of the whole diet (Leterme *et al.*, 2010). Azolla can be fed to pigs as green fresh or ensiled before feeding. *Azolla* can replace 10 per cent and 20 per cent of the protein (from soybean) in diets of growing and finishing pigs, respectively and it is readily consumed by pig especially during pregnancy. Water hyacinth reduced organic matter digestibility and digestible energy content without affecting feed intake. Men *et al.* (2006) found that it can supplement diet of finisher pig with high return and less back fat meat. *Eichhornia crassipes* is available all over India round the year and it contains moderate crude protein, high nitrogen free extract and total ash. Water spinach leaves can constitute 47 per cent of pig diets and has better nutrients digestibility and N balance than cassava leaves (Le, 1999). Fresh duckweed is highly palatable and pig consumed more dry matter of fresh duckweed and it had stimulating effect on body weight gain (Hang, 1998). Pig fed fresh duckweed grew 35 per cent faster than conventional diet.

In India, succulent forage *viz.* maize and pearl millet feeding to younger and finisher pigs have been found beneficial (Yadav *et al.*, 1990 a; Yadav *et al.*, 1990 b; Yadav & Gupta, 1994) Pigs also chew succulent guinea grass (grown up to 50 cm height). After chewing and ingesting juices from the forage, the residue fibre is extruded from the mouth by the pig (Singh *et al.*, 1995). Rameswar Deka *et al.* (2007) found common feed resources are Colocasia (*Colocasia esculenta*), locally known as Kachu, and some vegetables including pumpkin, gourd, papaya banana and water hyacinth in in the Northeast region. Mahadevappa *et al.* (2012) reported that feeding *azolla* and *salvina* as part of the basal diet of pig is limited due to their low digestible energy.

Antinutrtional factor *viz.* saponins contents in various forage plant species *viz., Brachiaria decumbens, B. brizantha, Amaranthus hypochondriacus, Chenopodium quinoa, Atriplex hortensis,* and *Medicago sativa* showed conflicting results in pigs (Hauptli & Lovatto, 2006). It had positive effects on both sows and piglets (Hauptli & Lovatto, 2006), compared with fewer stillborn piglets but the growth performance was significantly reduced (Ilsley & Miller, 2005). Phytate is another antinutritinal factor affecting major pool of phosphates in plants, resulting in low availability of the bound phosphorus to nonruminants. The intestinal apparent digestibility of phytate in pigs varies widely, between zero and 25 per cent (Rubio *et al.,* 2006). Moreover, phytate significantly reduce the availability of minerals such as zinc, calcium, and magnesium (Rimbach *et al.,* 2008).

Studies on feeding forage to pigs had shown that after an adaptation period of at least two months, the digestibility of the fibre component improves in the large intestine and the pigs absorb more nutrients from the forages (Howell & John, 1993, Kenneth, 2010). Economy of pigs on the pastures and forage varies considerably. It depended on the type of pasture, age of pigs, and management systems. Forages in the form of pasture, or as part of a complete feed, can be successfully used in pig production if it is fed to a limited extent mainly to young pigs and lactating sows. Pasture also help in better isolation and disease control due to improved sanitation and air quality, decreases waste management problems and cannibalism. The pastures can be recommended for the breeding herd especially to provide exercise and nutrients needed for boars and sows (Howell & John, 1993; Kenneth, 1998; Kenneth, 2010). Pigs on the pasture get more exercise than those housed in pens or enclosed facilities, hence, have a higher energy requirement. Limitations of feeding on forage and the pasture are, the pig must consume more material to get the same amount of nutrients found in grain or in complete feed as forages are generally having low dry matter than the sole feed and also not available round the year, hence, the feeding program has to be modified from one season to another season, more labourers are required for handling, feeding, watering and farrowing, greater problems of internal parasites and the forage and pasture based feeding takes slightly longer time to attain marketable body weight. Generally, graminaceous pastures can take considerably longer time for growing pig and they often become too fat for markets. Hence, use of the pasture containing higher proportion of forage legumes are recommended which can maintain high level reproductive capacity of boars, and in many cases increase the litter size as compared to intensively raised sows. Use of good pasture containing alfalfa, ladino clover, and grass also help to maintain high level of reproductive capacity of boars as well as in many cases increased litter size as compared to confinement rearing of pigs (Howell & John, 1993). However, it requires better environmental management skills during extreme weather.

Processed feed can be well utilized under different pig feeding system. Among processed forage, when forage is incorporated into a complete feed, the energy density of the feed is decreased because of the low concentrations of metabolizable energy in the forages. Silage made from maize or high quality alfalfa-grass silage, is a practical way of using more forage in the piggery. Leaf meal of *Leucaena leucocephala* treated with either acetic acid or NaOH raised the rate of nitrogen retention in pigs (Echeverria *et al.*, 2002). Process of ensiling also helps to reduce antinutritional factors. Ensiling the succulent shoots of *Leucaena leucocephala* reduced mimosine content from 07 per cent to 02 per cent (Liu and Wang, 1990) as well as the cassava leaves lost 77 per cent of their hydrogen cyanide, in addition digestibility increased in growing pigs (Borin *et al.*, 2005). Chhaya *et al.*, 2014 found that the ensiled mixture of taro (*Colocasia esculenta*) foliage and banana stem was more palatable than rice bran and dietary nutritional value (dry matter intake, apparent digestibility of dry matter, crude protein and nitrogen retention) increased when the ensiled mixture of taro and banana stem replaced rice bran. Manh *et al.* (2002) also found that water hyacianth at a 15 per cent DM content was successfully preserved by adding 4.5 per cent of molasses on a fresh matter basis. Due to its low cost, ensiled water hyacianth could be economically acceptable by small holder pig farms. The digestibility of dry matter, organic matter (OM) and CP of ensiled sweet potato leaves was found high in growing pigs, but that of crude fibre and lysine contents were low. Without addition of lysine to the sweet potato leaves diet, daily live-weight gain was found to be low (An, 2004; An, 2005; Mora *et al.*, 1992; González*et al.*, 2003). Pig fed diets containing *Leucaena* leaf meal up to 10 per cent of the basal diet gained significantly faster body weight than the control diets but further increase in leaf meal to basal diet had negative effects on live weight gain (Malynicz, 1974). Chen *et al.* (1981) found not any differences in the feed conversion ratios between the control diet and diets containing up to 16 per cent leaf meal in the diet. Sun-drying was found more effective in reducing hydrogen cyanide in cassava leaves (Bui Huy *et al.*, 2000). Leaf meal of cassava has been accorded with a favourable metabolizable energy value for graded inclusion of the leaf meal as a replacement for soyabean meal and maize in pig diets. Similarly, Ravindran *et al.* (1987) showed that performance of growing pigs was unaffected by cassava leaf meal inclusion up to 26.7 per cent of total diet if the leaf meal was substituted for coconut oil meal. Obua & Davidson, 2011 found sundried *Microdesmis puberula* leaf meal fed to pig beneficial. The leaf meal of *Desmanthus virgatus* has low nutritive value as revealed by low organic matter, NDF and N digestibility (Ly & Samkol (2001). The cowpea (*Vigna unguiculata*) leaf meal was well accepted by the pigs up to 30 per cent of the total diet (Sarria *et al.*, 2010). The inclusion of leaf meal of cowpea in the total diet increased dry matter intake by 08 per cent without affecting the apparent digestibility of dry matter and gross energy in the diet. Inclusion of cassava leaf meal up to 30 per cent level in a maize/ soybean meal diet for growing pigs resulted in high feed efficiency and gains in the body weight (Le, 1999). Cyanogenic glycosides were found to get reduced by chopping, crushing followed by sun drying and ensiling (Le, 1999). Banana leaf meal was found to supply up to 50 per cent of the ration for lactating sows without any significant difference in litter size with some body weight loss and could also replace up to 15 per cent of total dietary dry matter for growing pigs with satisfactory feed conversion (Garcia *et al.,* 1991). Weaned pigs tend to prefer pellets as compared to leaf meal (Laitat *et al.,* 2000, 2004). Further, silage of taro (*Colocasia esculenta*) mixed with water spinach is a good source of protein for feeding to growing pigs as the apparent digestibility of crude protein (>80 per cent) and the N retention as percentage of N digested (70.5 per cent) were high in the mixed diets and no adverse effect was noticed on reproductive performance measured as numbers of piglets born alive and weaned in Cambodia (Giang & Preston, 2011; Malavanh *et al.,* 2008). The interval from weaning to estrus was also found satisfactory. Azolla had beneficial effects on the carcass characteristics of crossbred Large White Yorkshire pigs (Cherry *et al.,* 2014). Inclusion of sun dried azolla upto 20 per cent in the diets in the diets of pig helped to obtain higher dressing percentage and pork with less fat percentage when compared to that of pigs fed with conventional concentrates. It did not appear to affect the haematological and carcass parameters of pigs.

Rabbit

Conventionally, forages and collected native grasses constitute the major feed resources for the small backyard rabbit production system. Rabbits could utilize forages as pseudo-ruminant unlike other monogastric livestock (Asuquo, 1997). Rabbit can be maintained on green leaves alone without any supplementation of the concentrate, but for the best growth, forage has to be supplemented with concentrate feed. Basically, cellular structure of tropical grasses is such that it resists degradation in digestive tract due to high contents of vascular tissue, parenchyma bundle sheaths and epidermis, poorly digested constituents and low contents of mesophyll cells i.e easily digested. The niche that rabbit production can occupy is in the utilisation of fibrous feeds (Cheeke, 2003). The strategy of using tropical forages as a supplement to a homemade feed can be a potential way to improve the growth performance of rabbit's with economic viability under various rabbit production systems (Bamikole & Ezenwa, 1999; Cheeke, 2003; Onwudike, 1995; Raharjo & Cheeke, 1985, Raharjo *et al.,* 1986; Raharjo, 1987; Tangendijaja *et al.,* 1990; Toung *et al.,* 1994). Tropical forages can supply a major part of the protein and fibre requirements of rabbits but for adequate growth and lactation performance, supplementation with sources of energy is needed. The energy sources may be obtained from roots and tubers (cassava, sweet potatoes), byproducts obtained from fruits (bananas, plantains, mangos, *etc.*), rice bran and other grain processing by-products, sugar cane products (whole cane, derinded cane, molasses, juice) and fats. Supplementation with roots (cassava, sweet potato) and fruit (banana) may allow the animals to meet their water requirements as well as increasing energy intake (Jin *et al.,* 1990). The adequacy of good quality forage centred feeding system in rabbits needs to be established by feeding trials conducted under tropical environments. If forage constitutes major part of the diet, the high palatability of forage needs to be ensured to provide a major part of the daily nutrient intake (Raharjo & Cheeke, 1985; Raharjo, 1987).

Among 96 temperate and tropical forage plants pecies studied for feeding of rabbits, 32 forage legumes *i.e., Arachis glabrata, Arachis hypogea, Arachis pintoi, Cajanus cajan, Calapogonium mucunoides, Cassia rotundifolia, Cassia tora, Centrosema pubescens, Clitoria ternata, Crotalaria juncea, Crotalaria ochroleuca, Desmanthus virgatus, Desmodium distortum, Desmodium heterophyllum, Desmodium scripiums, Hedysarum coronarium, Lablab purpureus, Lupinus albus, Macroptilium lathyroides, Macroptillium atropuropureum, Medicago sativa, Neonotonia wightii, Psophocarpus tetragonolobus, Pueraria phasloloides, Stylosanthes* species, *Stylosenthes guianensis, S. hamata* cv. Verano, *Trifolium species, Trifolium alexandrinum, Trifolium pratense, Vicia faba, Vicia sativa, Vigna umbellata* and *Vigna unguiculata*; 14 leguminous fodder tree fodder plant species, *Acacia albida, A. saligna, Albizia falcata, Bauhinia variegata, Calliandra* species, *Calliandra calothyrsus, Erythrina lithosperma, Gliricidia sepium, Leucaena leucocephala, Moringa oleifera, Robinia pseudoacacia, Sesbania aegyptica, Sesbania formosa* and *Sesbania sesban;* eight non-leguminous tree fodder plant species, Ficus *auriculata, Grevia oppositifolia, Grewia optiva, Morus alba, Musa paradisiaca, Populus tremuloides, Schleichera oleosa* and *Trichantera gigantea;* 18 forage grasses, *Avena sativa, Brachiaria brizantha, Bromus catharticus, Chloris gayana, Cynodon dactylon, Cynodon plectostachyus, Dactylis glomerata, Festuca arundinacea, Hordeum vulgare, Lolium perenne, Panicum maximum, Paspalum plicatulum, Penisetum purpureum, Poa pratene, Setaria splenda, Thysanolaena maxima, Triticum aestivum,* and *Zea mays;* 16 forage forb plant species, *Amaranthus hypochondriachus, Beta vulgaris, Boehmeria nivea, Brassica species, Commelina benghalensis, Helianthus annuus, Hibiscus rosasinensis, Ipomoea batatas, Leunae cornuta, Lotus corniculatus, Luffa egyptiaca Manihot esculenta, Maranta arundinacea, Musa paradiosica, Tridax procumbens* and *Telfaria occidentalis* and eight aquatic forage plant species, *Azolla species i.e. A. filiculoides, A. pinnata, Eichhornia crassipes, Ipomoea aquatica, Lemna minor, Pistia stratiotes, Potamogeton natans L. var. fluitans* and *Ulva fasciata* (Table 10.3). Among 96 forage plant species studied, feeding of 68 forage plant species had enhanced growth and reproductive performance or was equal to concentrate based diet, as a part substituted in the total basal diet. Among remaining 14 forage plant species, six forage plant species *Medicago sativa, V. unguiculata, Amaranthus hypochondriachus,* leaves of *Manihot esculenta, Leucaena leucocephala* and *Eichhornia crassipes* had reported positive effects on growth and performance with few negative effects (Table 10.3). Feeding of eight forage plant species, *Psophocarpus tetragonolobus, Calliandra* species, *Festuca arundinacea, Helianthus annuus, Hibiscus rosasinensis, Bauhinia variegata, Grevia oppositifolia* and *Robinia pseudoacacia* had negative effects on the growth and performance.

Table 10.3: Effects of Forages Incorporation in Rabbit Feeding Systems

Forage Plant Species	*Form*	*Acceptable Level in the Total Diet*	*Substituted Mainly to*	*Effects Growth/ Maintenance/ Other Effects on Poultry/ Duck*	*Reference*
Forage Legumes					
Cajanus cajan, Centrosema pubescens, Desmodium heterophyllum, Pueraria phasloloides, Stylosenthes guyanensis , Leucaena leucocephela, Gliricidia sepium	Fresh forage fed in Nigeria	—	—	—	Medugu *et al.*, 2012
Arachis glabrata (perennial groundnut)	Fresh forage	100 per cent	Lucerne	(+) Growth	Gomez-de-Varela *et al.*, 1983
Arachis hypogea (Groundnut)	Hay of haulms	27-50 per cent/sole feeding	Berseem meal	(+) Growth	Rao *et al.*, 1986; Ngodigha *et al.*, 1994; *Nwagu et al., 2010*
Arachis pintoi (pinto peanut)	Forage hay	10-15 per cent	Basal diet	(+) Growth	Huang *et al.*, 1998; Nieves *et al.*, 1998
A. pintoi	Fresh forage	30 per cent	Basal diet	(+) Growth	Nieves *et al.*, 1996; Nieves *et al.*, 2009
Arachis pintoi + Pennisetum purpureum	Fresh forage	adlibitum	Raw ingredients studied as only feed	(+) Growth	Nieves *et al.*, 1996
Cajanus cajan (Pigeopea)	Hay	100 per cent	*Cynodon dactylon*	(+) Growth	Moura *et al.*, 1992
C. cajan	Hay	28 per cent	Lucerne	(+) Growth	Crespi *et al.*, 1992a
Calapogonium mucunoides	Leaf meal	10-20 per cent	Basal diet	(=)Growth	Aderinola *et al.*, 2008; Nweze 2005, Ukpe *et al.*, 2009
Cassia tora	Fresh	100 per cent	Lucerne	(+) Growth	Cheeke *et al.*, 1983
Centrosema pubescens	Leaf meal/ Pelleted forage	20 per cent/≤ 75	Basal diet	(+) Growth	Prawirodigdo *et al.*, 1989; Nweze, 2005
C. pubescens	Leaf meal	10 per cent CP of basal diet	150g concentrate ad libitum	(=) Growth	Alikwe *et al.*, 2014

Forage Plant Species	*Form*	*Acceptable Level in the Total Diet*	*Substituted Mainly to*	*Effects Growth/ Maintenance/ Other Effects on Poultry/ Duck*	*Reference*
C. pubescens	Leaf meal	20 per cent	Basal diet	(=) Growth	Aderinola *et al.*, 2008
Clitoria ternata	Fresh	100 per cent	Lucerne	(+) Growth	Cheeke *et al.*, 1983
Crotalaria juncea (Sunhemp)	Dried leaves	27 per cent	Berseem meal	(+) Growth	Rao *et al.*, 1986
Crotalaria ochroleuca	Air dried	15-30 per cent	Sunflower meal	(+) Growth	Laswai *et al.*, 2000
Desmanthus virgatus (Hedge lucerne)	Fresh	27 per cent	Berseem meal	(+) Growth	Rao *et al.*, 1986
Desmodium distortum	Fresh	40-100 per cent	Lucerne	(+) Growth	Harris *et al.*, 1981b; Cheeke *et al.*, 1983
Desmodium scripiums	fresh	—	Basal diet	(+) Growth	Aduku *et al.*, 1986
Hedysarum coronarium (Sulla)	Fresh	35 per cent	Lucerne	(+) Growth	Cucchiara, 1989
Lablab purpureus	Fresh	*Adlibitum*	leaf meal of sunflower	(+) Growth	*Omole et al., 2007*
Lupinus albus (Lupin)	Green forage	20 per cent	Basal diet	(+) Growth	El-Gendy, 1999, Oduozo & Adegbola, 1992
Macroptilium lathyroides	Green forage	100 per cent	Lucerne	(+) Growth	Cheeke *et al.*, 1983
Macroptilium atropurpureum	Green forage	100 per cent	Basal diet	(+) Growth	Aduku *et al.*, 1986
Medicago sativa (Lucerne)	Dried forage	40-100 per cent	Basal concentrate diet	(+) Growing or pregnant	Fernandez-Carmona *et al.* ,1998; Reddy, 1987; Perez *et al.*, 1994; Pascual *et al.*, 2002; Payne *et al.*, 1983; Cheeke *et al.*, 1980
M. sativa	Dried forage	(99.5 per cent +NaCl)	Basal concentrate diet	(+) Growth	Perez *et al.*, 1998
M. sativa	Hay	92-96 per cent	Barley and soybean meal	(+) Breeding rabbit/fairly good growth rate with the addition of fat of carcasses and substantial decrease in mortality rate	Fernandez-Carmona *et al.*, 1998; Fernandez-Carmona *et al.*, 2000

Contd...

Forage Plant Species	*Form*	*Acceptable Level in the Total Diet*	*Substituted Mainly to*	*Effects Growth/ Maintenance/ Other Effects on Poultry/ Duck*	*Reference*
M. sativa	Hay	100 per cent	Barley and soybean meal	(-) Breeding rabbit	Pascual *et al.*, 2000
M. sativa	Dehydrated after ensiling	100 per cent	Lucerne	(-) Growth	Perez *et al.*, 1990
M. sativa	With molasses	100 per cent	Basal diet	(+) Growth	Linga & Lukefahr, 2000; Linga *et al.*, 2003
M. sativa	Dried with high saponins concentration,	Best with 0.6 per cent saponin in the diet	Lucerne low concentration of saponin	(+) Growth	Auxilia *et al.*, 1983
Neonotonia wightii (Perennial soybean)	Hay	38 per cent	Lucerne	(+) Growth	Crespi *et al.*, 1992b
Psophocarpus tetragonolobus	Hay	100 per cent	Lucerne	(-)Growth	Harris *et al.*, 1981b
P. tetragonolobus	Hay	22 per cent	Lucerne	(+) Growth	Grandi *et al.*, 1985
Stylosanthes species	Fresh	100 per cent	Lucerne	(-) Growth	Harris *et al.*, 1981b
Stylosanthes species	Dried	50 per cent stylo/ 75 per cent concentrate/ sole feeding	Basal diet	(+) Growth	Sanni *et al.*, 2005; *Nwagu et al., 2010*
S. guianensis	Fresh	*Adlibitum/20 per cent*	Leaf meal of sunflower	(+) Growth	El- Gendy, 1999; *Omole et al., 2007*
S. hamata cv. Verano	Hay	50 per cent	Concentrate	(+) Growth	Bamikole & Ezenwa, 1999; Bamikole *et al.*, 1999
Trifolium species (Clover)	Hay	20 per cent	Basal diet	(+) Growth	Marai *et al.*, 1979
Trifolium alexandrinum (Berseem)	Fresh	*Adlibitum*	—	(+) Growth of adult	Deshmukh *et al.*, 1989; Deshmukh *et al.*, 1990

Forage Plant Species	*Form*	*Acceptable Level in the Total Diet*	*Substituted Mainly to*	*Effects Growth/ Maintenance/ Other Effects on Poultry/ Duck*	*Reference*
T. alexandrinum (Berseem)	Hay	100 per cent	Lucerne	(+) Growth	El-Adawy *et al.*, 2000
T. alexandrinum	Silage	100 per cent	Fresh berseem	(+) Growth	El-Ayouty *et al.*, 2000
T. pratense (Red clover)	Hay	30 per cent	Lucerne	(+) Growth	Grandi *et al.*, 1988
Vicia faba (Horse bean)	Sun dried leaves	75 per cent	Soybean meal	(+) Growth	Magouze *et al.*, 1998
V. sativa (Common vetch)	Sun dried	60 per cent	Lucerne hay	(+) Breeding rabbit	Lopez *et al.*, 1996; Sese *et al.*, 1999
Vigna umbellata (Ricebean)	Fresh green	~50 per cent	Basal diet	(+) Growth	Gupta *et al.*, 1996b
V. unguiculata (Cowpea)	Dried testa of pod	60 per cent	Poultry mash	(+) Growth	Sese *et al.*, 1999
V. unguiculata	sun dried leaves	100 per cent	Soybean meal	(-) Growth	Magouze *et al.*, 1998
Forage Grasses					
Brachiaria brizantha, Chloris gayana, Panicum maximum, Paspalum plicatulum, Penisetum purpureum Setaria splenda	Fresh forage fed in Nigeria	—	—	—	Medugu *et al.*, 2012
Avena sativa (Oat)	Green, pre-flowering	—	—	(+) Growth	Deshmukh *et al.*, 1989
A. sativa	Whole plant meal	25 per cent	Basal diet	(+) Growth	Bhatt *et al.*, 2001
A. sativa) + *Vicia sativa*	Silage	100 per cent	Same fresh forage	(+) Growth	Kennou *et al.*, 1990
Bromus catharticus	Hay (first cut)	24 per cent	Lucerne	(+) Growth	Grandi, 1993
Chloris gayana (Rhodes grass)	Fresh	—	—	(+) Growth	Raharjo *et al.*, 1986a; Mutetikka *et al.*, 1990

Contd...

Forage Plant Species	*Form*	*Acceptable Level in the Total Diet*	*Substituted Mainly to*	*Effects Growth/ Maintenance/ Other Effects on Poultry/ Duck*	*Reference*
Cynodon dactylon (Bermuda grass)	Fresh	100 per cent lucerne	Lucerne	(-) Growth	Champe *et al.*, 1983; Deshmukh *et al.*, 1989; Deshmukh *et al.*, 1993a
C. dactylon	Fresh	100 per cent	Lucerne	(+) Growth	Daniels *et al.*, 1985
C. plectostachyus (Star grass)	Fresh	100 per cent	Basal diet	(+) Growth	Ramchurn *et al.*, 2000
Dactylis glomerata (Orchard grass)	Fresh	45 per cent	Lucerne	(+) Growth	El-Sayaad *et al.*, 1992
Festuca arundinacea (Tall fescue)	Hay	100 per cent	Basal diet	(-) Growth	Bhatt *et al.*, 2001
Hordeum vulgare (Forage Barley)	Hay	28-35 per cent	Meadow hay/ Cereal meal	(+) Growth	Corino *et al.*, 1982; Polidori *et al.*, 1984; Auxilia *et al.*, 1977
Lolium perenne (Rye grass)	Fresh	14 per cent	Lucerne	(+) Growth	Grandi, 1983
Lolium species	Fresh	20-50 per cent	Basal diet	(+) Growth	Payne *et al.*, 1983; Aderibigbe *et al.*, 1992
Panicum maximum (Guinea grass)	Hay/ fresh	10 per cent	Concentrate	(+) Growth	Bamikole *et al.*, 1999; Aderinola *et al.*, 2009; Ukpe *et al.*, 2009
Pennisetum purpureum	Fresh	—	Basal diet	(=/+) Growth	Raharjor *et al.*, 1988
Poa pratensis (Kentucky bluegrass)	Fresh	~40 per cent	Lucerne	(+) Growth	Harris *et al.*, 1981a
Thysanolaena maxima (Broom grass)	Leaves	40 per cent	Ricebean/Concentrate	(=/+) Growth	Singh *et al.*, 1995; Rohilla & Bujarbaruah, 2000; Rohilla *et al.*, 2000

Forage Plant Species	*Form*	*Acceptable Level in the Total Diet*	*Substituted Mainly to*	*Effects Growth/ Maintenance/ Other Effects on Poultry/ Duck*	*Reference*
T. maxima	Fresh leaves	15 per cent of the diet	Forage rice bean	(=)Growth	Singh *et al.*, 1995
Triticum aestivum (Forage wheat)	Dried whole plant	28 per cent	Cereal	(+) Growth	Auxilia *et al.*, 1977
Zea mays (Forage Maize)	Silage/ dried whole plant	40-100 per cent	Fresh berseem	(+) Growth	El-Ayouty *et al.*, 2000
Z. mays	Dried whole plant	28-40 per cent	Cereal meal/Maize grain/ Hay	(+) Growth	Auxilia *et al.*, 1977; Auxilia *et al.*, 1979; Auxilia & Masoero 1980; Auxilia *et al.*, 1980; Masoero *et al.*, 1979; Polidori *et al.*, 1982; Polidori *et al.*, 1984
Z. mays	Dried whole plant	70 per cent	Wheat grain and straw	(+) Growth	Seroux *et al.*, 1980
Forage Forbs/ Others					
Amaranthus hypochon-driachus	Whole plant meal (vegetative stage)	15 per cent	Lucerne	(+) Growth	Alfaro *et al.*, 1987
A. hypochondriachus	Sun dried forage hay	100 per cent	Lucerne	(-) Growth	Harris *et al.*, 1981a
Beta vulgaris (fodder beet)	Fresh leaves	~40 per cent	Concentrate	(+) Growth	Bassuny *et al.*, 1999
Boehmeria nivea (ramie)	Hay	25 per cent	Basal diet	(+) Growth	Mendes *et al.*, 1980
Brassica species (Forage rape)	Fresh	21 per cent	Mixed cereals	(+) Growth	Fedeli Avanzi *et al.*, 1976
Commelina benghalensis	Fresh leaves and stem	40 per cent	Basal diet	(+) Growth	Mtenga *et al.*, 1994

Contd...

Forage Plant Species	*Form*	*Acceptable Level in the Total Diet*	*Substituted Mainly to*	*Effects Growth/ Maintenance/ Other Effects on Poultry/ Duck*	*Reference*
Helianthus annuus (Sunflower)	Leaves	100 per cent	Lucerne	(-) Growth	Harris *et al.*, 1981a
Hibiscus rosasinensis	Fresh leaves	100 per cent	Basal diet	(-) Growth	Gidenne, 1985b
Ipomoea batatas (Sweet potato)	Vines	80 per cent	Sweet potato roots	(+) Growth	Abu *et al.*, 1999; Raharjo, 1987; Mutetikka *et al.*, 1990; Nieves *et al.*, 2009; Tam *et al.*, 2009; Luyen & Preston 2012
Leunae cornuta (Wild lettuce)	Stems and leaves	40 per cent	Basal diet	(+) Growth	Mtenga *et al.*, 1994
Lotus corniculatus (Birdsfoot trefoil)	—	32 per cent	Lucerne	(+) Growth	Grandi *et al.*, 1988
Luffa egyptiaca,	Fresh	*Adlibitum*	Basal diet	(+) Growth	Aduku *et al.*, 1986
Manihot esculenta (Cassava)	Leaves and stems hay	10-20 per cent	Basal diet	(+) Growth	Scapinello *et al.*, 2000
M. esculenta	Leaves and stems meal	50 per cent	Clover hay	(+) Growth	Toson *et al.*, 1999
M. esculenta	Leaves	100 per cent	Lucerne	(-) Growth	Harris *et al.*, 1981b; Reed *et al.*, 1982; Raharjo, 1987
M. esculenta	Leaves meal (dried)	42 per cent	Poultry offals and wheat bran	(+) Growth	Fotso *et al.*, 2000
M. esculenta	Leaves meal (dried	40 per cent	copra meal	(+) Growth	Ravindran *et al.*, 1986
Maranta arundinacea (Arrowroot)	Pelted Foliage	60-80 per cent	Basal diet	(+) Growth	Erdman, 1986
Musa paradiosica	Leaves	—	Basal diet	(+) Growth	Aduku *et al.*, 1986
Telfaria occidentalis	Leaf meal	5 per cent CP of basal diet	150g concentrate *ad libitum*	(=) Growth	Alikwe *et al.*, 2014

Forage Plant Species	*Form*	*Acceptable Level in the Total Diet*	*Substituted Mainly to*	*Effects Growth/ Maintenance/ Other Effects on Poultry/ Duck*	*Reference*
Trichantera gigantea	Leaf meal	30 per cent	Basal diet	(=)	Nieves *et al.* (2009)
Tridax procumbens	—		Basal diet	(+/=)	Aduku *et al.*, 1986
Tree Fodder (Legume and Non-legume)					
Acacia albida	Pods	20 per cent	Basal diet	(+) Growth	Igwebuike *et al.*, 1999
A. saligna	Leaves	40 per cent	Concentrate	(+) Growth	Abdel-Samee *et al.*, 1992
A. saligna	Dried leaves	30 per cent	Basal diet/ Berseem hay	(+) Growth	El-Gendy, 1999a; El-Gendy, 1999b
Bauhinia variegata	Fresh leaves	~15 per cent DM	Basal diet of angora	(-) Growth	Negi *et al.*, 1985
Ficus *auriculata*	Fresh leaves	15 per cent of the diet	Forage rice bean	(=)Growth	Singh *et al.*, 1995
Gliricidia sepium	Leaf meal/ Dried leaves	15-27 per cent	Berseem meal	(+) Breeding rabbit	Herbert, 1998; Rao *et al.*, 1986;
G. sepium	Fresh or pelleted	70 per cent	Basal diet	(+) Growth	Onwudike, 1995
G. maculata (*G. sepium*)	Fresh leaves	10 per cent DM	Concentrate	(+) Growth	Onwudike, 1995
Grevia oppositifolia	Leaves	100 per cent	Barley	(-) Growth	Sastry *et al.*, 1982
Grewia optiva	Fresh leaves to appetite	~15 per cent DM	Basal diet	(+) Growth of Angora rabbit	Negi *et al.*, 1985; Deshmukh *et al.*, 1989
Leucaena leucocephala	Hay/ Leaves (treated) : heated, silage, sundried/ Dried leaves	15-30 per cent	Lucerne/ Basal diet/ hard wheat bran/l ucerne	(+) Growth	Aduku *et al.*, 1986; Scapinello *et al.*, 2000; Awosanya *et al.*, 2000; El-Galil *et al.*, 2001; Mtenga *et al.*, 1994; Parigi-Bini *et al.*, 1984; Gupta *et al.*, 1996a; Ghazalah *et al.*, 1998
L. leucocephala	Leaf meal	40-100 per cent	Untreated *Leucena* leaves	(-) Growth	Harris *et al.*, 1981, Ekpenyong, 1985, Tangendjaja *et al.*, 1990; Scapinello *et al.*, 2000.

Contd...

Forage Plant Species	*Form*	*Acceptable Level in the Total Diet*	*Substituted Mainly to*	*Effects Growth/ Maintenance/ Other Effects on Poultry/ Duck*	*Reference*
L. leucocephala	Whole plant	100 per cent	-	(-) Growth	Nieves *et al.*, 1995;
L. leucocephala	Foliage	30 per cent	Basal diet	(+/=) Growth	Nieves *et al.*, 1998; Nieves *et al.*, 2009
L. leucocephala	Dried leaves treated with 1.2 per cent FeCl3	50 per cent	Basal diet	(+) Growth	Gupta *et al.*, 1996a
L. leucocephala	Fresh or pelleted	50 per cent fresh or 7 per cent of feed dry matter	Basal diet	(+) Growth	Onwudike, 1995
L. leucocephala	Fresh leaves	15-40 per cent DM	Wheat bran/ Mash concentrate/ Basal diet	(+) Growth	Nieves *et al.*, 1998; Muir *et al.*, 1992; Onwudike *et al.*, 1992; Rohilla *et al.*, 1999
L. leucocephala	Fresh leaves	100 per cent DM	Concentrate	(-) Growth	Onwudike, 1995
L. leucocephala	Fresh leaves	100 per cent	Other forages	(-) Breeding rabbits	Muir *et al.*, 1995
L. leucocephala	Mimosine extracted from seeds	0.26 per cent	Basal diet	(+) Growth	Szyszka *et al.*, 1985
Moringa oleifera	Leaf meal	45 per cent	Basal diet	(+) Growth	Abubakar *et al.*, 2011
Morus alba (Mulberry)	Fresh leaves	50 per cent	Basal diet	(+) Growth	Meena *et al.*, 1999,
M. alba	Leaves	*29-38 per cent*	Pelleted feed	(+) Growth	Singh *et al.*, 1984
M. alba	Fresh leaves	69 gm dry matter daily intake	Basal diet	(+) Growth	Deshmukh *et al.*, 1989; Deshmukh *et al.*, 1993
M. alba	Foliage	30 per cent	Basal diet	(=) Growth	Nieves *et al.*, 2009
M. alba	pelleted	29-38 per cent	Basal diet	(+) Growth	Singh, *et al.*, 1984
Musa paradisiaca (Banana)	Fresh leaves	40 per cent DM	Basal diet	(+) Growth	Rohilla *et al.*, 2000a

Forage Plant Species	*Form*	*Acceptable Level in the Total Diet*	*Substituted Mainly to*	*Effects Growth/ Maintenance/ Other Effects on Poultry/ Duck*	*Reference*
Populus tremuloides (Poplar)	Leaves	40 per cent	Lucerne	(+) Growth	Ayers *et al.*, 1992b
Robinia pseudoacacia (Black locust)	Leaves meal	100 per cent	Lucerne	(-) Growth	Harris *et al.*, 1984
R. pseudoacacia	Leaves	0 per cent	Lucerne	(-)Growth	Cheeke *et al.*, 1984; Raharjo *et al.*, 1990
R. pseudoacacia	Leaves	100 per cent	Concentrrate+ Forage *Pennisetum*	(+) Growth	Singh *et al.*, 1997; Singh *et al.*, 1999
R. pseudoacacia	Partially sun dried leaves	~ 17 per cent DM	Basal diet	(+) Growth	Sanjiv *et al.*, 2000
R. pseudoacacia	Dried leaves	75 per cent	Basal diet	(+) Growth	Bhatt *et al.*, 2000
Schleichera oleosa (Kusum tree)	Leaf meal	10 per cent	Basal diet	(+) Growth	Sreemannarayana *et al.*, 2001
Sesbania aegyptica	Dried leaves	15 per cent	Lucerne	(+) Growth	Ghazalah *et al.*, 1998
Aquatic Forage					
Azolla species	Sun dried hay	24 per cent	Clover hay	(+) Growth	Abou-Zeid *et al.*, 2001
A. filiculoides	Dried meal	50 per cent	Soybean meal	(+) Growth	Gualtieri *et al.*, 1988
A. pinnata	Sun dried hay	20 per cent	Lucerne	(+) Growth	Wittouck *et al.*, 1992
Azolla pinnata	Dried	1.5-3.0 per cent	Basal diet	(=) Growth	Anitha *et al.*,2016
Eichhornia crassipes (Water haycinth)	Dried	100 per cent	Lucerne	(-) Growth	El-Adawy *et al.*, 2000
E crassipes	Dried	10-30 per cent	Lucerne	(+) Growth/ breeding rabbit	Grandi, 1981a; Grandi *et al.*, 1983a; Moreland *et al.*, 1991a; Moreland *et al.*, 1991b
E. crassipes	Dried pre-bloom stage	10-30 per cent	Barley/ Basal diet	(+) Growth	Zeweil *et al.*, 1993; Biobaku *et al.*, 1991; Eleraky *et al.*, 1996

Contd...

Forage Plant Species	*Form*	*Acceptable Level in the Total Diet*	*Substituted Mainly to*	*Effects Growth/ Maintenance/ Other Effects on Poultry/ Duck*	*Reference*
Ipomoea aquatica (Water spinach)	Fresh	10 per cent	Basal diet	(+) Growth	Toung *et al.*, 1994; Becerra *et al*, 1995; Amornsak *et al.*, 2004; Hongthong Phimmasan *et al.*, 2004; Pok Samkol *et al.*, 2006; Tam *et al.*, 2009; Tam *et al.*, 2013; Hang *et al.*, 2011; 2013; Huyen*et al*, 2012,
Ipomoea aquatica +Panicum maximum	Fresh	10 per cent + 20 per cent of the body weight	Sole *Ipomoea aquatica* 10 per cent of body weight	(+) Growth	Khuc Thi Hue & Preston, 2006
Ipomoea aquatica + Stylosanthes sp.	Fresh	10 per cent + 20 per cent of the body weight	Sole *Ipomoea aquatica*	(+) Growth	Khuc Thi Hue & Preston, 2006
Pistia stratiotes (Water lettuce)	Dried	30 per cent	Basal diet	(+) Growth	Biobaku *et al.*, 1991
Potamogeton natans L. var. fluitans	Dried	8 per cent	Lucerne, bran and barley	(+) Growth	Grandi, 1978
Ulva fasciata (Seaweed)	Dried	15 per cent	Concentrate	(+) Growth	Raju & Screemanna, 1995; Raju *et al.*, 1995
U. fasciata	Dried	5 per cent	Basal diet	(+) Growth	Sreemannarayana *et al.*, 1995,

(+) positive effects; (–) negative effects; (+/–) both positive and negative effects; (=) equal effects compared to control.

Source: Siriwan *et al.*, 2012 and others as in reference.

Table 10.4: Digestibility of some Tropical Forage Resource Fed in Fresh (Wilted) in Rabbits

Forage	*Per cent Digestibility*	
	Gross Energy	*Crude Protein*
Tree Fodder (Legumes)		
Albizia falcata	70.3	73.4
Calliandra calothyrsus	51.4	49.8
Leucaena leucocephala	69.5	75.9
Sesbania formosa	65.8	64.2
Sesbania sesban	77.5	83.9
Forage Legumes		
Cassia rotundifolia	40.1	57.5
Centrosema pubescens	54.2	72.9
Desmodium heterophyllum	48.7	52.1
Nenotonia wightii	39.8	56.6
Pueraria phasoloides	44.3	65.6
Stylosanthes guianensis	55.1	58.9
Forage Grasses		
Brachiaria brizantha	24.5	17.8
Chloris gayana	36.3	32.4
Panicum maximum	10.7	13.0
Paspalum plicatulum	33.7	21.2
Pennisetum purpureum	45.2	64.7
Setaria spendida	9.4	6.2
Other Forages		
Cassava leaves and stems	47.0	42.0

Source: Raharjo *et al.*, 1986.

In general, digestibility of forage legumes including the leaves of many tree legumes is comparable to alfalfa meal (Table 10.4). The tannins in tree leaves are one factor that limits the intake and digestibility. The leaves and twigs of tree legumes *viz. Leucaena leucocephala, Gliricidia sepium* and *Calliandra* species have high tannins (Raharjo & Cheeke, 1985). Generally, tropical legumes *viz. Gliricidia sepium, Leucaena leucocephala* (even though it contains the toxic amino acid mimosine), *Erythrina lithosperma* are preferred over forage grasses and agricultural by-products. Apart from being a potential protein source, inclusion of forage legume is important to the farmers because of its cost effectiveness. It is therefore, recommended that farmers be encouraged to include a certain percentage (at least 10 per cent) into the feed (Alikwe *et al.,* 2014). Some of the tropical forage legumes contain toxic amino acids or alkaloids (*e.g. subabool* contains mimosine, while leaves of cassava contains cyanogens), hence, they are not good for sole feeding. By the use of a mixture of forages, the concentration of specific toxins can be kept to non-hazardous levels and the quality can be improved by adjusting close to voluntary consumption by rabbit (Lukefahr & Cheeke, 1990). Again, by using a mixture of forages, effects of toxins can be minimized. Forage being high in nutritional quality is also associated with some constraints for feeding of rabbits. Noted few constraints are seasonality of production, inherent antinutritional factors in some forage, will have have negative effects on absorption of minerals and rumen microbial activity, being high in fibre may result caecal impaction and limit energy intake, locality and collection in relation to area of use and processing (Adebowale, 1983; Champe and Maurice, 1983; Waghorn *et al.,* 1994; Nunez-Hernandez *et al.,* 1991; Abeke *et al.,* 2003; Abeke & Otu, 2008). Some toxins such as cyanide in *Manihot esculenta* and mimosine in *Leucaena leucocephala* are often present which could cause growth depression and mortality (Cheeke & Shull, 1985). Forage grasses and legumes could be used as feed for rabbits at different ages with respect to their inclusion levels (Medugu *et al.,* 2012). For effective utilization of these forages, it may be necessary to develop processing methods to reduce or completely eliminate anti-nutritional factors that are present in them. Processing methods for rabbit feed production must be simple, economical and inexpensive. Sometimes, single processing method may not effectively remove the antinutritional factors in forages; therefore, combination of two or more processing methods may be required.

Most of the grasses were found less palatable than the legumes. Tree leaves can be used in many areas to provide forage in the dry season. Digestibility of some tropical forage resources have been reported by Raharjo *et al.,* 1986 (Table 10.4). Raharjor *et al.* (1988) fed some tropical grasses (*Chloris gayana, Panicum maximum,* and *Pennisetum purpureum*) and legumes to rabbits and observed that digestibility of the grasses were lower than those of the tropical legumes and return on the investment, thus cost of feeding grower rabbit could be lowered by supplementing concentrate diets with *Stylosanthes species.* Nyako (2001) suggested that tropical forage grasses could only be used as source of indigestible fibre. Possibility of utilizing tree fodder and forage grass by mono-gastric livestock has been demonstrated in rabbit under a silvipasture system (Singh *et al.,* 1995). Rabbit fed on tree leaves of nevaro (Ficus *auriculata*) and the leaves of broom grass (*Thysanolaena maxima*) from the nevaro + broom grass silvipasture established on nonarable lands showed the gain in body weight marginally lower than the feeding of the forage rice bean (Table 10.5) but added benefit are availability round the year unlike rice bean which is available only during the monsoon and can be grown only on the cultivated lands competing with food crops. More research is required to reveal the potential and develop silvipasture based rabbit feeding systems in mountain and plain area of India.

Table 10.5: Body Weight Gain (kg) in Rabbits Fed Green Forage and Concentrates under Nevaro (*Ficus auriculata*) + Broom Grass Silvipasture System

Age of Rabbits (Days)	*Leaves of Broom Grass*	*Leaves of Ficus auriculata*	*Forage Ricebean*
42	0.575	0.575	0.577
56	0.770	0.808	0.875
70	0.989	1.065	1.161
84	1.189	1.257	1.464
98	1.400	1.498	1.691
112	1.643	1.661	1.884

Source: Singh *et al.,* 1995.

Chapter 11

Organic Forage Production

Increased purchasing power and rapid urbanisation have influenced life style and health consciousness in India. Now the organized organic way of livestock rearing is gaining popularity particularly in peri-urban areas. Peri urban-organic farming is an indicator of fast changing economic scenario in organic farming sector in India. The demand for organic foods particularly fruit, vegetables and milk is increasing rapidly. The organic farming movement is commonly agreed to have begun in the 1940s in England with the writings of Sir Albert Howard, who learned about organic practices in India during the 1920s. It is important to understand that organic farming is a system of production, a set of goal-based regulations that allow farmers to manage their own particular situations individually, while maintaining organic integrity. The objective of organic farming is to optimize productivity of interdependent communities of soil, plants, animal and the people for overall wellbeing of human beings (FAO, 2001). Speedy (2001) defined Organic farming as a system of farm management to achieve sustainable production without synthetic inputs *viz.,* chemical fertilizer and pesticides. Production of organic milk and meat products largely depend upon development of certified organic forage production systems as per prescribed guideline and standard for organic farming. Organic livestock products are produced by feeding organically produced feed and forage to the livestock along with prescribed treatment of livestock as per organic standard. The characteristics of certified animal husbandry involve organic operations *i.e.* biodiversity, diversification and integration, sustainability, natural plant nutrition, natural disease and pest management, health and integrity, animal welfare and ecosystem services. The Codex Alimentarius Commission (FAO, 2001), a joint FAO and WHO Food Standards Programme, began in 1991 elaborating guidelines for the production, processing, labelling and marketing of organically produced food. The IFOAM Basic Standards define how organic products are grown, produced, processed and handled. These are presented as general principles, recommendations and basic standards and provide a framework for certification bodies and standard setting organization worldwide to develop their own certification standards. The country specific certification standards take into account specific local conditions and provide more specific requirements than IFOAM Basic Standards. The requirements in these Codex Guidelines are in line with IFOAM Basic Standards and the EU Regulation for organic food. This Codex document is an important step in the harmonization of international rules to build up consumer trust. The global organic production cover approximately 26 million hectare which follows 61 standards having 364 certification agencies (Bhattachrya & Chakraborty, 2005). The four pillars of organic farming are organic standards, certification/regulatory mechanism, technology packages and network of markets. In India, National Programme on Organic Production (NPOP) was established in October 2001 under Ministry of Commerce and Industry, Government of India to provide information on all aspects of organic production in India. Keeping in mind Indian requirements, the standards and procedures of NPOP have been drawn in harmony with international standards of Codex and IFOAM. A National Steering Committee (NSC) comprising Ministry of Commerce, Ministry of Agriculture, APEDA, Spices Board, Coffee Board, Tea Board and various other Government and private organisations associated with the organic movement is monitoring the overall organic activities under the National Programme for organic production (NPOP). In the year 2005, NPOP standard has got equivalency with the standard of ED commission. Now, Indian Standard is acceptable in European Countries. For organic milk, Codex Standards are followed taking into account several pesticide residues in milk and recommended Maximum Residue Limits (MRL) for 83 pesticide residues in milk. Standards for seven biofertilizers and guidelines for certification have been issued in Gazette of India dated 24th March, 2006, (Palsaniya & Maity, 2017). NPOP has prescribed National Standards for Organic Production covers processes of conversion, crop production, animal husbandry, food processing and handling, labelling and storage and transport. The whole farm including livestock is converted to organic management according to the prescribed organic standard over a period of three years and a detailed management plan of organic forage production including ingredients of concentrate drawn and implemented. Organic certification is based on continuance of practices for organic plan management.

In organic forage production system, certified organic forage seed should be organically produced as per IFOAM guideline. Preferably forage seed should be produced organically. In case of unavailability, untreated conventional forage seed can be used if organic seeds are documented as not being commercially available. Documentation must be kept to verify that any non-organic seed used has not been treated with synthetic fungicides/insecticides, is not genetically modified, nor inoculated with a GMO nitrogen-fixing bacterial inoculant. In organic livestock farming, forage production should be planned in advance considering combinations of leguminous and non-leguminous forage crop under integrated forage production system, recommended forage varieties and forage crop rotations for rain fed and round the year forage cropping system for at least 50 per cent irrigated lands

Table 11.1: Requirements of Area for Organic Forage Production

Dairy Unit	*Total ACU*	*Annual Green Forage Requirement (q)*	*Area Required (ha)*	*Dry/ Bhoosa Require (q)*	*Area Required (ha)*	*Annual Concentrate Mix Requirement (q)*	*Area Required (ha)*	*Total Area Required (ha)*
10	12	730	0.5	146	0.5	146	2.0	3.0
20	23	1405	1.0	277.5	1.0	277.5	4.0	6.0
50	58	3540	2.0	708.0	2.0	708.0	10.0	14
100	113	6880	6.5	1372.5	6.5	1372.5	20.0	33

In the given area Rabi season will be for grain production and Kharif for legume hay as supplement.

Source: Maity *et al.* (2016).

Table 11.2: Mean Forage Yield (Dry matter tonnes/ha) under Organic and Conventional Forage Production Systems over Six Years

Cropping System	*Organic Production*	*Conventional (Use of Fertilizers)*	*Per cent Increase in Yield Over Conventional Forage Production Management*
Irrigated Land			
Maize + Cowpea – Berseem – Maize	39.1	33.2	17.8
Sorghum – Oat – Sorghum	38.9	32.1	21.2
Guinea grass + Cowpea (*Kharif*) – Berseem (*rabi*) – Cowpea (summer)	17.5	16.4	6.7
Rainfed land			
Sorghum – Barley – Sorghum	19.8	15.6	26.8
Sesame + Mustard	0.45 + 0.90*	0.38 + 0.73*	18.4 Sesame seed and 28.6 Mustard seed

*: Seed

Source: Maity *et al.* (2016).

in an agro climatic zone. Minimum two years conversion period is required for land to change from conventional to organic production systems. As per the NOP regulation "A producer must select and implement tillage and cultivation practices that maintain or improve the physical, chemical, and biological condition of the soil and minimize soil erosion" (USDA, 2000; 7 CFR 205.203(a)). It allows the use of predator species for pest control, development of natural habitat for predator species, and non-synthetic controls such as traps and repellants. It allows the control of weeds through mulching with fully biodegradable materials, mowing, mechanical cultivation, and grazing. All soil fertility inputs, pest control or weed control materials must be listed in the OSP and approved by the certifier prior to use. Records of use should be maintained and product labels or purchase records should be kept and made available during the organic inspection. Composted biomass of plant and animal origins generated on the organic farm itself should be sufficient to meet the nutrient requirement of forage production by promoting maximum microbial activity in the soil. The nitrogen source may be organic manure produced on same farm or forage legume as companion crop. Bio fertilizers can be used safely under all agro ecosystems and in all the forage crops. Accumulation of heavy metals and other pollutants should be prevented in the soil. As per the National Standard, crop residues, green manure, plant residues compost, seaweed compost on farm unit of plant origin are permitted but saw dust restricted. Dung and urine of livestock and poultry manure on farm unit as well as microbial bio fertilizers are permitted but FYM and slurry on and out farm and blood, meat, bone and leather meal are restricted. There is restriction on minerals such as rock phosphate, trace elements, basic slag, industrial by-products, organic house products (NPOP, 2005). Prevention methods such as disruption, elimination of pest habitat, mechanical, physical and biological methods are allowed (Maity *et al.*, 2016, Palsaniya & Maity, 2017). Permitted pesticides are used as per given in the standards. Management of weed, pests and disease should be done by soil solarisation, deep summer tillage, cultural practices such as crop rotation, intercropping, stale bed preparation, mulching, balance manure application, mechanical control and breaking pest development cycle in the field. Botanical pesticides are permitted but not the growth regulators, genetically engineered organisms, products and dyes. Soil and water resources should be appropriately used to prevent soil erosion, salinization and pollution of surface and ground water. Forage resource can be certified organically when the national standards requirements have been met during a conversion period of at least two years before sowing or in the case of perennial crops other than the grassland, at least three years (thirty-six months) before the first harvest of forage which can be appropriated by certifying agency (Palsaniya and Maity, 2017).

The percentage of organically produced feed and forage fed to livestock for organic livestock products is 100 per cent in Argentina, Australia and North America while under IFOAM, only 80 per cent organic feed and forage must be fed to livestock in Asian and European Union countries. There should be natural grasslands as a part of organic forage production system. Standards of organic production in these countries has requirement that organic livestock be managed organically from the last third of gestation of the mother, or at least from birth. However, Asian and European organic standards currently allow livestock to come from non-organic sources at different ages, depending on the animal species. Some national organic certification agencies prohibit the use of antibiotics while others specify a withholding period to be certified as organic products. Vaccinations are generally allowed, with some restrictions. Codex Maximum Residue Limits (MRLs in ppb) of veterinary drugs in milk are 0.05 for 2,4,5-Trichlorophenoxy acetic acid, 4.0 for Benzylopenecillin, 100 for Ceftiofur, 200 for Dihydrostreptomycin, 100 for Oxytetracyclin, 200 for Spectinomycin, 25 for Sulfadimine and 0.1 for Thiabendazole. The Maximum Residue Limits (MRL) of heavy metals in water is 1000 ppm for magnesium and aluminium; 100 ppm for arsenic and iodine; 40 ppm for barium, cobalt, copper, lead, and molybdenum.

Maity *et al.* (2016) developed standard process and protocol for organic milk production in three Indian cattle *Sahiwal, Tharparkar* and *Gir*. Conventional land was converted to certified organic land within 30 months from the month of starting organic conversion following NPOP standards by a certifying agency SGS India, Gurgaon. The requirements of area were computed as per size of a dairy unit (Table 11.1). The forage cropping systems identified practiced and average forage productivity obtained over six year under irrigated and rain fed situation are given in Table 11.2. Enrichment of the soil by *in situ* produced organic manure and biofertilizers increased system forage productivity under organic management. Under irrigated condition, annual forage based cropping systems recorded 18-21 per cent and in annual + perennial forage based cropping systems recorded seven per cent higher production in organic system as compared to conventional systems. Under rain fed condition, the increase in forage production was 27 per cent in forage yield, 18.4 per cent in sesame seed and 29 per cent in mustard seed under organic management as compared to conventional production systems. Neem oil was used as botanical pesticide. Application of *in situ* produced organic manure @ 20 tonnes/ha, and application of vermicompost at the rate one tonnes per hectare, as per standard nutrient requirement for the cropping system resulted in the improvement of two to three per cent in soil organic carbon and fertility status (NPK) over six years of organic farming. Organic forage fed to the indigenous cattle breed resulted in better reproductive efficiency over the study period. The analysis of blood samples of these livestock showed better blood glucose, urea and calcium strata in the livestock. Singh *et al.* (2012) reported that feeding organic forage to buffalo had no effect on milk yield and its composition; however, buffaloes on organic diet have higher fibre digestibility and low urinary N loss as compared to feeding of conventional forage.

Increasing reliance on pasture and forage-based feeds on dairy farms has as such considerable potential to improve the fatty acid profile of milk and dairy products (Couvreur *et al.,* 2006). Although both conventional and organic dairies can benefit from grazing and forage-based feeds, it is far more common on certified organic farms in the United Kingdom and United Stated of America to account or pasture and forage-based feeds share in the dry matter intake of dairy livestock. Benbrook *et al.* (2013) found that organic milk contained 25 per cent less ω-6 fatty acids and 62 per cent more ω-3 fatty acids than conventional dairy milk, yielding a 2.5-fold higher ω-6/ω-3 ratio in conventional compared to organic milk (5.77 vs. 2.28). All individual ω-3 fatty acid concentrations were higher in organic milk *i.e.* α-linolenic acid (by 60 per cent), eicosapentaenoic acid (32 per cent), and docosapentaenoic acid (19 per cent) as was the concentration of conjugated linoleic acid (18 per cent). In United Kigdom, the milk from organic dairy livestock consuming organic forage had higher concentrations of beneficial fatty acids (FA) than conventional milk, including total polyunsaturated fatty acids (PUFA); 39.4 vs. 31.8 g/kg of the total fat), conjugated linoleic acid cis-9, trans-11 (CLA 9; 7.4 vs. 5.6 g/kg of fatty acids), and α-linolenic acid (α-LN; 6.9 vs. 4.4 g/kg of fatty acids) as compared to conventionally produced milk (Butler *et al.,* 2008; Butler *et al.,* 2011). The organic milk contained higher concentrations of α- linoleic acid (ALA), Conjugated linoleic acid (CLA), and the long-chain omega omega ω-3s Eicosapentaenoic Acid (EPA) and Docosapentaenoic Acid (DPA), compared to cows lacking routine access to pasture and fed substantial quantities of grains had less linoleic Acid (LA) and other omega ω-6 fatty acids (O'Donnell *et al.,* 2010; Slots *et al.,* 2009; Molkentin, 2009). Conjugated linoleic acid (CLA) is dependent on grazing immature nutrient rich pasture (Couvreur *et al.,* 2006; Stergiadis *et al.,* 2012). The Conjugated linoleic acid loss in milk occurred due to feeding of mechanically harvested and stored forage. Season had strong interactions with the forage management practices. The differences in the fat composition between organic systems were greater for summer compared with winter organic milk. In India, the compositions of organic milk obtained from indigenous cattle *Sahiwal, Tharparkar* and *Gir* showed that main fatty acid by weight were palmitic acid ranging 25.2 to 26.6 per cent followed by oleic acid (18.11-20.98 per cent) and myristic acid (10.05-11.00 per cent) in organic milk (Maity *et al.,* 2016). More, ω-6 fatty acids *i.e.* linoleic acid (1.48±0.06) and linolenic acid (0.94±0.04) were reported in organic milk than the conventionally produced milk which had lower linoleic acid (1.34± 0.06) and linolenic acid (0.82± 0.06). Milk protein, milk sugar and total fat composition were similar in organic and conventional milk.

Organic production system reduces the environmental pollution in food production as it has lower eutrophication potential per tonne of milk produced due to lower or no use of chemical fertilizer and pesticides for forage production but the studies are very few to quantify these benefits (Boer, 2003). In general, organic milk production inherently increases methane emission, therefore, can considerably reduce global warming by reducing carbon dioxide and nitrous oxide. However, acidification potential of milk production, for example, is 78–97 per cent due to volatilisation of ammonia, which is not reduced necessarily by changing from conventional to organic milk production system.

Oruganti (2011) narrated the challenges and problems in the organic milk and organic forage production. Most important challenges faced by organic livestock producers were certification; paper work and high compliance cost; sourcing organic inputs particularly organic seed of suitable varieties of grains and forages; high costs of production and maintaining animal health and sometimes inadequate market for costly organic milk and meat. Besides, the organic standards also require abiding a rigorous system for inspection, certification and verification of organic production processes. Meeting the certification requirements and quality standards for an external export market could be extremely demanding. Expansion of organic livestock markets has challenges *viz.* establishment of a low cost certification agency so that small farmers can afford certification, improvisation of organic standards according to regional agro climatic conditions, development of a strong domestic organic market and organic livestock production and advancement in organic livestock production through research and training (Pathak *et al.,* 2003). The expansion of organic milk and meat production is largely in the hands of consumers growing demands in preferring organic over conventional livestock products. The growing scientific consensus on the nutritional benefits of organic foods and wider availability in major markets of the cities could push the demand further.

Glossary of Technical Terms

Accumulation: The increase in forage mass per unit area over a specified time, representing the balance growth, senescence, decomposition of dead organs and consumption by animals.

Acid detergent fibre: The fibrous component represents the least digestible fibre portion of forage or other roughage. This highly indigestible part of forage includes lignin, cellulose, silica and insoluble forms of nitrogen but not hemicellulose. Forages with higher ADF are lower in digestible energy than forages with lower ADF, which means that as the ADF level increases, digestible energy levels decrease. During laboratory analysis, ADF is the residue remaining after boiling a forage sample in acid detergent solution. ADF is often used to calculate digestibility, total digestible nutrients (TDN) and/or net energy for lactation (NEL).

Aftermath: Forage that growth following a harvest.

Actual production history (APH): APH yield is used to determine the production guarantee. The APH yield is based on up to 10 years of actual and/or assigned yields. Unit — The insurable acreage used to determine the APH yield, the production guarantee, and any indemnity (loss payment).

Agroforestry: Land-use system in which trees are used for forest products (*e.g.* timber, pulp, fruits, rubber, syrup and browse) combined with agricultural crops including forage crops and/or animal production. **Agro-silvipasture**: It incorporates agricultural crops, potentially including forage crops and livestock production, where trees may produce timber, pulp, fruits, rubber, syrup or browse for grazing animals. **Silvipasture:** It refers to the exclusive use of land for forest products and animal production by browsing of shrubs and trees and/or grazing of co-existing forage crops.

Alternate stocking: A method of repeated grazing and resting of forage using two paddocks in succession.

Animal unit: One mature, non-lactating bovine (middle-third of pregnancy) weighing 500 kg and fed at a maintenance level for zero gain (8.8 kg dry matter per day), or the equivalent, expressed as (weight) 0.75, in other kinds or classes of animals. **Tropical Livestock Unit** (TLU): One cattle with a live weight of 250 kg (FAO, 2010). An animal unit is based on the assumption that metabolic requirements are related to metabolic weight and provide the basis for comparing among different kinds and classes of animals.

Animal unit day: The amount of dry forage consumed by one animal unit per 24 hour period (8.8 kg). Animal unit day is used to express the quantity of forage intake by one animal unit or one forage intake unit for a 24-hour period of time and may be extrapolated to other time periods, such as a week, month, or year.

Anti-herbivory: Chemical or structural forage characteristics that inhibit or limit selection and consumption of the plant by livestock and wildlife (Anti-quality).

Antioxidants: They are chemical substances that help protect against cell damage from free radicals. Well-known antioxidants include vitamin A, vitamin C, vitamin E, carotenoids, and flavonoids

Anti-quality (Anti-nutritional): A description of any chemical factor in forage (such as lignin, an alkaloid, a phyto-hormone or a toxin) that negatively affects the animal including their animal physiology, health and well-being, reproduction and intake, or the degree to which the forage meets nutritional requirements of a specific kind and class of animal. **Anti-quality factor:** Apart from nutrients, forages may contain various harmful compounds that can adversely affect animal performance and cause sickness or even death. These compounds are called anti-quality factors and include tannins, nitrates, alkaloids, cyano-glycosides, estrogens and mycotoxins. The occurrence and/or severity of these factors depend on the forage and weed plant species present, season, environmental conditions and sensitivity of the animal. High-quality forages should be free from harmful levels of anti-quality components.

Ash: The non-organic, mineral component of plant material. The residue that remains after complete combustion of organic matter.

Balanced ration: Complete feed formulated to provide a specific animal species and class with appropriate amounts of all nutrients required for maintenance and a given level of performance.

Biomass: The total dry weight of vegetation per unit area of land above a defined reference level, usually ground level, at a specific time. It may be specifically forage biomass or herbage biomass or browse biomass as per type of biomass. Biomass and forage mass are instantaneous measures. To describe them over time, a series of instantaneous measures of biomass/forage mass are averaged.

Change over time or the accumulated biomass/forage mass are determined by the difference between an end point and an initial point. Biomass can include both forage and non-forage vegetation per unit area while forage mass is specific to forage plants. Depending on requirements, biomass and forage mass should be specified as above-ground at a specified cutting height, or below-ground as described by measurement method. It should be specified whether the biomass or forage mass is alive or dead, at the time of harvesting, and the proportions of each if both are included. Dry weight of biomass is defined as dried at 105°C to a constant weight unless otherwise noted.

Bite weight: The total weight of forage (dry matter basis) in one bite taken by an animal.

Biting rate: The number of bites taken during a specified time; usually as bites per minute or bites per day. Biting jaw movements should be separated from jaw movements devoted to gathering forage for biting and to chewing forage prior to swallowing.

Bloat: Excessive accumulation of gases in the rumen of the animals. It is due to the formation of stable foam within the rumen which traps the fermentation gases. The main foaming agent is now known to be a plant cytoplasmic protein with a sedimentary velocity of 18 km Svedberg units (18S) also called as Fraction –I. It occurs to the extent of 4-5 per cent in bloat producing forages, such as lucerne, but its concentration is <1 per cent in nontoxic forages.

Botanical composition: The relative proportions of the plant components (species and morphological units) in a canopy above a defined sampling height, preferably ground level. Botanical composition may be calculated based on forage mass, cover, density or frequency. Cover is measured in units of proportion or percentages. Frequency is presence or absence count data that follow the binomial distribution. Density is the number of individuals per unit area.

Browse: Leaf and twig growth of shrubs, woody vines, trees, cacti and other non-herbaceous vegetation that can be ingested by herbivores. In another word available to be consumed browse in situ by animals. **Browsing:** Feeding on the leaves that are on trees, bushes and shrubs higher up rather than the eating the grass and plants on the ground.

Bunchgrass/tussock grass: Grasses producing tillers, but not stolon's or rhizomes, and having an erect and clumped growth form (Examples: Tufted grass, Caespitose grass).

By-pass protein: The portion of intake protein that has a slow rate of degradability in the rumen. It is fed so that it may escape digestion in the rumen, reach the lower gastrointestinal (GI) tract essentially intact and be digested directly in the small intestine as it would be in non-ruminants. This can provide a balance of amino acids unaltered by microbial digestion and synthesis. By-pass protein is also known as undegradable intake protein (UIP), rumen undegradable protein (RUP) or escape protein.

Camp: An area where animals choose to rest or are confined by herders.

Canopy: The above-ground parts of a population or community of forage plants. It may include both herbaceous and woody vegetation. **Canopy architecture**: The spatial distribution and arrangement of the constituent parts of the canopy. **Canopy cover**: It is the proportion of the ground area covered by the canopy when viewed vertically. **Canopy density:** The bulk density of the canopy (mass unit volume). **Canopy height**: The surface height of an undisturbed canopy or the compressed height of a canopy, normally measured from ground level.

Capybara: Capybara (*Hydrochoerus hydrochaeris*) is herbivores that graze mainly on grasses and aquatic plants, as well as fruit and tree bark. As with other grazers, they can be very selective and feed on the leaves of one species, disregarding other species surrounding it. They eat a greater variety of plants during the dry season, as fewer plants are available. While they eat grass during the wet season, they have to switch to more abundant reeds during the dry season. The capybara's jaw hinge is not perpendicular and therefore they chew food by grinding back-and-forth rather than side-to-side. Capybara are coprophilous, as a source of bacterial gut flora, to help digest the cellulose in the grass that forms their normal diet, and to extract the maximum protein and vitamins from their food. They may also regurgitate food to masticate again, similar to cud-chewing by a cow. As with other rodents, the front teeth of capybara grow continually to compensate for the constant wear from eating grasses; their cheek teeth also grow continuously.

Carrying capacity: The maximum stocking rate that will achieve a target level of animal performance, in a specified grazing system that can be applied over a defined time without deterioration of the grazing land. In general terms, carrying capacity is a useful concept when based on adequate historical data and experience but is a number in a constant state of change. Carrying capacity includes the effects of variables that are not easily measured and whose impacts are not readily anticipated and for which it may be difficult or impossible to adjust (example: weather). Thus, carrying capacity is both site-specific and varies from season to season and year to year. The 'average' carrying capacity refers to the long term carrying capacity averaged over years, whereas the 'annual' carrying capacity refers to a specific year. It may also be defined over parts of years. While the definition above applies to the objective of animal production, there are increasing numbers of land uses that can be the primary objective of carrying capacity. These include economic, environmental and ecological objectives and those relating to biodiversity, ecotourism, global climate change and recreation. Because of the multifunctional uses of grazing lands, the carrying capacity can differ from one objective to another.

Concentrate: Concentrates refer to animal feeds that are rich in energy and/or protein but low in fibre, such as corn, soybean meal, oats, wheat, molasses, *etc.*

Conservation: The process of saving forage for future use. Forage can be conserved in situ (*e.g.* stockpiling) or harvested, preserved and stored (*e.g.* hay, silage, haylage).

Continuous stocking (Continuous gazing): A method of grazing livestock on a specific unit of land where animals have unrestricted and uninterrupted access throughout the time when grazing is allowed. The length of the stocking period should be defined and in context with the rationale and season of use (Example: Grazing stockpiled forage from late autumn to late winter).

Coprophagy: Rabbits are herbivores that feed by grazing on grass, forbs, and leafy weeds. They graze heavily and rapidly for about the first half-hour of a grazing period (usually in the late afternoon), followed by about half an hour of more selective feeding. If the environment is relatively non-threatening, the rabbit remains outdoors for many hours, grazing at intervals. Their diet contains large amounts of cellulose, which is hard to digest. Rabbits solve this problem by using a form of hindgut fermentation. They pass two distinct types of feeces: hard droppings and soft black viscous pellets, the latter of which are known as caecotrophs and are immediately eaten (*coprophagy*). Rabbits reingest their own droppings (rather than chewing the cud as do cows and many other grazer) to digest their food further and extract sufficient nutrients.

Creeping grass: Grasses that spread by stolon's, rhizomes or both.

Creep stocking: A method to allocate unrestricted quantities of high-quality forage to maximize intake by juvenile animals while restricting forage intake to meet but not exceed the nutritional requirements of their dams. This method allows juvenile animals to graze in areas that their dams cannot access at the same time to optimize animal performance through highly selective grazing without competition from the dams.

Crop: The cultivated produce of the land.

Crop land: Land devoted to the production of cultivated crops. It may be used to produce forage crops.

Crude fibre (CF): This older proximate method was used to divide carbohydrates into digestible and indigestible fractions. When CF content is higher, the energy content of the feed is lower because crude fibre is considered indigestible. Measuring crude fibre was one part of the original system of analysing the "digestible" fraction in feedstuffs. The CF extract was once used as a standard analysis for fibrous parts or the indigestible portion of carbohydrates in feeds. However, some of these substances are partially digestible by microorganisms in the rumen. Crude fibre accounts for most of the cellulose but only a portion of the lignin and no ash, so it underestimates true fibre and is less than acid detergent fibre (ADF). Thus, CF is not a good indicator of digestibility in ruminant animals, and the use of this parameter in feeds for ruminants is declining. Even though CF is not a very useful parameter for quantifying forage fibre where lignin content is substantial, the CF is a reasonable estimate of the fibre in grains because of their low lignin content. Thus, it is still commonly used for analysis of feeds for non-ruminants or monogastric animals (*i.e.*, those that do not have a chambered stomach or rumen; for example, horses and pigs). Crude fibre is still used today as the legal measurement of fibre in grains and finished feeds.

Crude protein: Crude protein does not distinguish between true protein and non-protein nitrogen and does not describe crude protein digestibility or quality. Proteins are organic compounds composed of building blocks called amino acids. They are a major component of vital organs, tissue, muscle, hair, skin, milk and enzymes. Protein is required on a daily basis for maintenance, lactation, growth and reproduction. The crude protein content of a feed sample represents the total nitrogen (N) in the diet, which includes not only true protein but also non-protein nitrogen (*e.g.*, urea and ammonia in a feed; nitrate is not included in non-protein nitrogen). Because N is an integral part of any amino acid, non-protein nitrogen has the potential to be utilized for protein synthesis by rumen microorganisms. In laboratory analysis, total N present in a feed sample is first determined and then the total amount of protein is calculated by multiplying the total N by a factor. This factor is 6.25 for forages because leaf and stem tissue proteins generally contain 16 per cent nitrogen, or one part nitrogen to 6.25 parts protein. For seeds, this factor is different (*e.g.*, 5.70 for wheat and 5.90 for other cereal grains). Unless otherwise stated, protein values given in laboratory reports, feed tables and feed tags are crude protein. Because the protein content of forages, silages or grains used in animal feeding are sometimes inadequate to meet the needs of the animal class, protein supplements become essential. Consequently, analysis for total protein or crude protein in a feed sample is important. Crude protein in feeds for ruminants can be further fractionated according to their rate of breakdown in the rumen, as discussed below for neutral detergent fibre insoluble crude protein (NDFICP) and discussed previously for acid detergent fibre insoluble crude protein (ADFICP). No doubt, CP is an important indicator of the protein content of a forage crop, and even estimates of non-protein nitrogen are important in evaluating nutritive value. However, it is a false perception that protein is always the most limiting nutrient in the animal's diet and CP is the ultimate measure of a forage quality. In fact, the energy value of forages is often the most limiting attribute for meeting an animal's requirements in most forage-based feeding. An over emphasis on CP may cause one to fail to pay due attention to meeting energy requirements. Furthermore, CP is merely an estimate of nitrogen content (N per cent × 6.25 = CP, per cent) and must be considered in context of plant maturity, species, fertilization rate and many other characteristics. For example, a high nitrate concentration in the forage will result in an artificially high CP level.

Decomposition: The processes of biodegradation of dead plant material, including detachment from the plant, movement to the litter layer and incorporation into soil organic matter.

Deferment: The postponement or delay of grazing or harvesting to achieve a specific management objective. A strategy aimed at providing time for plant reproduction, establishment of new plants, restoration of plant vigour, a return to environmental conditions appropriate for grazing or the accumulation of forage for later use.

Deferred rotation: Deferred rotation "involves at least two pastures with one not grazed until after seed-set". By using deferred rotation, grasses can achieve maximum growth during the period when no grazing occurs.

Deferred stocking: It is a method to defer grazing on the land units that may or may not be in a systematic rotation with other land units. A key concept of deferred stocking is that the deferment is a conservation practice for restoring and maintaining the desired condition of the grazing land. It is not a practice to increase livestock production within a stocking season. However, along with other management strategies, such as reseeding, weed control and prescribed burning, deferred stocking can improve the response of desired vegetation and, over time, increase animal production potential.

Defoliation; It is removal of plant tissue by grazing animals or machinery. While 'defoliate' is derived from 'foliage' which means 'leaves,' defoliation under grazing or mechanical harvest removes leaves, stems and inflorescences in varying proportions.

Desert land: Land on which vegetation is sparse or absent and is characterized by an arid climate. Deserts may be classified as hot or cold deserts depending on latitude and elevation.

Digestibility: The proportion of the dry matter, organic matter or nutrients absorbed during passage though the digestive tract. True digestibility is the actual digestibility of the dry matter, organic matter or specific nutrient consumed and excludes endogenous excretions in the faeces. Apparent digestibility is the difference between dry matter, organic matter or nutrient consumed and dry matter or nutrient excreted in the faeces expressed as a proportion but does not account for endogenous excretions in the faeces.

Digestible crude protein (DCP): It is the most common way of expressing the protein value and requirements for the ruminants. DCP of feed stuff is obtained by multiplying the crude protein content by digestibility coefficient of the protein in the forage.

Digestible dry matter: DDM (or DMD) is the portion of the dry matter in a feed that is digested by animals at a specified level of feed intake. There is no direct laboratory method for measuring DDM/DMD. It is often estimated by measuring *in vitro* or *in situ* digestibility. Both of these analyses are rather expensive and laborious. So, *in vitro* digestibility is frequently estimated by near infrared reflectance (NIR) analysis and/or estimated from the acid detergent fibre. The DDM can be calculated as follows:

Per cent DDM = 88.9 – [0.779 × per cent ADF (on a dry matter basis)].

Digestible energy (DE): The energy apparently absorbed from the digestive tract (energy in the food minus energy lost in the faeces). Digestible energy provides an indication of the actual amount of energy from a feed that can be available for use by the animal. It is estimated by subtracting energy lost in the faeces (faecal energy or FE) from the gross intake energy (GE), (*i.e.*, DE = GE – FE). Digestible energy is commonly used to evaluate poultry and horse feed. For poultry feed, DE is considered as an appropriate measure of feed quality, because FE is almost the sole form of energy loss during digestion. However, in horses, given that FE only partially accounts for the energy losses (considerable losses also occur via urine and gases) in the process of the utilization of nutrients, DE may over estimate low quality feeds relative to high quality feeds.

Dry matter basis: Dry matter basis indicates the nutrient levels in a feed sample based on its dry matter content (*i.e.*, excluding its water content). This is also referred to as "Dry Basis," "Dry Results" or "Moisture-free Basis." As there is considerable variation in the water content of forages, excluding the water or expressing the nutrient levels on a dry matter basis eliminates the dilution effect of the water, thereby providing the essential common basis for direct comparison of the nutrient contents across different forages.

Dry-matter intake: The amount of forage consumed by an animal on a dry-matter basis. Dry matter intake is the amount of (or prediction of the amount of) dry matter consumed by the animal and is a central concept to any discussion of animal nutrition. Typically, intake increases as the digestibility of the forage increases. However, anti-quality components such as tannins and alkaloids in feeds and forages may decrease intake. Scientists have consistently observed that as the per cent of neutral detergent fibre (NDF) increases in the feed, animals consume less (*i.e.*, DMI is less). This relationship, along with estimates of NDF digestibility, is used to estimate DMI for grasses and legumes using the following equations:

$DMI_{Grass} = -2.318 + 0.442 \times CP - 0.0100 \times CP^2 - 0.0638 \times TDN + 0.000922 \times TDN^2 + 0.180 \times ADF - 0.00196 \times ADF^2 - 0.00529 \times CP \times ADF$

Where: DMI_{Grass} is expressed as per cent of BW, and CP (Crude Protein), ADF (Acid Detergent Fibre), and TDN (Total Digestible Nutrient) are expressed as per cent of DM.

$DMI_{Legume} = (120 \div NDF) + ([\{NDFD - 45\} \times 0.374] \div 1350) \times 100$

Where DMI_{Legume} is expressed as per cent of BW, NDF (Neutral Detergent Fiber) as per cent of DM, and NDFD (48-hour *in vitro* NDF digestibility) as per cent of NDF with NDFD adjustment proposed by Oba and Allen (1999)]. Though these calculations have been proven to provide reasonable estimates of DMI, the estimates are not perfect. Dry matter intake is affected by the condition of the animal (*e.g.*, age, body weight, pregnancy status, level of milk production, *etc.*), feed factors (*e.g.*, palatability, balance of the diet, anti-quality factors in the feed, *etc.*) and the feeding environment (*e.g.*, temperature, humidity, *etc.*).

Dry-matter intake (Organic-matter intake): The amount of forage consumed by an animal on an organic-matter basis.

Energy: The potential to do work. Usually expressed as mega joules (MJ) kg^{-1} [mega calories (Mcal) kg^{-1}]. Forage dry matter where one calorie is standardized to be equal to 4.184 joules.

Ether extract: Ether extract is a portion of dry matter extracted with ether. It is a laboratory test to approximate the total fat (or crude fat) content of a feed and includes some waxes, pigments and other lipids to a minor degree in addition to true fats.

Fat: Chemically, fats are "triglycerides of fatty acids" that are a high-density source of energy for animals. Fat is rich in energy; it contains 2.25 to 2.8 times the energy found in carbohydrates and is highly digestible. Fat is added to rations to boost energy levels when intake may be limited due to poor animal health, less palatable feed or environmental stress. Some concentrates (such as soybean meal) contain relatively high levels of fats and oils. Fats are composed of building blocks called *fatty acids*. Fats in feed samples are typically determined through ether extraction (EE). In addition to fat, EE may solubilize some other compounds like plant pigments, esters and aldehydes. This is why the measurement of fat through EE is called crude fat. True fat can be measured by determining the content of fatty acids or it can be estimated in forages as ether extract minus one.

Fibre (US: Fibre): A nutritional entity that is relatively resistant to digestion and is slowly and only partially degraded by herbivores. Fibre is considered to be composed of structural polysaccharides, cell-wall proteins and lignin but chemical composition varies with type of plant cell wall and method of determination. It is a biological unit rather than a distinct chemical entity.

Field: A defined area of land used for cultivating crops or growing forages.

First-last stocking: A method of utilizing two or more groups of animals, usually with different nutritional requirements, to graze sequentially on the same land area. If more than two groups of animals graze sequentially, this would be described as 'first, second and last stocking.' The objective of this stocking method is to allocate nutrition among different groups of animals with different nutritional requirements such as lactating dairy cows and dry cows. Higher selective grazing and greater forage mass present during the period of occupation by lactating cows can contribute to meet their higher nutrient requirements, compared with dry cows that are the second group to occupy the paddock. It may also include the objective of increasing total forage use such as grazing cattle or sheep as the second group of grazing animals behind horses as the first group.

Flora: All plant species occurring within a site or a region. The flora of a site, region or country may be described in a systematized list of the collective species.

Fodder: Harvested forage fed intact or after chopping to livestock. It includes fresh, conserved and dried forage.

Forage: Edible parts of plants, other than separated grain, that can provide feed for grazing animals or that can be harvested for feeding. It includes browse, herbage, and mast.

Forage Crop: A crop of cultivated plants, other than separated grain, produced to be grazed or harvested for use as feed for animals.

Forage allowance: The relationship between forage mass and animal live weight per unit area of the specific unit of land being grazed at any one time; an instantaneous measurement of the forage-to-animal relationship. It is inverse of grazing pressure. This definition can be appropriately altered to be specific to herbage or browse by substituting these terms in place of forage. Forage allowance is expressed as a ratio of forage mass (kg per hectare) to animal live weight (kg per hectare) at a specific time. To describe forage allowance over time, a series of instantaneous measures are averaged. Where needed, it may be expressed as a ratio of forage mass to animal units or forage intake units per unit area at a specific time.

Forage intake: The forage consumed by an animal. Intake is expressed per unit of time such as intake per day, intake per month), intake per year or intake per stocking season.

Forage intake unit: A unit to measure the rate of forage consumption by grazing animals where one forage intake unit is equivalent to the consumption of 8.8 kg dry matter per day. By definition, one forage intake unit has a dry-matter intake rate of 8.8 kg per day; thus, any animal at any age or stage of production may be represented as a certain proportion or multiple of the forage intake unit, based solely on its rate of forage dry-matter intake per day. The intake rate of an animal that is larger or smaller than 8.8 kg dry matter per day will have a forage intake-unit-equivalent, which is a proportion or multiple of one forage intake unit.

Forage quality: Forage quality refers to the ability of forage to support desired levels of animal performance (*e.g.*, daily gain or milk production). It is a function of voluntary intake and nutritive value (nutrient content and digestibility).

Forb: Any herbaceous, dicotyledonous broad-leaved plant.

Forest land: Land on which the vegetation is dominated by trees or, if trees are lacking, the land bears evidence of former forest and has not been converted to other vegetation or land use.

Forward creep stocking: A method of creep stocking where dams and offspring rotate through a series of paddocks with offspring as first grazers and dams as last grazers. A specific form of First-last stocking.

Frontal stocking: A method that allocates forage within a land area by means of a sliding fence that livestock can advance to gain access to ungrazed forage.

Fruit: The seeds of plants and pulpy surrounding tissues.

Graminivory: Graminivory is a form of grazing involving feeding primarily on grass (specifically "true" grasses in the Poaceae). Horses, cattle, capybara, hippopotamuses, grasshoppers, geese, and giant pandas are graminivores. Giant pandas (*Ailuropoda melanoleuca*) are obligate bamboo grazers, 99 per cent of their diet consisting of sub-alpine bamboo species.

Grass: Plant or plant species of the *Poaceae* family. Given the importance of grasses in grazing lands, other terms have been used to further define specific characteristics.

Grassland: The term 'grassland' is synonymous with pasture land when referring to an imposed grazing-land ecosystem. The vegetation of grassland in this context is broadly interpreted to include grasses, legumes and other forbs, and at times woody species may be present (Native or Natural Grassland). There are many descriptive terms for pasture land/ grassland that take into account their age and stability. **Annual pasture land/ Grassland**: Forage is established annually, usually with annual plants, and generally involves soil disturbance, removal of existing vegetation, and other cultivation practices. **Cultivated pasture land/ grassland:** Forage is established with domesticated introduced or indigenous species that may receive periodic cultural treatment such as renovation, fertilization or weed control. **Permanent pasture land/ grassland:** Land on which vegetation is composed of perennial or self-seeding annual forage species which may persist indefinitely. It may include either naturalized or cultivated forages. **Temporary pasture land/ grass land/ Ley farming:** Land on which vegetation is composed of annual, biennial, or perennial forage species kept for a short period of time (usually only a few years). Temporary pasture land/grassland can be regularly resown or can be integrated in a crop rotation (ley). It is usually composed of simple mixtures of grasses, grass/legume or legume species. Ley is integrated in a crop rotation. **Naturalized pasture land/ grass land:** Forage species present are primarily introduced from other geographical regions that have become established and have persisted under the existing conditions of environment and management over a long time and used mainly for grazing by livestock and wildlife. There are many types of natural grasslands, with vegetation characteristics determined by climate and soil conditions, by grazing animals and by fire. **Semi-natural pasture land:** Managed ecosystem dominated by indigenous or naturally occurring grasses and other herbaceous species (Native grassland). **Meadow:** Natural or semi-natural grassland often associated with the conservation of hay or silage. A meadow may exist as a result of discontinuous features of hydrology, landscape position, or soil characteristics that differ from the surrounding landscape and vegetation. Descriptive terms include 'mountain meadow,' 'alpine meadow,' 'wet meadow,' and 'hay meadow.' 'Flower meadows' are kept for aesthetic interest and can also provide feeding or bedding. **Campos:** Grassland consisting mainly of grasses, along with herbs, small shrubs and occasional trees on undulating and hilly landscape, with variable soil fertility. It differs from Cerrado in having a longer and more severe winter and a relative abundance of native legumes. **Cerrado**: Savanna with varying amounts of trees and shrubs along rivers and in valley bottoms. It is characterized by a tropical climate with alternating wet and dry seasons. The wet season lasts usually six months. **Llanos**: Extensive system of grasslands, seasonally flooded, with infertile and acidic soils. The tropical climate is characterized by alternating wet and dry seasons. **Pampa**: Tree less grasslands on flat and fertile plains. The Pampa is a temperate grassland or a sub-tropical steppe. The climate is humid to arid; summers are warm and winters are mild. **Prairie:** Nearly level or rolling grassland, originally tree less or with a few scattered trees, and usually on fertile soils. It may be characterized as a short-grass, intermediate-grass, or tall-grass prairie depending on the influence of a continental climate and variation in total summer precipitation, rate of evapo-transpiration, periodic fire and soil depth. Soil depth and precipitation generally increase from west to east and vegetation changes from short-grass prairie in the west to tall-grass prairie in the east. **Sahelian steppe:** Discontinuous vegetation dominated by annual C_4 plants, especially grasses, and scattered shrubs. The arid or semi-arid tropical climate with alternating wet and dry seasons is characterized by a strong variability in rainfall patterns and one short rainy season. The soils are generally poor. **Savanna:** Grassland characterized by precipitation between 375 and 1500 mm year, variable proportions of trees or large shrubs, especially in tropical and sub-tropical regions. It is often a transitional vegetation type between grassland and forest land. Tropical savannahs are characterized by a climate with alternating wet and dry seasons. The wet season usually ranges between five and nine months. Sub-tropical savannas have a wet climate with warm summers and mild winters. **Steppe:** Semi-arid, sparse to rolling grassland characterized by short to medium-height grasses occurring with other herbaceous vegetation and occasional shrubs. **Forest steppe:** soils are black or brown-earth with high to medium contents of organic matter and high mineral contents **Veld:** Indigenous vegetation used as grazing and/or browsing which may be composed of any of a number of plant growth forms (predominantly C_4 grasses and *Acacia* or broad-leaf trees) and need not necessarily be climax vegetation. **Marsh land:** Flat, wet, treeless wetland usually covered by shallow water and dominated by marsh grasses, bushes, sedges, other grass-like plants and forbs. **Shrub land:** Land on which the vegetation is dominated by low-growing woody plants. **Tundra:** Land areas in arctic and alpine regions devoid of large trees, varying from bare ground to various types of vegetation consisting of grasses, sedges, forbs, dwarf shrubs and trees, mosses and lichens.

Grazers-scrapers: Water animals that feed for example on algae found on stones are called grazers-scrapers. Grazers-scrapers feed also on microorganisms and dead organic matter on various substrates.

Grazing: The eating or partial defoliation of any kind of standing vegetation by domestic livestock or other animals is called as grazing. Conservation grazing is the use of grazing animals to help improve the biodiversity of a site. Due to their hardy and thrifty nature, rare and native breeds are often used in conservation grazing. In some cases, to re-establish traditional hay meadows, cattle such as the English Longhorn and Highland (which happen to be rare breeds) are used to provide low intensity grazing. Grazing is a method of feeding in which a herbivore feeds on plants such as grasses, or other multi cellular organisms such as algae. In agriculture, grazing is one method used whereby domestic livestock are used to convert grass and other forage into meat, milk and other products.

Grazing event: It is the activity of grazing (including biting and masticating but not ruminating) without stopping (stocking period). When monitoring grazing activity, it is necessary to define a specific interval that will be the separation point of grazing events from other activities.

Grazing land: It's any vegetated land that is grazed or has the potential to be grazed by animals (domestic and wild). This term is all-inclusive and covers all kinds and types of land that can be grazed. **Graze:** To consume predominantly herbaceous forage in situ by animals. **Ungrazed:** It describes the status of grazing land that is not grazed by animals and the status of plants or plant parts that are not grazed by animals.

Grazing land management: The manipulation of the soil–plant–animal complex of the grazing land in pursuit of a desired result. The definition may be applied to specific kinds of grazing land by substituting the appropriate term such as grassland, pastureland or rangeland in place of grazing land.

Grazing management: It is manipulation of grazing in pursuit of a specific objective or set of objectives. **Extensive grazing management:** Grazing management that uses relatively large land areas per animal and a relatively low level of labour, resources or capital. **Intensive grazing management:** Grazing management that uses relatively high levels of labour, resources or capital to increase production per unit area or per animal, through a relative increase in stocking rates, grazing pressure and forage utilization. Intensive grazing management is not synonymous with rotational stocking. Grazing management can be intensified by substituting any one of a number of stocking (grazing) methods that increase production and efficiency of resource use. **Sequence grazing:** It is the grazing of two or more land units in a succession that differ in forage species composition. Sequence grazing takes advantage of differences among forage species and species combinations, grown in separate areas for management purposes, to extend grazing seasons enhance forage quality and/or quantity, or achieve some other management objective. **Strip grazing:** Confining animals to an area of grazing land to be grazed in a relatively short period of time, where the paddock size is varied to allow access to a specific land area. Strip grazing may or may not be a form of rotational stocking, depending on whether or not specific paddocks are utilized for recurring periods of grazing and rest. **Mob grazing:** In the management of a grazing unit, grazing by a relatively large number of animals at a high stocking density for a short time period. **Mixed grazing:** two or more types of livestock, such as sheep and cattle, are grazed on the same pasture.

Grazing management unit: The entire grazing land area used to support grazing animals over a defined time, generally a year. A grazing management unit may be a single area or it may have a number of subdivisions (Paddock, Pasture). It may also include periods of grazing and periods of rest depending on the stocking method(s) used or migrations of wildlife and does not imply continuous occupation by grazing animals during the defined time.

Grazing pressure: It is the relationship between animal live weight and forage mass per unit area of the specific unit of land being grazed at any one time; an instantaneous measurement of the animal-to-forage relationship. Grazing pressure may also be expressed as a ratio of animal units or forage intake units per unit forage mass to compare across differences in animal species or stages of production. Expressing grazing pressure in units of animal demand provides the basis for calculation of Grazing Pressure Index. To describe grazing pressure over time, a series of instantaneous measures are averaged. This differs from Grazing pressure index, which is a ratio of the integration of forage consumption to the integration of forage growth rate over time. This definition can be appropriately altered to be specific to herbage or browse by substituting these terms in place of forage.

Grazing pressure index: An animal-to-forage relationship measured in terms of integrated forage consumption by the animal (kg per day) over initial forage mass plus integrated forage growth rate (kg per day) over time. Grazing pressure index differs from grazing pressure in that grazing pressure is an instantaneous measurement of the animal-to-forage relationship.

Grazing station: It is the position from which an animal takes multiple bites without moving its feet.

Grazing system: A defined, integrated combination of soil, plant, animal, social and economic features, stocking (grazing), method(s) and management objectives designed to achieve specific results or goals. A grazing system is site-specific because it integrates specific biotic and abiotic components and their environments, management objectives and social factors. System behaviour is a consequence of the interrelationships among the parts of the system. When a system component is managed in isolation away from the influence of the rest of the system, it is no longer under the same influences and may behave differently. Thus, when managed within a system, responses and behaviour of plants and animals may differ from those observed when managed alone or in another system. Grazing system should include at least the following information: number, size, kind, slope, erosion status and soil classification of land units; number, kind, gender, size and age of livestock; duration of use and non-use periods for each unit in the system; stocking method(s); type(s) of forage; and geographical location and elevation; type of climate, mean annual and seasonal temperatures, and precipitation amounts and distribution. Grazing systems generally can be grouped into categories. **Nomadic grazing systems:** Systems based on extensive movement of herds and flocks in search of forage, led by human family units with no permanent home base. **Semi-sedentary grazing systems:** Systems based on a village permanently occupied by women and children from which herds and flocks, usually tended by men and boys, are absent for extended periods of time in search of forage. **Transhumance grazing systems:** Systems that differ from semi-sedentary systems in that the grazing is cyclical beginning at the end of winter with flocks and herds leaving lowland grazing areas where permanent villages are located and moving to mountain pastures for grazing during summer. Latitudinal transhumance occurs in tropical climates where the cyclic movement is under the influence of alternating wet and dry seasons. **Sedentary grazing systems:** Grazing systems managed at a particular location(s) by resident management. May be managed by either or both extensive and intensive grazing managements and can include rangeland, pasture land, crop land and forest land within the grazing system. **Short duration grazing**: A rotational grazing system employing high stocking density, one herd, commonly 5–12 pasture units, grazing periods of 3–10 days, and two to several grazing period cycles per year; also the common 'rotation grazing' of improved pasture.

Grazing time: The total amount of time devoted to grazing during a specified time; usually 24 hours.

Grazing value: The worth of a plant or cover for livestock determined by palatability and nutritional value as excellent, fair, good or poor.

Growth: The production of new plant tissue by forage plants. The term forage growth is used specifically to define the rate of production of new plant tissue. It can also be used as a general descriptor of plant development and structural change over time. The measurements may be reported in terms of fresh weight or (preferably) dry matter or organic matter weight per unit area (g/m^2; kg/

ha; tonnes/ha) and per unit time (day, annum), but may also be expressed per plant or plant unit.

Gross energy: Gross energy refers to the total energy in a feed before accounting for losses due to normal digestive, metabolic and productive functions. It is determined by measuring the amount of heat produced when a feed is completely oxidized in a bomb calorimeter. It is not a very useful measure since the gross energy in most common feeds is about the same, but they do not result in similar animal performance.

Habitat: The term habitat is used to describe a specific area where a particular animal lives, within an environment. Many animals have adapted to requiring specific conditions which can only be found in their natural habitat such as those animals that live in the Polar Regions that have longer, thicker body fur to keep them warm.

Harvest: Forage defoliated by a single grazing or cutting, or over a series of grazing's or cuttings. It may be reported as daily amount, a single harvest, or seasonal or annual totals. The terms **'Forage production'** and **'Forage yield'** are often used as alternatives to the term 'Forage harvested' but may be misleading, particularly in the context of dynamic tissue flow measurements.

Hay: Hay is feed preserved by dehydrating harvested green forage to a moisture content of 15 per cent or less, so that the biological processes do not proceed rapidly enough to build-up heat.

Haylage: Harvested forage ensiled at a moisture content of 35-50 per cent moisture under anaerobic conditions for future use.

Hemicellulose: Like cellulose, hemicellulose is a carbohydrate that exists in almost all plant cell walls along with cellulose. Whereas cellulose is composed only of glucose, hemicellulose is composed of many other sugars (*e.g.*, glucose, xylose, mannose, galactose, arabinose, *etc.*) in chains of 500 to 3,000 sugar units. Hemicellulose is a branched polymer. In contrast, cellulose is not branched. As hemicellulose content increases in animal feed, the voluntary feed intake typically decreases.

Herbage: The above-ground biomass of herbaceous plants, other than separated grain. Grasses, grass-like species, herbaceous legumes and other forbs collectively; the foliage and edible stems of herbs, generally above ground but including edible roots and tubers.

Herbaceous: Herbaceous refers to the non-woody above-ground parts of grass, grass-like legume and forb plants. Herbaceous species are differentiated from woody species by not having perennial woody stems.

Herbivore: Herbivore is an animal that only eats plant material, algae and bacteria in order to gain its nutrition. Those animals that are herbivorous have adapted to digest plant material specifically, such as elephants, donkeys and rabbits.

Holistic resource management (Holistic management): It refers to a management process/philosophy that frequently involve the use of rotational stocking with relatively high stocking densities. Holistic management is not an individual method but is a philosophy and should not be used as a name for any particular stocking method.

Hortipasture: In which land is managed by growing fruit trees and the interspaces utilized for growing understory forage crops for feeding livestock.

Ingestive behaviour: The behaviour of the animal involved in grazing including time devoted to searching for, selecting, apprehending and consuming forage, usually on a daily basis. Ingestive behaviour is usually described in terms of quantifiable activities like bite weight, bight rate, grazing event and grazing time. Bite weight, biting rate and grazing time are components of ingestive behaviour. Intake can be estimated as the product of these three components (Intake = bite weight × biting rate × grazing time). Ingestive behaviour is influenced by animal, plants and soil factors, the environment, time of day, season, precipitation, management and other factors.

***In situ* digestibility:** *In situ* digestibility is determined by incubating a ground forage sample in a porous nylon bag placed within the rumen via a fistula or port in the animal's side (*in situ*) for a fixed time period.

Intensive early stocking: A method of using high grazing pressure during an initial restricted period of the stocking season followed by total removal of livestock for the remainder of the season to allow rest and recovery by the forage. This method, designed for use with native rangelands dominated by warm-season species, provides a way to maximize use of forage during the early part of the stocking season when digestibility is generally highest and to overcome low forage digestibility during late summer.

Intermittent stocking: A method that imposes grazing on a particular management unit or area of land for indefinite periods at irregular intervals.

Invader– Plant that pushes out native plant species and decreases their population by increasing its own population

***In vitro*:** *In vitro* (Latin for "within the glass") generally refers to the technique of performing a given biological procedure in a controlled environment outside of a living organism. In other words, it is a process that is carried out in a test tube. In feed testing, *in vitro* refers to a feed sample that is digested in test tubes or tested outside the animal.

***In vitro* digestibility:** *In vitro* digestibility of a feed is determined by incubating a ground feed sample with rumen fluid in a beaker or test tube for 24 to 48 hours, followed either by addition of acid and pepsin and further incubation for 24 hours or by boiling in neutral detergent fibre solution.

Legume: Plant or plant species of the *Fabaceae* (*Leguminosae*) family with a wide range of physical characteristics from herbaceous forbs to shrubby and tree forms. The ability to form a symbiotic relationship with bacteria to fix atmospheric nitrogen is found in many species of legumes.

Leaf area index (LAI): The area of green leaf (one side only per unit area) of ground. It refers to leaf only or to lamina plus half the surface area of exposed sheaths and petioles.

Ley farming: A rotation of arable crops requiring annual cultivation and artificial pastures occupying field for two years or longer in a rotation cycle.

Lignin: Lignin is a complex carbohydrate compound, a major structural component of mature plants, contained in the fibrous portion of plant stems, leaves, cobs and hulls. It is undigestible and hence has a negative impact on cellulose digestibility. As the lignin content in a feed increases, digestibility of its cellulose decreases, thereby lowering the amount of energy potentially available to the animal. The term "lignin" was derived from the Latin word "lignum," meaning wood, since it most commonly occurs in the woody tissues of plant materials.

Lipid: Lipids are substances found in plant and animal tissues that are insoluble in water but soluble in benzene or ether; lipids include glycolipids, phosphoglycerides, fats, oils, waxes and steroids.

Litter: An accumulation of dead detached plant material at the soil surface. When appropriate, litter may be more narrowly defined, for example as tree litter (large woody material including branches, leaves and dead trees). Mulch is frequently used as a term for herbaceous material.

Macro-minerals: Macro-minerals, also called major minerals, are the elements present in the animal body in relatively larger amounts than micro minerals (thus the name). Therefore macro-minerals are required in relatively large amounts, generally in gram (g) quantities per head per day, if the animal is to live and function properly. Macro-minerals perform specific roles in the body's structure and functions; for example, they help to build body tissues (*e.g.* bone) or to regulate metabolic activities. Macro minerals interact with each other and must be supplied in proper quantities and ratios to maintain appropriate animal function. The seven macro-minerals essential to animals are: calcium (Ca), phosphorus (P), sodium (Na), magnesium (Mg), potassium (K), sulphur (S) and chlorine (Cl). The total mineral content of the body is approximately 4 per cent of body weight. Calcium makes up approximately 1.75 per cent of total body weight, phosphorus makes up approximately 1.10 per cent of total body weight, and magnesium makes up approximately 0.04 per cent of total body weight.

Mast: Fruit and seed of shrubs, woody vines, trees, cacti and other non-herbaceous vegetation available for consumption by animals.

Meadow: A tract of grassland where productivity of indigenous or introduced forage is modified due to characteristics of the landscape position or hydrology.

It may be characterized as: hay meadow, native meadow, mountain meadow, wet meadow, or other designations.

Metabolism: A mixture of chemical processes that occur within the body of an animal to either release energy (breaking down food) or to consume it (muscle movement).

Metabolizable energy (ME): The energy available for metabolism by an animal (energy in the food minus faecal energy, urinary energy and gaseous energy losses).

Micro-minerals (Micronutrients): Micro-minerals, or trace minerals, are present in the animal body tissues in extremely low concentrations. They are nutrients required in small amounts, generally in milligram (mg) or microgram (μg) amounts per head per day, but play critically important roles in animal nutrition. There are 10 micro-minerals recognized in animal nutrition: iron (Fe), manganese (Mn), copper (Cu), zinc (Zn), selenium (Se), cobalt (Co), iodine (I), chromium (Cr), molybdenum (Mo) and nickel (Ni). However, copper is particularly toxic in sheep and selenium can also be toxic. Deficiencies and toxic levels are quite regional.

Migration: The migration of animals is generally connected with the seasons and involves with travelling between one place and another, often along a well-known route.

Minerals: In feed analysis, mineral refers to inorganic feed elements essential for life.

Mixed stocking: A method of stocking two or more species of grazing or browsing animals on the same land unit, not necessarily at the same time but within the same stocking season. Objectives of mixed stocking include increased forage utilization, altering botanical composition, weed control and interruption of parasite cycles. Mixed stocking may be a form of first-last stocking where one animal species is followed by a second animal species with different grazing behaviour with the objective of increasing total forage use. In wildlife systems, many animal species can occupy the same land area either simultaneously or intermittently. Mixed stocking on rangelands is sometimes referred to as 'common use.'

Mob stocking: A method of stocking at a high grazing pressure for a short time to remove forage rapidly as a management strategy.

Monogastric: Animals with a monogastric digestive tract can get cellulose from their diet and digest some of it. However, unlike ruminants, which are animals with more than one stomach compartment, gathering energy from cellulose is not as efficient. There is symbiotic bacteria in the stomach that aids in cellulose digestion. Anatomical adaption over time is what makes the partial digestion of cellulose possible. Monogastric animals have a stomach that has a single compartment. Humans have a monogastric stomach, as do pigs, chickens, rabbits and horses.

Mycotoxins: Mycotoxins are toxic to animals, and are produced on plants by fungi, particularly during weather stress during the growing or harvest seasons or during feed storage (*e.g.*, vomitoxin, zearalenone, aflatoxin and T-2, *etc.*).

Net energy (NE): Metabolizable energy minus energy lost as heat of fermentation and heat of nutrient metabolism. The net increase in useful animal product expressed per unit increase in food intake. **Net energy for maintenance**: Change in retained energy per unit of feed intake measured between an intake of zero and intake when retained energy equals zero. **Net energy for gain (product deposition)**: Change in retained energy per unit of feed intake measured in a growing animal consuming an intake those results in a retained energy greater than zero. **Net energy for lactation**: Change in lactation energy (*i.e.*, energy in milk) per unit of feed intake measured under conditions in which retained energy remains constant.

Neutral Detergent fibre (NDF): NDF is the residue or insoluble fraction left after boiling a feed sample in neutral detergent solution. The NDF contains plant cell wall components except for some pectin's. The NDF is considered a close estimate of the total fibre constituents of feedstuffs since it measures cellulose, hemicellulose, lignin, silica, tannins and cutins. The hemicellulose, cellulose and lignin represent the fibrous bulk of the forage. Because they give the plant rigidity and enable it to support itself as it grows, these three components are classified as structural carbohydrates. Though lignin is indigestible, hemicellulose and cellulose can be (in varying degrees) digested by microorganisms in animals with either a rumen (*e.g.*, cattle, goats or sheep) or hind-gut fermentation (*e.g.*, horses, rabbits, guinea pigs) as part of their digestive tract. NDF concentration is negatively correlated with dry matter intake (*i.e.*, as NDF in the forage increases, animals will consume less forage). As a result, NDF is often used in formulas to predict the dry matter intake.

Neutral Detergent fibre digestibility: NDFD is the 48-hour *in vitro* digestible fraction of NDF expressed as percentage of the Neutral Detergent Fibre (NDF) content of a feed sample.

Nitrate (NO_3): Nitrate concentrations in forages and other feeds are generally low. When the rate of nitrate uptake (*e.g.*, uptake per day) by the plant from the soil exceeds its rate of conversion to protein, nitrates will accumulate in plants. Nitrates can accumulate in the forage crop due to excessive nitrogen fertilization and excessive moisture stress or other factors that limit growth. Excessive concentrations of nitrate in the animal's diet can induce nitrate toxicosis in the animal (*e.g.*, reduced weight gain, failure to rebreed, weakness, staggering and, in severe cases, death). The concentration of nitrate that is considered toxic varies considerably from one class of animal to another. The amounts of nitrate in the animal's water supply should also be considered.

Nitrate toxicity: Toxicity by nitrate accumulation (0.43 per cent) in forage which can cause acute poisoning in the livestock.

Non-fibrous carbohydrate (Neutral Detergent soluble carbohydrate): NFC or NDSC represents all forms of digestible carbohydrates that are solubilized after boiling a feed sample in neutral detergent solution. These are all forms of non-cell-wall carbohydrates and include starch, sugar, pectin and fermentation acids, which are digestible and serve as energy sources for the animal. Because NFC includes some other digestible compounds in addition to starch and sugars included in NSC, NFC generally shows a higher value than non-structural carbohydrate (NSC) on the feed analysis report. NFC is calculated from the following equation: NFC per cent = 100 – [per cent NDF + per cent CP + per cent Fat + per cent Ash], where EE per cent is the ether extract per cent or Fat per cent.

Non-grass grazing: Although grazing is typically associated with mammals feeding on grasslands, or more specifically livestock in a pasture, ecologists sometimes use the word in a broader sense, to include any organism that feeds on any other species without ending the life of the prey organism. Use of the term varies even more than this, for example a marine biologist may describe herbivorous sea urchins that feed on kelp as grazers, even when they kill the organism by cutting the plant at the base. Malacologists sometimes apply the word to aquatic snails that feed by consuming the microscopic film of algae, diatoms and detritus, a biofilm, which covers the substrate and other surfaces underwater.

Non-protein nitrogen (NPN): NPN refers to nitrogen in a feed sample that is not in the form of protein but can be used by the microbial population in the rumen or gastrointestinal tract to synthesize amino acids and proteins. Common forms of NPN are urea and ammonia.

Non-selective stocking: A method that uses high grazing pressures that increase the consumption of less-preferred forage species by grazing animals. Non-selective stocking is generally attempted by using mob stocking with a high animal-to-forage ratio during short time periods. In practice, stocking to overcome preference is achieved rarely.

Non-structural carbohydrate (NSC): NSCs are simple carbohydrates, such as starches and sugars, stored inside the cell that can be rapidly and easily digested by the animal. Hence, NSC is considered to serve as a readily available energy source.

Nutrient requirement: Nutrient requirement refers to the minimum amounts of nutrients (energy, protein, fat, minerals and vitamins) necessary to meet an animal's real needs for maintenance, growth, reproduction, lactation or work (but does not include a safety margin in ration formulation).

Nutritive value: Nutritive value refers to a feed's protein, mineral and energy composition, availability of energy, and efficiency of energy utilization. The predicted animal response based on chemical composition, digestibility and nature of digested products, as estimated by in vitro or in vivo chemical analyses.

Paddock: A grazing area that is a sub-division of a grazing management unit and is enclosed and separated from other areas by a fence or barrier.

Palatability: It is the acceptability of forage to an animal, estimated as free-choice selection from one forage over another or by degree of defoliation of one forage relative to another when the animal has access to both. However, when not given the opportunity to selectively feed the animal may perform just as well on the less palatable forage. This is a subjective term. The preference for one species of forage plant over another can be measured while factors affecting palatability are not generally quantifiable. Palatability is affected by the feed's odour, texture, moisture, physical form and temperature. For a forage plant species to be considered "high-quality," it generally must be highly palatable because quality includes intake and palatability is required for high levels of intake.

Particle size: Particle size refers to the diameter of granular feed materials (*e.g.*, grains, pellets, mineral particles) and/or the length and sometimes width of roughage or forage fragments. Particle size can affect mixing of feed ingredients and digestion rate.

Pasture: A type of grazing management unit enclosed and separated from other areas by fencing or other barriers and devoted to the production of forage for harvest primarily by grazing. Although pasture and pasture land are often used as synonyms, pasture refers to the place whereas pasture land refers to land devoted to the production of forage for harvest primarily by grazing and provides the basis for a land-use mapping unit. Animals eat forage growing in the pasture. Animals do not eat the pasture. The pasture is a grazing management unit of pasture land. The pasture may be sub-divided into two or more paddocks for management purposes. **Pasture land:** Land devoted to the production of indigenous or introduced forage for harvest primarily by grazing. Pasture land generally must be managed to arrest successional processes. **Pasture renovation:** Improvement of a pasture by tillage, reseeding, fertilization and liming.

Patch-burn grazing: Patch-burn grazing burns a third of a pasture each year, no matter the size of the pasture. This burned patch attracts the grazers (cattle or bison) which graze the area heavily because of the fresh grasses that grow in. The other patches receive little to no grazing. During the next two years the next two patches burn consecutively and then the cycle begins a new. In this way, patches receive two years of rest and recovery from the heavy grazing. All this results in a diversity of habitats that different prairie plants and birds can utilize, mimicking the effects of the pre-historical bison/fire relationship where bison heavily graze one area and other areas have opportunity to rest.

Period of occupation: The length of time that a specific land area is occupied, whether by one group of animals or by two or more groups of animals in succession (First-last stocking, Forward creep stocking, Period of stay.

Period of stay: The length of time that a particular group of animals occupies a specific land area. 'Period of occupation' and 'period of stay' differentiate between the total time a specific land area is utilized and the time that a particular group of animals is using said land area. The term is useful in describing stocking methods such as first-last stocking. The 'period of occupation' is the total time that a specific land area is utilized and may involve several different groups of animals moving through in sequence, as in first-last stocking or in migrations. It differs from stocking period in that grazing may or may not be involved (example: feeding hay on pasture in winter when ice prevents grazing). The 'period of stay' defines the fractional part of the 'period of occupation' that any one of the two or more animal groups occupies the specified land area and is only a part of the period of occupation.

Pod: The seed case of a leguminous plant.

Preference: A measure of relative intake of alternative forages or forage constituents, where access to forage is unrestricted. Preference describes an animal response but makes no assumptions about mechanisms determining the response. Preference is a relative expression that requires the opportunity for choice between two or more components. This is an objective measure of selective behaviour and is preferred to the more subjective term 'palatability'. While palatability attempts to describe the perceived acceptability of a single forage type without comparison to an alternative choice, it is subject to various interpretations, can be confounded with preference and is not a recommended term.

Protein: Protein is an essential nutrient. Proteins are composed of long chains of various kinds of amino acids. Animals meet their protein needs by breaking down plant and microbial protein (formed in the rumen) and reassembling them as animal proteins.

Proximate analysis: Proximate analysis is a chemical method of quantitative analysis that separates, identifies and quantifies the major categories of compounds in a mixture. In feed and forage analysis, it serves as a tool for assessing and expressing the broad nutritional value of a sample. The proximate system for routine analysis of animal feed and forage was devised in the mid-nineteenth century. It is important to remember that proximate analysis is not a nutrient analysis. Rather, it is a partitioning of both nutrients and non-nutrients into categories based on common chemical properties. The technique was developed to provide a top level, very broad classification of forage and feed components. The system consists of the consecutive steps of analytical separations and determinations of six categories of components and expressing the percentage of each that is present in a feed sample:

- ☆ Water/moisture (or dry matter)
- ☆ Ash (minerals)

- ☆ Total or crude protein (total nitrogen×6.25)
- ☆ Total or crude fat (or ether extract)
- ☆ Crude fiber (incompletely digested carbohydrates)
- ☆ Nitrogen-free extract (readily digestible carbohydrate)

Pseudo ruminant: The hippopotamus is a large, semi-aquatic, mammal inhabiting rivers, lakes and mangrove swamps. During the day, they remain cool by staying in the water or mud; reproduction and childbirth both occur in water. They emerge at dusk to graze on the grasses. While hippopotamuses rest near each other in the water, grazing is a solitary activity. Their incisors can be as long as 40 cm and the canines up to 50 cm, however, the canines and incisors are used for combat and play no role in feeding. Hippos rely on their broad, horny lips to grasp and pull grasses which are then ground by the molars. The hippo is considered to be a pseudo ruminant, it has a complex three- or four-chambered stomach but does not "chew cud".

Put-and-take stocking: A method of using variable animal numbers during a stocking period or stocking season, with a periodic adjustment in animal numbers in an attempt to maintain desired management criteria, *e.g.*, a desired quantity of forage, degree of defoliation, or grazing pressure.

Quality: A description of the degree to which forage meets the nutritional requirements of a specific kind and class of animal. Quality is a relative term. Nutritional requirements and anatomy vary among different kinds and classes of grazing animals; thus, what constitutes 'high quality' forage for one animal may be 'low-quality' forage for another. Quality should be quantified in terms of animal response.

Rangeland: Extensive land on which the indigenous vegetation (climax or sub-climax) is predominantly grasses, grass-like plants, forbs or shrubs that are grazed or have the potential to be grazed, and which is managed as a natural ecosystem for the production of grazing livestock and wildlife. Rangelands may include natural grasslands, savannahs, shrub lands, many deserts, steppes, tundra's, alpine communities and marshes.

Rangeland management: Management of a rangeland to produce maximum forage for sustained use without jeopardizing the other resources or uses of land.

Rangeland rejuvenation: Accomplishment of favourable environment to forage plants by soil-plant- animal management practices.

Ration: Ration refers to the 24-hour feed allowance for an individual animal.

Ration stocking: A method of confining animals to an area of grazing land to provide the daily allowance of forage per animal.

Relative feed value (RFV): An index for ranking cool-season grass and legume forages based on the intake of digestible energy calculated from acid detergent fibre (ADF) and neutral-detergent fibre (NDF) concentrations in forage.

Relative forage quality (RFQ): An index for ranking all forages based on the intake of total digestible nutrients calculated by summative equations after estimating digestible portions of protein, fatty acids, fibre and non-fibrous carbohydrate.

Resembling a grass: Herbaceous monocots, which are usually a member of the *Cyperaceae* (sedges) or *Juncaceae* (rushes) or which are families that are similar to grasses in appearance.

Residue: Forage remaining on the land after harvest.

Rest: To leave an area of grazing land ungrazed or unharvested for a specific time, such as a year, a growing season or a specified period required within a particular management practice. **Rest period:** The length of time that a specific land area is not stocked between stocking periods (Recovery period).

Rest rotation: Rest rotation grazing "divides the range into at least four pastures. One pasture remains rested throughout the year and grazing is rotated amongst the residual pastures." This grazing system can be especially beneficial when using sensitive grass that requires time for rest and regrowth.

Riparian area grazing management: Riparian area grazing is used more towards improving wildlife and their habitats. It uses fencing to keep livestock off ranges near streams or water areas until after wildlife or water fowl periods, or limiting the amount of grazing to a short period of time.

Rotational grazing: Rotational grazing "involves dividing the range into several pastures and then grazing each in sequence throughout the grazing period". Utilizing rotational grazing can improve livestock distribution while incorporating rest period for new forage.

Rotational stocking: A method that utilizes recurring periods of grazing and rest among three or more paddocks in a grazing management unit throughout the time when grazing is allowed. The lengths of the grazing and rest periods should be defined. Words such as 'controlled' or 'intensive' are sometimes used in an attempt to describe the degree of grazing management applied to this stocking method. These words are not synonyms for rotational stocking. Rotational grazing stocking is not a grazing system but it is a stocking method.

Roughage: Roughage refers to bulky and coarse feed high in fiber (greater than 18 per cent crude fibre) but lower in energy than most concentrates. For example, forage, hay, silage and haylage are sometimes called roughage.

Rumen: The rumen is the foregut (or fore stomach) of ruminant animals such as cattle, sheep and goats. The rumen is a large, hollow muscular organ that is the site of most of the fibre digestion that occurs in ruminant animals. This digestion is largely performed by microorganisms (bacteria, protozoa and fungi) that inhabit the rumen.

Ruminal microbes: Ruminal microbes include the whole community of microorganisms present in the rumen of ruminant animals. They accomplish the digestion or fermentation of feed. An estimated 150 billion microorganisms per teaspoon are present in the contents of the rumen. This microbial community consists of bacteria, protozoa and fungi.

Ruminants: Ruminants are a class of animals that have multiple organs working together to accomplish digestion. The digestive tract consists of the *reticulum* (involved in rumination and in passage from the rumen to the omasum), *rumen* (large compartment used for fermentation), *omasum* (once called the manyplies, it removes excess liquid and nutrients moving out of the reticulo-omasal orifice), and *abomasum* (acid-pepsin digestion similar to a monogastric). By comparison, monogastric animals (*e.g.*, swine and human being) have a simple or single-chambered stomach that utilizes an acid-pepsin digestion to extract nutrition from the ingested food.

Seasonal stocking: A method to restrict use of a land unit(s) to one or more specific seasons of the year.

Seed: Mature (ripened) ovules consisting of an embryonic plant and a store of food (stored in the endosperm, in some species), all surrounded by a protective seed coat.

Senescence: Applied to plants or organs, it is the process of remobilization and transfer of soluble constituents from mature to immature plant tissues that occurs with advancing age of plant parts, or through drought stress or depredation by pests, and is usually accompanied by chlorosis and subsequent death of mature tissue. Senescence is accompanied by a loss of dry matter as cell contents of the live tissue are metabolized and/or translocated.

Sequence (sequential) stocking: It is grazing of two or more land units in succession that differ in forage species composition. Sequence stocking takes advantage of differences among forage species and species combinations, grown in separate areas for management purposes, to extend stocking seasons to enhance forage quality and/or quantity or to achieve some other management objective.

Set stocking: A method that allows a specific, non-variable number of animals on a specific, non-variable area of land during the time when grazing is allowed.

Shrub: A woody plant having multiple stems arising at or near the base. Mature height is generally ally greater than 5 m. Coppice growth on a tree may have multiple basal stems.

Shrub land: Land on which the vegetation is dominated by shrubs.

Silage: Forage harvested and preserved at high moisture contents (generally >500 g per kg by organic acids produced during partial anaerobic fermentation (Syn. Ensilage). Silage refers to the feed preserved by an anaerobic fermentation process (*e.g.*, corn silage, haylage, high moisture corn) in which lactic acid and volatile fatty acids (produced by fermentation) lower the pH of the silage. The low pH preserves the silage. **Ensiling**: To produce silage by the process of fermentation of forages. **Silo**: The container used in the preservation of forage as silage. **Silage additive**: Silage additives refer to the substances added during the ensiling process to enhance production of lactic acid and/or a rapid decrease in pH of the feed.

Silvipasture: It is a land management system in which forests or trees are managed for the production of wood as well as understory for growing forage grasses and or legumes for the grazing/feeding livestock.

Soilage: Forage crops harvested and fed to livestock while in the fresh condition.

Steppe: Vast semi-arid grass covered plain, usually thinly wooded with shrubs and trees.

Stocking cycle: The time elapsed between the initiations of successive stocking periods on a specified grazing land area is called stocking cycle, usually in a regular cycle of use. One stocking cycle includes one stocking period plus one rest period. Stocking cycles may be variable or fixed time.

Stocking density: It is the relationship between the number of animals and the specific unit of land being grazed at any one time; an instantaneous measurement of the animal-to-land area relationship. Where needed, stocking density may be expressed as animal units or forage intake units per unit of land area (animal units at a specific time/unit of land area currently grazed). It is an animal-to-land relationship at a specific time and provides no information about forage mass, Canopy characteristics or the degree of forage use. This is a subjective term that can best be described in terms of stocking method, stocking period, rest, grazing pressure and Forage mass.

Stocking method: Procedure or technique to manipulate animals in space and time in order to achieve a specific objective(s) is called stocking method. In most cases, the term 'stocking' is preferred to 'grazing' (*i.e.* 'stocking method' vs. 'grazing method') because grazing refers to the consumption of standing forage, whereas it is the method of stocking grazing animals that allows the manipulation of how, when, what and how much the animals graze. One or more stocking methods are used within grazing systems to accomplish objectives. A stocking method is not site-specific. The objective of a stocking method may be 1) to allocate nutrition among varying classes of livestock [examples: creep stocking (grazing), first-last stocking]; 2) to improve efficiency of forage use [examples: frontal stocking (grazing), mixed stocking]; 3) to reduce negative effects on the soils or plants [examples: rotational stocking, deferred stocking]; 4) to extend stocking seasons [example: sequence stocking (grazing)]; or 5) to accomplish an experimental objective [example: put-and-take stocking]. One or more stocking methods can be used within a grazing system. It is important to select the correct stocking method to accomplish the intended objective(s). When describing the use of a stocking method, it is important to describe the context or the overall grazing system in which the method is applied.

Stocking period: The length of time that grazing livestock or wildlife occupies a specific pasture or paddock is called stocking period.

Stocking rate: The relationship between the number of animals and the total area of the land in one or more units utilized over a specified time; an animal to-land relationship over time is called stocking rate. Unless otherwise specified, this includes the total area of land in the grazing system supporting the total number of animals, including land areas being deferred or cropped (if crops are included in the system), and not just the areas actually grazed during the stated time. Where needed, it may be expressed as animal units or forage intake units per unit of land area over time (animal units over a described time/total system land area).

Stocking season: (i) stocking season is the time during which grazing can be practised normally each year or portion of each year. (ii) On US public lands, an established period for which grazing permits are issued (Syn. grazing season). The stocking season may be the whole year or a very short time span and is normally a function of forage mass and climate. In this context, the vegetative growing season may be only a part of the stocking season. Likewise, the stocking season may be only a part of the vegetative growing season.

Stockpiled forage: Forage which is allowed to accumulate for grazing at a later time.

Stover: Matured, cured stalks of food crops as maize, sorghum, millets *etc.* from which grains have been removed. It is a type of roughage.

Strip stocking: A method that confines animals to an area of grazing land to be grazed in a relatively short time, where the paddock size is varied to allow access to a specific land area. Strip stocking and ration stocking may or may not be a form of rotational stocking, depending on whether or not specific paddocks are utilized for recurring periods of grazing and rest.

Structural carbohydrate: Structural carbohydrates are the complex carbohydrates that form the plant cell wall and include cellulose, hemicellulose, lignin and pectin. They are typically measured in the laboratory as neutral detergent fibre (NDF).

Stubble: The basal portion of stems and leaves of herbaceous plants left standing after harvest.

Succulage: The root crops such as turnip carrot and fodder beets which provide succulent livestock feed are called succulage. These crops are rich in carbohydrates and make good source for mixing with poor quality roughages.

Supplement: A supplement feed or feed mixture is used to improve the nutritional value of the ration complementing the nutrients in the base feed. A supplement is rich in one or more of protein, energy, vitamins or minerals, and, in combination with the base feeds, produces a more complete feed.

Supplemental pasture: Additional pasture for use in adverse weather conditions, usually annual forage crops for dry periods or winter time.

Sward: A population or a community of herbaceous plants characterized by a relatively short habit of growth and relatively continuous ground cover, including both above- and below-ground parts.

Top feed: Leaves of trees/shrubs which are used for feeding livestock are called top feeds and they are often termed as top feed plant species.

True digestibility: Actual digestibility or availability of feed, forage or nutrient as represented by the balance between intake and faecal loss of the same ingested material. Thus the true digestibility of feed by animal is the apparent digestibility of the feed plus unavoidable metabolic excretions.

Total digestible nutrients (TDN): A general measure of the nutritive value of a feed calculated from the intake of digestible nutrients, with adjustment for the energy value of fat.

Total Mixed Ration: A total mixed ration is a homogeneous mixture of mechanically mixed ration ingredients that typically combine roughages (forages) and concentrates such as grains to optimize animal performance. TMRs are commonly used in large dairy or beef feedlot operations.

Total nomadism: In this livestock management system, livestock owners do not have a permanent place of residence. They do not practise regular cultivation and their families keep on moving with their herds for grazing the livestock.

Toxicity: Toxicity refers to the extent to which a substance can exert a poisonous effect on animals.

Utilization: It is extent to which grazing animals have cropped the current growth of pasture within their reach expressed in percentage of height or weight of forage removed or in such general terms as excessive, close, moderate or light.

Utilized Metabolizable Energy (UME): It is a measure of output from a forage system based on the estimated energy requirements of animals and the energy value of forage. The UME may be expressed either per animal on a daily basis [mega joules (MJ) per head] or per unit area over a specified time [gigajoules (GJ) per hectare]. It is calculated as the sum of ME requirements for maintenance and production of the classes of livestock involved less the ME value of all feed supplements provided from outside the area and augmented by the ME value of any conserved herbage made on the area but unused.

Variable stocking: The practice of allowing a variable number of animals on a fixed area of land during the time when grazing is allowed.

Vegetation: Plant life in general on a piece of land.

Vitamins: Vitamins are organic compounds that typically function as parts of enzyme systems essential for many metabolic functions.

Voluntary intake: The amount of forage consumed by an animal unrestricted by the quantity available.

Water soluble carbohydrate: WSCs are the carbohydrates that can be solubilized and extracted in water. WSCs include monosaccharaides, disaccharides and some short chain polysaccharides, mainly fructans, which are a major storage carbohydrate in some cool season grasses (*e.g.*, timothy).

Note: Modified and sourced from Allen *et al.* (2011), Allen (1991), Barnes *et al.* (2007), Edwards (1973), FAO (1965), Oregon State University (1992), Saha*et al.* (2013).

References

Abdel-Samee, A. M., El-Gendy, K. M. and Ibrahim, H. (1992). Growth performance and some related physiological changes in rabbits as affected by feeding acacia under subtropical conditions. *Egyptian Journal of Rabbit Science* 2:13-22.

Abebe Shiferaw, Ranjitha, Puskar, Azage Tegegne and Dirk Hoekstra (2011). Innovation in forage development: empirical evidence from Alaba Special District, southern Ethiopia. *Development in Practice* 21(8): 1138-1151.

Abou-Elezz, F. M. K., Sarmiento-Franco, L., Santos-Ricalde, R. and Solorio-Sanchez, F. (2011). Nutritional effects of dietary inclusion of *Leucaena leucocephala* and *Moringa oleifera* leaf meal on Rhode Island Red hens' performance. *Cuban Journal of Agricultural Science* 45(2): 163-169.

Abou-Zeid, A. E., Mohamed, F. F. and Radwan, M. S. M. (2001). Assessment of the nutritive value of dried azolla hay as a possible feed ingredient for growing NZW rabbits. *Egyptian Journal of Rabbit Science* 11:1-21.

Abu, O. A., Tewe, O. O. and Bakare, J. (1999). Performance, nutrient digestibility and carcass characteristics of rabbit fed sweet potato based diets. *International Journal of Animal Sciences* 14:197-201.

Adams, D. C. and Short, R. E. (1988). The role of mineral nutrition on productivity in range environment. *Fort Koegh Research Symposium*, Miles City, Mont, pp. 37-43.

Adegbola, T. A., Tibi, E. U. and Asugwa, D. C. (1985). Feed intake and digestibility of rabbits on all-forage, forage plus concentrate and all-concentrate diets. *J. Anim. Prod. Res.* 5: 185-191.

Aderinola, O. A., Ojebiyi, O. O., Rafiu, T. A., Akinlade, J. A. and Adepoju, L. O. (2008). Performance evaluation of growing rabbit fed diets containing varying inclusion of *Centrosema pubescens* or *Calopogonium mucunoides* in the savannah zone of Nigeria, *9 th World Rabbit Congress* – June 10-13, 2008, – Verona, Italy, pp. 525-528.

Agnihotri, N. P. (1999). *Pesticide Safety Evaluation and Monitoring*, New Delhi, All India Coordinated Research Project on Pesticide Residues, Indian Agricultural Research Institute, New Delhi, pp. 104-107.

Agrawal, R. K., Kumar Sunil, Tripathi, S. B., Sharma, R. K. and Singh, K. A. (2008). Agro-economic analysis of fodder-based cropping Systems. *Indian J. of Fertilizers* 4 (4): 121-131.

AICRPAF (2006). *Annual Report*, All India Cordinated Project on Agroforestry, NRCAF, Jhansi.

Akande, K. E., Doma, U. D., Agu, H. O. and Adamu, H. M. (2010). Major antinutrients found in plant protein sources: their effect on nutrition. *Pakistan Journal of Nutrition*, 9: 827–832.

Al Faïz, C., Thami Alami, I., Moussadek, R. and El Oumri, M. (2010). Some expected climatic changes impacts on the pasture and forage species in Morocco, eds. Porqueddu, C. and Ríos, S., *"The contributions of grasslands to the conservation of Mediterranean biodiversity"*, Zaragoza: CIHEAM. CIBO/FAO/SEEP, Zaragoza: CIHEAM/CIBIO/FAO/SEEP (Options Méditerranéennes: Série A. Séminaires Méditerranéens; n. 92), 2010, pp. 175-179.

Alalade, O. A. and Iyayi, E. A. (2006). Chemical Composition and the feeding value of Azolla (*Azolla pinnata*) Meal for Egg-Type Chicks. *International Journal of Poultry Science* 5(2) : 137-141

Albert Nii Armah Tagoe (2012). The response of broiler chikens to dietary inclusion of *Alternanthera sessilis* leaf meal (ASLM). *Thesis.* Master of Philosophy in Animal Nutrition, Kwame Nkrumah University of Science and Technology, Kumasi, Ghana.

Alfaro, M. A., Ramirez, R., Martinez, A. and Bressani, R. (1987). Evaluation of different amounts of amaranth meal (vegetative parts) to replace lucerne leaf meal in diets for growing rabbits. *Archivos Latinoamericanos de Nutricion* 37: 174-185.

Alhassan, W. S. and Odoi, F. (1982). Use of cassava leaf meal in diets for pigs in the humid tropics. *Tropical Animal Health and Production* 14: 216–218.

Alikwe, P. C. N., Dambo, L. B. and Elijah I. Ohimain (2001). Growth performance of rabbit fed *Telfairia occidentalis* and *Centrosema pubescens* as protein supplement. *International Journal of Research In Agriculture and Food Sciences* 1(6): 11-15.

Allan, H. H. (1947). The use and misuse of shrubs and trees as fodder. *Joint Publication of Imperial Bureau of Pastures and Field Crops,* Aberystwyth, UK. pp. 40-46.

Allard, V., Soussana, J. F., Falcimange, R., Berbigier, P., Bonnefond, J. M., Ceschia, E. *et al.* (2007). The role of grazing management for the net biome productivity and greenhouse gas budget (CO_2, N_2O and CH_4) of semi-natural grassland. *Agric. Ecosyst. Environ.* 121: 47–58.

Allen, V. G. (1991). Terminology for grazing animals. *Proceedings of Grazing Livestock Nutrition Conference. In* F. T. McCollum and M. B. Judkins (ed.) *"Proceedings of the 2nd Grazing Livestock Nutrition Conference"*, 2–3 August, 1991, Steamboat Springs, Oklahoma Agricultural Experiment Station, Stillwater Colorado, pp. 103–110.

Allen, V. G., Batello, C, Berretta, E. J, Hodgson, J., Kothmann, M., Li, X., McIvor, J., Milne, J., Morris, C., Peeters, A. and Sanderson, M. (2011). The Forage and Grazing Terminology Committee - An international terminology for grazing lands and grazing animals. *Grass and Forage Science*, 66: 2–28, doi: 10. 1111/j. 1365-2494. 2010. 00780.

Amare, Haileslassie, Blummel, M., Clement, F., Ishaq, S. and Khan, M. A. (2011). Adapting livestock water productivity to climate change. *International Journal of Climate Change: Strategies and Management* 3: 156-166

Amezquita, M. C., Ibrahim, M., Llanderal, T., Buurman, P. and Amezquita, E. (2005). Carbon sequestration in pastures, silvopastoral systems and forests in four regions of the Latin American tropics. *J. Sustainable Forestry* 21: 31-49.

Amlaku, A., Sölkner, J., Puskur, R. and Wurzinger, M. (2012). The impact of social networks on dairy technology adoption: evidence from Northwest Ethiopia. *International Journal of AgriScience* 2(11): 1062-1083.

Amodu, J. T. (2004). *Stylosanthes*: A promising legume for Africa Chapter 22, In: Chakraborty, S. (Ed.), *"High-yielding anthracnose-resistant Stylosanthes for agricultural systems."* Australian Centre for International Agricultural Research Canberra, Monograph No. 111: pp. 225-234.

Amornsak, N., Thy San and Preston, T. R. (2004). Duckweed (*Lemna minor*) and water spinach (*Ipomoea aquatica*) as protein supplements for ducks fed broken rice as the basal diet. *Livestock Research for Rural Development, Vol. 16, Art. #16.* Retrieved August 14, 2016, from http://www. lrrd. org/lrrd16/3/ amo16016. htm

An, L. V. (2004). Sweet potato leaves for growing pigs: Biomass yield, digestion and nutritive value. *Doctor's Dissertation.* ISSN 1401-6249, ISBN 91-576-6750-0

Anderson, R. H. (1947). The use and misuse of shrubs and tree fodder, *Joint Publication of Imperial Bureau of Pastures and Field Crops,* Aberystwyth, UK, pp. 13-19.

Angthong, W., Cheva-Isarakul, B., Promma, S. and Cheva-Isarkul, B. (2007). Beta-carotene, mimosine and quality of leucaena silage kept at different duration. *Kasetsart Journal, Natural Sciences* 41: 282– 287.

Anita Fanish and Sathyapriya, R. (2013). Review of benefits of agroforestry system. *International J. Education and Research* 1(1): 1-12.

Anitha, K. C., Rajeshwari, Y. B., Prabhu, T. M., Patil1. M. V., Shilpa Shree, J. and Anupkumar, P. K. (2016). Effect of supplementary feeding of azolla on growth performance of broiler rabbit. *ARPN Journal of Agricultural and Biological Science* 11(1): 30-36

Annual Report (1989-90 and 1990-91). Survey of feed and fodder resources of Mithun. *Annual Report*, 1989-90 and 1990-91, ICAR - National Research Centre on Mithun, Jharnapani, Medziphema, Nagaland, India. pp. 24-38.

Annual Report (1999-2000). Survey of feed and fodder resources of Mithun. *Annual Report*, 1999-2000, ICAR - National Research Centre on Mithun, Jharnapani, Medziphema, Nagaland, India. pp. 26-40.

Anonymous (1908). *Imperial Gazetteer of India: Provincial series. Madras*. Supt. of Govt. Printing, Calcutta, 78p.

Anonymous (2010-14). Fodder availability during lean period through silvipasture system. *ICAR-IGFRI*, Jhansi *Annual Report, 2010-14.*

Anonymous (2015). *Scenario of breeder seed production*. In "*Annual Breeder Seed Review Report* 2013-14", Crop Science Division, ICAR, New Delhi, pp. 2-3.

Anthra (1999). *Community-based Research on Local Knowledge Systems*: The Anthra Project on Ethnoveterinary Research', in Evelyn Mathias, D. V. Rangnekar and C. M. McCorkle (eds.), Ethnoveterinary Medicine: Alternatives for Livestock Development. *Proceedings of an International Conference*, held in Pune, India, 4-6 Nov. 1997. Vol 1: Selected Papers. BAIF Development Research Foundation, Pune.

Anthra (2000). *Summary Observations of the Internal Impact Study of IKAH Program in Andhra Pradesh and Maharashtra,* Unpublished Working Paper, Anthra, Hyderabad,.

Anthra (2005-08). Pastoral Livelihood Reports – training reports, Anthra Publications, Hyderabad, India.

Anthra (2008). Bridging the Knowledge Divide: Livestock and Livelihoods in the Emerging Context. Anthra Publications, Hyderabad, India.

Anwar Hussain and Abdul Haleem (1956). Feeding Cottonseed Cake (Whole Pressed) and Berseem Hay Meal to Laying Pullets in Pakistan. *World's Poultry Science Journal* 12(4): 280-285. DOI: http://dx. doi. org/10. 1079/ WPS19560033, Published online: 01 September 2007.

Archana, S. and Sharma Varun. (2009). *Securing community tenure over common lands*. South Asia Pro Poor Livestock Policy Programme, New Delhi, India.

Armstrong Donkoh, Atuahen, C. C., Poku-Prempeh, Y. B. and Twum, I. G. (1999). The nutritive value of chaya leaf meal (*Cnidoscolus aconitifolius* (Mill.) Johnston):

Studies with broiler chickens. *Animal Feed Science and Technology* 77(1):163-172, DOI: 10. 1016/S0377-8401(98)00231-4.

Arnold, J. E. M. and Stewart, W. C. (1990). *Common Property Resource Management in India.* Oxford Forestry Institute. Oxford (UK), 72 p.

Ash, A. J. and Peteia, L. A. (1992). Nutritional value of *Sesbania grandiflora* leaves for ruminants and monogastrics. *Tropical Agriculture (Trindad)* 69: 223-228.

Asuquo, B. O. (1997). Nutritional potentials of *Ipomea, Centrosema, Pueraria,* Emilia and Tridax forages in mixed feeds for weaner rabbits. *Niger. J. Anim. Prod.* 24(1): 46–50.

Auxilla, M. T. and Masoero, G. (1980). Emploi du maïs-fourrage deshydraté dansd l'alimentation des lapins. *2nd World Rabbit Congress,* Bracelona 2: 147-156.

Auxilla, M. T., Bergoglio, G., Masoero, G., Mazzocco, P., Ponsetto, P. D. and Terramoccia, S. (1983). Feeding meat rabbits. Use of lucerne with different saponin content. *Coniglicoltura* 20: 51-58.

Auxilla, M. T., Masoero, G. and Costantini, F. (1977). Meals of dried whole maize, barley and wheat plants cut at the waxy stage of maturity in feeds for fattening rabbits. *Annali dell'Istituto Sperimentale per la Zootecnia* 10: 201-212.

Auxilla, M. T., Masoero, G. and Terramoccia, S. (1979). Use of dehydrated whole maize plant in diets for growing rabbits. *Annali dell'Istituto Sperimentale per la Zootecnia* 12: 43-50.

Awasthi, M. D. and Ahuja, A. K. (1995). Organochlorine insecticide residue in bovine milk and commercial baby milk powder. *Journal of Food Science Technology* 32: 420–422

Awosanya, B. and Akinyode M. A. (2000). Treatment effect of leucaena leaf meal on the carcass characteristics of rabbits. *Nigerian Journal of Animal Production* 27: 90- 94.

Ayers, A. C., Cheeke, P. R. and Patton, N. M. (1992). Evaluation of hybrid poplar leaves as a feedstuff for rabbits. *Journal of Applied Rabbit Research* 15:1033-1042.

Ayssiwede, S. B., Chrysostome, C. A., Zanmenou, J. C., Dieng, A., Houinato, M. R., Dahouda, M., Akpo, Y., Hornick, J. L. and Missohou, A. (2011). Growth Performances, Carcass and Organs Characteristics and Economics Results of Growing Indigenous Senegal Chickens Fed Diets Containing Various Levels of *Leuceana leucocephala* (Lam.) Leaves Meal. *International journal of Poultry Science* 10(9): 734-749.

Ayssiwede, S. B., Dieng, A., Bello, H., Chrysostome, C. A., Hane, M. B., Mankor, A., Dahouda, M., Houinato, M. R., Hornick, J. xL. and Missohou, A. (2011). Effects of *Moringa oleifera* (Lam.) Leaves Meal Incorporation in Diets on Growth Performances, Carcass Characteristics and Economics Results of Growing Indigenous Senegal Chickens. *Pak. J. Nutrition* 10: 1132-1145.

Bagchi, Sumanta and Ritchie Mark, E. (2010). Biodiversity, carbon dioxide, climate change, ecosystem function, ecosystem services, human natural ecosystems, land use change, livestock production, rangelands, wildlife conservation. *Ecology Letters* 13(*8*): 959-968.

Bai Changjun, Liu Guodao, Wang Dongjun, Daida Krishna, Qudratullah, S, Prasad, V. L. K., Rama Rao, S. V., Parthasarthy Rao, P., Ramesh, C. R., Balagopal, R. and Gopalan, A. (2004). Stylosanthes leaf meal for animal industries in China and India Chapter 24. In: Chakraborty, S. (Ed.), "*High-yielding anthracnose-resistant Stylosanthes for agricultural systems.* " Australian Centre for International Agricultural Research Canberra, Monograph No. 111: 243-252.

BAIF (2003). *Development of Community Silvipasture through People's Participation in Bhilwara District of Rajasthan*. BAIF Development Research Foundation, Pune, India, 5p.

BAIF (2006). *Community Pasture Development in Rajasthan.* BAIF Development Research Foundation, Pune, India, 2 p.

Baig, M. J., Sultan Singh, Rai, S. K., Singh, J. B., Misra, A. K., Biradar, N. B., Verma, O. P. S. and Anil Kumar (2007). *Water Productivity in Forage Crops and Livestock-Prospects and Opportunity for Improvement*, IGFRI, Jhansi (India), 88p.

Balaji, K., Jalaludeen, A., Churchil, R. R., Peethambaran, P. A. and Senthil Kumar, S. (2009). Effect of dietary inclusion of Azolla (*Azolla pinnata*) on production performance of broiler chicken. *Indian Journal of Poultry Science* 44 (2):195-198

Balaraman, N., Singh, K. A. and Gupta, H. K. (1988). Performance of goat on nevaro + nandi setaria grass based Silvi-pastoral system in Sikkim. *J. Hill Research* 2:43-48.

Balaraman, N. (1981). *Feed and fodder resources for livestock in Sikkim, Research Bulletin* No. 13, ICAR Research Complex for NEH Region, Sikkim centre, Tadong, Gangtok.

Bamikole, M. A. and Ezenwa, I. (1999). Performance of rabbits on Guinea grass and Verano stylo hays in the dry season and effect of concentrate supplementation. *Animal Feed Science and Technology* 80: 67-74.

Barnes, R. F., Miller D. A., and Nelson, C. J. (eds.) (1995). Forages, Fifth Edition, Volume I: An Introduction to Grassland Agriculture, Iowa State University Press, Ames, IA, USA: 516 p.

Barnes R. F., Miller D. A., and Nelson, C. J. (eds.) (1995). Forages, Fifth Edition, Volume II: The Science of Grassland Agriculture, Iowa State University Press, Ames, IA, USA, 357 p.

Barnes, R. F., Nelson, C. J., Moore, K. J. and Collins, M. (2007). *Forages: the science of grassland agriculture. Glossary*, 6th edn. Vol II. Ames, Blackwell Publishing Professional. Iowa, USA.

Baruah, K. K., Saikia, B. N., Bhattacharya, M., and Sarma, K. (1995). The nutritive value of para grass for mithuns (*Bos frontalis*), *Indian Veterinary Journal*, 72 (12): 1333-1335.

Basak, B., Pramanik, A. H., Rahmnan, M. S., Taradar, S. U. and Roy, B. C. (2002). Azolla (*Azolla pinnata*) as a feed ingredient in broiler ration. *Int. J. Poult. Sci.* 1: 29-32.

Bassny, S. M., Sarhan, M. A. and El-Adawy, M. M. (1999). Nutritional studies on some green forage in Egypt. 6. Effect of partial replacement of concentrates by fodder beet leaves and roots in growing rabbit diets. *Egyptian Journal of Rabbit Science* 9: 215-227.

Battu, R. S, Singh, B. and Kang, B. K. (2004). Contamination of liquid milk and butter with pesticide residues in Ludhiana. *Ecotoxicol Environ Saf.* 59: 324-331.

Battu, R. S., Singh, B. and Kalra, R. L. (1996). Seasonal variation in residues of DDT and HCH in dairy milk in Punjab, India. *Pesticide Res. J.* 8: 32-37.

Battu, R. S, Gupta, S. C., Chawla, R. P. and Kalra R. L. (1984). Residues of DDT and BHC in market samples of meat of various animals. *Ind. J. Ecol.* 11 (2): 177-182.

Becerra, M., Preston, T. R. and Ogle, R. B. (1995). Effect of replacing whole boiled soya beans with azolla in the diets of growing ducks. *Livestock Research for Rural Development*, 7(3): http://www. cipav. org. co/lrrd/lrrd7/3/7. htm

Belete Shenkute Gemeda and Abubeker Hassen (2014). *In vitro* fermentation, digestibility and methane production of tropical perennial grass species. *Crop and Pasture Science* 65(5): 479-488 http://dx. doi. org/10. 1071/CP13450.

Bello, H., Dieng, A., Ayssewede, S. B., Chrysostomy, C. A. and Hane, M. B. (2011). Effects of *Moringa oleifera* leaves meal incorporation in diets on growth performance, carcass characteristics and economic results of growing indigenous Senegal chickens. *Pakistan Journal of Nutrition* 10(12): 1132-114.

Benbrook, C. M, Butler G., Latif, M. A., Leifert C. and Davis, D. R. (2013). Organic Production Enhances Milk Nutritional Quality by Shifting Fatty Acid Composition: A United States–Wide, 18-Month Study. PLoS ONE 8(12): e82429. https://doi. org/10. 1371/*journal. pone.* 0082429

Berwal, R. S., Lohan, O. P., Rathee, C. S. and Yadav, K. K. (1993). Effect of various chemicals on post compression expansion and durability of crop residue blocks. *Indian Journal of Animal Nutrition* 10: 119–122.

Bhagmal, Singh, K. A., Roy, A. K., Ahmad, Shahid and Malviya, D. R. (2009). Chapter on Oat, *In "Handbook of Agriculture",* ICAR, New Delhi pp. 1054-1065.

Bhagmal, Singh, K. A., Roy, A. K., Ahmad, Shahid and Malviya, D. R. (2009). Chapter on Forage Crops and Grasses, *In "Handbook of Agriculture",* ICAR, New Delhi pp. 1353-1417.

Bhatia, I. and Chowdhury, G. N. (1996). Lead poisoning of milk – the basic need for foundation of human civilization. *J Anim Health* 40: 24-26.

Bhatnagar, Y. V., Rinchen, W., Herbert, H., Prins, T. Spike, E., Van Wieren, S. E. and Mishra, C. (2006). Perceived Conflicts Between Pastoralism and Conservation of the Kiang Equus kiang in the Ladakh Terans-Himalaya, India, *Nature Conservation Foundation and International Snow Leopard Trust-India Program 3076/5* IV Cross, Gokulam Park, Mysore, Karnataka, India - 570002.

Bhatt, R. S (2000). Replacement of maize with rice polishes in the feeding regime of weaner rabbits. *7th World Rabbit Congress*, Valencia vol. C: pp. 125-130.

Bhatt, R. S. and Sharma, S. R. (2001). Nutrient utilisation and growth performance of broiler rabbits fed oat plant meal and tall fescue hay. *Asian-Australasian Journal of Animal Sciences* 14: 1228-1232.

Bhattacharya, P. and Chakraborty, G. (2005). Current status of organic farming in India and other countries. *Indian Journal of Fertilisers* 1 (9): 111-123.

Bhattacharyya, A., Majumdar, S., Bhanja, S. K. and Mandal. A. B. (2008). Green berseem feeding on growth performance and blood biochemical profiles in Turkey. *The Indian veterinary journal* 85(5): 523-525.

Bimbao, M. A., Lopez, T. and Lightfoot, C. (1995). *Learning about sustainability*. ILEIA Newsletter 11 (2): 28-30.

Binti Ma'rifah, Atmomarsono, U. and Suthama, N. (2013). Nitrogen Retention and Productive Performance of Crossbred Native Chicken Due to Feeding Effect of Kayambang (*Salvinia molesta*). *International J. of Sci. and Eng.* 5(1):19-24.

Biobaku, W. O. and Ekpenyong, T. E. (1991). Effect of feeding graded levels of water lettuce and water hyacinth on the growth of rabbits. *Journal of Applied Rabbit Research* 14: 98-100.

Biradar, N. B. Sridhar, K., Suryavanshi, V., Karthikeyan, S. Vinod Kumar and Kulkarni, N. (2010). 'Hasiru Hullu Beladu Hesarad Raitharu (Farmers who became successful by growing fodder crops), IGFRI, Jhansi, India

Biradar, N. B., Sridhar, K. and Karthikeyan, S. (2008). *Manual on Forage Crops for peninsular India,* IGFRI, Jhansi, India.

Biradar, N. B., Sridhar, K. and Karthikeyan, S. (2009). The saga of Mr Malleshappa Hakalad from labour to successful farmer. *IGFRI News Letter* 15 (1): 5.

Birthal, P. and Rao. P. P. (2002). Economic contributions of livestock sub-sectors in India. In: P. S. Birthal and P. P. Rao (eds). *Technology Options for Sustainable Livestock Production in India.* NCAEPR, New Delhi and ICRISAT, Hyderabad: pp. 12-19.

Blummel, M., Anandan, S. and Prasad, C. S. (2009). Potential and limitations of byproduct based feeding system to mitigate green house gases for improved livestock productivity. *In "Proceedings of 13th biennial conference of Animal Nutrition Society of India"*, Bangalore, pp. 68–74.

Bluthgen, A. H., Burt, R. and Heeschen, W. H. (1997). Heavy metals and other trace elements. *Monograph on Residues and Contaminants in Milk and Milk Products* International Dairy Federation, Brussels, Belgium. pp. 65-73.

Bluthgen, A. H. (2000). Contamination of milk from feed. Bull. *Int. Dairy Federation* 356, 43-47.

Boer, I. J. M. (2003). Environmental impact assessment of conventional and organic milk production. *Livestock Production Science* 80: 69–77 www. elsevier. com/ locate/livprodsci.

Bolton, J. L., Goplen, B. P. and Baenziger, H. (1972). World distribution and historical developments. Alfalfa Science and Technology (ed. C. H. Hanson), American Society of Agronomy pp. 1-34. Madison, WI, USA.

Bond, W. J. and Midgley, G. F. (2000). A proposed CO_2-controlled mechanism of woody plant invasion in grasslands and savannas. *Glob. Change Biol.* 6: 865–869. (doi:10. 1046/j. 1365-2486. 2000. 00365. x).

Bond, W. J., Woodward F. I. and Midgley G. F. (2005). The global distribution of ecosystems in a world without fire. *New Phytol.* 165: 525–538. (doi:10. 1111/j. 1469-8137. 2004. 01252. x).

Booysen, D. V. (1967). Grazing and grazing management terminology in Southern Africa. *Proceedings of the Grassland Society of Southern Africa* 2: 45–57.

Borgohain, B. (1983). A decade of forage research in Assam, *Research Bulletin* No. 7, AAU, Jorhat (Assam).

Bounejmate, M. (1992). Distribution des luzernes annuelles spontanées au Maroc en relation avec certains facteurs climatiques et édaphiques. Dans: *Al Awamia*, 79, pp. 35-49.

Bounejmate, M., Loss, S. P. and et Robson, A. D. (1994). Effects of temperature and frost on gentotypes of *Medicago truncatula L.* and *Medicago aculeata L.* from contrasting climatic origins. *J. Agronomy and Crop Science* 172 (4): 227-236.

Bowman, D. M., Murphy, B. P. and Banfai, D. S. (2011). Has global environmental change caused monsoon rainforests to expand in the Australian monsoon tropics? *Landscape Ecol.* 25: 1247–1260. (doi:10. 1007/s10980-010-9496-8).

Brady, N. C. (1974). *The nature and properties of soils.* 8th edition, McMillan Publishing Co. New York.

Bridgewood, L. (2003). Hydroponics: Soilless gardening explained. Ramsbury, Marlborough, The Crowood Press Limited, Wiltshire.

Briggs, J. M., Knapp, A. K., Blair, J. B., Heisler, J. L. and Hoch, G. A. (2005). An ecosystem in transition: causes and consequences of the conversion of mesic grassland to shrub land. *Bioscience* 55: 243–254 (doi:10. 1641/0006-3568(2005)055[0243:AEITCA]2. 0. CO;2).

Brithal, P. S. and Jha, A. (2005). Review of emerging trends in India's livestock economy: implications for development policy. *Indian Journal of Animal Sciences* 75: 1227-1232.

Brower, B. and Dennis, A. (2000). Yak grazing and forest dynamics in Sagarmatha National Park, Nepal. *Proceeding Integrated ecological studies of pasture problems in the Tibetan Autonomous,* China. Eds. H. Jianlin, C. Richard, O. Hanotte, C. McVeigh and J. E. O. Rege, held in Lhasa, P. R. China, 4–9 September 2000, Lhasa, P. R China.

Brown, G. L., Barnes, D. A., Rezende, S. A. and Klasing, K. C. (1987). Yield, composition and feeding value of irrigated *Sesbania sesban. Animal Feed Science and Technology* 18: 247–255.

Bruinsma, J. (2003). World agriculture: towards 2015/2030, an FAO perspective. *Earthscan,* FAO, Rome, Italy.

Brum, K. B., Haraguchi, M., Garutti, M. B., Nobrega, F. N., Rosa, B. and Fioravanti, M. C. S. (2009). Steroidal saponin concentrations in *Brachiaria decumbens* and *B. brizantha* at different developmental stages. *Ciencia Rural* 39: 279–281.

Brunori, G., Jiggins, J., Gallardo, R. and Schmidt, O. (2008). *The Second SCAR Foresight Exercise, Synthesis Report, 'New challenges for agricultural research: climate change, food security, rural development, (knowledge systems',* EU Commission Standing Committee on Agricultural Research (SCAR), 105p.

Buchanan, F. (1807). *A journey from Madras through the countries of Mysore, Canara and Malabar.* Vol II., Blumar and Co., London,

Bui Huy, N. P., Ogle, B. and Lindberg, J. E. (2000). Effect of replacing soybean protein with cassava leaf protein in cassava root meal based diets for growing pigs on digestibility and N retention. *Animal Feed Science and Technology* 83: 223–235.

Bui Xuan Men, Ogle, R. B. and Preston, T. R. (1995). Use of duckweed (*Lemna spp*) as replacement for soya bean meal in a basal diet of broken rice for fattening ducks. *Livestock Research for Rural Development.* 7(3): http://www. cipav. org. co/lrrd/lrrd7/3/2. htm

Bui Xuan Men, Ogle, R. B. and Preston, T. R. (1996). Duckweed (*Lemna spp*) as replacement for roasted soya beans in diets of broken rice for fattening ducks on a small scale farm in the Mekong delta. *Livestock Research for Rural Development 8(3):* http://www. cipav. org. co/lrrd/lrrd8/3/men831. htm

Buley, J. (1985). List of recommended species of trees and shrubs for Nepal *IN* "Global needs and problems of the collection, storage and distribution of multipurpose tree germplasm. *Science and Practices of Agroforestry* 2: 59-62.

Burke, I. C., Lauenroth, W. K., Vinton, M. A., Hook P. B., Kelly, R. H. (1998). Plant-soil interactions in temperate grasslands. *Biogeochemistry* 42: 121–143. doi: 10. 1007/978-94-017-2691-7.

Butler, G., Nielsen, J. H., Slots, T., Seal, C. and Eyre, M. D. (2008). Fatty acid and fat-soluble antioxidant concentrations in milk from high- and low-input conventional and organic systems: seasonal variation. *J. Sci. Food Agric* 88: 1431–1441.

Butler, G., Stergiadis, S., Seal, C., Eyre, M. and Leifert, C. (2011). Fat composition of organic and conventional retail milk in North East England. *J. Dairy Sci.* 94: 24–36.

Carter, A. S. and Paudiyal, B. R. (1987). Fodder trees – propagation by cuttings, *Tech. Note* 1/87, Australian Forestry Project, New Baneswar, Kathmandu, Nepal.

CGIAR (1989). *Sustainable Agriculture Production: Final Implication for International Agricultural Research*, FAO, Rome, Italy.

Chakoma, C., Franzel, S., Hove, I., Matimati, I. and Maarsdorp, B. V. (2004). The adoption of fodder trees in Zimbabwe: Smallholder farmers Experiences and Innovations. Workshop proceedings: May 2002. World Agro forestry Centre (ICRAF), Nairobi, Kenya.

Champe, K. A. and Maurice D. V. (1983). Response of early weaned rabbits to source and level of dietary fiber. *Journal of Animal Science* 56: 1105-1114.

Chander, M. (2011). Chaff cutters and fodder chaffing: A simple technology worth adoption. In H. P. S. Makkar, eds. *"Proceedings of FAO E- Conference on, Successes and failures with animal nutrition practices and technologies in developing countries"*, 1–30 Sept. 2010, pp. 133–136.,.

Chander, M., Subrahmanyeswari, B., Mukherjee, R. and Kumar, S. (2011). Organic livestock production: an emerging opportunity with new challenges for producers in tropical countries. *Revue scientifiqueet technique (International Office of Epizootics)* 30(3): 969-983.

Chandra, A., Pathak, P. S and Bhatt, R. K. (2006). *Stylosanthes* research in India: prospects and challenges ahead. *Current Science* 90 (7): 915-921.

Changjun, B., Guodao, L., Donjun, W., Daidakirishna, Qudratullah, S, Prasad, V. L. K., Ramarao, S. V., Rao, P. P., Ramesh, C. R., Blagopal, R. and Gopalan, A. (2004). *Stylosanthes* leaf meal for animal industries in China and India, In "High yielding anthracnose resistant *Stylosanthes* for agricultural systems", eds. S. Chakraborty, ACIAR, CSRIO, Canberra, pp. 243-252.

Chaplin-Kramer, R. and George, M. R. (2013). Effects of Climate Change on Range Forage Production in the San Francisco Bay Area. PLoS ONE 8(3): e57723. doi:10. 1371/journal. pone. 0057723.

Chaves, A. V., Thompson, L. C., Iwaasa, A. D., Scott, S. L., Olson, M. E., Benchaar, C., Veira, D. M. and McAllister T. A. (2006). Effect of pasture type (alfalfa vs. grass) on methane and carbon dioxide production by yearling beef heifers. *Canadian Journal of Animal Science* 86: 409–418.

Chawla, N. K. Kurup, M. P. G. and Sharma, Vijay Paul. (2002). *State of the Indian Livestock Farmers and the Indian Livestock Sector: A Status Paper.* Indian Institute of Management, Ahmadabad.

Cheeke, P. R. and Carlsson, R. (1978). Evaluation of several crops as sources of leaf meal: composition, effect of drying procedure, and rat growth response. *Nutrition Reports International* 18: 465–473.

Cheeke, P. R. and Patton, N. M. (1980). Alfalfa (lucerne) utilization by rabbits. *Commercial Rabbit* 8: 8-9.

Cheeke, P. R., Harris, D. J. and Patton, N. M. (1983). Utilization of tropical forages and alfalfa meal by rabbits. *South African Journal of Animal Science* 13: 25-26.

Cheeke, P. R., Harris, D. J. and Patton, N. M. (1984). Utilization of black locust (*Robinia pseudoacacia*) leaf meal by rabbits. *Nitrogen Fixing Tree Research Reports* 31.

Cheeke, P. R., Harris, D. and Patton, N. M. (1983). Utilization of tropical forages and alfalfa meal by rabbits (Abstract). *Nutri. Abst. Rev. Series* B53: 812

Cheeke, P. R. (2003). Feeding systems *for* tropical rabbit production emphasizing root, tubers *and* bananas. http://www. *fao*. org/DOCEP/003/T0554E/T0554E16. htm.

Chen, M. T., Chang, T. and Lin, L. S. (1981). Feeding efficiency of *Leucaena* leaf meal used in pig rations. *Leucaena Research Reports* 2: 46.

Cherry, D. M., Prasad, R. M. V., Rao, S. J., Jayalaxmi, P. and Rao, B. E. (2014). Effect of inclusion of *Azolla pinnata* on the haematological and carcass characteristics of crossbred large white Yorkshire pigs, *Veterinary World* 7(2): 78-82.

Chhabra, A. (2004). Contaminants in milk products and measures to minimize their level. *Indian Dairyman* 56: 115-125.

Chhay Ty, Borin K. and Chiv Phiny (2007). A note on the effect of fresh mulberry leaves, fresh sweet potato vine or a mixture of both foliages on intake, digestibility and N retention of growing pigs given a basal diet of broken rice. *Livestock Research for Rural Development. Volume 19, Article #136.* Retrieved May 5, 2017, from http://www. lrrd. org/lrrd19/9/chha19136. htm.

Chhay Ty, Kimseang, A., Borin, K. and Preston, T. R. (2014). Replacing rice bran by an ensiled mixture of taro (*Colocasia esculenta*) foliage and banana stem increases feed intake, diet digestibility and N retention in growing pigs. *Livestock Research for Rural Development. Volume 26, Article #71.* Retrieved September 5, 2016, from http://www. lrrd. org/lrrd26/4/chha26071. html.

Choi, C. Q. (2014). *Hunters to Herders: Ancient Civilization Made Rapid Switch.* Live Science Contributor | April 28, 2014.

Choubey, R. N., Roy, A. K., Sai Prasad, S. V., Zadoo, S. N. and Bhaskar, R. B. (2002). *IGFRI Oat Germplasm Catalogue*, IGFRI, Jhansi, India.

Choubisa, S. L., Modasiya, V., Bahura, C. K., Sheikh, Z. (2012). Toxicity of fluoride in cattle of the Indian Thar desert, Rajasthan, India. *Research Report Fluoride* 45(4): 371–376.

Christin, P. A., Besnard G., Samaritani E., Duvall M. R., Hodkinson T. R., Savolainen V.,

Christopher Dean and Johnson, G. W. W. and Harper, R. J. (2012) Carbon management of commercial rangelands in Australia: Major pools and fluxes. *Agriculture, Ecosystems and Enviornment* 135 (1-2): 15-23.

Christos Makridis, Christos Svarnas, Nikolaos Rigas, Nikolaos Gougoulias, Loukia Roka and Stefanos Leontopoulos (2012). Transfer of Heavy Metal Contaminants from Animal Feed to Animal Products. *Journal of Agricultural Science and Technology* A 2: 149-154.

Cincotta, R. and G. Pangare. (1993). *Pastoralism and Pastoral Migration in Gujarat.* Institute of Rural Management, Anand (India), 55p.

Cincotta, R., van Soest, P, Robertson, J., Beall, C. and Goldstein, M. (1991). Foraging ecology of livestock on the Tibetan Chang Tang: a comparison of three adjacent grazing areas. *Arctic and Alpine Research* 23(2): 149-161.

Clarke, Graham (1998). Development, Society and Environment in Tibet. Papers Presented at a Panel of *the 7th International Association for Tibetan Studies,* Graz 1995. Berlag de Osterreichischen Adademie der Wissenschaften, Wien.

Cleland, E. E., Chiariello, N. R., Mooney, H. A., Field, C. B. (2006). Diverse response of phenology to global changes in a grassland eco-system. *Proceedings of National Academy of Science,* USA 103: 13740-13744.

Close, W. H. (1993). Fibrous diets for pigs. In M. Gill, E. Owen, G. E. Pollot, and T. L. G. Lawrence (Eds.), Animal production in developing countries,. *British Society of Animal Production Occasional Publication No 16,* London, pp. 107–115.

Conant, R. T., Paustian, K. and Elliott, E. T. (2002). Grassland management and conversion into grassland: effects on soil Carbon. *Ecol. Appl. :* 11: 343–355.

Conger, B. V. (1991). Biotechnology in forage grass breeding. In: *Fodder Crops Breeding: Achievements, novel strategies and biotechnology. Proceedings of the 16th Meeting of Eucarpia Fodder Crops Section,* Wageningen, the Netherlands, pp. 147-153.

Corah, L. (1996). Trace mineral requirement of grazing cattle. *Animal Feed Science and Technology* 59: 61-70.

Corino, C., Dell'orto, V., Pedron, O., Polidori, F., Rigoni, M. and Castrovilli, C. R. (1982). Estimation of the nutritive value and the effects of different amounts of dried whole barley in the diet on performances of meat rabbits. *Zootecnica e Nutrizione Animale* 8: 289-299.

Couvreur, S., Hurtaud, C., Lopez, C., Delaby, L. and Peyraud, J. L. (2006). The linear relationship between the proportion of fresh grass in the cow diet, milk fatty acid composition, and butter properties. *J. Dairy Sci.* 89: 1956–1969.

Crespi, M. P., De., A. L., Coll, J. F. C., Itagiba, M. D., Sousa, G. O. R., De, J. C. D., Gomes, D., Da, C. and Goncalves, A. S., (1992). The use of guandu hay (*Cajanus cajan*) as a fibre and protein source in the diet of growing rabbits. *Revista da Sociedade Brasileira de Zootecnia* 21: 28-32.

Crespi, M. P., De., A. L., Coll, J. F. C., Sousa De, J. C. D., Gomes, A. V., Da, C. and Goncalves A. S. (1992). Use of perennial soya (*Neonotonia wightii*) as a source of fibre and protein for growing rabbits. *Revista da Sociedade Brasileira de Zootecnia* 21: 23-27.

Crouch, E. K. (1964). Dehydrated Aquatic Plant material for Cattle Feeding. *Feedstuffs* 36: 43.

Cucchiara, R. (1989). Sulla in the nutrition of meat rabbits. *Rivista di Coniglicoltura* 26: 39-42.

D' Mello, J. P. F. (1995). Leguminous leaf meals in non-ruminant nutrition. In: *Tropical Legumes in Animal Nutrition*, (Eds.) D' Mell, J. P. F. and Devendra, C., CAB Internatioonal Wallingford, UK, pp. 247-282.

D'Mello, J. P. E. (1992). Nutritional Potentialities of fodder trees and fodder shrubs as protein sources in mongastric nutrition. In "*Legume trees and other fodder trees as protein sources for livestock*", eds., Speedy, A and Puglies, P. H., 1st Edition, Food and Agriculture Organization, Rome. pp. 115-127.

D'mello, J. P. F. (1992). Chemical constraints to the use of tropical legumes in animal nutrition. *Animal Feed Science and Technology* Volume 38, Issues 2–3, pp. 237-261.

Dabadghao, P. M. and Shankarnarayanan, K. A. (1973). *Grass Cover of India*. Indian Council of Agricultural Research, New Delhi, India.

Daghir, N. J. (1995). Feedstuffs used in hot regions. *In "Poultry production in hot climates"* Daghir, N. J. (ed): 1st Edition CAB International Wallingford UK., pp. 125-154.

Daniels, L. B., Shriver, L. A. and Nelson, T. S. (1985). Evaluation of bermudagrass in diets of domestic rabbits. *Arkansas Farm Research* 34: 2.

Darabant, A., Rai, P. B., Tenzin, K., Roder, W. and Gratzer, G. (2007). "Cattle grazing facilitate tree regeneration in a conifer forest with palatable bamboo understory," *Forest Ecology and Management* 252(1–3): 73–83.

Das, K. C., Haque, N. and Rajkhowa, C. (2009). Tree leaves and forest grasses based feed block for Mithun in North Eastern region of India. In T. K. Walli, eds. "*Proceedings of the national symposium on fodder block technology*", ILDEX India–2009, New Delhi, pp. 80–83.

Das, M. (1990). Evaluation of dairy based farming on steep hills in meghalaya. Annual Report 1990-91, ICAR Research Complex for NEH Region, barapani, Meghalaya, pp. 212-213.

Datta, B. K., Bhar, M. K., Patra, P. H., Majumdar, D., Dey, R. R., Sarkar, S., Mandal, T. K. and Chakraborty, A. K. (2012). Effect of Environmental Exposure of Arsenic on Cattle and Poultry in Nadia District, West Bengal, India. *Toxicology International* 19(1): 59–62.

Davtyan, A. and Manukyan, V. (1987). Effect of grass meal on fertility of hens. *Ptitsevodstvo*. 6: 28-29.

Dawd, A. G., Gezmu, T. B. and Haki, G. D. (2012). Essential and toxic metals in cow's whole milk from selected sub-cities in Addis Ababa, Ethiopia. *Online International Journal of Food Science* 1(1): 12-19.

Dei, H. K., Alhassan, M. and Borkloe, G. (2012). *Evaluation of Albizia jullibrissin leaf meal as an egg yolk colourant in laying hens.* University for Development studies, Tamale-Ghana, pp. 21-28.

Delgado, C., M. Rosegrant, H. Steinfeld, S. Ehul and C. Courbois. (1999). *Livestock to 2020: The Next Food Revolution (Discussion Paper 28*). IFPRI, Washington DC; FAO Rome and ILRI, Nairobi, 72p.

Deng, M. X. J. and Gu, A. (1992). *Rangelands ofXizang (Tibet),* Science Press, Beijing, China.

Deng, W. D., Wanapat, M., Ma, S. C., Chen, J., Xi, D. M., He, T. B., Yang, Z. F. and Mao, H. M. (2007). Phylogenetic analysis of 16S rDNA sequences manifest rumen bacterial diversity in Gayal (*Bos frontalis*) fed fresh bamboo leaves and twigs (*Sinarumdinaria*). *Asian-Aust. J. Anim. Sci.* 20(7):1057-1066.

Deng, W. D., Wang, L. P., Ma, S. C., He, T. B., Yang, Z. F. and Wanapat, M. (2007). Comparison of Gayal (*Bos frontalis*) and Yunnan Yellow Cattle (*Bos taurus*): rumen function, digestibilities and nitrogen balance during feeding of pelleted lucerne (*Medicago sativum*). *Asian-Aust. J. Anim. Sci.* 20: 900- 907.

Derner, J. D. and Schuman, G. E. (2007). Carbon sequestration and rangelands: a synthesis of land management and precipitation effects. *J. Soil Water Conserv.* 62: 77–85.

Deshmukh, S. V. and Pathak, N. N. (1989). Voluntary intake and dry matter digestibility of green fodders and tree leaves *in* New Zealand White Rabbits. *Cherion* 18: 223- 225.

Deshmukh, S. V. and Pathak, N. N. (1990). Voluntary intake, digestibility and nutritive value of green berseem (*Trifolium alexandrinum*) in rabbits. *Indian Journal of Animal Nutrition* 7: 233-234.

Deshmukh, S. V., Pathak, N. N., Randhe, S. R. and Deshmukh, S. S. (1993). Voluntary intake, digestibility and nutritive value of coastal Bermuda grass (*Cynodon dactylon*) employed as sole feed for rabbits. *World Rabbit Science* 1: 109-111.

Deshmukh, S. V., Pathak, N. N., Takalikar, D. A. and Digraskar, S. U. (1993). Nutritional effect of mulberry (*Morus alba*) leaves as sole ration of adult rabbits. *World Rabbit Science* 1: 67-69.

DFID (2002). *Better livelihood for the poor people: The role of land policy*, UK.

Dhumal, M. V., Siddiqui, M. V., Siddiqui, M. B. A. and Avai, P. E. (2009). Performance of broilers fed on different levels of Azolla meal. *Indian Journal of Poultry Science* 44 (1): 65-68.

Dikshit, A. K. and Birthal, P. S. (2010). India's Livestock feed demand: Estimates and Projections. *Agricultural Economics Research Review.* 23: 15-28.

Diwedi, P. N., Goyal, R. K. and Singh, K. K. (2003). Preparation and evaluation of densified feed blocks in growing buffaloes. *Indian Journal of Animal Nutrition* 20: 202–205

Dixon, R. K., Smith, J. and Guill, S. (2003). Life on the edge: vulnerability and adaptation of African ecosystems to global climate change. *Mitigation and Adaptation Strategies for Global Change* 8: 93–113.

D'Mello, J. P. F. (1995). Leguminous leaf meals in non-ruminant nutrition. In "Tropical Legumes in Animal Nutrition", eds. J. P. F. D'Mello and C. Devendra,. CAB International, Wallingford, UK.

D'Mello, J. P. F. (1992). Nutritional potentialities of fodder trees and fodder shrubs as protein sources in monogastric nutrition, *IN "FAO Animal Production and Health paper* 102 (*Proceedings of the FAO Expert Consultation held at the Malaysian Agricultural Research and Development Institute* (*MARDI*) *in Kuala Lumpur, Malaysia*), *14–18 October 1991* Eds. Andrew Speedy and Pierre-Luc Pugliese, FAO, Rome, 1992, pp. 144- 158.

D'Mello, J. P. F. and Acamovic, T. (1989). *Leucaena leucocephala* in poultry nutrition: A Review. *Animal Feed Science and Technology* 26: 1–28.

Donkoh, A., Atuahene, C. C., Poku-Prempeh, Y. B. and Twnm, I. G. (2002). The nutritive value of chaya (*Cnidoscolus aconitifolius* (*Mill.*) Johnson) leaf meal: Studies with broiler chickens. *Anim. Feed Sci. Tec.* (*Netherlands*) 77: 163-172.

Dougall, H. W. and Bogdan, A. V. (1966). The chemical composition of some leguminous plants grown in the herbage nursery at Kitale, Kenya. *East African Agricultural and Forestry Journal* 32: 45-49.

Down to Earth (2010). Strained Milk. Cover Story, June, 16-30, pp. 23-29.

Drake, B. G., Gonzalez-Meler, M. A. and Long, S. P. (1997). More efficient plants: a consequence of rising atmospheric CO_2. *Annual Review Plant Physioloy, Plant Molecular Biology* 48: 609-639 (doi:10. 1146/annurev. arplant. 48. 1. 609).

Du Guo, Z. zhen and Zhang Degang. (2000). *Final Report of ICIMOD Regional Rangeland Programme—MaquCounty, Gansu* (*Phase I*). Submitted to ICIMOD (International Centre for Integrated MountainDevelopment), Kathmandu, Nepal.

Durunna, C. S., Chiaka, O. E., Ebenebe, A. B. I., Udedibie, M. C., Uchegbu, O., Durunna, N. (2011). Value of Bitter Leaf (*Vernonia amygdalina*) Meal as Feed Ingredient in the Diet of Finisher. *International Journal of Agriculture and Rural Development 14(2)*: 722-726

Dwivedi, R. N. and Pandey, J. K. (1987-92). Up to date Report on Forage Crops., ICAR Res. Complex for NEH Region, Arunachal Pradesh Centre, Basar (Arunachal Pradesh).

Echeverria, V., Belmar, R., Ly, J. and Santos-Ricalde, R. H. (2002). Effect of *Leucaena leucocephala* leaf meal treated with acetic acid or sodium hydroxide on apparent digestibility and nitrogen retention in pig diets. *Animal Feed Science and Technology* 101: 151– 159.

Edwards, P. J. (1973). Definitions of some pasture terms. *Proceedings of the Grassland Society of Southern Africa* 8: 133–135.

Eggum, B. O. (1970). The protein quality of cassava leaves. *Feeding pigs in the tropics* Br. J. 04/11/2011. *D:/cd3wddvd/NoExe/.../meister10. htm 339/427 Nutr.* 24(3):761–768.

Ehleringer, J. R., Cerling, T. E. and Helliker, B. R. (1997). C_4 photosynthesis, atmospheric CO_2 and Climate. *Oecologia* 112: 285-299. (doi:10. 1007/ s004420050311.

Eichie, F. O., Akinyemi1, D. E. and Adegbenro, M. (2015). Effect of Replacement of Soybean Meal with *Leucaena leucocephala* Leaf Meal on Performance, Haematology, Carcass Measures and Organ Weight in Broiler Chickens. *American Journal of Experimental Agriculture* 8(5): 327-334.

Ekpenyong, T. E. (1985). Toxic effects of feeding *Leucaena* leaves to growing rabbits. *Nutrition Reports International* 31: 345–350.

El-Adawy, M. M., Borhami, B. E. and Abdel-Hamid, A. E. Y. (2000). Utilization of sugar beet pulp and fibrous residues of berseem and water hyacinth in feeding growing New Zealand White rabbits. *Egyptian Journal of Rabbit Science* 10: 1-17.

El-Ayouty, S. A., Abdel-Khalek, A. E., El-Ghany, A. I. A. and Shatifa, M. A. (2000). Effect of diets containing silage on growth performance, digestibility and carcass traits of growing rabbits. *Egyptian Journal of Nutrition and Feeds* 3: 43-56.

Eleraky, W. A. and Mohamed, W. E. (1996). Growth performance, carcass traits and some related physiological changes of growing rabbits fed on *Acacia* and water hyacinth. *Egyptian Journal of Rabbit Science* 6: 87-98.

El-Galil, K. A., Khalil, F. S. and El-Ganzoury, E. H. (2001). Utilization of *Leucaena* leaf meal by growing rabbits under the recently reclaimed areas. *Egyptian Journal of Rabbit Science* 11: 151-165.

El-Gendy, K. M. (1999). Effect of dietary inclusion of *Acacia* leaves meal (*Acacia saligna*) on digestibility, growth performance and blood constituents of growing rabbits. *Egyptian Journal of Rabbit Science* 9: 271-283.

El-Gendy, K. M. (1999). Nutritional studies on some green forage in Egypt. 7. Utilization of sweet lupin (*Lupinus albus*) as green forage for feeding rabbits. *Egyptian Journal of Rabbit Science* 9: 257-269.

El-Sayaad, G. A. E., Luick, B. R. and Cheeke, P. R. (1992). Preliminary evaluation of seed screening of orchard grass as a feed stuff for rabbits. *Egyptian Journal of Rabbit Science* 2: 73-80.

Emechebe, A. (2010). IAR4D approach for strengthening partnerships among stakeholders to encourage collective action towards post harvest issues. Power point presentation in panel discussion on the theme of CORAF/WECARD Science week, May 24-29, 2010, Cotonou, Benin Republic.

Erdman, M. D. (1986). Growth performance of rabbits fed arrowroot (*Maranta arundinacae*) forage and processing waste. *Nutrition Reports International* 33: 1029-1034.

Ergano, K. A., Duncan, A., Adie, A., Tedla, G., Woldewahid, Z., Ayele, G., Berhanu and Alemayehu, N. (2010). Multi-stakeholder platforms strengthening the selection and use of fodder options in Ethiopia: Some lessons. *Presentation to the Ethiopian Fodder Roundtable on Effective Delivery of Input Services to Livestock Development,* 22 June 2010, Addis Ababa, Ethiopia.

Ergano, K., Duncan, A., Adie, A., Tedla, A., Woldewahid, G., Ayele, Z., Berhanu, G. and Alemayehu, N. (2010). Implementation challenges of innovation systems perspective in fodder production in Ethiopia'. Paper presented at the *"Innovation and Sustainable Development in Agriculture and Food Symposium".* 28 June-1 July 2010. Montpellier, France.

Erika, S., Zavaleta, M., Rebecca Shaw, Nona, R., Chiariello, Brian, Thomas, D., Elsa E. Cleland, Christopher B. Field, and Mooney, H. A. (2003). Grassland responses to three years of elevated temperature, CO_2, precipitation and N deposition. *Ecological Monographs* 73:585–604. http://dx. doi. org/10. 1890/02-4053.

Eruvbetine, D., Tajudeen, I. D., Adeosun, A. T. and Olojede A. A. (2003). Cassava (*Manihot esculenta*) leaf and tuber concentrate in diets for broiler chickens. *Bio Resource Tech.*, 86: 277–281.

Esonu, B. O., Iheukwumere, F. C., Iwuji T. C., Akanu, N. and Nwugo, O. H. (2003). Evaluation of *Microdesmis puberula* Leaf meal as feed ingredient in broiler starter diets. *Nigerian Journal of Animal Production* 30(1): 3-8.

Faber-Langendoen, D. and Josse, C. (2010). World Grasslands and Biodiversity Patterns. : *A Report to IUCN Ecosystem Management Programme Nature Serve,* Arlington, VA.

FAO (1965). *Glossary of pasture and fodder terms.* FAO, Rome, Italy.

FAO (1992). Legume tree and other fodder trees as protein sources for livestock, *FAO Animal Production and Health Paper* FAO, Rome, Italy, 102:14-18.

FAO (1992). Roots, tubers, plantain and bananas in animal feeding. (D. Machin and S. Nyvold, editors). Animal Production and Health paper No. 95. FAO, Rome, pp 289.

FAO (1993). Tropical Feeds by B. Göhl. Computerized version 4. 0 edited by A. Speedy, Rome, Italy.

FAO (1994). *Tropical Feeds and Feeding Systems- Azolla, eds* Andrew Speedy, 1984-1993. reference 634, FAO, Rome, Italy.

FAO (1994). *Tropical Feeds Database Extract-* 1983-94 - *Prosopis juliflora,* Andrew Speedy eds, FAO, Rome, Italy, reference: 563.

FAO (1997). Feeding pigs in the tropics. *FAO Animal Production and Health Paper* 132, FAO, Rome, Italy, 289p.

FAO (1998). Tropical feeds. Data Base- nutrient requirements of swine by Andrew Speedy and Nick Waltham, version 8. 156 24. National Research Council, Tenth edition, FAO, Rome, Italy.

FAO (2001). Codex Alimentarius- Organically Produced Foods. FAO, Rome, Italy.

FAO (2002). *Mulberry for Animal Production,* eds. M. D. Sanchez, *Animal Production and Health Paper* 147, FAO, Rome, Italy.

FAO (2003). The Yak, *Second edition,* Regional Office for Asia and the Pacific, Maliwan Mansion Phra Atit Road, Bangkok 10200, Thailand.

FAO (2005). Terms and definition FRA 2005. FRA working paper 73, FAO, Rome, Italy.

FAO (2007). *Livestocks Long Shadow.* Food and Agriculture Organization of the United Nations, Rome, Italy.

FAO (2012). Crop residue based densified total mixed ration – A user-friendly approach to utilise food crop by-products for ruminant production, by T. K. Walli, M. R. Garg and Harinder P. S. Makkar. *FAO Animal Production and Health Paper No.* 172. FAO, Rome, Italy.

FAO. (2006). *World agriculture: towards 2030/2050. Interim report, Global Perspective Studies Unit,* Food and Agriculture Organization of the United Nations, Rome, Itally

Faruqui, S. A. and Sunil Kumar (2009). Forages for Lean Period in East and North East Region, AICRP on Forage Crops, IGFRI, Jhansi, India, p. 20.

Faruqui, S. A., Kumar Sunil and Singh, D. N. (eds.) (2009). *Napier Bajra Hybrid: An Excellent Perennial Forage,* AICRP on Forage Crops, IGFRI, Jhansi, India p. 20.

Favier, C., De Namur, C. and Dubois, M. A. (2004). Forest progression modes in littoral Congo, Central Atlantic Africa. *J. Biogeogr.* 31, 1445–1461. (doi:10. 1111/j. 1365-2699. 2004. 01094. x)

Fazaeli, H., Golmoihammadi, H. A., Shoayee, A. A., Montaiebi, N. and Mosharraf, S. H. (2011). Performance of feedlot calves fed hydroponic fodder barley. *J. Agriculture Science Technology* 13: 365-375.

Fazaeli, H., Golmoihammadi, H. A., Tabatabayee, S. N. and Asghari-Tabrizi, M. (2012). Productivity and Nutritive Value of Barley Green Fodder Yield in Hydroponic System. *World Applied Sciences Journal* 16(4): 531-539.

Fedeli Avanzi, C. and Jannella, G. G. (1976). Growth and thiocyanate level in blood and urine of rabbits fed with carrots, rape and cabbage. *First World Rabbit Congress, Dijon: communication* 28, 1-4.

Fernandez-Carmona, J., Bernat, F., Cervera, C. and Pascual, J. J. (1998). High lucerne diet for growing rabbit. *World Rabbit Science* 6(2): 237-240.

Fernandez-Carmona, J., Santiago, S., Alquerda, A. I., Cervera, C. and Pascual, J. J (2000). Effect of lucerne-based diets on the reproductive performance of rabbit at high environmental temperatures. *7th World Rabbit Congress*, Valencia vol. C: 203-208.

FGTC (The Forage and Grazing Terminology Committee) (1991). *Terminology for Grazing Lands and Grazing Animals*. Blacksburg, Virginia, USA: Pocahontas Press, Inc., *Journal of Production Agriculture* 5: 191–201

Fotso, J. M., Fomunyam, R. T. and Ndoping, B. N. (2000). Protein and energy sources for rabbit diets in Cameroon. 1 - Protein sources. *World Rabbit Science* 8: 57-60.

Foulkes, D., Espejo, S., Marie, D., Delpeche, M. and Preston, T. R. (1977). The banana plant as cattle feed: composition and biomass production. *Trop. Anim. Prod.*, 3 (1): 45-50

Foundation for Ecological Security. (2009). *Common Land and Poor Livestock-keepers: Experiences from Common Land Development in the States of Rajasthan and Madhya Pradesh in India. South Asia Pro Poor Livestock Policy Programme*, ILRI, NASC, New Delhi.

Gabino, A. T. (2012). *The Effect of Chopped Banana Trunk and Sweet Potato Leaves and Vines as Feed Supplement on the Performance of Growing- Finishing Pigs.* Benguet State University, La Trinidad, Benguet.

Gad, S. D. and Shyama, S. K. (2009). Studies on the food and feeding habits of Gaur (*Bos gaurus* H. Smith (Mammalia: Artiodactyla: Bovidae) in two protected areas of Goa". *Journal of Threatened Taxa* **1** (2): 128–130. doi:10. 11609/ jott. o1589. 128-30.

Gadzirayi, C. T. and Mupangwa, J. F. (2014). The Nutritive Evaluation and Utilisation of *Moringa oleifera Lam.* in Indigenous and Broiler Chicken Production: A Review. *Greener Journal of Agricultural Sciences.* 4(1): 015- 021, http://dx. doi. org/10. 15580/GJAS. 2014. 1. 012914078.

Galindo, W., Rosales. M., Murgueitio, E. and Larrahondo, J. (1989). Sustancias antinutricionales en las hojas de Guamo, Nacedero y Matarratón. *Livestock Research for Rural Development.* Volume 1, Number 1: 36-47.

García, A., Duarte, F. and Magaña, A. (1991). Crecimiento y finalizacón de cerdos con diferentes nivles de vinaza. *Livest. Res. Rur. Dev.* 3(1): 41–46

Garg, M. R, Bhanderi, B. M. and Serasia, P. L. (2003). Trace mineral status of feeds and fodder in Dahod and Panchmahal district of Gujarat. *Animal Nutrition and Feed Technology* 3: 27-36.

Garg, M. R. and Bhanderi, B. M. (2011). Enhancing livestock productivity through balanced feeding. In S. S. Kundu, C. Dutt, V. Mani and A. Kumar. *"Proceedings of the 14th Biennial Conference"*, Animal Nutritional Society of India, GBP University of Agriculture and Technology, Pantnagar, Uttra Khand, India, pp. 11–21.

Garg, M. R., Bhanderi, B. M. and Sherasia, P. L. (2005). Assessment of adequacy of macro and micro mineral content of feed stuffs for dairy animals in Semi arid zone of Rajsthan. *Anumal Nutrition and Feed Technology* 5: 9-20.

Gee, E. P. (1964). *The wild life of India*, Collins S T, James Place, London.

Gerlitz, J. V., Banerji, S., Hoermann, B., Hunzai, K., Macchi, M. and Tuladhar, S (2014). *Poverty and Vulnerability Assessment – A survey instrument for the Hindu Kush Himalayas*, ICIMOD, Kathmandu (Nepal), p102.

Ghazalah, A. A., El-Shahata, A. A and, El-Yamny, A. T. (1998). Evaluation of some tropical forage for nutrition and meat production of rabbits. *Egyptian Journal of Rabbit Science* 8: 127-139.

Ghotge, N. S. and Ramdas, S. R. (1998). 'Women in Livestock Production: Towards Gender Sensitive Policies, Technologies and Development Strategies', in *Second Pan Commonwealth Veterinary Conference: Scientific Proceedings*, Vol. 1, Commonwealth Veterinary Association and Veterinary College, University of Agricultural Sciences, Bangalore.

Giang, N. T. and Preston, T. R. (2011). Taro (*Colocacia esculenta*) silage and water spinach as supplements to rice bran for growing pigs. *Livestock Research for Rural Development. Volume 23, Article #45.* Retrieved September 5, 2016, from http://www. lrrd. org/lrrd23/3/gian23045. htm.

Giang, N. T. and Preston T. R (1997). Evaluation of protein in duckweed (*Lemna spp*) using a duckling growth assay. *Livestock Research for Rural Development.* 9(2): http://www. cipav. org. co/lrrd/lrrd9/2/anh92. htm

Gidenne, T. (1985). Feeding value of hibiscus for young rabbits reared under tropical or temperate climatic conditions. Preliminary results. *Cuni-sciences* 3: 23-28.

Gilgen, A. K., Constant Signarbieux, Urs Feller and Nina Buchmann (2010). Competitive advantage of *Rumex obtusifolius* L. might increase in intensively managed temperate grasslands under drier climate. *Agriculture, Ecosystems and Enviornment* 135 (1-2): 15-23.

Giridhar, K. and Rajendran, D. (2013). Cultivation and usage of azolla as supplemental feed for dairy cattle. In: Value addition of feed and fodder for dairy cattle, *NIANP*, June 27-July 6, 2013, 32-34.

Gliessman, K. A. Sr. (2005): Pastoral nomadism, a sustainable system for grazing land management in arid areas. *Journal of Sustainable Agriculture* 5(4): 113-131. DOI: 10. 1300/J064v25n04_09.

Gohi, B. (1981). Tropical feeds, *FAO Animal Production and Health Series.* 12:529, FAO, Rome, Italy.

GOI (1987). *Report of the Task Force to Evaluate Impact of Sheep and Goat Rearing in Ecologically Fragile Zones.* Ministry of Agriculture, Government of India, New Delhi, 67p.

GOI (1999). *Common Property Resources in India (NSS 54th round)*. National Sample Survey Organization, Department of Statistics and Programme Implementation, Government of India, New Delhi, 290p.

GOI (2001). *National Action Programme to Combat Desertification (Volume 1).* Ministry of Environment and Forests, Government of India, New Delhi, 352p.

GOI (2001). *Report of Task Force on Greening India for Livelihood Security and Development.* Planning Commission, Government of India, New Delhi, 231p.

GOI (2002). *National Report to the Third Session of the United Nations Forum on Forests.* Ministry of Environment and Forests, Government of India, New Delhi, 19p.

GOI (2002). *Watersheds at work: success stories of watershed programmes.* Government of India, 263p.

GOI (2006). Standing Parliamentary Joint Committee on the Scheduled Tribes, (Recognition of Rights) Bill, 2005, Report of the Joint Committee, May 2006.

GOI (2007). *Report of the Task Force on Grassland and Deserts.* Planning Commission, Government of India, New Delhi, 34p.

GOI (2009). *Manual for Drought Management*, Department of Agriculture and Coperation, Ministry of Agriculture, Government of India, New Delhi.

GOI (2009). Ministry of New and Renewable Energy Resources (2009) www. mnre. gov. in/relatedlinks/biomassrsources.

Goldstein, M., Beall, C. and Cincotta, R. (1990). Traditional Nomadic Pastoralism and Ecological Conservation on Tibet's Northern Plateau. *National Geographic Research*, 6(2): 139-156.

Gomez-De-Varela, A., Harris, D. J., Cheeke, P. R. and Patton, N. M. (1983). Evaluation of perennial peanut (*Arachis glabrata*) and kudzu (*Pueraria phaseoloides*) as feedstuffs for rabbits. *Journal of Applied Rabbit Research* 6: 97-98.

González, G., Ivonne Díaz, Vecchionacce, H. and Ly, J. (2003). Performance traits of pigs fed sweet potato (*Ipomoea batatas L.*) foliage ad libitum and graded levels of protein. *Livestock Research for Rural Development* 15 (9), 2003. http:// www. lrrd. org/lrrd15/9/gonz159. htm.

Gonzalez-Estrada, E., Rodrigues, L. C., Walen, V. K., Naab, J. B., Koo, J., Jones, J. W., Herrero, M. and Thornton, P. K. (2008). Carbon sequestration and farm income in West Africa: identifying best management practices for smallholder agricultural systems in northern Ghana. *Ecological Economic,* 67: 492–502

Good Practice Code (2012). INGP 28 - Building Fodder Security in Rural Areas: Validation of Traditional Knowledge on Fodder and Its Reintegration into Livelihoods, published on SAPPLPP website.

Gordon King (2017). Evolution of Animal Agriculture, Department of Animal and Poultry Science, University of Guelph, Ontario, Canada. www. aps. uoguelph. ca/~gking/Ag_2350/evolve. htm download 03-05-2017.

Gorksi, B. (2000). Nutritional Analysis of Pastured Poultry Products. *APPPA GRIT, American Pastured Poultry Producers Association* Vol. 11, P. 1-3.

Gowda, N. K. S., Prasad, C. S., Ramana, J. V. and Shivaramaiah, M. T. (2002). Assessment of mineral status in hilly and central dry zone of Karnataka and ways to supplement them. *Indian J. Animal Science* 72: 165-170.

Grandi, A. and Battaglini, M (1988). *Trifolium pratense L.* hay in diets of growing rabbits. *4th World Rabbit Congress*, Budapest 3: 123-131.

Grandi, A. (1978). Aquatic plants in the feeding of rabbits. The hydrophyte *Potamogeton natans* L. var. fluitans Roth. in feeds for rabbits. *Coniglicoltura* 15: 31-33.

Grandi, A. (1981). Use of water hyacinth (*Eichhornia crassipes*) in diets for rabbits. *Coniglicoltura* 18: 43-48.

Grandi, A. (1993). Use of *Bromus catharticus* in the nutrition of rabbits. *Rivista di Coniglicoltura* 30: 41-43.

Grandi, A. and Cagiotti, M. R. (1985). *Psophocarpus tetragonolobus* (winged bean) meal in diets for growing rabbits. *Zootecnica e Nutrizione Animale* 11: 427-436.

Grandi, A., Marzetti, P. and Blasi, F. (1983). Water hyacinth (*Eichhornia crassipes*) meal for feeding rabbits: digestibility, nutritive value and productive performances. *Zootecnica e Nutrizione Animale* 9: 297-309.

Grigg, D. B. (1974). The Agricultural Systems of the World, an Evolutionary Approach, Cambridge University Press, Cambridge, UK, 358pp.

Groenefeld, E. and Glodek, P. (2000). Animal Breeding and Animal Genetic Resources. Landbauforschung Völkenrode, FAL Agricultural Research, Special Issue 228, Braunschweig, 94p.

Gualtieri, M., Rapaccini, S. and Balloni, W. (1988). Effects of the inclusion of *Azolla filiculoides* meal in feeds for growing rabbits. *Rivista di Coniglicoltura* 25: 55-58.

Guodao, L., Changjun, B., Donjun, W., Ramesh, C. R., and Rao, P. P. (2004). Leaf meal production from *Stylosanthes* in China and India, In "High yielding anthracnose resistant *Stylosanthes* for agricultural systems", eds S. Chakraborty, ACIAR, CSRIO, Canberra, pp. 253-256.

Gupta, B. N. and Singh, R. B. (1983). Chemical composition and nutritive value of *Styloshanthes guyanensis* (HBK) hay (Abstract). *Nutr. Abst. Rev. Series* B53: 28.

Gupta, H. K. and Atraja, P. P. (1996). Influence of mimosine and 3-hydroxy 4(1H) pyridone fed through *Leucaena leucocephala* leaf meal on growth and nutrient utilization in meat rabbits. *6th World Rabbit Congress*, Toulouse 1: 191-193.

Gupta, H. K., Balaraman, N. and Yadav, B. P. S. (1990). Feeding value of some important fodder trees of agroforestry systems in eastern Himalayan region. *Proc. Int. Symp. "Water Sedimentation and Resource Management*", 9-13 October, 1990, CSWRTI, Dehradun, 453-57.

Gupta, P. D, Goyal, R. K. and Chattopadhyay, P. S. (1998). Performance evaluation of IGFRI densifying machine for grasses. *Journal of Agricultural Engineering,* 35: 33–38.

Gupta, P. K., Sahai, S., Singh, N., Dixit, C. K., Singh, D. P., Sharma, C., Tiwari, M. K., Gupta, R. K. and Garg, S. C. (2004). Residue burning in rice–wheat cropping system: Causes and implications. *Current Science.* 87(12): 1713-1717

Gupta, S. K. and Roy, M. M. (2009). *Luecaena- A Promising Forage Resource.* IGFRI, Jhansi, India.

Gura, S. (2008). Industrial livestock production and its impact on smallholders in developing countries. *Consultancy report to the League for Pastoral Peoples and Endogenous Livestock Development,* Germany. *http://www. pastoralpeoples. org.*

Ha, N. N. and Phan, P. T. (1995). Vegetative propagation capacities and effect of fertilization on biomass production of *Trichanthera gigantea. Livestock Research for Rural Development.* Volume 7, Number 1 (10kb), from http:// www. lrrd. org/lrrd9/1/nhan91. htm.

Haile, S. G., Nair, P. K. R. and Nair, V. D. (2008). Carbon storage of different soil size fractions in Florida silvopastoral Systems. *J. Enviorn. Quality* 37: 1789-1797.

Hall, A. (2007). Challenges to strengthen agricultural innovation systems: where do we go from here? *UNU-MERIT Working Paper Series.* 2007-038. Available from http://www. merit. unu. edu/publications/wppdf/2007/wp2007-038. pdf (accessed on 15 March 2012).

Hall, A., Dijkman, J. and Sulaiman, R. V. (2010). Research into Use: Investigating the Relationship between Agricultural Research and Innovation. *UNU-MERIT Working Paper Series* #2010-044.

Hall, A., Rasheed Sulaiman V and Peter Bezkorowajnyj. (2007). *Reframing Technical Change: Livestock Fodder Scarcity Revisited as Innovation Capacity Scarcity. A Conceptual Framework.* Published by the Fodder Innovation Project Phase 2, International Livestock Research Institute (ILRI), Hyderabad, India, 52p.

Haneef, S. S., Swarup, D., Kalicharan and Dwivedi, S. K. (1995). The effect of concurrent lead and cadmium exposure on cell mediated immune response in goats. *Vet Human Toxicol* 37: 428-429.

Hang, B. P. T., Lam, V. and Preston, T. R. (2011). Effect of different sources of carbohydrate as supplements to basal diets of water spinach and cabbage leaves on the growth rate in rabbits. Livestock Research for Rural Development. Volume 23, Article #227. http://www. lrrd. org/lrrd23/11/hang23227. htm.

Hang, B. P. T., Lam, V. and Preston, T. R. (2013). Effect of paddy rice on growth performance of rabbits fed water spinach as the sole forage or combined with leaves of water hyacinth. *Livestock Research for Rural Development. Volume 25, Article #45.* Retrieved, from http://www. lrrd. org/lrrd25/3/hang25045. htm

Hang, D. T. (1998). Ensiled cassava leaves and duckweed as protein sources for fattening pigs on farms in Central Vietnam. Livestock Research for Rural Development. 10(3): http://www. cipav. org. co/lrrd/lrrd10/3/hang2. htm.

Hanson, J. G. and Baker, B. B. and Bourdon, R. M. (1993). Comparison of the effects of different climate change scenarios on rangeland livestock production. *Agricultural Systems,* 41: 487–502.

Hardin, G. (1968). 'The tragedy of commons'. *Science,* 162(Dec. 13): 1243-1248.

Harris, D. J., Cheeke, P. R. and Patton, N. M. (1981). Effect of feeding amaranthus, sunflower leaves, Kentucky bluegrass and alfalfa to rabbits. *Journal of Applied Rabbit Research* 4: 48-50.

Harris, D. J., Cheeke, P. R. and Patton, N. M. (1984). Evaluation of black locust leaves for growing rabbits. *Journal of Applied Rabbit Research* 7: 7-9.

Harris, D. J., Cheeke, P. R., Telek, L., and Patton, N. M. (1981). Utilization of alfalfa meal and tropical forages by weanling rabbits. *Journal of Applied Rabbit Research* 4: 4-9.

Harris, R. B. and Bedunath, D. J. (2001). Sheep vs argali and livestock in western China. *Final Report, National Geographical Society,* University of Montana, Missoula, USA, 52p.

Hartley, W. (1963). The development of forage crops. *Span* 6, 68-71.

Hasan, M. R. and Chakrabarti, R. (2009). Floating aquatic macrophytes: water hyacinths. *In* **Hasan**, M. R. ; Chakrabarty, R. (Eds). Use of algae and aquatic macrophytes as feed in small scale aquaculture - a review. *FAO Fisheries and Aquaculture Technical Paper*. No. 531. FAO, Rome, 123p.

Hassan, H. M. A., El-Moniary, M. M., Hamouda, Y., Eman, F. El-Daly, Amani, W. Youssef, A. W. and El-Azeem, N. A. A. (2016). Effect of different levels of *Moringa oleifera* leaves meal on productive performance, carcass characteristics and some blood parameters of broiler chicks reared under heat stress conditions. *Asian Journal of Animal and Veterinary Advances* 11(1): 60-66 DOI: 10. 3923/ ajava. 2016. 60. 66.

Hauptli, L. and Lovatto, P. A. (2006). Feeding sows in gestation and lactation with diets containing saponins. *Ciencia Rural* 36: 610–616.

Haynes, R. J. (*1980*). Competitive aspects of the grass-legume association. *Advances in Agronomy* 33: 227-256.

Hazarika, A., Borah, U., Borah, M. M., Bordoloi, L. and Hirendra N. Sarma, H. N. (2010). Documentation of ecology and food plant of *Bos frontalis* (Mithun) in Arunachal Pradesh (India). *Journal of Environmental Research and Development* 4 (3): 687.

Hazra, C. R. (1995). Soil and water conservation in relation to land use and biomass production. In: R. P. Singh (ed). *Forage Production and Utilization*. Indian Grassland and Fodder Research Institute, Jhansi, India.

Heady, H. F. (1970). Grazing systems: terms and definitions. *Journal of Range Management* 23: 59–61

Hegde, N. G. (1998). Dairy Development for Poverty alleviation and Environmental protection. Integrated Rural Development for Sustainable Livelihood. BAIF, Pune, India, pp. 33-39.

Hegde, N. G. (2006). *Livestock Development for Sustainable Livelihood of Small Farmers*. CLFMA Souvenir, pp. 50-63.

Hegde, N. G. (2014). Forage resource development in India. *The Indian journal of animal sciences* 84(7): 715-722.

Heinritz, S. N., Hoedtke, S., Martens, S. D., Peters, M. and Zeyner, A. (2012). Evaluation of ten tropical legume forages for their potential as pig feed supplement. *Livestock Research for Rural Development. Volume 24(1), Article #7.* Retrieved September 5, 2016, from http://www. lrrd. org/lrrd24/1/ hein24007. htm.

Herbert, U. (1998). Reproductive performance of rabbit does feed diets containing *Gliricidia* leaf meal from conception through weaning of kits. *Nigerian Journal of Animal Production* 25: 163-168.

Herrero M., *et al.* (2009). *Drivers of change in crop–livestock systems and their potential impacts on agro-ecosystems services and human well-being to 2030.* Study commissioned by the CGIAR Systemwide Livestock Programme (SLP). ILRI, Nairobi, Kenya:

Higgins, S. I. *et al.* (2007). Effects of four decades of fire manipulation on woody vegetation structure in savanna. *Ecology* 88: 1119–1125. (doi:10. 1890/06-1664).

Ho Thanh Tham (2015). Utilisation of Water Hyacinth as Animal Feed. *Nova Journal of Engineering and Applied Sciences* 4(1), 2015:1-6 PII: S229279211500004-4 www. novaexplore. com.

Hoffmann, W. A., Adasme, R., Haridasan, M., de Carvalho, M. T., Geiger, E. L., Pereira, M. A. B., Gotsch, S. G. and Franco, A. C. (2009). Tree topkill, not mortality, governs the dynamics of savanna–forest boundaries under frequent fire in central Brazil. *Ecology* 9: 1326–1337. (doi:10. 1890/08-0741. 1).

Holzner, W., Kriechbaum, M., Ramble, C. and Diemberger, H. (2000). Yak grazing and forest dynamics in Sagarmatha National Park, Nepal. Integrated ecological studies of pasture problems in the Tibetan Autonomous, *Proceedings of the Third International Congress on Yak,* eds. H. Jianlin, C. Richard, O. Hanotte, C. McVeigh and J. E. O. Rege, held in Lhasa, P. R. China, 4–9 September 2000, Lhasa, P. R. China.

Holzner, W., Kriechbaum, M., Ramble, C. and Diemberger, H. (2000). Integrated ecological studies of pasture problems in the Tibetan Autonomous Region, P. R. China, *Proceedings of the Third International Congress on Yak,* eds. H. Jianlin, C. Richard, O. Hanotte, C. McVeigh and J. E. O. Rege, held in Lhasa, P. R. China, 4–9 September 2000, Lhasa, P. R. Cina.

Hongthong Phimmasan, Siton Kongvongsay, Chhay Ty and Preston, T. R. (2004). Water spinach (*Ipomoea aquatica*) and Stylo 184 (*Stylosanthes guianensis* CIAT 184) as basal diet for growing rabbits. *Livestock Research for Rural Development. Vol. 16, Art. # 34.* Retrieved from http://www. lrrd. org/lrrd16/5/hong16034. htm.

Hopkins A. (ed.) (2000). *Grass, Its Production and Utilization,* Third Edition, Blackwell Science Ltd., Oxford, UK, 440p.

Hopkins, A. Del Prado, A. (2007). Implications of climate change for grassland in Europe: impacts, adaptations and mitigation options: a review. *Grass and Forage Science* 62: 18–126.

Hove, L. (2000). Influence of season/year and species on chemical composition and in vitro digestibility of five Indigofera accessions. *Animal Feed Science and Technology* 136(3-4): 312 – 322.

Hoveland, Carl S. (1987). *The Growth, Culture and Multiplication of Grassland Forage Plants.* pp. 3-13. In: D. B. Hannaway and H. W. Youngberg (eds.) Grassland Production and Management Short Course. Beijing Agricultural University, Beijing, People's Republic of China.

Howell, N. W. and John, C. R. (1993). Forages for Swine. Extension G2360, University of Missouri, USA.

Hozhabr Naghshi, Sasan Khojasteh and Masoud Jafari (2014). Investigation the Effect of Different Levels of Azolla (*Azolla pinnata*) on Performance and Carcass Characteristics of Cobb Broiler Chicks. *International Journal of Farming and Allied Sciences* 3(1): 45-49, www. ijfas. com.

Huang, X., Zheng, Z., Fang, J., Wu, J. and Fu, C. (1998). Nutrition improvement for rabbits by feeding orchard-growing forage on hilly red soils. *Journal of Fujian Academy of Agricultural Sciences* 13: 102-107.

Huyen, N. T. D., Trach, N. X. and Preston, T. R. (2012). Effects of supplementation of paddy rice and/or rice grain and/or rice husk to sweet potato (*Ipomoea batatas*) vines as basal diet on growth performance and diet digestibility in rabbits. *Livestock Research for Rural Development. Volume 25*, Article #19. http://www. lrrd. org/lrrd25/1/huye25019. htm.

Ibraham, K. (1975). *Glossary of terms in pasture and range survey research, ecology and management*, FAO, Rome, Italy.

ICAR -IGFRI (1990). *Forage Production Technology*, Indian Grassland and Fodder Research Institute, Jhansi, India, p. 12.

ICAR -IGFRI (1991). *Research Highlights,* Indian Grassland and Fodder Research Institute, Jhansi, India 1990-91. pp. 27 and 37.

ICAR-IGFRI (2008). *Farming System Approach for Livelihood Improvement Vision and activities*. IGFRI, Jhansi, India.

ICAR-IGFRI (2008). Production and Utilization of Value Aided Densifying Products. IGFRI, Jhansi, India.

IDF - 9701 (1997). *Monograph on residues and contaminants in milk and milk products., Proceedings of a Seminar in Santander* (*Spain*), 22-24 May, 1989

Ige, A. O., Odunsi, A. A., Akinlade, J. A, Ojedapo, L. O., Ameen, S. A., Aderinola, O. A. and Rafiu, T. A. (2006). Gliricidia leaf meal in layer's diet: Effect on performance, Nutrient digestibility and Economy of production. *Journal of Animal Veterinary Advances* 5: 483–486.

Igwebuike, J. U., Anugwa, F. O. I. and Mustapha, I. (1999). Utilization of *Acacia albida* pods for rabbit feeding. 1. Effects on growth and organ weights of growing rabbits. *International Journal of Animal Sciences* 14: 231-236.

Iisley, S. E. and Miller, H. M. (2005). Effect of dietary supplementation of sows with quillaja saponins during gestation on colostrum composition and performance of piglets suckled. *Animal Science* 80: 179– 184.

ILRI (2000). Yak production in central Asian highlands, ILRI Proceedings, *Proceedings of the Third International Congress on Yak,* eds. H. Jianlin, C. Richard, O. Hanotte, C. McVeigh and J. E. O. Rege, held in Lhasa, P. R. China, 4–9 September 2000, Lhasa, P. R. China.

ILRI (2006). *Livestock production - an effective use of water in developing countries. 'TOP STORIES': Better lives through livestock.* WWW. ILRI. ORG, posted on 22 August 2006.

ILRI (2010). *Improvement of Fodder markets and identification of crop varieties with improved fodder characteristics in selected disadvantaged areas of India.* Unpblished Project Report Submitted to OPEC Fund for International Development. August, 2010,. ILRI, Nairobi, Kenya.

ILRI (International Livestock Research Institute) (2000). *ILRI strategy to 2010: Making the livestock revolution work for the poor.* ILRI, Nairobi, Kenya.

Imungi, J. K. and Potter, N. N. (1983). Nutrient contents of raw and cooked cowpea leaves. *Journal of Food Science* 48: 1252–1254.

Inder Dev, Singh, V. and Bimal Misri (2003). Socioeconomic profile of migratory grazier and participatory appraisal of forage production and utilization of an alpine pasture of of North-west Himalaya, *ENVIS Bulletin: Himalayan Ecology* 11(2): 52-62.

Inthapanya, S. and Preston, T. R. (2009). Effect of supplementation with sweet potato root and paddy rice on growth performance of local rabbits fed water spinach (*Ipomoea aquatic*) and paper mulberry (*Broussonetia papyrifera*) as basal diets. *Livestock Research for Rural Development. Volume 21*, Article #176. http://www. lrrd. org/lrrd21/10/sang21176. htm.

IPCC (2007). "Climate Change 2007: The Physical Science Basis. Summary for Policymakers. " *Contribution of Working Group I to the Fourth Assessment Report of the Intergovernmental Panel on Climate Change*, 5 February 2007.

IUCN (International Union for Conservation of Nature and Natural Resources) (2002). *IUCN Red, List of Threatened Species.* IUCV/UNEP/WWF (1991). *Caring for the earth- A strategy for sustainable living, land,* Switzerland, pp. 198-201.

Jackson, J. K. (1987). *Manual of Afforestation in Nepal.* Forest Survey and Res. Office, Kathmandu, Nepal.

Jadav, G. D. (1986). DDT and BHC residues in milk samples from Marathawada region. In *"Pesticide residue and environmental pollution",* eds. S. C. Goel., Muzzafarnagar, India, pp. 72–85.

Jadhav G D. (1986). DDT and BHC residues in milk samples from Marathwada region. In *"Pesticide Residues and Environmental Pollution"*, eds. S. C. Goel., Muzzafarnagar, India, pp. 86-92.

Jain, R. K. and Nagar, R. P. (2005). *IGFRI Clusterbean Germplasm Catalogue.* IGFRI, Jhansi, India.

Jain, R. K. and Nagar, R. P. (2005). *IGFRI Pearl Millet Germplasm Catalogue.* IGFRI, Jhansi, India.

Jakhmola, R. C. (2009). Densified feed blocks - prospects of this technology in India. In T. K. Walli, eds. *"Proceedings of the national symposium on fodder block technology"*, ILDEX India–2009, New Delhi, pp. 74–79.

Jeroch, H., Flachowsky, G. and Weissbach, F. (1993). Futtermittelkunde. Jena, Stuttgart.

Jia, B. R., Zhou, G. S., Wang, Y. H., Wang, F. Y. and Wang X (2006). Effects of temperature and soil water-content on soil respiration of grazed and ungrazed *Leymus chinensis* steppe, Inner Mongolia. *Journal of Arid Environments* 67: 60–76. Doi: 10. 1016/j. jaridenv. 2006. 02. 002

Jodha, N. (1990). *Rural Common Property Resource: a Growing Crisis.* International Institute of Economic Development, London, 16p.

Jodha, N. (1995). Common property resource and the dynamics of rural poverty in India's dry regions. *Unasylva* 46: 23-29.

Jodha, N. (2008). Rural commons and livelihood strategies in dry regions of India. *European J. Development Research* 20: 507-611.

Jones, J. W., Hoogenboom, G., Porter, C. H., Boote, K. J., Batchelor, W. D. Hunt, L. A., Wilkens, P. W., Singh, U., Gijsman, A. J. and Ritchie, J. T. (2003). The DSSAT cropping system model. *European Journal of Agronomy,* 18: 235–265.

Joshi, V., Tyagi, V., Kak Anjali and Arjun Lal (2009). A Review of Status of Genetic Resources of Forage Crops in India:, Indian Journal of Plant Genetic Resources http://www. indianjournals. com/ijor. aspx?target=ijor:ijpgr and volume=22 and issue=3 and article=014.

Kala, C. P. (2009). Grassland ecosystem and grazing policy. *Curr. Sci.* 96: 326.

Kalra, R. L., Chawla, R. P., Kapur, S. K., and Battu, R. S. (1979). Status report on pesticide contamination of bovine milk in India. Paper presented at the IDA's *Seminar on Quality Control of Milk in Delhi*, 9–10 November, 1979.

Kalra, R. L., Chawla, R. P., Sharma, M. L., Battu, R. S. and Gupta, S. C. (1983). Residues of DDTA and HCH in butter and ghee in India 1978–81. *Environmental Pollution* (Series B), 6, pp. 195–206.

Kamra, D. N., Singh, R., Jakhmola, R. C. and Srivastava, R. V. N. (1983). Effect of wilting and the additives, straw, molasses and urea on the fermentation pattern of maize silage. *Animal Feed Science and Technology* 9(3): 185-196.

Kan, C. A. and Meijer, G. A. L. (2007). The risk of contamination of food with toxic substances present in animal feed. *Animal Feed science and Technology* 133 (1-2): 84-108.

Kaphalia, B. S. (1982). Organochlorine pesticide contamination in some species of fodder grasses. *Environmental Pollution* (*Series B*) 3: 31–37.

Kaplinski, R. and Morris, M. (2001). *A Handbook for Value Chain Research.* IDRC, Ottawa.

Karsten, H. P., Patterson, Stout, R. and Crews, G. (2010). Vitamins A, E, and fatty acid composition of the eggs of caged hens and pastured hens. *Renewable Agriculture and Food Systems.* 25(1): 45-54.

Kaur, A., Gupta, S. R. and Singh, G. (2002). Carbon storage and nitrogen cycling in silvopastoral systems on a sodic soil in northwestern India. *Agroforestry Systems* 54: 21-29.

Kaushal, P., Malaviya, D. R. and Roy, A. K. (2008) *Biotechnological Approaches for Forage Crop Improvement: Prospects, Achievements and Road Map.* IGFRI, Jhansi, India.

Keating, B. A., Carberry, P. S., Hammer, G. L., Probert, M. E., Robertson, M. J., Holzworth, D.,. Hutha, N. I, Hargreaves, J. N. G., Meinke, H., Hochman, Z., McLean, G., Verburg, K., Snow, V., Dimes, J. P., Silburn, M., Wang, E., Brown, S., Bristow, K. L., Asseng, A., Chapman, McCown, S. R. L., Freebairn, D. M. and Smith, C. J. (2003). An overview of APSIM, a model designed for farming systems simulation. *European Journal of Agronomy*, 18: 267–288.

Kenneth, B. K. (1998). Forages For Swine, Pennsylvania State University Gilbert R. Hollis, University of Illinois D. Murray Danielson, University of Nebraska, Extension Bulletin E- 2284, October 1998.

Kenneth, B. K. (2010). Forages for Swine, PIH-126, eXtension. org, Pennsylvania State University; Gilbert R. Hollis, University of Illinois; D. Murray Danielson, University of Nebraska, USA.

Kennou, S. and Lebas, F. (1990). Growth performances of local Tunisian rabbits fed on diets with fresh or ensiled green forage. *Cuni-Sciences* 6: 31-39.

Khan, M. I. (2003). NASSD background paper: Communication for sustainable development. Gilgit: *NACS Support Project*, IUCN, Pakistan.

Khan, M. K. A., Akbar, M. A., Khaleduzzaman, A. B. M. and Rahman, M. M. (2009). Utilization of *Leucaena* and *Sesbania* leaf meals as protein supplements in broiler ration. *Bang. J. Anim. Sci.* 38(1 and 2): 123–131.

Khan, T. A. and Sunil Kumar (2010). Modelling understorey pasture production in a hortipastoral system. *Indian J. of Agroforestry* 12 (1): 46-48.

Khan, T. A. and Sunil Kumar (2009). *Physical Hortipastoral Models, their Management and Evaluation.* IGFRI, Jhansi, India.

Khandekar, S. S., Noronha, A. B. C. and Banerjee, S. A. (1981). Organochlorine pesticide residues in egg and milk available in Bombay markets. *Science and Culture* 47: 137.

Khatta, V. K., Kumar, N., Gupta, P. C. and Sagar, V. (1987). Effect of ensiling at different intervals on mimosine content of subabul (*Leucaena leucocephala*). *Indian Journal of Animal Sciences* 57: 340–342.

Khuc Thi Hue and Preston, T. R. (2006). Effect of different sources of supplementary fibre on growth of rabbits fed a basal diet of fresh water spinach (*Ipomoea aquatica*). *Livestock Research for Rural Development. Volume 18, Article #58.* Retrieved September 17, 2016, from http://www. lrrd. org/lrrd18/4/ hue18058. htm.

Kirby, K. R. and Potvin, C. (2007). Variation in carbon storage among tree species: implication for the management of a small scale carbon sink project. *Forest Ecology and Management* 246: 208-221.

Klein, J., Xinquan, Zh. and Harte, J. (2000). Climatic and grazing controls on vegetative aboveground biomass: Implications for the rangelands on the north-eastern Tibetan Plateau, In "*Yak production in central Asian highlands, ILRI* Proceedings, *Proceedings of the Third International Congress on Yak*", eds. H. Jianlin, C. Richard, O. Hanotte, C. McVeigh and J. E. O. Rege, held in Lhasa, P. R. China, 4–9 September 2000, Lhasa, P. R. China.

Knapp, Brian. (1991). *What Do We Know About Grasslands*?. Simon and Schuster Young Books, Hempstead, England.

Kobu, K., Prakash, B., Das, K. C., Vidyarthi, V. K., Mech, A. and Rajkhoa, C. (2007-08). Effect of feeding paddy straw, concentrate and *Lagerstroemia specicosa* leaves based ration on nutrient intake and utilization in adult mithun. *Annual Report*, 2007-08, National Research Centre on Mithun, Jharnapani, Medziphema, Nagaland, India, pp. 20.

Koehler, I. H., Poulton, P. R., Auerswald, K. and Schnyder, H. (2010). Intrinsic water-use efficiency of temperate seminatural grassland has increased since 1857: an analysis of carbon isotope discrimination of herbage from the Park Grass Experiment. *Glob. Change Biol.* **16:** 1531–1541. (doi:10. 1111/j. 1365-2486. 2009. 02067. x).

Kohler-Rollefson, I. (2007). *Keppers of Genes – The Interdependence between Pastoralist, Breeds and Access to Common Lands and Livelihoods.* LIFE Network, Pali (Rajasthan), India, 62 p.

Krishna, D., Qudratullah, S., Prasad, V. L. K. and Rama Rao, S. V. (2008). Nutritive value and feasibility studies of *Stylosanthes* leaf meal in broiler diets. *Indian Journal of Poultry Science* 43 (1): 39-44

Kumar, A., Natarajan, S., Biradar, N. B. and Trivedi, B. K. (2011). Evolution of sedentary pastoralism in south India: case study of the Kangayam grassland. Pastoralism (2011) 1: 7. doi:10. 1186/2041-7136-1-7

Kumar, Anil, Kumar, Sunil, Pandey, K. C. and Singh, K. A. (2007). *IGFRI Profile. ICAR News* 13(3): 6-10.

Kumar, B. M. (2006). Agroforestry: the new old paradigm for Asian food security. *J. Tropical Agriculture* 44(1-2): 1-14.

Kumar, R. (1992). Anti-nutritional factors. The potential risks of toxicity and the methods to alleviate them. *In* A. W. Speedy, and P. L. Pugliese (Eds.), *Legume trees and other fodder trees as protein source for livestock. FAO Animal Production and Health Paper No.* 102 FAO, Rome, Italy, pp. 145–160.

Kumar, S., and Faruqui, S. A. (2009). Production potential and economic viability of food forage based cropping system under irrigated conditions. *Indian J. Agronomy* 54(1): 36-41.

Kumar, Sumil, Sudhir Kumar and Chaubey B. K. (2009). Aonla-based hortipastoral system for soil nutrient buildup and profitability. *Annals of Arid Zone* 48(2):153-157.

Kumar, Sunil and Shukla A. K. (2011). Fruit, forage productivity and economics of grownup aonla -based hortipastoral system under semi-arid condition *Indian J. Agroforestry* 13 (1): 22-26.

Kumar, Sunil and Choubey, B. K (2008). Performance of Aonla (*Emblicaofficinalis*) –based hortipastoral system in semi-arid region under rainfed situation. *Indian J. of Agril. Sci.* 78 (9): 748-51.

Kumar, Sunil and Ram, S. N. (2009). Effect of pruning intensities and pastures combination on productivity of *Ber*- based hortipastoral system. *Ann. of Arid Zone* 48 (1): 63-66.

Kumar, Sunil and Shukla, A. K. (2010). Productivity, economics and soil nutrient buildup in grownup aonla based hortipastoral system under semi-arid condition. *Range Mgmt. and Agroforestry. Symposium Issue* (A): 29-30.

Kumar, Sunil, Shukla, A. K. and Singh H. V. (2013). Intercropping of pearlmillet + Cowpea as rainfed fodder crops with aonla based Agri-horti-system. *International Journal of Horticulture* 3(7): 28-34.

Kurukulasuriya, P. and Rosenthal, S. (2003). Climate change and agriculture: a review of impacts and adaptations. *Climate Change Series Paper No. 91*, World Bank, Washington, DC.

Lacassagne, L., Francesch, M., Carre, B. and Melcion, J. P. (1988). Utilization of Tannin-Containing and Tannin-Free Faba Beans (*Vicia faba*) by Young Chicks - Effects of Pelleting Feeds on Energy, Protein and Starch Digestibility. *Animal Feed Science and Technology* 20: 59–68.

Lailer, P. C., Dahiya S. S., Lal, D. and Chauhan, T. R. (2006). Growth performance of buffalo heifers fed complete feed blocks. In *"Proceedings of the national symposium on buffalo for rural upliftment"*, Indian Society for Buffalo Development, Mumbai, India, pp. 52.

Lailer, P. C., Dahiya, S. S., Sharma, M. L. and Lal, D. (2009). Effect of feeding complete feed block on milk yield, nutrient digestibility and economics of lactating Murrah buffaloes. In T. K. Walli, eds. *"Proceedings of the national symposium on fodder block technology"*, ILDEX India–2009, New Delhi, pp. 35–47.

Laitat, M., de Jaeger, F., Vandenheede, M. and Nicks, B. (2004). Factors influencing feed ingestion and performance of weaned pigs: perception and characteristics of diets. *Annales de Medecine Veterinaire* 148: 15- 29.

Laitat, M., Vandenheede, M., Désiron, A., Canart, B. and Nicks, B. (2000). Pellets or meal after weaning: the choice of piglets. 32èmes Journées de la Recherche Porcine en France, Paris, France, 1, 2 et 3 février 2000. *Journées de la Recherche Porcine en France*, 32: 157–162.

Lajide, L., Oseke, M. O. and Olaoye, O. O. (2008). Vitamin C, fibre, lignin and mineral contents of some edible legume seedlings. *Journal of Food Technology* 6: 237–241.

Lambers, H. and Poorter, H. (1992). Inherent variation in growth rate between higher plants: a search for physiological causes and ecological consequences. *Adv. Ecol. Res.* **23:** 187–261. (doi:10. 1016/S0065-2504(08)60148-8).

Lambert, D. (1988). *Our World Grasslands*. Silver Burdett Press, Englewood Cliffs, New Jersey.

Laswai, G. H., Lugembe, K. K. M., Mosha, R. D. and Kimanbo, A. E. (2000). Estimate of nurtrients digestibility, growth performance, slaughter traits and blood parameters in rabbit fed diets containing various levels of *Cortalaria ochroleuca*. *7th World Rabbit Congress*, Valencia Vol. C: 305-311.

Laswai, G. H., Ocran, J. N., Lekule, F. P. and Sundstol, F. (1997). Effects of dietary inclusion of leucaena leaf meal with and without ferrous sulphate on the digestibility of dietary components and growth of pigs over the weight range 20-60 kg. *Animal Feed Science and Technology* 65: 45–57.

Le Dividich, J., Geoffroy, F., Canope, I. and Chenost, M. (1976). Using waste bananas as animal feeds. *World Animal Review No.* 20: 22–30.

Le, T. M. (1999). Evaluation of water spinach (*Ipomoea aquatica*) for Baxuyen and Large White sows and fattening crossbred pigs. Master in Science Thesis. Swedish University of Agricultural Sciences. Uppsala pp. 71.

Le, T. M., Bui Hong Van, Mai Thi Chinh and Preston, T. R. (1997). Effect of dietary protein level and duckweed (*Lemna spp*) on reproductive performance of pigs fed a diet of ensiled cassava root or cassava root meal. *Livestock Research for Rural Development*. 9(1): http://www. cipav. org. co/lrrd/lrrd9/1/lemen911. htm.

Leng, R. A. (1997). Tree foliage in ruminant nutrition. *FAO Animal Production and Health Paper*, FAO, Rome, Italy, pp. 48-87.

Leng, R. A. (2008). The potential of feeding nitrate to reduce enteric methane production in ruminants. *A Report to the Department of Climate Change Commonwealth Government of Australia*, Canberra, ACT, Australia. www. penambulbooks. com.

Leterme, P., Londono, A. M., Ordonez, D. C., Rosales, A., Estrada, F., Bindelle, J. and Buldgen, A. (2010). Nutritional value and intake of aquatic ferns (*Azolla filiculoides Lam.* and *Salvinia molesta Mitchell.*) in sows. *Animal Feed Science and Technology* 155: 55– 64.

LIFE (2007). *Livestock Keepers and Management of Animal Genetic Resources – Roles, Rights and Responsibilities.* League for Pastoral People and Endogenous Livestock Development, New Delhi, 10p.

Linden, J. C., Henk, L. L., Murphy, V. G., Smith, D. H., Gabrielsen, B. C. and Tengerdy, R. P. (1987). Preservation of potential fermentables in sweet sorghum by ensiling. *Biotechnol Bioeng* 30: 860–867.

Linn, J. G., Goodrich, R. D. and Meiske, J. C. (1973). Aquatic Plants from Minnesota Part 4 - Nutrient Composition, *WRRC Bulletin* 56. University of Minnesota.

Littlewood, R. W. (1936). *Livestock of southern India.* Government Press, Madras, India.

Liu Guodao, Bai Changjun, Wang Dongjun, Ramesh, C. R. and Parthasarthy Rao, P. (2004). Leaf meal production from *Stylosanthes* in China and India, chapter 25. In: Chakraborty, S. (Ed.), "*High-yielding anthracnose-resistant Stylosanthes for agricultural systems.*" Australian Centre for International Agricultural Research Canberra, *Monograph* No. 111, pp. 253-256.

Liu, G. D. and Wang, D. J. (1990). Detoxication of leucaena juvenile shoot's. *Leucaena Research Reports* 11: 112–113.

Liu, Y. S., Pan, Q. M., Liu, H. D., Bai, Y. F. and Simmons, M. (2011). Plant responses following grazing removal at different stocking rates in an Inner Mongolia grassland ecosystem. *Plant and Soil* 340: 199–213. doi: 10. 1007/s11104-010-0458-3.

Lloyd, J., Bird, M. I., Vellen, L., Miranda, A. C., Veenendaal, E., Djagbletey, G., Miranda, H. S., Cook, G. and Farquhar, G. D. (2008). Contributions of woody and herbaceous vegetation to tropical savanna ecosystem productivity: a quasi-global estimate. *Tree Physiol.* 28, 451. (doi:10. 1093/treephys/28. 3. 451).

Loehle, C., Li, B. L., Sundell, R. C. (1996). Forest spread and phase transitions at forest-prarie ecotones in Kansas, USA. *Landscape Ecol.* 11: 225–235. (doi:10. 1007/BF02071813)

Long, C. and Alterman, T. (2007). Meet real free-range eggs. *Mother Earth News,* October/November, p. 4.

Long, E. A., Ainsworth, A., Leakey, D. B., Nösberger, J. and Ort, D. R. (2006). Food for thought: lower-than-expected crop yield stimulation with rising CO_2 concentrations. *Science* 312: 1918–1921.

Long, R. J. *et al.* (1999). Feed value of native forages of the Tibetan Plateau of China. *Animal Feed Science and Technology* 80: 101-113.

Lopez, E., Pro, A., Becerril, C., Perez, P. and Cuca, A. M. (1996). Common vetch (*Vicia sativa*) for feeding does. *6th World Rabbit Congress*, Tooulouse 1: 227-230.

Lopez, J. L. and Tapia, L. (2003). Foliage of legumes as forage for swine: nutritive value and anti-nutritional factors. *Ensaios e Ciencia: Serie Ciencias Biologicas, Agrarias, e da Saude* 7: 349–358.

Lopez-Bote, J. A., Sans, A., Rey, A., Castano, B. I. and Thos, J. (1998). Effect of free range feeding on n-3 fatty acid and a-tocopherol content and oxidative stability of eggs. *Animal Feed Science Technology* 72: 33-40.

Lovett, D. K., McGillowway, D., Bortolozzo, A., Hawkins, M., Callan, J., Flynn, B. and O'Mara, F. P. (2006). *In vitro* fermentation patterns and methane production as influenced by cultivar and season of harvest of *Lolium perenne* L. *Grass and Forage Science* 61(1): 9–21.

Lu, C. D. (1988). Grazing behaviour and diet selection of goats. *Small Rumin. Res.* 1: 205-216.

Lu, J., Fu, Z. F and Yin, Z. (2008). Performance of a water hyacinth (*Eichhornia crassipes*) system in the treatment of wastewater from a duck farm and the effects of using water hyacinth as duck feed. *Journal of Environmental Sciences* 20: 513-519.

Lukefahr, S. D. and Cheeke, P. R. (1990). Rabbit project planning strategies for developing countries. *Livestock Research for Rural Development* 2. 2., International Small Livestock Research Center, Department of Food Science and Animal Industries, Alabama A and M University, PO Box 264, Normal, Alabama 35762 (USA).

Luyen, L. T. and Preston, T. R. (2012). Growth performance of New Zealand White rabbits fed sweet potato (*Ipomoea batatas*) vines supplemented with paddy rice or Guinea grass supplemented with commercial concentrate. Livestock Research for Rural Development. Volume 24, Article #127. http://www. lrrd. org/lrrd24/7/luye24127. htm

Ly, J. (1990). The physiological and biochemical basis for feeding pigs and poultry in the tropics (part 1). *Livestock Research for Rural Development*, 2 (2).

Ly, J., Hean Pheap, Keo Saeth and Pok Samkol (2002). The effect of DL-methionine supplementation on digestibility and performance traits of growing pigs fed broken rice and water spinach. *Livestock Research for Rural Development 14(5):* http://www. cipav. org. co/lrrd/lrrd14/5/ly145b. htm.

Ly, J., Samkol, P. and Preston, T. R. (2001). Nutritional evaluation of tropical leaves for pigs: Pepsin/pancreatin digestibility of thirteen plant species. *Livestock Research for Rural Development. Volume 13(5), Article #48.* Retrieved September 5, 2016, from http://www. lrrd. org/lrrd13/5/ly135. htm.

Ly, J. and Samkol, P. (2001). Nutritional evaluation of tropical leaves for pigs; Desmanthus (*Desmanthus virgatus*). *Livestock Research for Rural Development.*

Volume 13, Article #37. Retrieved September 5, 2016, from http://www. lrrd. org/lrrd13/4/ly134a. htm.

Lyon, C. K. (1985). Degradation of mimosine during ensiling of Leucaena. *Journal of the Science of Food and Agriculture* 36: 936–940.

Maarse, L., Dhamankar, M. and Pica-Gamerra, U. (2008). *Drawing Policies, Programmes and Institutional Lessons out of Good Practices: Common Property Resources – Livestock.* South Asia Pro Poor Livestock Policy Programme, New Delhi, 7 p.

Maass, B. and Sawkins, M. (2004). History, relationships and diversity among *Stylosanthes* species of commercial significance, In "High yielding anthracnose resistant *Stylosanthes* for agricultural systems", eds. S. Chakraborty, ACIAR, CSRIO, Canberra. Pp. 9-26.

Madan Pal, Karthikeyapandian, K., Vanita Jain, Srivatava, A. C., Anupam Raj and Sengupta, U. K. (2004). Effect of CO_2 enrichment on growth and nutritional quality of berseem. *Agriculture Ecosystems and Enviornment* 101: 31-38.

Madan Pal, Vanita Jain and Sengupta, U. K. (1997). Effect of CO_2 enrichment on growth and nutritional quality of berseem. *Indian J. Plant Physiology* 2(4): 300-303.

Magouze, F. I., Mahmoud, S. A., El-Kelawy, H. M., Homouda, I. A. and Alla, S. A. (1998). Productive and reproductive performance of rabbits fed diets containing different agricultural by-products. 1. Productive performance of growing rabbits. *Egyptian J. Rabbit Sci.* 8 (1): 49-60

Mahadevappa, D. G., Sanganal, J. S., Gopinath, C. R. and Kalibavi1, C. M. (2012). Importance of *Azolla* as a sustainable feed for livestock and poultry - a review. *Agriculture Reviews* 33(2): 93-102.

Mahanta, S. K., Singh, S., Kumar, A. and Pachauri, V. C. (1999). Subabool leaf meal as replacement of mustardcake in lamb diets. *Small Ruminants Research* 32: 37-42.

Mahesh, M. S. (2013). Integrated organic farming and organic milk production: Opportunities and challenges in India. *Indian Dairyman* 65(5): 56-60.

Maity, S. B., Anil Kumar, Sunil Kumar, Sultan Singh, Das, S. K., Nag, S. K., Rai, A. K. and Kushwaha, B. P. (2016). Organic Forage and its impact on milk production of indigenous dairy cattle (*Personal communication*).

Maity, S. B., Kumar, A, Tiwari, S. K., Singh, S., Nag, S. K., Rai, A. K. and Kushwaha, B. P. (2009). *Organic feeding system for milk and meat production-vision and activities.* IGFRI, Jhansi, India.

Maity, S. B., Singh, U. P., Anil Kumar and Kushwaha, B. P. (2007). Performance of *Tharparkar* cattle under organic and inorganic feeding system. *Range Mgmt and Agroforestry* 28(2B): 228-230.

Makkar, H. P. S. (1991). Antinutritional factors in animal feedstuffs - mode of actions. *International Journal of Animal Sciences* 6: 88–94.

Makumba, W. Akinnifesi, F. K., Janssen, B. and Oenema, O. (2007). Long term impact of a gliricidia – maize intercropping system on carbon sequestration in southern Malawi. J. *Agri. Eco. Enviorn.* 118: 237-243.

Malavanh, C., Preston, T. R. and Ogle, B. (2008). Effect of replacing soybean meal by a mixture of taro leaf silage and water spinach on reproduction and piglet performance in Mong Cai gilts. *Livestock Research for Rural Development. Volume 20, supplement.* Retrieved September 5, 2016, from http://www. lrrd. org/lrrd20/supplement/mala2. htm.

Malviya, D. R. and Kaushal, P. (2005). IGFRI *Guinea Grass Germplasm Catalogue.* IGFRI, Jhansi, India.

Malviya, D. R., Roy, A. K., Kaushal, P., Singh, U. P. and Bhaskar, R. B. (2001). *IGFRI Berseem Germplasm Catalogue,* IGFRI, Jhansi, India.

Malynicz, G. (1974). The effect of adding *Leucaena leucocephala* meal to commercial rations for growing pigs. *Papua New Guinea Agricultural Journal* 25: 12–14.

Mangurkar, B. R. and Ravi Kumar, C. (2001). *Preliminary Desk Study on Livestock Environment Interactions in Watersheds in India,* SDC/IC, October 2001.

Manh, L. H., Dung, N. N. X., Yamasaki, S. and Takada, R. (2002). Replacement of concentrate by water hyacinth (*Eichhroria crassipes*): Effects on digestibility feed intake and live weight gain in pig production. In: Yamasaki S, Kamakawa A, eds. *"Proceedings of the 2002 annual workshop of JIRCAS Mekong Delta Project - Development of new technologies and their practice for sustainable farming systems in the Mekong Delt",* November 26- 28, 2002, Can Tho University, Can Tho city, Vietnam2002. pp. 152-8.

Manjunath, B. L. and Singh, S. P. (2005). System approach for higher productivity and profitability, *Technical Bulletin No.* 6, ICAR Research Complex for Goa, Panaji, India.

Marai, I. F. M., Radwan, M. A. H., Nowar, M. S., Allam, M. E. A. and Aboul-Ela, S. S. (1979). Feeding rabbits on different levels and sources of roughages and concentrates. *Zagazig Journal of Agricultural Research* 6: 153-168.

Marcel Dekker (2011). Forages collection. Centro Internacional de Agricultura Tropical, CIAT, Newyork. URL http://isa. ciat. cgiar. org/urg/foragecollection.

Masahiko Hirata (2009). The Role of Food, Agriculture, Forestry and Fisheries in Human Nutrition, eds. Victor R. Squires – Vol. I- Forage Crop Production ©*Encyclopedia of Life Support Systems (EOLSS) Forage Crop Production,* Faculty of Agriculture, Miyazaki University, Miyazaki, Japan.

Masoero, G., Terramoccia, S. and Auxilia, M. T. (1979). The digestibility of diets containing whole dried maize meal for growing rabbits. *Annali dell'Istituto Sperimentale per la Zootecnia* 12: 145-153.

Mayland, F. F., Greene, L. W, Robinson, D. L. and Wilkinson, S. R. (1990). Grass Tetany: A review of Mg in soil-plant-animal continuum. In *"Proc. Pacific Northwest Animal Nutrition Conference"*, Vancouver, BC. pp. 29-41.

Mc Dowell, L. R. (1997). *Mineral for grazing ruminants in tropical region*, 3rd ed. Institute of Food and Agriculture Sciences, University of Florida, Gainesville, USA.

McIntire, J., Bourzat D., and Pingali P. (1992). *Crop-livestock interactions in sub-Saharan Africa.* World Bank, Washington, DC.

McIntire, J. and Powell, J. M. (1993). African semi-arid tropical agriculture cannot grow without external inputs. *AGRIS (International Information System for the Agricultural Science and Technology,* FAO, Rome, Italy.

McSherry, M. E. and Ritchie, M. E. (2013). Effects of grazing on grassland soil carbon: a global review. *Global Change Biology* 19: 1347–1357. doi: 10. 1111/ gcb. 12144.

Medugu, C. I., Mohammed, G., Raji, A. O., Barwa, E. and Andi Zhinma, A. (2012). Utilization of different forages by growing rabbits. *International J. Advanced Biological Research* 2(3): 375-381

Meena, S., Farooq, M., Satyanarayan, K., Rajeshwari, Y. B. and Singh, K. C. (1999). Nutritional evaluation of mulberry leaves in broiler rabbits. *Indian Veterinary Journal* 76: 625-629.

Mello, J. P. and Devendra, D. C. (1995). *Tropical legumes in animal nutrition*, IDRC, Singapore, 350p.

Men, L. T., Yamasaki, S., Cadwell, J. S., Yamada, R., Takada, R. and Taniguchi, T. (2006). Effect of farm household income levels and rice-based diet or watwer hyacinth (*Eichhornia crassipes*) supplementation on growth/cost performances and meat indexes of growing and finishing pigs in the Mekong Delta of Vietnam. *Animal Science Journal* (Japan) 77: 320 - 330.

Mendes A. A., Funari S. R. C., Nunes J. R. V. and Spers A. (1980). Increasing levels of ramie hay in diets for growing rabbits. *Revista Latino-Americana de Cunicultura* 1: 27-35.

Menhi Lal and Shukla, N. P. (1985). Evapotranspiration, growth, and forage yield of MP Chari in relation to to growth stages and meteorological conditions. *Indian J. Agronomy* 30: 287-291.

Menhi Lal and Shukla, N. P. (1987). Studies on irrigation scheduling in mixed stand of forage oat and legumes. *Indian J. Agronomy* 32: 21-23.

Menhi Lal and Shukla, N. P. (1988). Studies on evapotranspiration and water use in forage oat in relation to growth phases and weather conditions. *Indian J. Agronomy* 30: 117-120.

Menhi Lal, Shukla, N. P. and Saxena, D. C. (1984). Studies on water use and irrigation scheduling in mixed cropping of forages involving different canopy surface. *Indian J. Agronomy* 29: 68-71.

Milchunas, D. G. and Lauenroth, W. K. (1993). Quantitative effects of grazing on vegetation and soils over a global range of environments. *Ecol. Monog*raph 63: 327–366.

Milchunas, D. G., Lauenroth, W. K. and Burke, I. C. (1998). Livestock grazing: animal and plant biodiversity of shortgrass steppe and the relationship to ecosystem function. *Oikos* 83: 65–74.

Miller, D. J. (1998). *Fields of Grass: Portraits of the Pastoral Landscape and Nomads of the Tibetan Plateau and Himalayas.* International Centre for Integrated Mountain Development, Kathmandu, Nepal.

Miller, D. J and Craig, S. R. (1996). Rangeland and Pastoral Development in Hindu-Kush Himalayas. *Proc. of a Rangeland Expert Meeting.* (*November 5-7, 1996*), Kathmandu, Nepal.

Miller, D. J. (1995). Herds on the move: Winds of change among pastoralists in the Himalayas and on the Tibetan Plateau. *Discussion Paper Series MNR 95/2.* ICIMOD (International Centre for Integrated Mountain Development), Kathmandu, Nepal, 27p.

Miller, D. J. (2005). The Tibetan. Steppe. Chapter. 8., *FAO Grassland of the World,* IN J. M. Suttie, Stephen G. Reynolds, Caterina Batello - *Business and Economics, FAO Plant Production Series 34,* FAO, Rome, Italy. pp. 305-337.

Mishra, C. (2001). High altitude survival: conflicts between pastoralism and wildlife in the Trans-Himalaya. *Dissertation,* Wageningen University, Wageningen, The Netherlands.

Mishra, C., Prins, H. H. T. and Van Wieren, S. E. (2001). Overstocking in the Trans-Himalayan rangelands of India. *Environmental Conservation* 28: 279–283.

Mishra, C., Spike, E., Van Wieren, Ketner, P., Ignas, M., Heitkong, A., Herbert, H. and Prins, T. (2004). Livestock–wildlife competition. *Journal of Animal Ecology* 73: 344 –354.

Mishra, C., Van Wieren, S. E. and Prins, H. T. (2003). Diversity, risk mediation, and change in a Trans-Himalayan agro pastoral system. *Human Ecology* 31: 595–609.

Mishra, C., Van Wieren, S. E., Heitkönig, I. M. A. and Prins, H. H. T. (2002). A theoretical analysis of competitive exclusion in a Trans-Himalayan large herbivore assemblage. *Animal Conservation* 5: 251–258.

Mishra, U. S., Bhat Vishnu, Katiyar D.S., Venkatesh Bhat, B., Shah N. K. and Saxena P. (2002). IGFRI *Cenchrus* Germplasm Catalogue. IGFRI, Jhansi, India.

Mishra, U. S., Roy A. K., Singh U.P., Verma O.P.S., Shah N. K., Saxena P., Kaushal P., Karnani L. K. and Malaviya D. R. (2009). IGFRI Deenanath Grass Germplasm Catalogue (*Pennisetum pedicellatum Trin.*). IGFRI, Jhansi, India.

Misra, A. K., Mishra, A. S. Tripathi, M. K. and Jakhmola, R. C. (2003). Scope and future role of non-conventional feeds in ruminant feeding. *In*: *Sustainable Animal production (Eds: Jakhmaola, R. C. and Jain, R. K.*). Pointers Publishers, S. M. S. Highway, Jaipur-302003 (Rajasthan), India, pp. 376-397.

Misra, A. K., Mahanta, S. K. and Singh, K. K. (2008). *Guidelines for feeding of farm livestock in peninsular India,* IGFRI, Jhansi, India.

Misra, A. K., Pailan, G. H., Singh, K. K., Das, M. M. and Khan, T. A. (2007). Variability in straw quality of wheat cultivars in India. In "*Proc. of International Tropical Animal Nutrition Conference*" NDRI, Karnal, (October 4-7, 2007), pp. 42.

Misra, A. K., Singh, K. K., Das, M. M., Pailan, G. H., Nag, S. K. and Khan, T. A. (2007). Variability in straw quality of rice cultivars in India. In *"Proc. of National symposium on grassland and forage resources in changing socio-economic scenario.* " IGFRI, Jhansi, (November, 15-17, 2007), India, pp. 230.

Misri, B., Radotra, S. and Inder Dev (2007). *Tryst with Destiny- Research Initiatives for Fodder Resource Development in North-Western Himalaya (Jammu and Kashmir and Himachal Pradesh),* IGFRI, Jhansi, India.

Molina Botero, I. C., Manuel Cantet, J., Sebastián Montoya, Antonio Correa, G. and Barahona Rosales, R. (2013). In vitro methane production from two tropical grasses alone or in combination with *Leucaena leucocephala* or *Gliricidia sepium. Revista CES Medicina Veterinaria y Zootecnia* 8(2): 15-31.

Molkentin, J. (2009). Authentication of organic milk using $\delta^{13}C$ and the α-linolenic acid content of milk fat. *J. Agric. Food Chem.* 57: 785–790.

Mora, L., Dominguez, P. L., Calderon, R. and Quintano, J. (1992). Notes on the use of sweet potato (*Ipomoea batatas*) foliage in diets for weaned pigs. *Zootechnia de Cuba* 2: 65 - 90.

Moreland, A. F., Collins B. R., Hansen, C. A. and O'Brien, N R. (1991). Wastewater grown waterhyacinth as an ingredient in rabbit food. *Journal of Aquatic Plant Management* 29: 32-39.

Morgan, J. A., Milchunas, D. G., Le Cain, D. R., Mark West and Mosier, A. R. (2007). Carbon dioxide enrichment alters plant community structure and accelerates shrub growth in the short grass steppe. *Proceedings of the National Academy of Sciences of The United States of America* 104(37): 14724–14729.

Morgan, J. A., Pataki, D. E., Komer, C., Clark, H., Del Grosso, S. J., Grunzweig, J. M., Knapp, A. K., Mosier, A. R., Newton, P. C. D., Niklaus, P. A., Nipper, J. B., Nowark, R. S., Parton, W. J., Polley, H. W. and Shaw, M. R. (2004). Water relations in grassland and desert ecosystems exposed to elevated atmospheric $CO_{2.}$ *Oecologia* 140 (1): 11-25.

Morgan, J. A., Skinner, R. H. and Hanson, J. D. (2001). Nitrogen and CO_2 affect regrowth and biomass portioning differently in forage of three functional groups. *Crop Science* 41: 78-86.

Moura, S. A. Costa, M. T., Claudinei, C. and Parre, C. (1992). Comparison of hay of *Cajanus cajan* and *Cynodon dactylon* cv. Coast Cross for growing rabbits. *Veterinaria e Zootecnia* 4: 69-75.

Mtenga, L. A. and Laswai, G. D. (1994). *Leucaena leucocephala* as feed for rabbits and pigs: detailed chemical composition and effect of level of inclusion on performance. *Forest Ecology and Management* 64:249-257.

Muhamed Arif, Mustaq Ahmed Mia and Durrani, F. R. (2000). Effect of different dietary levels of Egyptian clover on broiler performance. *Pakistan J. biological Sciences* 3(7): 1086-1087.

Muir, J. P. and Massaete, E. (1996). Seasonal growth in rabbits fed wheat and maize bran with tropical forages Instituto de Produção Animal, CP 1410, Maputo, Moçambique *Livestock Research for Rural Development* 8: 1 http://www. lrrd. org/lrrd25/3/hang 82045. htm.

Muir, J. P. and Massaete, E. S (1992). Growth response in rabbits to various levels of *Leucaena leucocephala* fed fresh with a wheat bran diet. *Journal of the Zimbabwe Society for Animal Production* 4: 131-134.

Muir, J. P. and Massaete, E. S. (1995). Reproductive performance of rabbits fed wheat bran with tropical forages *or Leucaena leucocephala. World Rabbit Science* 3: 91-93.

Mulliniks, J. T., Jennifer Muscha, Lodge-Ivey, S. L. and Petersen, M. K. (2010). Predicted mineral intake utilizing both water and forage analysis varies by source and location of livestock water in Montana. *Proceedings, Western Section, American Society of Animal Science* 61: 182-185.

Munro, D. G. (1931). *Grass Farming in Coimbatore*. Bulletin no. 23, Government Press, Madras, India

Mutetikka, D. B., Carles, A. B., and Wanyoike, M. M. (1990). The effect of level of supplementation to diets of Rhodes grass (*Choris gayana*) hay, maize (*Zea mays*) leaves and sweet potato (*Ipomea batatas*) vines on performance of grower rabbits. *Journal of Applied Rabbit Research* 13 (Personal Communication).

Nag, S. K. and Raikwar, M. K. (2006). *Final Report of Lal Bahadur Shastri Young Scientist Scheme 'Monitoring of pesticide residues in animal feed, fodder and milk with special reference to Bundelkhand region'* submitted to ICAR, New Delhi.

Nag, S. K. and Raikwar, M. K. (2008). Organochlorine pesticide residues in bovine milk. *Bull Environ Contam Toxicol* 80: 5-9.

Nag, S. K., Mahanta, S. K., Raikwar, M. K. and Bhadoria, B. K. (2007). Residues in milk and production performance of goats following the intake of a pesticide (endosulfan). *Small Ruminant Research* 67: 235-242.

Nagpal, A. K. and Arora, M. (2002). Utilization of guar (*Cyamopsis tetragonolobus*) phalgati and groundnut haulms based complete feeds in camel calves. *Indian Journal of Animal Nutrition* 19: 69–72.

Nagpal, A. K., Arora, M. and Singh, G. P. (2005). Nutrient utilization of gram straw (*Cicer arietinum*) based complete feed blocks in calves. *Indian Journal of Animal Sciences* 75: 64–68.

Naik, P. K., Swain, B. K. and Singh, N. P. (2015). Production and utilization of hydroponic fodder. *Indian J. Nutrition* 32(1): 1-9.

Naik, P. K. (2014). Hydroponics green fodder for dairy animals. In: *Recent Advances in Animal Nutrition* (M. P. S. Bakshi and M. Wadhwa; eds.). Satish Serial Publishing House, 403, Express Tower, Commercial Complex, Azadpur, Delhi-110 033, India.

Naik, P. K., Dhuri, R. B., Swain, B. K. and Singh, N. P. (2012). Nutrient changes with the growth of hydroponics fodder maize. *Indian J. Anim. Nutr.* 29: 161-163.

Naik, P. K., Dhuri, R. B., Karunakaran, M., Swain, B. K. and Singh, N. P. (2014). Effect of feeding hydroponics maize forage on digestibility of nutrients and milk production in lactating cows. *Indian J. Anim. Sci.* 84(8): 880-883.

Nair, P. K. R., Kumar, B. M. and Nair, V. D. (2009). Agroforestry as a strategy for carbon sequestration. *J. Plant, Nutr. Soil Sci.* 172: 10-23.

Namara, M., Hataba, N. A. and Abdel Wahed, M. (2010). The productive performance of growing fayoumi chicks fed restricted diets supplemented with free fresh Azolla. *Egypt. Poult. Sci.,* 30 (3): 747-762

Narayana, H. and Shetty, S. V. (1977). Studies on the incorporation of mulberry leaves (*Morus indica*) in layers mash on health, production and egg quality. *Indian Journal of Animal Science* 47: 212-215.

Narwal, R. P. (2006). *Annual progress report of AICRP micro and secondary nutrients and pollutant elements in soils and plants.* CCSHAU, Hissar, India, pp. 85.

Negi, S. S. and Goel, G. C. (1985). Voluntary intake and nutritive value of *Grewia optiva* and *Bauhinia variegata* tree leaves supplemented to a diet of Angora rabbits. *Indian Journal of Animal Sciences* 55: 502-504.

Nelson, K. E., Zinder, S. H., Hance, I., Burr, P. Odongo, D., Wasawo, D., Odenyo, A. and Bishop, R. (2003). Phylogenetic analysis of the microbial populations in the wild herbivore gastrointestinal tract: Insights into an unexplored niche. *Environ. Microbiol.* 5:1212-1220.

Newas, R., Tewari, R. K., Ajit and Yadav R. S. (2006). Aonla based Agroforestry: A promising System for rainfed areas. *Technical Bulletin No. 06*, National Research Centre for Agroforestry Jhansi, India.

Ngodigha, E. M. (1994). Incorporation of *Centrosema pubescens* in broiler diets. Effects on performance characteristics. *Bulletin Animal Health Africa* 42:159-161.

Nhan, N. T. H, Preston, T. R. and Dolberg, F. (1997). Use of *Trichantera gigantea* leaf meal and fresh leaves as livestock feed. *Livestock Research for Rural Development. Volume 9, Article #7.* Retrieved August 14, 2016, from http:// www. lrrd. org/lrrd9/1/nhan91. htm.

Nhan, N. T. H., Hon, N. V., Son, V. V., Preston, T. R. and Dolberg, F. (1996). Effect of shade on biomass production and composition of the forage tree *Trichanthera gigantea. Livestock Research for Rural Development.* 8(2): 93-98.

Nicholson, F. A. (1887). *Manual of the Coimbatore district in the Presidency of Madras.* Government Press, Madras, India.

Nieves, D., Farinas, S., Munoz, A., Torrealba, E. and Rodriguez, N. (1996). Use of *Arachis pintoi* and *Pennisetum purpureum* in the feeding of fattening rabbits. *Revista Unellez de Ciencia y Tecnologia, Produccion Agricola* 14: 82-91.

Nieves, D., Moncada, I., Terán, O., González, C., Silva, L. and Ly, J. (2009). Digestive parameters in fattening rabbits given tropical foliage based diets. *Bioagro* 21(1): 33–40.

Nieves, D., Quintero M. and Mendoza, M. (1995). Inclusion of a probiotic and non-conventional feedstuffs (*Cajanus cajan* and *Leucaena leucocephala*) in meal diets for fattening rabbits. *Revista Unellez de Ciencia y Tecnologia, Produccion Agricola* 13: 35-44.

Nieves, D., Rodriguez, J. and Carvajal, L. (1998). Inclusion of probiotic and non conventional feedstuffs in mash diets for fattening rabbits: *Leucaena leucocephala* and *Arachis pintoi. Revista Unellez de Ciencia y Tecnologia, Produccion Agricola* 16: 37-48.

Nieves, D., Santana, L. and Benavenata, J. (1996). Increasing levels of Arachis pintoi (Krap. and Greg.) in fattening rabbit's mash diets. *Revista Unellez de Ciencia y Tecnologia, Produccion Agricola* 14: 33-43.

Nitya Ghotge (2004). *Forest Policies and Livestock in India*, Lead working group paper No. 4, Anthra, Hyderabad, India.

Nkwocha, G. A., Anukam, K. U. and Nkwocha, V. I. (2008). Performance, carcass and haematological characteristics of pigs fed graded levels of *Gmelina arborea* leaf meal. *Animal Production Research Advances* 4(2): 148 - 151.

Noble, A. D., Orr, D. M., Middleton, C. H., and Rogers. L. G. (2000). Legumes in native pasture – asset or liability? A case study with stylo. *Tropical Grasslands* 34: 199-206.

NPOP (2005). *National Programme on Organic Production*, Department of Commerce, Ministry of Commerce and Industry, Government of India, New Delhi.

Nweze, B. O. (2005). Performance of young rabbits fed two grasses and two legumes species supplemented with compounded diet. *Trop. J. Anim. Sci.*, 8(2): 91–95.

Nworgu, F. C. and Fasogbon, F. O. (2007). Centrosema (*Centrosema pubescens*) Leaf Meal as Protein Supplement for Pullet Chicks and Growing Pullets. *International Journal of Poultry Science* 6 (4): 255-260.

Nworgu, F. C., and Egbunike, G. N. (2013). Nutritional Potential of *Centrosema pubescens Mimosa invisa* and *Pueraria phaseoloides* leaf meals on growth performance responses of broiler chickens. *American Journal of Experimental Agriculture* 3(3): 506-519.

Nworgu, F. C., and Fasogbon, F. O. (2015). Leaf Meal as Protein Supplement for broiler chicks production. *Journal of Experimental Biology and Agricultural Sciences* 3(5): 440-447 DOI: http://dx. doi. org/10. 18006/2015.

Nworgu, F. C., Faphunda, J. B. (2002). Performance of broiler chicks fed mimosa (*Mimosa invisa*) leaf meal supplements. *Proceeding of 27th Ann. Conf. of Nigerian Society of Animal Production (NSAP)*, eds., Aletor V. A. and Onibi G. E. held at the Federal University of Technology Akure, Nigeria on the 17th – 21st March, 2002. pp. 128–131.

Nworgu, F. C., Adebowale, E. A, Oredein, O. A. and Oni, A. (1999). Prospect and economics of broiler production using two plant protein sources. *Tropical Journal of Animal Science* 2(1): 159-166.

O'Donnell, A. M., Spatny, K. P., Vicini, J. L. and Bauman. D. E. (2010). Survey of the fatty acid composition of retail milk differing in label claims based on production management practices. *J. Dairy Sci* 93: 1918–1925.

Obua, B. E. and Davidson, E. S. (2011). Evaluation of sundried *Microdesmis puberula* leaf meal on the growth performance of weaner pigs. International *Journal of Agriculture and Rural Development* 14(1): 499 - 506.

Obua, B. E., McAlbert, F. U., Okoro, B. O. and Efrenie, S. (2012). Survey of the diversity of forage plants used in feeding pigs in small holders farms in Southeastern Nigeria. Int. *Journal of Agric. and Rural Dev.* 15 (3): 1310-1316.

Odunsi, A. A. (2003). Assessment of Lablab (*Lablab purpureus*) Leaf Meal as a Feed Ingredient and Yolk Colouring Agent in the Diet of Layers. *International Journal of Poultry Science* 2: 71–74.

Odunsi, A. A., Ogunleke, M. O., Alagbe, O. S. and Ajani, T. O. (2002). Effect of feeding *Gliricidia sepium* leaf meal on the performance and egg quality of layers. *International Journal Poultry Scienc,* 1: 26-28.

OECD and FAO (2010). Meat - OECD-FAO Agricultural Outlook (2010). OECD Publishing, pp. 147–158.

Ogle, R. B. (2006). Forages for pigs: nutritional, physiological and practical implications. Workshop/ Seminar "*Forages for Pigs and Rabbits*" MEKARNCelAgrid, 22-24 August, 2006. Phnom Penh, Cambodia. URL http:// www. mekarn. org/proprf/ogle. htm.

Ohler, J. G. (2014). Modern coconut management; palm cultivation and products. *Miscellaeous Published item – Food for All*, FAO, Rome, Italy.

Olff, H., Ritchie, M. E. and Prins, H. H. (2002). Global environmental controls of diversity in large herbivores. *Nature* 415: 901-904.

Olugbemi, T. S., Mutayoba, S. K. and Lekule, F. P. (2010). Effect of Moringa (*Moringa oleifera*) Inclusion in Cassava Based Diets Fed to Broiler Chickens. *International journal of Poultry Science* 9: 363-367.

Omeje, S. I., Odo, B. I. and Egwu, P. O. (1997). The effects of *Centrosema pubescens* as leaf meal concentrate for broiler birds. In" *Book of Abstract of the 22 Annual Conference of NSAP*", held at Abubakar Tafawa Balewa University, BauchiNigeria, pp. 136-137.

Omole, A. J., Adejuyigbe, A., Ajayi, F. T., Fapohunda, J. B. (2007). Nutritive value of *Stylosanthes guianensis* and *Lablab purpureus* as sole feed for growing rabbits. *African Journal of Biotechnology 6(18):* 2171-2173.

Onibi, G. E., Folorunso, O. R. and Elumelu, C. (2008). Assessment of Partial Equi-Protein Replacement of Soyabean Meal with Cassava and *Leucaena* Leaf Meals in the Diets of Broiler Chicken Finishers. *International journal of Poultry Science* 7: 408-413.

Onwudike, O. C. (1995). Use of the legume tree crops *Gliricidia sepium* and *Leucaena leucocephala* as green feeds for growing rabbits. *Animal Feed Technology and Science* 51(1-2): 153-163.

Onwuka, C. F. I., Adeliyi, G. O., Biobaku, W. O. and Adu, I. F. (1992). *Leucaena leucocephala* leaves in rabbit diets. *Leucaena Research Reports* 13: 65-67.

Oregon State University (1992). Terminology for Grazing Lands and Grazing Animals. *Journal of Production Agriculture* 5:191-201.

Oruganti, M. (2011). Organic Dairy Farming – A New Trend in Dairy Sector. *Veterinary* world 4(3): 128-130.

Osman, M. and Rao, J. V. (1999). *Alternate land use: Hortipastoral systems.* In "*Fifty years of dryland agriculture research in India*", Central Research Institute for Dryland Agriculture, Hyderabad, India, pp. 485–495.

Otte J., Roland-Holst D., Pfeiffer D., Soares-Magalhaes R., Rushton J., Graham J., Silbergeld, E. (2007). *Industrial livestock production and global health risks. Pro-Poor Livestock Policy Initiative Research Report.* FAO, Rome, Italy.

Ozdemir, S. Dundar, M. S., Sengorur, B. and Senol, A. S. (2009). Lead and zinc contents of cow's milk affected by varying traffic density. *Int. J. Env. Pollution* 6(4): 411-417.

PACA (2008). National Consultation on Conservation Agriculture, Background Paper, Professional Alliance for conservation of agriculture, 1st Floor, NASC Complex, DPS Marg, New Delhi, pp. 1-27.

Pachauri, V. C., Goyal, R. K., Mahnata, S. K. and Gupta, P. D. (1999). Effect of densification of dry grass on nutrient utilization and growth in crossbred calves. *Indian Journal of Animal Nutrition* 16: 244–247

Pagani, M. (2002). The alkenone–CO_2 proxy and ancient atmospheric carbon dioxide. *Phil. Trans. Royal. Soc. London A* **360**, 609–632. doi:10. 1098/rsta. 2001. 0959.

Pagani, M., Caldeira, K., Berner, R. and Beerling, D. (2009). The role of terrestrial plants in limiting atmospheric CO_2 decline over the past 24 million years. *Nature* 460: 85–89. doi:10. 1038/nature08133.

Pailan, G. H. and Pathak, P. S. (2005). *Chara Evam Pashudhan Utpadan-Vartman Evam Bhavishya*. IGFRI, Jhansi, India.

Pal, M., Singh, K. A., Saxena, J. P. and Singh, H. K. (1985). Energetics of Cropping Systems. *Indian J. Agron.* 30(2): 1-LXI.

Pal, T. D., Singh P., Prasad, C. S. and Sampath, K. T. (2009). Mineral content in feeds and fodders and their release in rumen. *Technical Bulletin No.* 4, NIANP, Bangalore.

Palsaniya, D. R. and Maity, S. B. (2017). Organic Fodder and Milk Production In "*Towards Organic Agriculture*, eds. *B. Gangwar and N. K. Jat* Today and Tomorrow's Printers and Publishers, New Delhi - 110 002, India, pp 451-464.

Pandey, H. N. and Pathak, N. N. (1991). Nutritional evaluation of artificially grown barley fodder in lactatingcrossbred cows. *Indian J. Anim. Nutr.* 8 (1): 77-78.

Pandey, K. C. and Roy A. K. (2011). *Forage Crop Varieties.* Indian Grassland and Fodder Research Institute, Jhansi, Uttar Pradesh, India, 84p.

Pandey, K. K. (1982). *Fodder Trees and Tree fodder in Nepal,* Swiss Federal Institute of Forestry Research, Birmensdert, Switzerland.

Pandey, K. K. and Nosberger, J. (1985). Growth and development of saplings of the fodder tree *Artopacpus lakkoccha* Roxb. and the altitude in Nepal. *The International Tree Crops Journal* 3: 125-134. http://dx. doi. org/10. 1080/01435698. 1985. 9752783.

Pandey, K. K. and Nosberger, J. (1985). Nutrients contents of the leaves of the fodder tree *Artocarpus lakoocha* Roxb. *Agroforestry Systems* 3: 297-303.

Pandeya, S. C. (2000). The grazing land resources and fodder cultivation in India. *Foundation Day Lecture.* IGFRI (Indian Grassland and Fodder Research Institute), Jhansi, UP, India.

Parigi-Bini, R., Cinetto, M. and Carotta, N. (1984). Digestibility and nutritive value of *Leucaena leucocephala* in growing rabbits. *3rd World Rabbit Congress,* Rome 1: 399-407.

Parigi-Bini, R., Xiccato, G., Dale Zotte, A. and Carazzolo, A. (1994). Influence of different levels of dietary fibre on digestibility, performance and carcass and meat quality of rabbits. 6e Journées Rech. *Cunicole en France* 2: 347-354.

Parihar, S. S., Kanodia, K. C. and Singh, K. A. (1984). Effect of pH on germination of range grasses. *Forage Res.* 10(1): 43-44.

Pariyar, D. (2000). Yak production in central Asian highlands, *ILRI Proceedings, Proceedings of the Third International Congress on Yak,* eds. H. Jianlin, C. Richard, O. Hanotte, C. McVeigh and J. E. O. Rege, organized by ILRI, held in Lhasa, P. R. China, 4–9 September 2000, Lhasa, P. R. China.

Parthasarathy Rao and Hall, A. J. (2003). Importance of crop residues in crop–livestock systems in India and farmers' perceptions of fodder quality in coarse cereals. *Field Crop Research* 48(1-2): 189-198.

Parthasarathy Rao, P. and Bhowmick, T. (2001). Database development, typology construction, and spatial temporal analysis of crop–livestock systems in India. *Progress Report 1. Submitted to System Wide Livestock Program (SLP),* ILRI, Ethiopia. International Crops Research Institute for Semi-Arid Tropics, Patancheru, Hyderabad, India, 56p.

Pascual, J. J., Cervera, C. and Fernandez-Carmona, J. (2002). A feeding programme for young rabbit does base on lucerne. *World Rabbit Science* 10: 7-13.

Pascual, J. J., Fonfria, M. J., Alquedra, I., Cervera, C and Fernandez-Carmona, J. (2000). Use of lucerne-based diet on reproductive rabbit does. *7th World Rabbit Congress, Valencia* vol. C: 379-384.

Paterson, R., Roothaert, R. and Kiruiro, E. (2010). Tropical Animal Health and Production: The Feeding of Leaf Meal of *Calliandra calothyrsus* to Laying Hens. 32: 51-61. doi:10. 1023/A:1005293019581.

Pathak, P. K, Chander, M. and Biswas, A. K. (2003). Organic Meat: an overview. *Asian—Aust. J. Anim Sci.* 16(8): pp. 1230-1237.

Pathak, P. S. (2002). Potential technological and management interventions for improving the productivity of grasslands. In "*Technology Options for Sustainable Livestock Production in India*, eds. P. S. Birthal and P. P. Rao, NCAEPR, New Delhi and ICRISAT, Hyderabad, India pp. 164-182.

Pathak, P. S. and Singh K. A. (1999). Sustainable Farming Systems in the Hilly Region of India with Special Reference to North Eastern Region. In *"Man and Forests",* eds. K. Dhir, D. R. Batish and D. K. Khurana, DNAES, IUFRO, ISTS, Botany Deptt. and Centre for Vocational Studies, Panjab University, Chandigarh, pp. 313-346.

Pathak, P. S., Chandra, A., Roy A. K., Bhatt, R. K. and Dwivedi., P. N (2005). *IGFRI Stylo Germplasm Catalogue*, IGFRI, Jhansi, India.

Pathak, P. S., Singh, J. P., Sharma, P., Singh, K. K., Dwivedi, P. N. and Singh, J. B. (2005). *Bundelkhand region: Agricultural perspectives (Status, constraints and prospects*). IGFRI, Jhansi, India, 96p.

Pathak, P. S., Suresh, G., Khan, T. A. and Gupta, S. K. (2006). *Subabul in the Rural Economy of Andhra Pradesh-A case study.* IGFRI Technical Bulletin, Jhansi, India, 37p.

Patiram and Singh, K. A. (1993). Effect of continuous application of manure and nitrogeneous fertilizers on some properties of acid *Inceptisol*. *J. Indian Soc. Soil Sci.* 41(3): 430-433.

Pattabhiraman, D. (1958). *The Kangayam breed of cattle*. Popular Education Publishing, Madras, Tamil Nadu, India.

Payne, M., Bryant, M. J., Owen, E., Capper, B. S., Wood, J. F., Machin, D. H. and Butcher, C. (1983). The effect of diets containing 50 per cent roughage on performance and digestibility in growing rabbits. *Tropical Animal Production* 8: 269-275.

Peden, D., Amede, T., Haileslassie, A and, Taddesse, G. (2008). Strategies for Improving Livestock Water Productivity. In "*Fighting Poverty Through Sustainable Water Use*": Volumes I, Proceedings of the CGIAR *Challenge Program on Water and Food 2nd International Forum on Water and Food*, Addis Ababa, Ethiopia, November 10—14, 2008.

Peden, D., G., Tadesse, and Misra, A. K. (2007). *Water and livestock for human development. In Water for food, water for life: A comprehensive assessment of water management in agriculture*, David Molden (eds.), London, International Water Management Institute., Colombo, 485-514.

Peden, D., Girma Tadesse and Amare Hialeslassie (2009). Livestock water productivity: implications for sub-Saharan Africa. *The Rangeland Journal* 31: 187–193.

Pedersen, M. W., Anderson, J. O., Street, J. C., Wang, L. C. and Baker, R. (1972). Growth response of chicks and rats fed alfalfa with saponin content modified by selection. *Poultry Science* 51: 458–463.

Pellant, M., Shaver, P., Pyke, D. A. and Herrick, J. E. (2005). *Interpreting Indicators of Rangeland Health, Version 4. Natural Resource Conservation Service*, US Department of Agriculture, Denver, USA.

Perez, J. M. (1994). Digestibility and energy value of dehydrated lucerne meal for rabbit: influence of chemical composition and technological process. *6e Journées Rech. Cunicole en France* 2: 355-364.

Perez, J. M., Lamboley, B. and Beranger, C. (1998). Digestibility and energy value for rabbits of individual or mixed batches of dehydrated alfalfa. *7e Journées Rech. Cunicole Lyon*: 129-132.

Perez, J. M., Lebas, F. and Lamboley, B. (1990). Feeding value of lucerne dehydrated after ensiling: digestibility, protein efficiency and utilisation by growing rabbits. 5e Journées Rech. *Cunicole en France 2: Communication* 58: 1-13.

Peter, F. F. and Eric Oberholzer (1985). Use of multipurpose trees in hill farming system in Nepal. *Agroforestry Systems* 2: 187-197.

Philip Grime, J., Jason, D., Fridley Andrew, P., Askew Ken Thompson, John, G. Hodgson, and Chris R. Bennet (2008). Long-term resistance to simulated climate change in infertile grassland *Proc. Natl. Acad. Sci. USA* 22 July 2008: 10028-10032. *Science* 2 February 2007: Vol. 315 no. 5812 pp. 640-642. *DOI:* 10. 1126/science. 1136401

Pillai, G. R., Rajasree, Girija Devi and Sheela, K. R. (1994). Coconut based forage production system compared with others in the humid tropical conditions in Kerela. In*"Agroforestry for Degraded Lands Volume I"*, eds. Panjab Siingh *et al.*, p. 450-457.

Planning Commision (2012). Report of the working group on "Research, capacity building and information management for environment and forest sector for 11th five year plan. ", Planning Commision, Government of India, New Delhi.

Planning Commission of India (2007). *Report of the working group on Animal Husbandry and Dairying for the Tenth Five Year Plan 2002-2007.* Govt. of India, New Delhi.

Planning Commission of India (2009). *Report of the Working Group on Advisory Committee on " Matter Relating to the development of Animal husbandry and Dairy in the States/UT's Subgroup Fodder Chairman Dr, K. A. Singh; Members- Dr. C. R. Hazra, Dr. Baljit Singh Sidhu, Dr. K. T. Sampat,.* Govt. of India, New Delhi.

Pok Samkol, Preston, T. R. and Leng, R. A. (2006). Effect of offering leaves or stems of water spinach on patterns of eating, consumption of caecotrophs, and excretion of faeces by growing rabbits. *Livestock Research for Rural Development. Volume 18, Article # 78.* Retrieved, October 16, 106, from http:// www. cipav. org. co/lrrd/lrrd18/06/samk18078. htm.

Pok Samkol, Preston, T. R. and Ly, J. (2006). Effect of increasing offer level of water spinach (*Ipomoea aquatica*) on intake, growth and digestibility coefficients of rabbits. *Livestock Research for Rural Development. Volume 18,* Article No. 25. http://www. lrrd. org/lrrd18/2/samk18025. htm.

Polidori, F., Dell'orto, V., Corino, C. (1982). Use of dried whole maize as a substitute for hay in diets for meat rabbits. *Coniglicoltura* 19: 25-27.

Polidori, F., Dell'orto, V., Corino, C., Pedron, O. and Bigli, A. (1984). Impiego di disidrati di cereali integrali in diete ad alta energia per coniglio da carne. *3rd World Rabbit Congress*, Rome 1: 339-346.

Polley, H. W., Mayeux, H. S., Johnson, H. B. and Tischler C. R. (1997). Viewpoint: atmospheric CO_2, soil water, and shrub/grass ratios on rangelands. *J. Range Manage.* 50: 278–284. doi:10. 2307/4003730.

Ponte, P., Prates, J., Crespo, D., Crespo, J., Mourão, S., Alves, R., Bessa, M., Chaveiro-Soares, L., Ferreira, L. and Fontes, C. (2008). Improving the Lipid Nutritive Value of Poultry Meat through the incorporation of dehydrated leguminous-based forage in the diet for broiler chicks. *Poultry Science* 87: 1587-1594.

Ponte, P., Prates, J., Crespo, D., Crespo, J., Mourão, S., Alves, R., Bessa, M., Chaveiro-Soares, L., Ferreira, L. and Fontes, C. (2008). Restricting the Intake of a Cereal-

Based Feed in Free-Range-Pastured Poultry: Effects on Performance and Meat Quality. *Poultry Science* 87: 2032-2042.

Post, W. M. and Kwon, K. C. (2000). Soil carbon sequestration and land use change. *Global change biology* 6: 317-328.

Pradeep Behari and Hazra, C. R. (1989). Forage productivity of sorghum and oat evapotranspiration studies. *Forage Research* 15(2): 162-166.

Pradhan, P. R. (1982). Time for planting and bare rooted planting trial. Pakhribas Agriculture Centre, Dhankuta, Nepal, *PAC Bulletin No.* 59, 13 p.

Prakash, B., and Rathore, S. S. (2007-08). A comprehensive study on survey, nutritional evaluation, nutrient requirement and development of improved agronomic practices for year round availability of green fodder for mithun under North Eastern Hill Region of India. *Annual Report,* 2007-08, National Research Centre on Mithun, Jharnapani, Medziphema, Nagaland, India, pp. 22-24.

Prakash, B., Dhali, A., Rathore, S. S., Das, K. C., Walling, I., Vupru, K., Mech, A., Baruah, K. K. and Rajkhoa, C. (2009). Chemical composition and nutritional evaluation of various foliages consumed by mithun (*Bos frontalis*). *Animal Feed Science and Technology* 150 (3-4): 223-229.

Prasad, R. N. and Singh K. A. (1994). Integrated Land Use Management for Eastern Himalayan Agro-ecosystems - Chapter 18, IN *"Agroforestry Systems for Sustainable Land uses",* Eds. Panjab Singh, P. S. Pathak and M. M. Roy, Oxford and IBH Co. Pvt. Ltd., New Delhi, pp. 228-236.

Prasad, R., Yadav, R. P., Agrawal, R. K., Agnihotri, Y., Mittal, S. P., Tiwari, A. K. and Samara, J. S. (2007). *Prospects of aonla cultivation on degraded Shivalik Lands,* Tech. Bulletin no. T-55/C-13, Central Soil and Water Conservation Research and Training Institute, Research Centre, Chandigarh (U. T.), p. 133.

Prasad, R. N., Singh, K. A. and Thakur, P. K. (1994). Approaches to agricultural development in the hills of North Eastern Region-A thought. *Proc. Planning Conference on "Perspective Development Studies on North Eastern Region"* organized by North Eastern Hills University at Shillong, 16-17 March, 1994, pp. 27-30.

Prine, G. M., Dunavin, L. S., Glennon, R. J. and Roush, R. D. (1986). Arbrook rhizpma -*Perennial Forage Legume,* IFAS, University of Florida, Gainesville, USA, 16p.

Prins, H. T. (1992). The pastoral road to extinction: competition between wildlife and traditional Pastoralism in East Africa. *Environ. Conserv.* 19: 117–123.

Purushothaman Seema, Viswanath, S. and. Swaminathan, L. P (2005). Managing Our Grazing lands: Need for a Rational Grazing policy, Azimpremji University, Bengaluru, India.

Puskur, R. (2011). Assessing impact of research for development initiatives that use innovation systems perspectives: Lessons from the Fodder Innovation Project in

India. Paper presented at the "*7th Conference of the Asian Society of Agricultural Economists*", Hanoi, Vietnam, 13-15 October,2012, URL: http://7thasae. ipsard. gov. vn/ppt/presentation/C3/C3_ASAE_Puskur per cent 20Ranjitha_India. pdf,

Puskur, Ranjitha, Carlos Quiros and Iain Wright. (2009). From research outputs to developmental outcomes: A framework for fostering innovation in livestock systems. Paper presented at the Innovation Asia-Pacific Symposium, Kathmandu, Nepal.

Putnam, D. H., Robinson, P. H. and Eric Lin (2013). *Does Hydroponic Forage Production Make sense? Alfalfa and Forage News,* News and information from UC Cooperative Extension about alfalfa and forage production, California, USA.

Rachuonyo, H. A., Allen, V. G. and McGlone, J. J. (2005). Behavior, preference for, and use of alfalfa, tall fescue, white clover, and buffalo grass by pregnant gilts in an outdoor production system. *Anim. Sci.* 83: 2225-2234.

Raharjo, Y. C. and Cheeke, P. R. (1985). Palatability of tropical tree legume forage to rabbits. *Nitrogen Fixing Tree Research Reports* 3, 31-32.

Raharjo, Y. C. (1987). Evaluation of tropical forages and by-product feeds for rabbit production. *Ph. D. Thesis*, Oregon State University, 251 pp.

Raharjo, Y. C., Cheeke, P. R. and Patton, N. M. (1986). Growth and productive performance of rabbits on a moderately low crude protein diet with or without methonine or urea supplementation. *Journal of Animal Science* 63: 795–803.

Raharjo, Y. C., Cheeke, P. R., Patt on, N. M. and Supriyati, K. (1986). Evaluation of tropical forages and by-product feeds for rabbit production. I. Nutrient digestibility and effect of heat treatment. *Journal of Applied Rabbit Research* 9: 56–66.

Rai, A. K., Ghosh, P. K., Ram, S. N., Singh, S. Sunil Kumar, Mahanta, S. K., Maity, S. B. Singh, J. P., Dixit, A. K., Tiwari, S. K. Roy, A. K., Kumar, R. V., Mishra, A. K. and Tripathi, S. B. (2013). *Carbon sequestration in forage based land use systems.* Indian Grassland and Fodder Research Institute, Jhansi, India. pp. 1-52.

Rai, M. P. and Gupta, S. K. (2007-13) Establishment and management of silvipasture for optimizing forage productivity and availability. *ICAR-IGFRI Annual Report,* 2007-13, Jhansi, India.

Rai, P. and Singh, K. A. (1991). Response of *Panicum antidotale* to nitrogen and phosphorus under rainfed condition. *Range Mgmt. and Agroforestry* 12(1): 61-65.

Rai, P., Kanodia, K. C., Patil, B. D., Velaudhan, K. C., Agarwal, R. (1980). Nitrogen equivalence of range legumes introduced in natural grassland. *Indian J. range Mgmt.* 1(2); 97-101.

Rai, P., Kanodia, K. C., Patil, B. D., Velayudhan, K. C., and Agarwal, R. (1985). Nitrogen equivalence of range legumes introduced in natural grassland. *Indian J. Range Management* 1(2): 97-101.

Rai, P., Ramana, D. B. V., Rao, G. R., Yadav, R. S., and Singh, U. P. (2001). Growth and production of *Anogeissus latifolia* based silvipastoral system under different pruning managements. *Indian J. Agroforestry* 3(2): 168-172.

Rai, P., Singh, K. A. and Kanodia, K. C. (1982). Effect of different levels of nitrogen, phosphorus and potassium on the productivity of *Macroptillium atropupureum* (DC) Urb Cv. Siratro. *Forage Res.* 8 (2): 105-109.

Rai, P., Solanki, K. R., Deb Roy, R. and Singh, R. (1998). Performance of lamb and kids on silvipastoral system and effect of grazing on constituent's vegetation. *Indian J. Animal Science* 68(9): 973-975.

Rai, P., Yadav, R. S., Solanki, K. R., Rao, G. R. and Singh, R. (2001). Growth and pruned biomass production of multipurpose tree species under silvipastoral systems on degraded lands in semiarid region of Uttar Pradesh, India. *Forest Tree and Livelihood* 11: 347-364.

Rai, R. N., Venugopal, K., Singh, K. A. and Singh, R. D. (1988). Studies on compatibility of grasses under natural forest of *utis* (*Alnus nepalensis*) and chillowance (*Schima wallichii*). *Indian J. Hill Farming* I A (1): 71-72.

Rajalahti, R., Janssen, W. and Pehu, E. (2008). *Agricultural Innovation Systems: From Diagnostic Towards Operational Practices Agricultural and Rural Development Discussion Paper 38*, World Bank, Washington DC, USA

Rajendra, P., Seghal, J. P., Patnayak, B. C. and Beniwal, R. K. (1998). Utilization of artificially grown barley fodder by sheep. *Indian J. Small Rumin.* 4 (2): 63-68.

Raju, K. V. S. and Sreemannarayana, O. (1995). Feeding of *Ulva fasciata* to rabbits - feed efficiency and carcass characteristics. *Indian Veterinary Journal* 72: 1331-1332.

Ram, S. N. and Shukla, A. K. (2008-13). Evaluation of *Hardwickia binata* based silvipasture systems under soil moisture conservation practices. *ICAR-IGFRI Annual Report* 2008-13, Jhani, India.

Ram, S. N., Kumar S., and Roy M. M. (2005). Performance of jujube (*Ziziphus mauritiana*) based hortipasture system in relation to pruning intensities and grass legume association under rainfed conditions, *Indian J. Agronomy* 50(3): 181-183

Ramamurthy, V. (1986). *History of Kongu: Prehistoric period to 1300 AD.* International Society for the Investigation of Ancient Civilization, Madras, India.

Ramana, D. B. V., Sultan Singh and Negi, A. S. (2000). Nutritive evaluation of some nitrogen and non-nitrogen fixing multipurpose tree species. *Animal Science and Feed Technology* 88(1-2): 103-111. DOI: http://dx. doi. org/10. 1016/ S0377-8401(00)00200-5.

Ramchurn, R., Dullull, Z. B., Ruggoo, A. and Raggoo, J. (2000). Effects of feeding star grass (*Cynodon plectostachyus*) on growth and digestibility of nutrients in the domestic rabbit. *Livestock Research for Rural Development* 12: 1-6.

Ramdas, S. R. (2003). Strategies For Livestock Development *In* "Watershed Interventions : A Final Report of A Study Commissioned By NABARD for the Indo-German Watershed Development Programme" Andhra Pradesh Study Team :Anthra (Anthra Team: Amala, Ashalatha, Digamber, Narsimlu, Rajamma, Rambabu, Sagari, Sanyasi Rao, Savitri, Subramanyam), Wassan Team: Janardhan, Padma, Ramachandrudu, Ravindra, Suresh, Vanaja, Hyderabad, Wassan, Hyderabad, ftp://ftp. fao. org/docrep/nonfao/lead/ae753e/ ae753e00. pdf

Ramdas, S. R. (2014). *Adivasis*, Pastoralists, and Forest Governance: Challenges and Opportunities. In "*Democratizing Forest Governance in India*", eds. Lele, S and Menon, A Oxford University Press, New Delhi), 63-99p.

Ramdas, S. R., Ashalatha and Sanyasi Rao, (2008). *Recognizing the Grazing Rights of Communities in Forests: Good Practice Note*, Anthra, December, 2008.

Ramdas, S. R., Yakshi, Girija Deepika (2001). Changing Livelihoods, Livestock and Local Knowledge Systems: Women Stake their Claim in Andhra Pradesh and Maharashtra, *Anthra,* Hyderabad. *Indian Journal of Gender Studies* 8(2): 175-194

Rameswar Deka, William Thorpe, M. Lucila Lapar and Anjani Kumar (2007). Karbi Anglong's pig sub-sector: current status, constraints and opportunities. International Livestock Research Institute CG Block, NASC Complex, DPS Marg, Pusa Campus New Delhi-110012, India.

Randhir, S., Sawal, R. K., Bhasin, V. and Bhatia, D. R. (1994). Nutrient utilization by rabbits on diets containing kudzu hay. *Indian Journal of Animal Nutrition* 11: 255-258.

Rangnekar, D. V. (2006). *Livestock in Livelihoods of the Underprivileged Communities in India – a Review*, International Livestock Research Institute, Nairobi (Kenya), 72p.

Rangnekar, D. V. (1999). Some observations on non- adoption of silage making in central and western India, Poster 1. 4, In "*Silage making in the tropics with particular emphasis on smallholders*", *FAO Electronic Conference on Tropical Silage,* 01 September- 15 December, 1999, Grassland and Pasture Crops Group, AGPC and Feed Resource Group, AGAP (Caterina Batello, Stephen Reynolds, Andrew Speedy, Len 'tMannetje, Max Shelton), FAO, Rome, Italy.

Rao, K. S., Reddy, M. R. and Reddy, G. V. N. (1986). Utilization of unconventional roughages in the formulation of complete feed for rabbits: effect on growth and carcass characteristics. *Indian Journal of Animal Nutrition* 3: 238-243.

Ravindran, V. (1990). Feeding value and digestibility of cassava leaf meal for growing pigs. *Proc. Fifth Australasian Animal Production Congress.* Vol. 3. Taipei, Taiwan, pp 20.

Ravindran, V., Kornegay, E. T. and Rajaguru, A. S. B. (1983). Utilization of the whole cassava plant as a swine feed. *World. Rev. Anim. Prod.* 19(1): 8–14.

Ravindran, V., Kornegay, E. T., Rajaguru, A. S. B. and Notter, D. R. (1987). Cassava leaf meal as a replacement for coconut oil meal in pig diets. *Journal of the Science of Food and Agriculture* 41: 45–53.

Ravindran, V., Kornegay, E. T., Rajaguru, A. S. B., Potter, L. M. and Cherry, J. A. (1986). Cassava leaf meal as a replacement for coconut oil meal in broiler diets. *Poultry Science* 65: 1720- 1727.

Ravindran, V., Rajdevan, P., Goonewardene, L. A. and Rajaguru, A. S. B. (1986). Effects of feeding cassava leaf meal on the growth of rabbits. *Agricultural Wastes* 17: 217-224.

Rebecca, M., Nelson, J. and Elizabeth, A. (2012). Effects of Climate Change on Pasture Production and Forage Quality. Dept of Plant and Soil Sciences, University of Kentucky, rebecca. mcculley@uky. edu

Reddy, G. V. N., Reddy, M. R. and Reddy, K. K. (1988). Nutrient utilization by milch cattle fed on rations containing artificially grown fodder. *Indian J. Anim. Nutr.* 5 (1): 19-22.

Reid, R. S., Thornton, P. K., McCrabb, G. J., Kruska, R. L., Atieno, F. and Jones, P. G. (2004). Is it possible to mitigate greenhouse gas emissions in pastoral ecosystems of the tropics? *Environment, Development and Sustainability* 6: 91–109.

Rekhate, D. H., Madavi, V. B. and Dhok, A. P. (2007). Performance of goats fed Arhar (*Cajanus cajan*) straw based pelleted complete ration. *Anim Nutr Feed Technol.* 7: 47–52.

Ren, H. Y., Zheng, S. X. and Bai, Y. F. (2009). Effects of grazing on foliage biomass allocation of grassland communities in Xilin River Basin, Inner Mongolia. *Chinese Journal of Plant Ecology* 33: 1065–1074.

Reynolds, J. F., Smith, D. M. S., Lambin, E. F., Turner, B. L. II, Mortimore, M., Batterbury, S. P. J. *et al.* (2007). Global desertification: building a science for dryland development. *Science:* 316:847.

Rich, K., Baker, D., Negassa, A. and Ross, R. B. (2009). Concepts, applications and extensions of value chain analysis to livestock systems in developing countries. Contributed Paper prepared for presentation at the "*International Association of Agricultural Economists Conference*", August 16-22, 2009, Beijing, China.

Riveros, F. (1992). The genus *Prosopis* and its potential to improve livestock production. In: Legume trees and other fodder trees as protein sources for livestock. Speedy, A. W. and Pugliese, P. L. (Eds.), *Animal Production and Health Paper*, FAO, Rome, Italy. pp. 257-276.

Robert Guisepi (2006). Dio's Rome Part II Agriculture and The Origins Of Civilization: The Neolithic Revolution, http://www. pgdp. net/

Roder, W. (2000). *Grazing resources for yak production systems in Bhutan*. Renewable Natural Resources Research Centre (RNRRC), Ministry of Agriculture, Jakar, Bumthang, Bhutan.

Rohilla, P. P. and Bujarbaruah, K. M. (1999). Effect of subabul (*Leucaena leucocephala*) feeding on growth and physiology of rabbits. *Journal of Hill Research* 12: 135- 137.

Rohilla, P. P. and Bujarbaruah, K. M. (2000). Effect of banana leaves feeding on growth of rabbits. *Indian Veterinary Journal* 77: 902-903.

Rohilla, P. P. and Bujarbaruah, K. M. (2000). Effect of feeding broom grass (*Thysanolaena maxima*) to rabbits. *Indian Journal of Animal Nutrition* 17: 87-89.

Rohilla, P. P. and Bujarbaruah, K. M. (2000). Effect of *Morus alba* feeding on growth and physiological parameters of rabbits. *International Journal of Animal Sciences* 15: 9-12.

Rosenzweig, C., Karoly, D., Vicarelli, M., Neofotis, P., Wu, Q., Casassa, G., Menzel, A., Root, T. L., Estrella, N., Seguin, B., Tryjanowski, P., Liu, C., Rawlins, C. andImeson, A. (2008). Attributing physical and biological impacts to anthropogenic climate change. *Nature* 453: http://dx. doi. org/10. 1038/nature06937.

Ross, E. and Enriquez, F. Q. (1969). The nutritive value of cassava leaf meal. *Poultry Science* 48: 846–853.

Rötter, R. and van de Geijn, S. C. (1999). Climate change effects on plant growth, crop yield and livestock. *Climatic Change* 43: 651–681.

Roy, A. K., Malaviya, D. R., Kaushal, P., Chandra, A., Singh, U. P. (2009). *Egyptian clover/Berseem* (*Trifolium alexandrium L.*), Catalogue IGFRI, Jhansi, India.

Roy, A. K., Malaviya, D. R., Kaushal, P., Chandra, A., Singh, U. P., Mahanta, S. K., Chauhan Roopali and Tiwari Rupali (2009). *Descriptors for Dichanthium Bothriochloa Complex.* IGFRI, Jhansi, India.

Roy, M. M. and Singh, K. A. (2008). The fodder situation in rural India: future outlook. *International Forestry Review* 10 (2): 217-234.

Roy, M. M. and Singh, K. A. (2009). Tropical Pasture Legumes and Their Role in Wasteland Management. *IN "Legumes for Ecological Sustainability"*, Eds. Masood Ali, Sanjeev Gupta, P. S. Basu, Naimuddin, Indian Society of Pulses Research and Development (Indian Institute of Pulses Research, Kanpur, pp. 480-490.

Roy, S., Saxena, P. and Roy, M. M. (2009). *Soil biodiversity under forage production systems.* IGFRI, Jhansi, India, 45p.

Rubio, L. A., Pedrosa, M. M., Cuadrado, C., Gelencser, E., Clemente, A., Burbano, C. and Muzquiz, M. (2006). Recovery at the terminal ileum of some legume non-nutritional factors in cannulated pigs. *Journal of the Science of Food and Agriculture* 86: 979–987.

Ruppert, J. C., Holm, A., Miehe, S., Muldavin, E. and Snyman, H. A. (2012). Meta-analysis of ANPP and rain-use efficiency confirms indicative value for degradation and supports non-linear response along precipitation gradients in dry lands. *Journal of Vegetation Science* 23: 1035–1050. doi: 10. 1111/j. 1654-1103. 2012. 01420. x

Russell, C. E. and Felker, P. (1987). The prickly-pears (*Opuntia* spp., Cactaceae): a source of human and animal food in semiarid regions. *Economic Botany* 41(3): 433-445.

Russell-Smith, J., Stanton, P. J., Edwards, A. C. and Whitehead, P. J. (2004). Rain forest invasion of eucalypt-dominated woodland savanna, Iron Range, north-eastern Australia: II. Rates of landscape change. *J. Biogeogr.* 31:1305–1316. (doi:10. 1111/j. 1365-2699. 2004. 01070. x.

Rybina, E. A. and Reshetova, T. A. (1981). Digestibility of nutrients and biochemical values of eggs in relation to the amount of lucerne and grass meal and the quality of supplementary fat in the diet of laying hens. *Trudy Uzbekskogo Nauchno-Issledovatel'skogo Instituta Zhivotnovodstva* 35: 148–152.

Sage, R. F. (2004). The evolution of C_4 photosynthesis. *New Phytology* 161: 341–370. doi:10. 1111/j. 1469-8137. 2004. 00974. x.

Saha, U., Leticia Sonon, Dennis Hancock, Nicholas Hill, Lawton Stewart, Gary Heusner and David E. Kissel (2013). *Common Terms Used in Animal Feeding and Nutrition*, College of Agricultural and Environmental Sciences, The University of Georgia, *Bulletin No.* 1367.

Sahay G., Malaviya D. R., Singh U. P., Roy S. and Bhaskar R. B. (2011). IGFRI *Cowpea catalogue*, IGFRI, Jhansi, India.

Sahay, K. M. and Singh, K. K. (1994). *Unit Operations of Agricultural Processing*, Vikas Publishing House Pvt. Ltd., New Delhi, India.

Salamin, N. (2008). Oligocene CO_2 decline promoted C_4 photosynthesis in grasses. *Curr. Biol.* 18**:** 37–43. doi:10. 1016/j. cub. 2007. 11. 058.

Samanta, A. K., Singh, K. K., Das, M. M. and Kundu S. S. (2004). *Complete Feed Block-A Balanced Ration for Livestock.* IGFRI, Jhansi, India.

Samanta, A. K., Singh, K. K., Das, M. M., Maity, S. B. and Kundu, S. S. (2003). Effect of complete feed block on nutrient utilization and rumen fermentation in barbari goats. *Small Ruminant Research* 48: 95–102.

Samuel, D. V. K. and Anap, G. R. (1980). Design criteria for portable hay drier. *Forage Res.* 6: 209-214.

Sanjiv, K. and Bhatt, R. S. (2000). Nutrient utilization and wool yield in Angora weaning rabbits fed on three levels of *Robinia* leaves. *Indian Journal of Animal Nutrition* 17: 171-174.

Sankaran, M. *et al.* (2005). Determinants of woody cover in African savannas. *Nature* 438*:* 846–849. doi:10. 1038/nature04070.

Sansamma, G. and Pillai, G. R. (2000). Effect of vermicompost on yield and economics of Guinea grass (*P. maximum*) growth as intercrop in coconut (*C. nucifera*) garden. *Indian Journal of Agronomy* 45: 693-97.

Sareen, S., Misri, B. and Dev, I. (2005). *White clover Germplasm Catalogue*. IGFRI, Jhansi, India. 81p

Sarmiento-Franco, L., McNab, J. M., Pearsson, R. A. and Belmar-Casso, R. (2002). Performance of broilers fed on diets containing different amounts of Chaya (*Cnidoscolus aconitifolius*) leaf meal. *Tropical Animal Health and Production* 34(3): 257-269.

Sarmiento-Franco, L., Mcnab, J. M., Pearson, A. and Belmar-Casso, R. (2003). The effect of chaya (*Cnidoscolus aconitifolius*) leaf meal and of exogenous enzymes on amino acid digestibility in broilers. *British Poultry Science* 44(3): 458-463. Published online: 28 Jun 2010 http://dx. doi. org/10. 1080/00071660310001598256.

Sarria, P., Montoya, C., Yusti, L. M., Orejuela, I., Guevara, M., Cruz, A. C., Arredondo, J., Londoño, A. and Peters, M. (2010). Nutritive value of leaf meal of Cowpea (*Vigna unguiculata* L. Walp.) for growing pigs. *Livestock Research for Rural Development*, 22 (6): 110 http://www. lrrd. org/lrrd22/6/sarr22110. htm.

Sastry, V. R. B. and Mahajan, J. M. (1982). Comparative adequacy of supplementation of white clover hay with tree leaves and whole grains on the performance of New Zealand White rabbits. *Indian Journal of Animal Sciences* 52: 901-905.

Savon, L. (2005). Tropical roughages and their effect on the digestive physiology of monogastric species - Alimentos fibrosos y su efecto en la fisiología digestiva de especies monogástricas. *Cuban Journal of Agricultural Science - Revista Cubana de Ciencia Agrí- cola* 39: 475–487.

Saxena, M. C. and Siddiqui, M. K. J. (1982). Pesticide pollution in India: Organochlorine pesticides in women, buffalo and goat milk. *J. Dairy Sci* 65, 430-434.

Saxena, P., Pandey, K. C., Padmavati, C., Sah, N. K., Faruqui, S. A., Sharmila Roy, Hasan, N., Bhaskar, R. B. and Azmi, M. I. (2002). *Forage Plant Protection, Bulletin* published by ICAR-IGFRI, Jhansi (UP), India, 38p.

Scapinello, C., Antunes, E. B., Melo, E. V. I., Furian, A. C. and Jobim, C. C. (2000). Nutritive value and utilization of Leucaena hays (*Leucaena leucocephala* cv. Cunnigham) for growing rabbits. *7th World Rabbit Congress*, Valencia vol. C: 423-428.

Scapinello, C., Falco, J. E., Furian, A. C. and Faria, H. G. DE (2000). Performance of growing rabbits feeding on different levels of cassava foliage hay (*Manihot esculenta*, Crantz). *Ciencia Rural* 30: 493-497.

Scarnecchia, D. L. (1990). Concepts of carrying capacity and substitution rations: a systems viewpoint. *Journal of Range Management* 43: 553–555.

Schaller, G. (1998). Wildlife of the Tibetan Steppe. University of Chicago Press, Chicago.

Schultze-Kraft, R. and Peters, M. (1997). *Tropical legumes in agricultural production and resource management: An overview.* Presented at the Tropentag JLU Giessen 22. -23. 5. 1997. Pp. 1–17.

Scott, M. L., Nesheim, M. C. and Young, R. J. (1982). *Nutrition of chicken. Scott ML, Ithaca, NY. (eds.*) 1::25-50.

Scurlock, J. M. O. and Hall, D. O. (1998). The global carbon sink: a grassland perspective. *Global Change Biology* 4(2): 229–233.

Seemannarayana, O., Raju, K. V. S. and Prasad, J. R. (2001). Utilisation of *Schleichera oleosa* leaf meal as rabbit feed. *Indian Journal of Animal Nutrition* 18: 93-95.

Seemannarayana, O., Raju, K. V. S., Ramaraju, G. V. A. N. S. and Prasad, J. R. (1995). The use of *Ulva fasciata*, a marine algaeas rabbit feed: growth and conversion efficiency. *Indian Veterinary Journal* 72: 989-991.

Seré, C. and Steinfeld, H. (1996). World livestock production systems: current status, issues and trends. *FAO Animal Production and Health Paper* 127, FAO, Rome, Italy.

Seroux, M., Franck, Y. and Mercier, P. (1980). Utilisation du maïs deshydraté par le lapin à l'engraissement. *2nd World Rabbit Congress*, Barcelona 2: 157-166.

Sese, B. T., Oruwari, B. M. and Berepubo, N. A. (1999). The value of cowpea testa in the diet of growing rabbits. *International Journal of Animal Sciences* 14: 209-214.

Shah, Amita. (2004). *Changing Interface between Agriculture and Livestock: A Study of Livelihood Options under Dryland Farming Systems in Gujarat.* Gujarat Institute of Development Research, Gujarat, India.

Shaimi, N., Kallida, R., Volaire, F., Saidi, N. and Al Faiz, C. (2009). Summer dormancy and drought survival of Moroccan ecotypes of *Dactylis glomerata* L. In: *Crop Science* 49: 1416-1424.

Shamna, T. P., Peethambaran, P. A., Jalaludeen, A., Leo Joseph and Muhammad Aslam, M. K. (2013). Broiler characteristics of Japanese quails (*Coturnixcoturnix japonica*) at different levels of diet substitution with *Azolla pinnata*. *Animal Sci. Rep.* 7(2): 75-80.

Shankar, V. and Singh, J. P. (1996). Grazing Ecology. *Tropical Ecology*. 37 (1): 67-78.

Shankarnarayan, K. A. and Shankar, V. (1984). *Grasses and legumes for forage and soil conservation.* Indian Council of Agricultural Research, New Delhi, India.

Sharma, S. K. (2004). Hortipastoral based land use systems for enhancing productivity of degraded lands under rainfed and partially irrigated conditions, *Uganda Journal of Agricultural Sciences* 9: 320-325

Sharma, V., Sharma, R. and Wadhwa, B. K. (2002). Status and control of pesticide residues in milk and milk products. *Indian Dairyman* 54: 59-63.

Sharma, V. P., Köhler-Rollefson, I. and Morton, J. (2003). Pastoralism in India: A scoping study. Centre for Management in Agriculture, IIM (Indian Institute of Management), Ahmedabad, India; League for Pastoral Peoples, Ober-Ramstadt, Germany; and NRI (Natural Resources Institute), University of Greenwich, UK. http://r4d. dfid. gov. uk/PDF/outputs/ZC0181b. pdf.

Sharrow, S. H. and Ismail, S. (2004). Carbon and nitrogen storage in agroforests, tree plantations and pastures in western Oregon, USA. *Agroforestry System* 60: 123-130.

Shoukat Araa, Adilb, S., Banday, M. T. and Manzoor A. Khan (2015). Feeding Potential of Aquatic Fern-Azolla in Broiler Chicken Ration. *Journal of Poultry Science and Technology Journal* homepage: www. jakraya. com/journal/jpst.

Shreshtha, N. P. and Pakhim, B. (1989). Effect of different tree fodders on milk production in buffaloes. PAC Tech. Paper No. 17, Pakhribas Agric. Centre, Kathmandu, Nepal.

Shreshtha, N. P., Neopane, S. P., Gurung, H. B., Pakhim, B. and Shrestha, P. (1990). Observation of the adaptability and seed production of forage legumes at Pakhribas Agriculture Centre, PAC Tech. Paper No. 126, Kathmandu, Nepal.

Shukla, A. K., Sunil Kumar, H. V. Singh, Watpade, S. G. and Pramanik, K. K. (2014). Bael (*Aegle marmelos*) based hortipastoral system with moisture conservation in semiarid condition. *J. Tree Science* 33: 7-11.

Shukla, R., Khare, P. K. (1998). Food habits of wild ungulates and their competition with live stock in Pench Wildlife Reserve Central India. *Journal of the Bombay Natural History Society* 95(3): 418–421.

Silva, L. C. R., Sternberg, L., Haridasan, M., Hoffmann, W. A., Miralles-Wilhelm, F. and Franco, A. C. (2008). Expansion of gallery forests into central Brazilian savannas. *Glob. Change Biol.* 14:2108–2118. doi:10. 1111/j. 1365-2486. 2008. 01637. x

Silvia, E. Valtorta (2002). *Animal production in a changing climate: impacts and mitigation*, National Council for Scientific and Technical Research, National Institute of Agricultural Technology, Rafaela Experimental Station 2300 – Rafaela, Santa Fe, Argentina.

Simply Hydro (2008). Basic hydroponic systems and how they work. Retrieved November 11, 2011, from http://www. simplyhydro. com/system. htm.

Singh, B., Goel, G. C. and Negi, S. S. (1984). Effect of supplementing muberry (*Morus alba*) leaves *ad libitum* to concentrate diets of Angora rabbits on wool production. *Journal of Applied Rabbit Research* 7(4): 156-160.

Singh, H. P., Ramakrishna, Y. S., Sharma, K. L. and Venkateswarulu, B. (1999). Fifty years of dryland agriculture research in India. Central Research Institute for Dryland Agriculture, Hyderabad, India.

Singh, J. B., Pradeep Behari, and Yadava, R. B. (2002). Evapotranspiration and crop coefficient of maize + cowpea intercropping system in semiarid region. *Range Mgmt. and Agroforestry* 23(2): 152-154.

Singh, J. P., Roy, M. M. and Radotra, S. (2009). *Grasslands of Himachal Pradesh.* IGFRI, Jhansi, India.

Singh, K. A. (2000). Potential of Agroforestry Interventions in the Humid Hilly Ecosystems of India. *IN "Agroforestry and Forest Products - Proceeding of International Workshop on Agroforestry and Forest Products",* 28-30th November, 2000, eds. L. K. Jha, D. Paul, U. Sahoo, R. P. Tiwari, S. N. Ramanujam, H. Lairamnghinglova and L. N. Singh, Pub: Linkman Publication, 45 Annapuna Devi Road, Titagarh, North 24 parganas-743 188 West Bengal, India, Vol. 1, pp. 87-106.

Singh, K. A. (2002). Resource Management Perspective of Arunachal Agriculture, Silver Jubilee Publication, ICAR Research Complex for NEH Region, Arunachal Pradesh Centre, Basar, 371p.

Singh, K. A. and Rai, P. (1984). Response of forage *Panicum* species to nitrogen fertilization on semiarid lands. *Indian J. agric. Sci.* 54(5): 382-386.

Singh, K. A. and Singh, N. P. (1983). Effect of sowing time on dry matter accumulation and nitrogen uptake in jute. *Indian J. Agron.* 28(4): 451-453.

Singh, K. A. and Yadav, B. P. S. (1995). Advances in Forage Resource Development and its Potential for Sustained Hill Agro-ecosystems in Eastern Himalayan Region, IN *"New Vistas in Forage Production",* Eds. C. R. Hazra and Bimal Misri, Publication and Information Directorate, New Delhi, pp. 162-183.

Singh, K. A., Satapathy, K. K., Ramchandra, Singh, A. and Arun Verma (2002). Micro-watershed Approach to Hill Slope Management in North East India *IN "Resource Management Perspective of Arunachal Agriculture"* eds. K. A. Singh, ICAR Research Complex for NEH Region, Arunachal Pradesh Centre, Basar, pp. 351-371.

Singh, K. A., Premchandran, M. N. and Handa, D. P. (1985). Studies of forage utilization based on height-weight relationship in *Panicum* species. *Indian J. Agron.* 30(4): 450-454.

Singh, K. A. Singh, A. and Satapathy, K. K. (2008). Tree based land use systems for optimization of bio-productivity and resource conservation in a humid subtropical climate. *Range Mgmt. and Agroforestry* Vol. 29 (2): 83-93.

Singh, K. A. (1984). Response of forage green panic and verano stylo intercropping to fertilization. *Indian J. Agron.* 29(3): 277-281.

Singh, K. A. (1985). Effect of phosphorus levels and verano stylo seed rates on the productivity of guinea grass and verano stylo intercropping. *Indian J. Agron.* 30(2): 260-262.

Singh, K. A. (1985). Influence of phosphorus levels on dry matter accumulation, nodulation and plant composition of *Glycine javanica* Linn. *Indian J. agril. Sci.* 55(9): 586-590.

Singh, K. A. (1985). Studies on salt tolerance in *Panicum* species. *Indian J. Agron.* 30(1): 94-97.

Singh, K. A. (1985). Grow forage *Panicum* of your choice. *Indian Farmers Digest* 18(3): 7-10.

Singh, K. A. (1986). Intercropping of grain legumes with blue panic *Indian Farming* 35(3):9.

Singh, K. A. (1987). Grow more fodder without nitrogen fertilizer. *Indian Farmers Digest* 20(11): 17-18.

Singh, K. A. (1994). Cutting management of oat varieties at the mid hills of Sikkim. *J. Hill Research* 7 (2): 153-154.

Singh, K. A. (1996). Growth, biomass productivity and coppicing behavior of multipurpose tree species at a mid hill of Sikkim. *Range Mgmt. and Agroforestry* 17(1): 93-97.

Singh, K. A. (1996). Effect of age, cutting interval, nitrogen and farmyard manure on root growth parameters of forage grasses in an acid *Inceptisol. Indian J. agric. Sci.* 66 (1): 38-44

Singh, K. A. (1998). Micro watershed approach to conservation farming in North Eastern Hills of India. *ICAR News Letter.* July-Sept., 1998 Vol. 4(3): 1-3.

Singh, K. A. (1998). Nutrient contents in the important fodder trees of eastern Himalaya as influenced by leaf position and the month of lopping. *J. Hill Research* 11 (1): 1-7.

Singh, K. A. (1999). Effect of nitrogen levels on yield, root biomass distribution, nitrogen recovery by forage grasses and changes in soil properties of an acid *Inceptisol. Indian J. agric. Sci.* 69(8): 551-554.

Singh, K. A. (1999). Golden timothy grass under integrated nutrient management on an acid *Inceptisol* under humid sub-tropical climate. *Trop. Agric.* (*Trindad*) 76(2): 104-108.

Singh, K. A. (1999). Nutrient contents in tree fodder and the leaves of bamboo used during scarcity period in eastern Himalaya. *Indian J. Anim. Nutrition* 16(2): 178-182.

Singh, K. A. (1999). Promising forage legumes for low to mid hills of north-eastern region of India. *Indian Farming* 49(2): 15-16.

Singh, K. A. (1999). Resource management and productivity enhancement through agroforestry in the eastern hilly agro-ecosystems of India. *Indian J. agroforestry* 1(1): 63-72.

Singh, K. A. (1999). Resource management and productivity enhancement through agroforestry in the eastern hilly agro-ecosystems of India. *Indian J. agroforestry* 1(1): 63-72.

Singh, K. A. (1999). Response of promising forage grasses to nitrogen. *Indian J. Agron.* 44(2): 272-76.

Singh, K. A. (2001). Leaf morphology and leaf area of fodder trees of NEH Region. *Range Management and Agroforestry* 22(1): 85-93.

Singh, K. A. (2005). Studies on forage production, soil-binding capacity of grasses and changes in soil properties on the terrace risers. *Range Mgmt. and Agroforestry* 26(2): 113-115.

Singh, K. A. (2005). Impact of forage resource development on sustained land use and environment in Northeastern hills region of India. *Range Mgmt. and Agroforestry* 26(1): 58-70.

Singh, K. A. (2006). Ecological perspective of forage based livestock production. *IGFRI Newsletter* 12(2) & (3).

Singh, K. A. (2006). Mixed farming and fodder based livestock production. *IGFRI Newsletter* 11(4) and 12(1).

Singh, K. A. (2006). Potential of forage production under non-competitive land use in north-eastern hills region. *Indian Farming* 56(5): 22-25.

Singh, K. A. (2006). Water productivity in livestock sector. *IGFRI Newsletter*12 (4): 3.

Singh, K. A. (2007). Flood, Forage and Livestock. *IGFRI Newsletter* 13(4): 2-3.

Singh, K. A. (2007). Grasslands and Grazing. *IGFRI Newsletter* 13(3): 2-3.

Singh, K. A. (2007). Healthy livestock for healthy people. *IGFRI Newsletter* 13(4): 2-3.

Singh, K. A. (2007). Challenges and priority in improvement of forage crops. *IGFRI Newsletter* 13(1): 2.

Singh, K. A. (2007). Ecological Perspective of Forage based Livestock production in India, ISA Seminar, 2-4 December, 2010.

Singh, K. A. (2008). Feed and fodder resource development in India. *IGFRI Newsletter* 14(3): 2-3.

Singh, K. A. (2008). Resource management perspective for forage production and agroforestry system development in eastern Himalayan region: a review. *Indian Journal of Agronomy* Vol. 53(4): 255-266.

Singh, K. A. (2008). Forage based feeding system for efficient livestock production. *IGFRI Newsletter* 14(2): 2-3.

Singh, K. A. (2008). Organizational structure for fodder development. *IGFRI Newsletter* 14(1): 2.

Singh, K. A. (2009). Role of Agro-forestry in sustaining crop productivity and soil health in eastern Himalayan region. Course Compendium of Training Programme under NAIP for NEH Region, NRCAF, Jhansi and NAIP, ICAR, New Delhi, pp. 14-19.

Singh, K. A. (2009). Grassland and carbon sequestration. *IGFRI Newsletter* 15(2 & 3): 3-4.

Singh, K. A. (2009). Investment and administrative reforms for accelerated fodder development. *IGFRI Newsletter* 15(4): 2-3.

Singh, K. A. (2009). Livestock and climate change. *IGFRI Newsletter* 15(1): 2-3.

Singh, K. A. (2010). Ecological perspective of forage based livestock production in India. *Souvenir of XIX National Symposium on "Resource Management Approaches Towards Livelihood Security"*, Indian Society of Agronomy, ICAR, UAS, Dharwad (karnataka) pp. 27-37.

Singh, K. A. (2010). Pasture development and management of grasslands through involvement of communities. *IGFRI Newsletter* 16(4):2.

Singh, K. A. (2010). Food Security for livestock. *Down to Earth, June 16-30: 28-29.*

Singh, K. A. (2010). Growth in Animal Husbandry Sector is dependent on availability of quality feed and fodder. *IGFRI Newsletter* 16(3): 2.

Singh, K. A. (2010). Policy Support and Programmes for Forage resource Management, *National Symposium on "Optimising foage production for Livelihood,* 12-14 November, 2010.

Singh, K. A. (2010). Report of fodder sub-group of the committee on animal husbandry of Planning Commission, *IGFRI Newsletter* 16(1 and 2): 2-3.

Singh, K. A. (2010). The goat gamble. *Down to Earth, December 28-35.*

Singh, K. A. (2011). *Diverse rangelands for a sustainable society*. eds. Feldman S R., Oliva G. E. and Sacido M. B. IRC 2011 Congress. Rosario, Argentina April 2-8, 2011.

Singh, K. A. and Anil Kumar (2012). Flood, Forage and Livestock in India. *Agriculture Year Book 2012,* published by *Agriculture Today,* New Delhi, pp114-118.

Singh, K. A. and Desai, A. R. (1996). *World Bibliography on Tree Response to Defoliation",* published by Director, ICAR Research Complex for NEH Region, Barapani, Meghalaya, 45p.

Singh, K. A. and Desai, A. R. (1998). Tree Response to Defoliation-A Review. *Agricultural Reviews,* 19(2): 105-119.

Singh, K. A. and Kanodia, K. C. (1981). Develop nutritious pasture. *Indian Farmers Digest* 14(9): 37-38.

Singh, K. A. and Mahendra Pal (1989). Nutrient balance sheet and agro-energetics of pigeon pea-wheat cropping systems. *Indian J. agric. Sci.* 59(4): 211-218.

Singh, K. A. and Pandey, J. K. (1995). Grow *Zizania* for green fodder on waterlogged lands in NEH Region. *Indian Farming* 45(1-2): 26.

Singh, K. A. and Patil, B. D. (1984). Resource use and plant interference in relation to nutrients in buffel grass and verano stylo association. *Indian J. Agron.* 29(3): 356-362.

Singh, K. A. and Pradhan, I. P. (1993). An appraisal of some indigenous farming systems with special reference to agroforestry in Sikkim (Eastern Himalaya). *Indian J. Hill Farming* 6(1): 113-122.

Singh, K. A. and Pradhan, I. P. (1995). Effect of cutting intervals on the forage yield and quality of some tropical grasses at a mid hill of Sikkim. *Range Mgmt. and Agroforestry* 16(2): 123-128.

Singh, K. A. and Rai, R. N. (1990). Important Fodder and Fuel Tree Species of Sikkim Hills. *Proc. Nat. Workshop on "Multipurpose Trees for Agroforestry",* eds. P. S. Pathak, R. Debroy, and Panjab Singh, organized by Range Management Society of India, 1-3 February, 1990, held at IGFRI, Jhansi (U. P.), pp 29-35.

Singh, K. A. and Singh, L. N. (1990). Intercropping of some temperate grasses at the mid hills of Sikkim. *Indian J. Agric Sci.* 60(7): 466-470.

Singh, K. A. and Singh, L. N. (1992). Performance of oat varieties at the mid hills of Sikkim. *Indian J. Hill Fmg.* 5(2): 133-134.

Singh, K. A. and Sunil Tiwari (2011). Livestock Deserves Fodder and Feed Security in India. *Agriculture Year Book 2011,* published by *Agriculture Today,* New Delhi, pp. 111-115.

Singh, K. A. and Thompson, F. B. (1995). Effect of lopping on water potential, transpiration, regrowth, ^{14}C-photosynthate distribution and biomass production in *Alnus glutinosa. Tree Physiology* 15: 197-202.

Singh, K. A. Anil Kumar and Premchandran, M. N. (1987). Performance of verano stylo intercropped and nitrogen fertilized *Panicum* species/varieties under rain fed condition. *Indian J. Agron.* 32 (4): 401-405.

Singh, K. A., Kanodia, K. C. and Rai, P. (1983). Stand geometry of *Cenchrus ciliaris* (L) Taub- effect on forage yield and quality. *Forage Res.* 9(1): 49-56.

Singh, K. A. Lal, M., Dwivedi, G. K. and Kanodia, K. C. (1984). Effect of N nutrition on dry matter, nutrient composition and nodulation in siratro. *Forage Res.* 10(2): 131-133.

Singh, K. A. Niranjan, K. P. and Anil Kumar (1985). Effect of row spacing on the productivity of blue panic intercropped with butterfly pea in semi-arid conditions. *Indian J. Agron.* 30(3): 349-352.

Singh, K. A. Patil, B. D. and Rai, P. (1984). Integration of forage and food production through intercropping in semiarid region. *Indian J. Agron.* 29(4): 510-515.

Singh, K. A. Patiram, Singh, L. N. and Rai, R. N. (1992). Effect of bamboo (*Bambusa nutans* Walls Ex. Munro) shade on the yield of some agricultural crops at the mid hills of Eastern Himalaya. *Indian Journal of Forestry* 15 (4): 339-341.

Singh, K. A., Anil Kumar and Niranjan, K. P. (1986). Bio-energetic and economics of nitrogen fertilized and legume intercropped forage systems in semiarid tropics. *Indian J. Agron.* 31(2): 139-146.

Singh, K. A., Balaraman, N., Bhusan, B. and Pradhan, I. P. (1995). Studies on nevaro (*Ficus auriculata*) based silvo-pastoral system in Eastern Himalaya. *Range Mgmt and Agroforestry* 16 (2): 115-121.

Singh, K. A., Hazarika, U. K. and Pradhan, I. P. (1995). Response of forage oat to nitrogen at mid hill of Sikkim. *J. Hill Research* 8(1): 120-122.

Singh, K. A., Kanodia, K. C. (1982). Blue panic - a grass for dry region. *Indian Farmers Digest* 15(3): p. 22.

Singh, K. A., Misri, B. K. and Roy, M. M. (2008). Sustainable Fodder Production for Livestock in High Altitude Regions of India IN "Advances in Agriculture Environment and Health", Eds. Shashi Bala Singh, O. P. Chaurasia, Ashish Yadav, Agnes M. Rimando and Thomas H. Terrill. Leh, India. pp. 373-383.

Singh, K. A., Pandey, J. K. and Dwivedi, R. N. (1996). Non-competitive land use strategies for fodder production in the hills of Eastern Himalayas. *Indian Farming* 45(10): 26-28.

Singh, K. A., Parthasarthy, V. A. and Pradhan, I. P. (1994). Lopped biomass prediction models for fodder trees of Eastern Himalaya. In "*Agroforestry for Degraded Lands Volume I*", eds. Panjab Singh *et al.,* pp. 435-443.

Singh, K. A., Pradhan, I. P. and Hazarika, U. K. (1995). Response of maize and teosinte to nitrogen as influenced by growing season at a mid hill of Sikkim. *J. Hill Research* 8(1): 79-82.

Singh, K. A., Prasad, R. N., Satapathy, K. K. and Sharma, U. C. (1993). Need for forage resource development in the hills of Eastern Hiamalayas- *Indian Farming* 43(8): 21-26.

Singh, K. A., Rai, R. N. and Balaraman, N. (1987). Performance of forage rice bean strains at mid hills of Sikkim. *J. Hill Res.* 1: 134-136.

Singh, K. A., Rai, R. N., and Pradhan, I. P. (1989). Grow amliso grass in the NEH Region. *Indian Farming* 38 (10): 43-45.

Singh, K. A., Rai, R. N. Balaraman, N. and Singh, L. N. (1995). Evaluation of forage grasses for the mid hills of Sikkim (Eastern Himalaya). *Range Mgmt. and Agroforestry* 16 (1): 39-45.

Singh, K. A., Rai, R. N., Gupta, H. K. and Singh, L. N. (1993). Cutting management of some temperate grasses in Sikkim. *Range Mgmt. and Agroforestry* 14 (1): 23-27.

Singh, K. A., Rai, R. N., Patiram and Bhutia, D. T. (1989). Large cardamom (*Amomum subulatum Roxb.*) plantation-an age-old agroforestry system in Eastern Himalayas. *Agroforestry Systems* 9: 241-257.

Singh, K. A., Singh, A., Satapathy, K. K., Yadav, B. P. S. and Ramchandra (2010). Micro-Watershed land use systems for optimization of Bio-productivity and resource conservation in a humid subtropical climate. *Journal of Sustainable Agriculture,* 34: 292-311.

Singh, K. A., Yadav, B. P. S. and Goswami, S. N. (1996). Forage Resource Management for Sustained Livestock Productivity in *Himalayan Agro-Ecosystems* Published by Director, ICAR Research Complex for NEH Region, Barapani, Meghalaya, 93p.

Singh, K. C. and Singh, R. P. (1999). Alternate Land Use Systems: Grassland Management. Pages 513–529 In *"Fifty years of dryland agriculture research in India"*, Central Research Institute for Dryland Agriculture, Hyderabad, India.

Singh, K. K., Das, M. M., Samanta, A. K., Kundu, S. S. and Sharma, S. D. (2004). Effect of complete feed block on nutrient utilization and growth in crossbred heifers. In *"Proceedings of XI biennial conference of India"*, Animal Nutrition Society of India, Jabalpur, India, pp. 79–80.

Singh, K. K., Nag, S. K., Garg, M. R. and Mojumdar, A. B. (2005). Mineral status of some commonly available trees and shrubs in rangeland. *Indian Journal of Animal Nutrition* 22: 55-58.

Singh, K. K., Nag, S. K., Garg, M. R. and Mojumdar, A. B. (2005). Mineral status of some commonly used feeds and forages vis-à-vis their requirements. *Animal Nutrition and Feed Technology* 5: 171-177.

Singh, K. K., Srinivas, B. and Ramchandra, K. S. (1997). Chemical changes with maturity and its impact on in sacco dry matter degradation of some range grasses and legumes. *Indian Journal of Animal Nutrition* 14: 254-257.

Singh, M. V. (2009). Micronutrient nutritional problems in soils of India and improvement for human and animal health. *Indian Journal of Fertilizers* 5: 11-16, 19-26 and 56.

Singh, O. P, Sharma, A., Singh, R. and Shah, T. (2004). Virtual water trade in dairy economy irrigation water productivity in Gujarat. *Economic and Political Weekly*. pp. 3492–3497.

Singh, P. (1986). Status of Himalayan rangelands in India and their sustainable management, eds. Daniel J. Miller and Sienna R. Craig, In *"Proc. Rangeland and Pastoral Development in Hindu Kush-Hiamalayas (November5-7, 1996)"*, Kathmandu, Nepal, pp. 13-22.

Singh, P., Chaudhary, L. C., Verma, A. K. and Pathak, N. N. (1997). Nutritive value of Robinia (*Robinia pseudoacacia*) leaves in growing Soviet Chinchilla rabbits. *World Rabbit Science* 5: 135-137.

Singh, R. A., Sharma, P., Meena, B. S., Singh, R. and Das, S. K. (2005). *Ex-Post Development Impact Analysis of Tejpura Watershed Project.* IGFRI, Jhansi, India.

Singh, R. B., Saha, R. C. and Misra, R. K. (2002). Arsenic profile in livestock feed and products in Ghentunga and Gontra districts of West Bengal. *Project Report, NDRI Eastern Regional Station*, Kolkatta.

Singh, R. P. and Osman, M. (1995). *Alternate Land Use Systems for Drylands in Sustainable Development of Dryland Agriculture in India.* Scientific publishers, Jodhpur, India, pp. 375–398.

Singh, S. (2004). *Metabolic and Rumen Microbial Activities in Sheep and Goats on grass-tree leaves/shrubs Dietary Regimens".* IGFRI, Jhansi, India.

Singh, S. (2008). The forest rights act 2007: implications for forest dwellers and protected areas. *International Forestry Review* 10: 325-330.

Singh, S., Kushwaha, B. P., Naga, S. K., Mishra, A. K., Singh, A. and Anele, U. Y. (2012). In vitro ruminal fermentation, protein and carbohydrate fractionation, methane production and prediction of twelve commonly used Indian green forages. *Animal Feed Science and Technology* 178, 2–11.

Singh, S., Nag, S. K. and Maity, S. B. (2012). Nutritional evaluation of organically grown fodders in lactating Murrah buffaloes (*Bubalus bubalis*). *Trop Anim Health Prod* 45: 251-257. doi:10. 1007/s11250-012-0198-x.

Singh, S., Singh, K. K., Das, M. M., Anil Kumar and Misra, A. K. (2008). *Livestock Rearing- An Ecological Perspective.* IGFRI, Jhansi, India.

Singh, S. K., Meena, H. R., Kolekar, D. V. and Singh, Y. P. (2012). Climate Change Impacts on Livestock and Adaptation Strategies to Sustain Livestock Production. *Journal of Veterinary Advances* 2(7): 407-412.

Singh, U. P., Roy, A. K., Faruqui, S. A. and Srivas, S. K. (2005). *IGFRI Forage Maize Germplasm Catalogue.* IGFRI, Jhansi, India, 74p.

Singhal, K. K. (2009). Scope for sugarcane bagasse based complete feed for ruminants. T. K. Walli, eds., In "*Proceedings of the national symposium on fodder block technology*", ILDEX India–2009, New Delhi, pp. 93–98.

Siriwan, D. M., Tassilo, T. T., Jérôme Bindelle, Michael Peters and Carlos, E. L. (2012). Alternative plant protein sources for pigs and chickens in the tropics – nutritional value and constraints: a review. *Journal of Agriculture and Rural Development in the Tropics and Subtropics* 113(2): 101–123, Feedipedia - Animal Feed Resources Information System. URL 116 S. D. Martens *et al.*, http://www. trc. zootechnie. fr.

Sirohi, S. and Michaelowa, A. (2007). Sufferer and cause: Indian livestock and climate change. *Climatic Change* 85: 285–298.

Skerman, P. J., Cameron, D. J. and Riveros, F. (1988). *Tropical forage legumes*, FAO, Rome, 692p.

Slots, T., Butler, G., Leifert, C., Kristensen, T. and Skibsted, L. H. (2009). Potentials to differentiate milk composition by different feeding strategies. *J. Dairy Sci* 92: 2057–2066.

Smit, I. P. J., Asner, G. P., Govender, N., Kennedy-Bowdoin, T., Knapp, D. E. and Jacobson, J. (2010). Effects of fire on woody vegetation structure in African savanna. *Ecol. Appl.* 20, 1865–1875. doi:10. 1890/09-0929. 1.

Smith P., *et al.* (2007). *Agriculture. Climate change 2007: mitigation. Contribution of working group III to the fourth assessment report of the intergovernmental panel on climate change.* (eds. Metz B., Davidson, O. R., Bosch, P. R., Dave, R., Meyer, L. A.), Cambridge University Press, Cambridge, UK.

Society for Range Management (1989). *A glossary of terms used in range management, 3rd edn.* Glossary Revision Special Committee, Publications Committee. Peter W. Jacoby, Chairman. Society for Range Management. Edison Press, Denver, Colorado, USA.

Solanki, K. R. and Ram Newaj (1999). Agroforestry: an alternate landuse system for dryland agriculture, In "*Fifty years of dryland agriculture research in India.* " Central Research Institute for Dryland Agriculture, Hyderabad, India, pp 463–473

Solarte, A. (1989). Development of feeding systems for rabbits and guinea pigs based on sugar cane juice and tree foliages. *Livestock Research for Rural Development* 1(1): Paper No. 8, Colombia.

Solomon Geberselassie, Peden, D., Amare Haileslassie and Mpairwe, D. (2009). Factors affecting livestock water productivity: animal scale analysis using previous cattle feeding trials in Ethiopia. *The Rangeland Journal*. 31: 251–258.

Son, V. V. and Trung, N. B. (2002). Utilization of water hyacinth (Eichhornia crassipes) as a supplement in diets of fattening pig. In: Yamasaki S, Kamakawa A, eds. " *Proceedings of the 2002 annual workshop of JIRCAS Mekong Delta Project - Development of new technologies and their practice for sustainable farming systems in the Mekong Delta*, November 26-28, 2002, Can Tho University, Can Tho city, Vietnam,. pp. 146-51.

Sonder, K., Astatke, A., Abiye. El Waked, Molden, D. and Peden, D. (2004). Strategies for increasing livestock water productivity in water stressed agricultural systems. http://cpwf theme1. irri. org/documents3. htm.

Soussana, J. F., Teyssonneyre, F., Picon-Cochard, C., Casella, E., Besle, J. M, Lherm, M. and Loiseau, P. (2002). Impact des changements climatiques et atmosphériques sur la prairie et sa production. In: "*Fourrage*, 169", p. 3-24.

Speedy, A. W. (2001). *FAO and pre-harvest food safety in the livestock and animal food,* FAO, Rome, Italy.

Spielman, D. J, Davis, K., Negash, M. andAyele, G. (2011). Rural innovation systems and networks: findings from a study of Ethiopian smallholders. *Agriculture and Human Values* 28(2): 195-212.

Spielman, D. J., Javier Ekboir and Kristin Davis (2009). The art and science of innovation systems inquiry: Applications to Sub-Saharan African agriculture. *Technology in Society* 31(4): 399-405.

Spielman, David and Dawit Kelemework (2009). Measuring Agricultural Innovation System Properties and Performance. IFPRI Discussion Paper 00851, Washington, D. C.

Sridhar, K., Biradar, N. B., Karthigeyan, S., Rao D. V. K. N. and Roy, A. K. (2008). *Tryst with Destiny- Research Initiatives for Fodder resource Development in Peninsular India*. IGFRI, Jhansi, India.

Srikosamatara, S. and Suteethorn, V. (1995). Populations of gaur and banteng and their management in Thailand, Natural History *Bulletin of the Siam Society* 43 (1): 55-83.

Stan, H. J. (1990). *Pesticide in principles and applications of Gas Chromatography in Food analysis.* Ellis Horwood Series in Food Science and Technology, Ellis Horwood Ltd Publication, Springer, USA, pp 59-85.

Stegman, E. C., Musick, J. C. and Stewart, J. I. (1990). Irrigation management *In* "Design and Operation of Irrigation Systems" (M. E. Jensen ed.), *ASAE Monogram* 3: 63-81.

Stergiadis, S., Leifert, C., Seal, C. J., Eyre, M. D., Nielsen J. H., *et al.* (2012). Effect of feeding intensity and milking system on nutritionally relevant milk components in dairy farming systems in North East of England. *J. Agric. Food Chem.* 60: 7270–7281.

Su, Y. Z., Zhao, H. L., Zhang, T. H. and Zhao X. Y. (2004). Soil properties following cultivation and non-grazing of a semi-arid sandy grassland in northern China. *Soil and Tillage Research* 75: 27–36. doi: 10. 1016/s0167-1987(03)00157-0.

Sujatha, T., Udhayakumari, D., Kundu, A., Jeyakumar, S., Jai Sundar and Kundu, M. S. (2013). Utilization of raw azolla (*Azolla pinnata*), as natural feed additive for sustainable production in Nicobari foul. *Animal Science Reporter* 7(4): 146-152.

Sulaiman, V. R. and Hall, A. (2002). Beyond Technology Dissemination: Reinventing Agricultural Innovation. *Outlook on Agriculture* 31(4): 225-233.

Sumil Kumar, Sudhir Kumar and Chaubey, B. K. (2009). Aonla-based hortipastoral system for soil nutrient buildup and profitability. *Annals of Arid Zone* 48(2): 153-157.

Sun, T., Liu, Z., Qin, L. and Long. R. (2012). Aspects of lipid oxidation of meat from free-range broilers consuming a diet containing grasshoppers on alpine steppe of the Tibetan Plateau. *Poultry Science*. 91: 224-231.

Sun, T., Liu, Z., Qin, L. and Long. R. (2012). Meat fatty acid and cholesterol level of free-range broilers fed on grasshoppers on alpine rangeland in the Tibetan Plateau. *Journal of the Science of Food and Agriculture* 92: 2239-2243.

Sunil Kumar and Ram, S. N. (2009). Effect of pruning intensities and pastures combination on productivity of *Ber*- based hortipastoral system. *Ann. of Arid Zone* 48 (1):63-66.

Sunil Kumar and Shukla, A. K. (2010). Productivity, economics and soil nutrient buildup in grown up aonla based hortipastoral system under semi-arid condition. *Range Mgmt. and Agroforestry. Symposium Issue* (A): 29-30.

Sunil Kumar, and Shukla, A. K. (2011). Fruit, forage productivity and economics of grownup aonla –based hortipastoral system under semi-arid condition *Indian J. of Agroforestry* 13 (1): 22-26.

Sunil Kumar, Shukla, A. K. and. Singh, H. V. (2013). Intercropping of pearlmillet + Cowpea as rainfed fodder crops with aonla based Agri-horti-system. *International Journal of Horticulture* 3(7): 28-34.

Sunil Kumar, Shukla, A. K., Rai, A. K. and Ahmed, A. (2014). Performance of guava based hortipasture system in rain fed situation with *insitu* soil moisture conservation. *IN "National Symposium on Climate Resilient Forage Production and Utilization"*, BCKV, Kalyani, West Bengal, pp. 7.

Sunil Kumar, Singh, R. and Shukla, A. K. (2014). Performance of aonla based hortipasture system in degraded land through *insitu* soil moisture conservation. *IN "National Symposium on Climate Resilient Forage Production and Utilization"*, BCKV, Kalyani, West Bengal, India, pp. 8.

Suresha, C., Ramachandra, B., Suresh, B., Prasad, N., Madhav, C. B. and Shetter, V. B. (2009) Evaluation of red gram and Bengal gram based pelleted complete diets in goats. *Indian J Anim Nutr.* 26: 349–353.

Surk, R. F. (1978). Selenium in nutrition, *World Record of Nutrition and Deities* 30: 88-106.

Suttle, K. B., Meredith, A., Thomsen Mary, and Power, E. (2007). Species Interactions Reverse Grassland Responses to Changing Climate. Science 2 February 2007: Vol. 315 no. 5812 pp. 640-642 DOI: 10. 1126/science. 1136401 accessed On 20-6-2013.

Swarup, D., Patra, R. C., Naresh, R., Kumar, P. and Shekhar, P. (2005). Blood lead levels in lactating cows reared around polluted localities; transfer of lead into milk, *The Science of the Total Environment,* 349: 67-71.

Szyszka, M., Posri, S. and Meulen U. Ter. (1985). Estimation of acceptable daily intake of mimosine for rabbits. *Zeitschrift fur Tierphysiologie, Tierernahrung und Futtermittelkunde* 54: 156-160.

Takatsuki, S. (2009). "Effects of sika deer on vegetation in Japan: a review," *Biological Conservation* 142(9): 1922–1929.

Takimoto, A. Nair, P. K. R. and Nair, V. D. (2008). Carbon stock and sequestration potential of traditional and improved agroforestry systems in the West African Sahel. *Agri. Eco. Enviorn.* 125: 159-166.

Takroo, R., Kaphalia, B. S. and Seth, T. D. (1985). Chlorinated pesticide residues in different brands of butter. *Journal of Food Science and Technology* 22: 57–59.

Tam, N. H., Hang, B. P. T., Lam, V. and Preston, T. R. (2013). Growth performance of rabbits fed components of paddy rice as supplements to *Operculina turpethum* foliage or *Operculina turpethum* mixed with water spinach. *Livestock Research for Rural Development. Volume 25, Article #49.* Retrieved August 26, 2016, from http://www. lrrd. org/lrrd25/3/tam25049. htm

Tam, N. H., Tuan, V. T., Lam, V., Hang, B. P. T. and Preston, T. R. (2009). Effects on growth of rabbits of supplementing a basal diet of water spinach (*Ipomoea aquatica*) with vegetable wastes and paddy rice. *Livestock Research for Rural Development. Volume 21,* Article #174. http://www. lrrd. org/lrrd21/10/hang21174. htm

Tangendijaja, B., Rahardjo, Y. C. and Lowry, J. B. (1990). *Leucaena* leaf meal in the diet of growing rabbits: Evaluation and effect of a low mimosine treatement. *Animal Feed Science and Technology* 29: 63–72.

Tangtaweewipat, S. and Elliott, R. (1989). Nutritional value of pigeon pea (*Cajanus cajan*) meal in poultry diets. *Animal Feed Science and Technology* 25: 123–135.

Tavendale, M. H., Meagher, L. P., Pacheco, D., Walker, N., Attwood, G. T. and Sivakumaran, S (2005). Methane production from in vitro rumen incubations with *Lotus pedunculatus* and *Medicago sativa*, and effects of extractable condensed tannin fractions on methanogenesis. *Animal Feed Science and Technology* 123 and 124: 403–419.

Taylor, A. H. (1987). "Culm dynamics and dry matter production of bamboos in the Wolong and Tangjiahe giant panda reserves, Sichuan, China," *Journal of Applied Ecology* 24(2): 419–433,

Teteh, A., Lawson, E., Tona, K. Decuypere, E. and Gbeassor, M. (2013). *Moringa Oleifera* Leave: Hydro-Alcoholic Extract and Effects on Growth Performance of Broilers. *International journal of Poultry Science* 12(7): 401-405.

Tews, J., Esther, A., Milton, S. J. and Jeltsch, F. (2006). Linking a population model with an ecosystem model: assessing the impact of land use and climate change on savanna shrub cover dynamics. *Ecological Modelling* 195: 219–228.

Thomas, H. (1980). Terminology and definitions in studies of grassland plants. *Grass and Forage Science*, 35: 13–23.

Thornton P. K. and Gerber P. (2010). Climate change and the growth of the livestock sector in developing countries. *Mitigation Adapt. Strateg. Glob. Change* 15:169–184.

Thornton, P. K. (2010). Livestock production: recent trends, future prospects Published 16 August 2010. DOI: 10. 1098/rstb. 2010. 0134

Thornton, P. K., Van de Streeg, J., Notenbaert, A. and Herrero, M. (2009). The impacts of climate change on livestock and livestock systems in developing countries: a review of what we know and what we need to know. *Agricultural systems* 101(3): 113-117.

Tietjen, B. and Jeltsch, F. (2007). Semi-arid grazing systems and climate change: a survey of present modelling potential and future needs. *Journal of Applied Ecology* 44: 425–434.

Timberlake, J. and Van Der Poel, P. (1979). *Glossary of terms used in range ecology, soil conservation, soil science and land use planning*, Division of Land Utilization, Department of Agriculture Field Services, Gaborone, Republic of Botswana.

Tiwari, K. M. (1970). Interim results of intercrops of miscellaneous tree species with main crop of Taungya plantation to increase the productivity. *Indian Forester* 96(a): 142–148.

Togun, V. A., Farinu, G. O. and Olabanji, R. O. (2006). Feeding Graded Levels of Wild Sunflower (*Tithonia diversifolia Hemsl. A. Gray*) Meal in Replacement of Maize at Pre-pubertal Age, Negatively Impacts on Growth and Morphormetric Characteristics of the Genitalia of Anak 2000 Broiler Cocks at Their Pubertal Age. *World Applied Sciences Journal* 1(2): 115-121.

Tomita, Y., Hayashi, K. and Hashizume, T. (1985). Palatability of pigs to sweet potato-silage and digestion trial by them. *Bulletin of the Faculty of Agriculture,* Kagoshima University No. 35, pp. 75–80.

Topp, C. F. E. and Doyle, C. J. (1996). Simulating the impacts of global warming on milk and forage production in Scotland. 1. The effects on dry matter yield of grass and grass-white clover stands. *Agricultural Systems* 52: 213-242.

Topp, C. F. E. and Doyle, C. J. (1996). Simulating the impacts of global warming on milk and forage production in Scotland. 2. The effects on milk yields and grazing management of dairy herds. *Agricultural Systems* 52: 243-270.

Toson, M. A., El-Latif, S. A. A. and Ibrahim, M. R. M. (1999). Growth performance of Newzealand rabbits fed diets containing cassava leaves and stems meal as a substitution of clover hay meal. *Egyptian Journal of Nutrition and Feeds* 2: 541-549.

Toung, N., Ogle, B. and Preston, T. R. (1994). Optimum protein supply and level of inclusion of water spinach (*Ipomea aquatica*) in sugar cane juice based diets for growing ducks. In *"Sustainable Livestock Systems in the Tropics." MSc thesis,* Swedish University of Agricultural Sciences, Uppsala.

Tripathi, S. B., Rai, A. K., Dixit, A. K. and Singh, K. A. (2009). Improving yield and quality of fodders through secondary and micronutrients. *Indian J. Fertilizers* 5(4): 81-88.

Trivedi, B. K. (2010). Grasses and Legumes for Tropical Pasture, IGFRI, Jhansi, India 152p.

Trivedi, R. K. and Gunasekaran, M. (2014). Seed Certification Standards for Forage Crops. *Indian Minimum Seed Certification Standards.* Department of Agriculture and Co-operation, Ministry of Agriculture, Govt. of India, pp. 135-168.

Tubiello, F. N., Amthor, J. S., Boote, K. J., Donatelli, M., Easterling, W., Fischer, G., Gifford, R. M., Howden, M., Reilly, J. and Rosenzweig, C. (2007). Crop response to elevated CO_2 and world food supply. A comment on "Food for Thought. " by Long *et al., Science* 312: 1918–1921.

Turner, R. L. (2004). *Livestock Production and Rural Poor in Andhra Pradesh and Orissa States, India (PPLPI Working Paper no. 9)*. FAO, Rome, 42 p.

Ty, C., Borin, K. and Phiny, C. (2007). A note on the effect of fresh mulberry leaves, fresh sweet potato vine or a mixture of both foliages on intake, digestibility and N retention of growing pigs given a basal diet of broken rice. *Livestock Research for Rural Development* 19 (9), *Article #136.* Retrieved May 6, 2017, from http://www. lrrd. org/lrrd19/9/chha19136. htm.

Tyagi, R. K and Singh, P. (1988). Grazing resources and grazing systems in India. In: P. Singh (ed.). *Pasture and Forage Crops Research*. Range Management Society of India, Jhansi, India pp. 17-34.

Tyagi, R. K. (1997). *Grassland and Fodder Atlas of Bundelkhand.* IGFRI, Jhansi, India.

Udar, S. A., Chopde, S. and Dhore, R. N. (2003). Mineral profile of soil, feed, fodder and buffaloes in Western Agro-climatic zone of Vidarbha. *Animal Nutrition and Feed Technology* 3: 165-172.

Udedibie, A. B. I. and Igwe F. O. (1989). Dry matter yield and chemical composition of pigeon pea (*C. cajan*) leaf meal and the nutritive value of pigeon pea leaf meal and grain meal for laying hens. *Animal Feed Science and Technology* 24 (1–2): 111-119

Ugwu, S. O. C. and Chukwuka, C. O. (2001). The effect of inclusion of dried Centro (*Centrosema pubescens*) leaves as source of protein in diet of growing pigs. Proceedings 6 th Annual Conf. of Animal Science Association of Nigeria, September, 2001, Maiduguri, Nigeria, pp. 8 1- 82

Vamsidhar Reddy, T. S., Hall, A. and Sulaiman, R. (2012). Locating research in agricultural innovation trajectories: Evidence and implications from empirical cases from South Asia', *Science and Public Policy* 2012; doi: 10. 1093/scipol/ scs034.

Vamsidhar Reddy, T. S., Puskur, R., Hall, A. and Sulaiman, R. (2013). Applying innovation system principles to fodder scarcity: Experiences from the Fodder Innovation Project. *CRISP Working Paper* 2013-001, Centre for Research on Innovation and Science Policy (CRISP), Hyderabad, India:

Van An, L., Hong, T., Ogle, B. and Lindberg, J. E. (2005). Utilization of Ensiled Sweet Potato (*Ipomoea batatas* (L.) Lam.) Leaves as a Protein Supplement in Diets for Growing Pigs. *Tropical Animal Health and Production* 37: 77-88.

Van Auken, O. W. (2009). Causes and consequences of woody plant encroachment into western North American grasslands. *J. Environ. Manage.* 90: 2931–2942. (doi:10. 1016/j. jenvman. 2009. 04. 023).

Van Keuren, R. W., Dudley, J. W., Matthias Stelly, David M. Kral and Linda C. Eisele (1974). *Systems Analysis in Forage Crops Production and Utilization*, Papers presented at the annual meeting of the Crop Science Society of America in Las Vegas, Nevada, November 1973 Crop Science Society of America, 677 South Segoe Road Madison, Wisconsin 53711, USA.

Vasey D. E. (1992). *An Ecological History of Agriculture*, 10,000 B. C. -A. D. 10,00, 363 pp. Ames, IA, USA: Iowa State University Press.

Vashishtha, B. B. (2005). Ber (Ziziphus mauritiana) based cropping systems in arid and semi-arid regions eds. Omanya, GO and Pasternak, D. (eds.) 2005. *"Sustainable Agriculture Systems for the Drylands, Proceedings of the International Symposium for Sustainable Dryland Agriculture Systems"*, 2–5 December 2003, Niamey, Niger. Niamey BP 12404, Niger: International Crops Research Institute for the Semi-Arid Tropics (ICRISAT). ISBN 92-9066-477-0, 336 p.

Verchot, L. V, Van Noordwijk, M., Kandji, S., Tomich, T., Ong, C., Albrecht, A., Mackensen, J., Bantilan, C., Anupama, K. V. and Palm, C. (2007). Climate change: linking adaptation and mitigation through agroforestry. *Mitigation and Adaptation Strategies for Global Change* 12: 901–918.

Vijayan, R., George, T., Beevi, S. N., and Kurien, M. O. (2006). Residues of organochlorine insecticides in meat of slaughtered animals in Kerala. *Pesticide Research Journal, 18*(2), 228–230.

Visalakshi, A., Beevi, S. N. and Mathew, T. B. (1993). Monitoring of insecticide residue in bovine milk from Trivandrum region. *Journal of Environmental Biology* 14: 225–229.

Vivekanandan, P. (2007). *Korangadu: A traditional pastureland farming system in the drylands of Tamil Nadu, South India*. Sustainable-agriculture and Environmental Voluntary Action, Madurai, Tamil Nadu, India.

Voelcker, J. A. (1893). *Report on the improvement of Indian agriculture*. Eyre and Spottiswoode Publisher, London.

Walli, T. K. (2009). Crop residue based densified feed block technology for improving ruminant productivity eds. Walli, T. K., *"Proceedings of the national symposium on fodder block technology"*, ILDEX India–2009, New Delhi, pp. 67–73.

Walli, T. K. (2009). Nutritional strategies for ruminant production in Afro-Asian tropics. In *"Proceedings of the World Animal Nutrition Conference"*, Animal Nutrition Association, NASC Complex, New Delhi, pp. 294–300.

Walli, T. K. (2010). Densified complete feed block technology and its role in enhancing ruminant production, including buffaloes. In *"Proceedings of the International Buffalo Conference"*, NASC Complex, New Delhi, pp. 294– 300.

Walli, T. K. (2011). A novel technology to provide complete feed to dairy animals for enhancing their productivity. Nandi, S. K., Sur, A. and Saha, D. eds., In *"Proceedings of the 39th Dairy Industry Conferenc"*, Indian Dairy Association, Kolkata, pp. 81–84.

Walli, T. K. (2011). Straw based densified complete feed block technology. In H. P. S. Makkar, eds. *"Proceedings of FAO E-Conference on successes and failures with animal nutrition practices and technologies in developing countries"*, 1–30 Sept. 2010, FAO, Rome, Italy, pp. 41–44.

Walter, W. M., Purcell, A. E. and McCollum, G. K. (1978). Laboratory preparation of a protein-xanthophyll concentrate from sweet potato leaves. *J. Agri. Food Chem.* 26 (5): 1222–1225.

Wan, H. W., Bai, Y. F., Schonbach, P., Gierus, M. and Taube, F (2011). Effects of grazing management system on plant community structure and functioning in a semiarid steppe: scaling from species to community. *Plant and Soil* 340: 215–226. doi: 10. 1007/s11104-010-0661-2.

Wang, R. Z. (2004). Responses of *Leymus chinensis* (Poaceae) to long-term grazing disturbance in the Songnen grasslands of north-eastern China. *Grass and Forage Science* 59: 191–195. doi: 10. 1111/j. 1365-2494. 2004. 00417.

Wangchuk, K., Wurzinger, M., Darabant, A., Gratzer, G. and Zollitsch, W. (2014). "The changing face of cattle rearing and forest grazing in the Bhutan Himalaya," *Mountain Research and Development* 34: 131–138.

Wangchuk, K., Darabant, A. and Rai, P. S. (2014). Morphological Responses Explain Tolerance of the Bamboo *Yushania microphylla* to Grazing. *Journal of Botany* Volume 2014, Article ID 573415, 7 pages http://dx. doi. org/10. 1155/2014/573415.

Werner, P. A., Walker, B. H. and Stott, A. (1991). An introduction to savanna ecology. In *"Savanna Ecology and Management"*. Blackwell Publishing Company, Oxford (UK): pp. 181-186.

Wharton, C. H. (1957). *An ecological study kouprey, Novibos Sauveli (Urban) Monograph of the Institute of Science and Technology*, Manilla, Philippines.

WHO (1992). Hexachlorocyclohexanes. *Health and Safety Guide* Environmental Health Criteria 123, Geneva.

William, J. B. and Midley, G. F. (2012). Carbon dioxide and the uneasy interactions of trees and savannah grasses. doi:10. 1098/rstb. 2011. 0182 *Phil. Trans. R. Soc. B* 19 Vol. 367 (1588): 601-612, http://rstb. royalsocietypublishing. org/.

Williams, R., Jones, C. A., Kiniry, J. R. and Spanel, D. A. (1989). The EPIC crop growth model. Transactions of the *ASAE* 32: 497–511.

Winicow, I. (1991). Charecterization of salt tolerant Alfalfa (*Medicago sativa*) plants regenerated from salt tolerant cell line. *Plant cell reports* **10**: 561-564.

Wittouck, P., Detimmerman, F., Petry, Y M. and Hove, C. Van (1992). Azolla as a food for rabbits in Africa. *Journal of Applied Rabbit Research* 15: 1058-1062.

World Bank (2001). *Livestock Development: Implications for Rural poverty, the Environment, and Global Food Security*. Washington D. C.

World Bank (2006). *"Enhancing Agricultural Innovation: How to Go Beyond the Strengthening of Research Systems. ", Economic Sector Work report*, World Bank, Washington, D. C.

World Bank (2007). *Enhancing Agricultural Innovations: How to go beyond the strengthening of research systems*, The International Bank for Reconstruction and Development/World Bank, Washington DC.

Xie, A. *et al.* (1996). The herbage yield and the nutrient variation in mountain meadow. *Journal of Qinghai Animal Husbandry and Veterinary Medicine* 26(2): 8-10.

Xie, D., Wanapat, M., Deng, W., He, T., Yang, Z. and Mao, H. *et al.* (2007). Comparison of Gayal (*Bos frontalis*) and Yunnan Yellow Cattle (*Bos taurus*): In Vitro Dry Matter Digestibility and Gas Production for a Range of Forages *Asian-Aust. J. Anim. Sci.* 20(8): 1208-1214.

Yadav, B. P. S. and Gupta, J. J. (1994). Perforamance of finisher pigs fed diets based on green maize fodder. *Indian J. Animal Science* 64(4): 420-22.

Yadav, B. P. S., Gupta, H. K. and Gupta, J. J. (1992). Growth pattern in goats on native pasture of north eastern India. *Int. Conf. on goat, 2-8 March, 1992, New Delhi.*

Yadav, B. P. S., Gupta, H. K. and Gupta, J. J. (1994). Nutritional evaluation of alder (*Alnus nepalensis*) leaves in goat. *"Agroforestry Systems for Degraded Lands"*, eds. Panjab Singh *et al.*, Oxford and IBH Co. Pvt. Ltd Vol I, pp. 365-66.

Yadav, B. P. S., Varma, A. and Gupta, J. J. (1990). Green maize fodder feeding to pigs. *Indian J. Nutr.* 7: 315-316.

Yadav, B. P. S., Varma, A. and Gupta, J. J. (1990). Utilization of green bajra fodder in pig ration. *Indian J. Animal Science* 60(09): 1127-28.

Yadav, B. P. S., Varma, A. and Gupta, J. J. (1991). Comparative studies on nutritional values of different pasture grasses for growing calves. *Indian J. Animal Science* 61(11): 1242-44.

Yadav, K. K., Rathee, C. S. and Lohan, O. P. (1991). Compaction behavior of straw based complete feed. *Indian Journal of Animal Sciences* 61: 980–984.

Yan, L, Zhou, G. and Zhang, F. (2013). Effects of Different Grazing Intensities on Grassland Production in China: A Meta-Analysis. *PLoS ONE* 8(12): e81466. doi:10. 1371/journal. pone. 0081466.

Yokoyama, S. and Shibata, E. (1998). The effects of sika-deer browsing on the biomass and morphology of a dwarf bamboo, *Sasa nipponica*, in Mt. Ohdaigahara, Central Japan. *Forest Ecology and Management* 103(1): 49–56.

Yu, Y. Q., Shi, J., Liu, C. G., Luo, N., Gu, Z. L. and Chu, H. (1999). Geographical distribution of argali in Xinjiang. *Chinese Biodiversity* 7: 270-276.

Zadražil, F., Permana, I. G. and Wiesche, C. (1999). Is the conversion of lignocellulose into feed with white rot fungi realizable? *"Practical problems of scale up and technology transfer"*. Deutscher Tropentag 1999, Berlin.

Zamora, R. G. and Veum, T. L. (1979). Whole soybeans fermented with *Aspergillus oryzae* and *Rhizopus oligosporus* for growing pigs. *J. Anim. Sci.* 48(1):63–68.

Zamora, R. G., Flores, L., Dantes, E., Villegas, M. and Villareal, L. (1985). Feeding value of banana meal reject for growing/finishing pigs. Phil. *J. Vet. Anim. Sci.* 11(2): 363–373.

Zanu, H. K., Mustapha, M. and Addo Nartey, M. (2012). Response of broiler chickens to diets containing varying levels of leucaena (*Leucaena leucocephala*) leaf meal. *Journal of Animal and Feed Research* 2(2): 108-112.

Zeweil, H. S., Nour, A. H. and Abd El-Rahman, S. A. (1993). Use of water hyacinth (*Eichhornia crassipes* Mart Sloms) in rabbit feeding. *Egyptian Journal of Rabbit Science* 3: 191-198.

Zhang, Y. (1999). Effect of plateau zokor on characters and succession of plant communities in alpine meadow. *Zoological Research* 20: 435–440.

Zhang, Y., Fan, N., Wang, Q. and Jing, Z. (1998). The changing ecological process of rodent communities during rodent pest managements in alpine meadow. *Acta Theriologica Sin* (Chin) 18: 137-143.

Zmudzki, J., Juszkiewicz, T., Niewiadowska, A., Szkoda, J., Semeniuk, S., Golebiowski A. and Szyposzynski, K. (1992). Chemical pollution of tissues from cattle and in milk and eggs in the Zgorzelec-Bogatynia district. Medycyna-Weterynaryjna 48, 213-215.

Zwierzykowski, Z. (1987). Interspecific and intergeneric hybridization in the *Lolium- Festuca* complex. *Proc. Intern. Conf. Hladke Zivotice, Czechoslovakia*, pp. 29-38. **Tietjen**, B. and Jeltsch, F. (2007). Semi-arid grazing systems and climate change: a survey of present modelling potential and future needs. *Journal of Applied Ecology* 44: 425–434.

Timberlake, J. and Van Der Poel, P. (1979). *Glossary of terms used in range ecology, soil conservation, soil science and land use planning*, Division of Land Utilization, Department of Agriculture Field Services, Gaborone, Republic of Botswana.

Tiwari, K.M. (1970). Interim results of intercrops of miscellaneous tree species with main crop of Taungya plantation to increase the productivity. *Indian Forester* 96(a): 142–148.

Togun, V.A., Farinu, G.O. and Olabanji, R.O. (2006). Feeding Graded Levels of Wild Sunflower (*Tithonia diversifolia Hemsl. A. Gray*) Meal in Replacement of Maize at Pre-pubertal Age, Negatively Impacts on Growth and Morphormetric Characteristics of the Genitalia of Anak 2000 Broiler Cocks at Their Pubertal Age. *World Applied Sciences Journal* 1(2): 115-121.

Tomita, Y., Hayashi, K. and Hashizume, T. (1985). Palatability of pigs to sweet potato-silage and digestion trial by them. *Bulletin of the Faculty of Agriculture,* Kagoshima University No. 35, pp. 75–80.

Topp, C.F.E. and Doyle, C.J. (1996). Simulating the impacts of global warming on milk and forage production in Scotland. 1. The effects on dry matter yield of grass and grass-white clover stands. *Agricultural Systems* 52, 213-242

Topp, C.F.E. and Doyle, C.J. (1996). Simulating the impacts of global warming on milk and forage production in Scotland. 2. The effects on milk yields and grazing management of dairy herds. *Agricultural Systems* 52: 243-270.

Toson, M. A., El-Latif, S. A. A. and Ibrahim, M. R. M. (1999). Growth performance of Newzealand rabbits fed diets containing cassava leaves and stems meal as a substitution of clover hay meal. *Egyptian Journal of Nutrition and Feeds* 2: 541-549.

Toung, N., Ogle, B. and Preston, T. R. (1994). Optimum protein supply and level of inclusion of water spinach (*Ipomea aquatica*) in sugar cane juice based diets for growing ducks. In *"Sustainable Livestock Systems in the Tropics." MSc thesis,* Swedish University of Agricultural Sciences, Uppsala.

Tripathi, S.B., Rai, A.K., Dixit, A.K. and Singh, K.A. (2009). Improving yield and quality of fodders through secondary and micronutrients. *Indian J. Fertilizers* 5(4): 81-88.

Trivedi, B.K. (2010.) Grasses and Legumes for Tropical Pasture, IGFRI, Jhansi, India 152p.

Trivedi, R. K. and Gunasekaran, M. (2014). Seed Certification Standards for Forage Crops. *Indian Minimum Seed Certification Standards.* Department of Agriculture and Co-operation, Ministry of Agriculture, Govt. of India, pp. 135-168.

Tubiello, F.N., Amthor, J.S., Boote, K.J., Donatelli, M., Easterling, W., Fischer, G., Gifford, R.M., Howden, M., Reilly, J. and Rosenzweig, C. (2007). Crop response to elevated CO_2 and world food supply. A comment on "Food for Thought." by Long *et al., Science* 312: 1918–1921.

Turner, R. L. (2004). *Livestock Production and Rural Poor in Andhra Pradesh and Orissa States, India (PPLPI Working Paper no. 9*). FAO, Rome, 42 p.

Ty, C., Borin, K. and Phiny, C. (2007). A note on the effect of fresh mulberry leaves, fresh sweet potato vine or a mixture of both foliages on intake, digestibility and N retention of growing pigs given a basal diet of broken rice. *Livestock*

Research for Rural Development 19 (9), *Article #136.* Retrieved May 6, 2017, from http://www.lrrd.org/lrrd19/9/chha19136.htm.

Tyagi, R. K and P. Singh. (1988). Grazing resources and grazing systems in India. In: P. Singh (ed.). *Pasture and Forage Crops Research.* Range Management Society of India, Jhansi, India pp. 17-34.

Tyagi, R.K. (1997). *Grassland and Fodder Atlas of Bundelkhand.* IGFRI, Jhansi, India.

Udar, S.A., Chopde, S. and Dhore, R.N. (2003). Mineral profile of soil, feed, fodder and buffaloes in Western Agro-climatic zone of Vidarbha. *Animal Nutrition and Feed Technology* 3: 165-172.

Udedibie, A.B.I. and Igwe F.O. (1989). Dry matter yield and chemical composition of pigeon pea (*C. cajan*) leaf meal and the nutritive value of pigeon pea leaf meal and grain meal for laying hens. *Animal Feed Science and Technology* 24 (1–2): 111-119

Ugwu, S.O.C. and Chukwuka, C.O. (2001).The effect of inclusion of dried Centro (*Centrosema pubescens*) leaves as source of protein in diet of growing pigs. Proceedings 6 th Annual Conf. of Animal Science Association of Nigeria, September, 2001, Maiduguri, Nigeria, pp. 8 1- 82

Vamsidhar Reddy, T. S., Hall, A. and Sulaiman, R. (2012). Locating research in agricultural innovation trajectories: Evidence and implications from empirical cases from South Asia', *Science and Public Policy* 2012; doi: 10.1093/scipol/scs034.

Vamsidhar Reddy, T.S., Puskur, R., Hall, A. and Sulaiman, R. (2013). Applying innovation system principles to fodder scarcity: Experiences from the Fodder Innovation Project. *CRISP Working Paper* 2013-001, Centre for Research on Innovation and Science Policy (CRISP), Hyderabad, India:

Van An, L., Hong, T., Ogle, B. and Lindberg, J.E. (2005). Utilization of Ensiled Sweet Potato (*Ipomoea batatas* (L.) Lam.) Leaves as a Protein Supplement in Diets for Growing Pigs. *Tropical Animal Health and Production* 37: 77-88.

Van Auken, O. W. (2009). Causes and consequences of woody plant encroachment into western North American grasslands. *J. Environ. Manage.* 90: 2931–2942. (doi:10.1016/j.jenvman.2009.04.023).

Van Keuren, R. W., Dudley J. W., Matthias Stelly, David M. Kral and Linda C. Eisele (1974). *Systems Analysis in Forage Crops Production and Utilization*, Papers presented at the annual meeting of the Crop Science Society of America in Las Vegas, Nevada, November 1973 Crop Science Society of America, 677 South Segoe Road Madison, Wisconsin 53711, USA.

Vasey D.E. (1992). *An Ecological History of Agriculture*, 10,000 B.C.-A.D. 10,00, 363 pp. Ames, IA, USA: Iowa State University Press.

Vashishtha, B.B. (2005) Ber (Ziziphusmauritiana) based cropping systems in arid and semi-arid regions eds. Omanya, GO and Pasternak, D. (eds.)

2005. *"Sustainable Agriculture Systems for the Drylands, Proceedings of the International Symposium for Sustainable Dryland Agriculture Systems",* 2–5 December 2003, Niamey, Niger. Niamey BP 12404, Niger: International Crops Research Institute for the Semi-Arid Tropics (ICRISAT). ISBN 92-9066-477-0, 336 p.

Verchot, L.V, Van Noordwijk, M., Kandji, S., Tomich, T., Ong, C., Albrecht, A., Mackensen, J., Bantilan, C., Anupama, K.V. and Palm, C. (2007). Climate change: linking adaptation and mitigation through agroforestry. *Mitigation and Adaptation Strategies for Global Change* 12: 901–918.

Vijayan, R., George, T., Beevi, S. N., and Kurien, M. O. (2006). Residues of organochlorine insecticides in meat of slaughtered animals in Kerala. *Pesticide Research Journal, 18*(2), 228–230.

Visalakshi, A., Beevi, S.N. and Mathew, T.B. (1993). Monitoring of insecticide residue in bovine milk from Trivandrum region. *Journal of Environmental Biology* 14: 225–229.

Vivekanandan, P. (2007). *Korangadu: A traditional pastureland farming system in the drylands of Tamil Nadu, South India.* Sustainable-agriculture and Environmental Volunt3ary Action, Madurai, Tamil Nadu, India.

Voelcker, J.A. (1893). *Report on the improvement of Indian agriculture.* Eyre and Spottiswoode Publisher, London.

Walli, T.K. (2009). Crop residue based densified feed block technology for improving ruminant productivity eds. Walli, T.K., *"Proceedings of the national symposium on fodder block technology"*, ILDEX India–2009, New Delhi, pp. 67–73.

Walli, T.K. (2009). Nutritional strategies for ruminant production in Afro-Asian tropics. In *"Proceedings of the World Animal Nutrition Conference"*, Animal Nutrition Association, NASC Complex, New Delhi, pp. 294–300.

Walli, T.K. (2010). Densified complete feed block technology and its role in enhancing ruminant production, including buffaloes. In *"Proceedings of the International Buffalo Conference"*, NASC Complex, New Delhi, pp. 294– 300.

Walli, T.K. (2011). A novel technology to provide complete feed to dairy animals for enhancing their productivity. Nandi, S. K., Sur, A. and Saha, D. eds., In *"Proceedings of the 39th Dairy Industry Conferenc"*, Indian Dairy Association, Kolkata, pp. 81–84.

Walli, T.K. (2011). Straw based densified complete feed block technology. In H.P.S. Makkar, eds. *"Proceedings of FAO E-Conference on successes and failures with animal nutrition practices and technologies in developing countries"*, 1–30 Sept. 2010, FAO, Rome, Italy, pp. 41–44.

Walter, W.M., Purcell, A.E. and McCollum, G.K. (1978). Laboratory preparation of a protein-xanthophyll concentrate from sweet potato leaves. *J. Agri. Food Chem.* 26 (5): 1222–1225.

Wan, H.W., Bai, Y.F., Schonbach, P., Gierus, M. and Taube, F (2011). Effects of grazing management system on plant community structure and functioning in a semiarid steppe: scaling from species to community. *Plant and Soil* 340: 215–226. doi: 10.1007/s11104-010-0661-2.

Wang, R.Z. (2004). Responses of *Leymus chinensis* (Poaceae) to long-term grazing disturbance in the Songnen grasslands of north-eastern China. *Grass and Forage Science* 59: 191–195. doi: 10.1111/j.1365-2494.2004.00417.

Wangchuk, K., Wurzinger, M., Darabant, A., Gratzer, G. and Zollitsch, W. (2014). "The changing face of cattle rearing and forest grazing in the Bhutan Himalaya," *Mountain Research and Development* 34: 131–138.

Wangchuk, K., Darabant, A.and Rai, P.S. (2014). Morphological Responses Explain Tolerance of the Bamboo *Yushania microphylla* to Grazing. *Journal of Botany* Volume 2014, Article ID 573415, 7 pages http://dx.doi.org/10.1155/2014/573415.

Werner, P. A., Walker, B. H. and Stott, A. (1991). An introduction to savanna ecology. In *"Savanna Ecology and Management"*. Blackwell Publishing Company, Oxford (UK): pp. 181-186.

Wharton, C.H. (1957). *An ecological study kouprey, Novibos Sauveli (Urban) mono -graph of the institute of Science and Technology monographs*, Manilla, Philippines.

WHO (1992). Hexachlorocyclohexanes. *Health and Safety Guide* Environmental Health Criteria 123, Geneva.

William, J.B. and Midley, G. F. (2012).Carbon dioxide and the uneasy interactions of trees and savannah grasses.doi:10.1098/rstb.2011.0182 *Phil. Trans. R. Soc. B* 19 Vol. 367 (1588): 601-612, http://rstb.royalsocietypublishing.org/.

Williams, R., Jones, C.A., Kiniry, J.R. and Spanel, D.A. (1989).The EPIC crop growth model.Transactions of the *ASAE* 32: 497–511.

Winicow, I. (1991). Charecterization of salt tolerant Alfalfa (*Medicago sativa*) plants regenerated from salt tolerant cell line. *Plant cell reports* **10**: 561-564.

Wittouck, P., Detimmerman, F., Petry, Y M. and Hove, C. Van (1992). Azolla as a food for rabbits in Africa. *Journal of Applied Rabbit Research* 15: 1058-1062.

World Bank (2001). *Livestock Development: Implications for Rural poverty, the Environment, and Global Food Security*. Washington D.C.

World Bank (2006), *"Enhancing Agricultural Innovation: How to Go Beyond the Strengthening of Research Systems."*, *Economic Sector Work report*, World Bank, Washington, D.C.

World Bank (2007). *Enhancing Agricultural Innovations: How to go beyond the strengthening of research systems*, The International Bank for Reconstruction and Development/World Bank, Washington DC.

Xie, A. *et al.* (1996). The herbage yield and the nutrient variation in mountain meadow. *Journal of Qinghai Animal Husbandry and Veterinary Medicine* 26(2): 8-10.

Xie, D., Wanapat, M., Deng, W., He, T., Yang, Z. and Mao, H. *et al.* (2007). Comparison of Gayal (*Bos frontalis*) and Yunnan Yellow Cattle (*Bos taurus*): In Vitro Dry Matter Digestibility and Gas Production for a Range of Forages *Asian-Aust. J. Anim. Sci.* 20(8): 1208-1214.

Yadav, B.P.S. and Gupta, J.J. (1994). Perforamance of finisher pigs fed diets based on green maize fodder. *Indian J. Animal Science* 64(4): 420-22.

Yadav, B.P.S., Gupta, H.K. and Gupta, J.J. (1992). Growth pattern in goats on native pasture of north eastern India. *Int. Conf. on goat, 2-8 March, 1992, New Delhi.*

Yadav, B.P.S., Gupta, H.K. and Gupta, J.J. (1994). Nutritional evaluation of alder (*Alnus nepalensis*) leaves in goat. *"Agroforestry Systems for Degraded Lands"*, eds. Panjab Singh *et al.,* Oxford and IBH Co. Pvt. Ltd Vol I pp365-66.

Yadav, B.P.S., Varma, A. and Gupta, J.J. (1990). Green maize fodder feeding to pigs. *Indian J. Nutr.* 7: 315-1.

Yadav, B.P.S., Varma, A. and Gupta, J.J. (1990). Utilization of green bajra fodder in pig ration. *Indian J. Animal Science* 60(09): 1127-28.

Yadav, B.P.S., Varma, A. and Gupta, J.J. (1991). Comparative studies on nutritional values of different pasture grasses for growing calves. *Indian J. Animal Science* 61(11): 1242-44.

Yadav, K.K., Rathee, C.S. and Lohan, O.P. (1991). Compaction behavior of straw based complete feed. *Indian Journal of Animal Sciences* 61: 980–984.

Yan, L, Zhou, G. and Zhang, F. (2013). Effects of Different Grazing Intensities on Grassland Production in China: A Meta-Analysis. *PLoS ONE* 8(12): e81466. doi:10.1371/journal.pone.0081466.

Yokoyama, S. and Shibata, E. (1998). The effects of sika-deer browsing on the biomass and morphology of a dwarf bamboo, *Sasa nipponica*, in Mt. Ohdaigahara, Central Japan. *Forest Ecology and Management* 103(1): 49–56.

Yu, Y.Q., Shi, J., Liu, C.G., Luo, N., Gu, Z.L. and Chu, H. (1999) Geographical distribution of argali in Xinjiang. *Chinese Biodiversity* 7: 270-276.

Zadražil, F., Permana, I.G. and Wiesche C (1999). Is the conversion of lignocellulose into feed with white rot fungi realizable? *"Practical problems of scale up and technology transfer"*. Deutscher Tropentag 1999, Berlin.

Zamora, R.G. and Veum, T.L. (1979). Whole soybeans fermented with *Aspergillus oryzae* and *Rhizopus oligosporus* for growing pigs. *J. Anim. Sci.* 48(1):63–68.

Zamora, R.G., Flores, L., Dantes, E., Villegas, M. and Villareal, L. (1985). Feeding value of banana meal reject for growing/finishing pigs. Phil. *J. Vet. Anim. Sci.* 11(2):363–373.

Zanu, H.K., Mustapha, M. and Addo Nartey, M. (2012). Response of broiler chickens to diets containing varying levels of leucaena (*Leucaena leucocephala*) leaf meal. *Journal of Animal and Feed Research* 2(2): 108-112.

Zeweil, H. S., Nour, A. H. and Abd El-Rahman, S. A. (1993). Use of water hyacinth (*Eichhornia crassipes* Mart Sloms) in rabbit feeding. *Egyptian Journal of Rabbit Science* 3: 191-198.

Zhang, Y. (1999). Effect of plateau zokor on characters and succession of plant communities in alpine meadow. *Zoological Research* 20: 435–440.

Zhang, Y., Fan, N., Wang, Q. and Jing Z. (1998).The changing ecological process of rodent communities during rodent pest managements in alpine meadow. *Acta Theriologica Sin* (Chin) 18: 137-143.

Zmudzki, J., Juszkiewicz, T., Niewiadowska, A., Szkoda, J., Semeniuk, S., Golebiowski A. and Szyposzynski, K. (1992). Chemical pollution of tissues from cattle and in milk and eggs in the Zgorzelec-Bogatynia district. Medycyna-Weterynaryjna 48, 213-215.

Zwierzykowski, Z. (1987). Interspecific and intergeneric hybridization in the Lolium- Festuca complex. Proc. Intern. Conf. Hladke Zivotice, Czechoslovakia, pp. 29-38.

Index

A

B

C

D

E

F

G

H

T

V

W

Z